# Statistical Mechanics, Kinetic Theory, and Stochastic Processes

# Statistical Mechanics, Kinetic Theory, and Stochastic Processes

C. V. HEER

*Ohio State University*



*ACADEMIC PRESS New York and London*

ACADEMIC PRESS, INC.
111 Fifth Avenue, New York, New York 10003

*United Kingdom Edition published by*
ACADEMIC PRESS, INC. (LONDON) LTD.
24/28 Oval Road, London NW1

LIBRARY OF CONGRESS CATALOG CARD NUMBER: 75-137591

PRINTED IN THE UNITED STATES OF AMERICA

# Contents

## IV INTERMOLECULAR INTERACTIONS



## V TRANSPORT PHENOMENA



## VI STATISTICAL THERMODYNAMICS



## VII MOLECULAR SYSTEMS AT LOW DENSITIES

## XI STOCHASTIC PROCESSES IN QUANTUM SYSTEMS

# Preface

A primary goal of this book is to present the statistical aspects of physics as a "living and dynamic" subject and to present the topics in a manner which will be useful to the research scientist. The material is arranged so that the earlier chapters can form the basis of an undergraduate course, while taken as a whole, the book is a graduate text.

In order to provide an elementary introduction to kinetic theory, physical systems in which particle–particle interaction can be neglected are considered first. Only wall interactions are important and the fascinating aspects of transport phenomena in the free-molecular flow region for gases and the transport of thermal radiation are discussed. Discrete random processes such as the random walk, the binomial and Poisson distributions, throwing of dice, and so forth are studied by means of the characteristic function, which permits a very simple counting procedure for mutually independent events. Conditional probability is introduced by considering the throwing of a set of dice with a single constraint and it is shown that this problem has many features which are similar to those encountered in statistical thermodynamics. The elementary aspects of binary collisions are developed and then used to study their effect on nuclear and chemical reaction rates, on laser kinetics, etc. This study is followed by the classical development of the Boltzmann equation and then the binary collision rates are related to the transport coefficients.

Following the methods developed for conditional probability, the canonical and grand canonical ensembles are introduced for the study of systems which are in thermodynamic equilibrium. Systems for which interactions between particles or between states are not important are considered first, and topics such as the chemical potential of gases, rotational and vibrational heat capacities, Bose and Fermi gases, the photon gas, paramagnetic ions and low temperature paramagnetism, and gases at extremely high temperatures are discussed. Imperfect gases for which binary interactions are important are studied by means of the cumulant or cluster expansion and are followed by a study of the very degenerate Fermi–Dirac gases which occur for electrons in metals, in very dense matter, and in neutron stars. The general aspects of the equilibrium properties of gases, liquids, and solids, of binary mixtures, etc. are developed and then used to study the Third Law of Thermodynamics as well as many special examples such as the $^3He$–$^4He$ dilution refrigerator. The combinatorial problem of placing interacting A and B particles on sites with a constraint on the total energy and number of particles is studied in some detail as an example of a many body problem. This introduces the fascinating topic of counting of pairs, polygons, and self-avoiding walks, the Ising model, and the dispersion in particle number and pairs which appears to be related to the experimentally observed critical point data.

Stochastic processes which occur for the simple harmonic oscillator and macroscopic particles are developed in considerable detail and are used to discuss such topics as Brownian motion and thermal noise in electrical circuits. Elastic scattering of waves by thermal fluctuations is used to study the single particle and pair interactions in physical systems. A nonlinear oscillator such as a laser and coupled nonlinear oscillators are studied as systems which approach a distinct operation point and which have unusual entrainment properties. The characteristic function is introduced for a more formal development of random processes. Stochastic processes in quantum systems is the last topic treated and starts with a simple discussion of transition rates, the time dependent density matrix, and the absorption or emission of energy by simple systems. Saturating pulses and photon echoes are given as an example for which the time proportional transition rate is not useful. A detailed derivation of the radiation emitted by an emitter in a simple harmonic oscillator potential is given and then used to discuss in detail the Mössbauer effect and the scattering of electrons, neutrons, and x rays by crystal lattices. The characteristic function is also used for the study of line shapes, spectral broadening in gases, liquid scattering, etc. Inelastic scattering is used to study the single particle and pair correlation aspects in gases, liquids, solids, binary mixtures, etc., and this very powerful tool provides a correlation between the scattering data and the measureable thermodynamic properties and the coefficients in the macroscopic flow equations.

The rationalized "meter, kilogram, second, and ampere" system of units is used throughout.

References to books and articles whose contents have influenced the author and which may be used for references to more detailed studies are given throughout. No attempt has been made to completely reference the topics and many excellent research papers do not appear among the references.

I wish to express my sincere appreciation to my wife Esther, to my daughters Barbara and Deborah, and to my son Daniel for their patience during the writing of this manuscript, and to my daughter Barbara for typing much of this manuscript.

# List of Symbols[1]

| | |
|---|---|
| $a$ | index |
| $a^+$, $a$ | creation and annihilation operators |
| $A$ | Helmholtz free energy |
| $A(\nu)$ | absorption coefficient |
| $b$ | index |
| $B$ | rotational constant |
| $\mathbf{B}$ | magnetic induction |
| $B(T)$ | second virial coefficient |
| $c$ | index; velocity of light or sound |
| $\mathbf{C}$ | intrinsic velocity |
| $C_V$ | heat capacity at constant volume, etc. for $C_p$, $C_B$ |
| $d\Omega$ | differential solid angle |
| $d\mathbf{r}$ | differential element of volume |
| $d\mathbf{v}$ | differential element of volume in velocity space |
| $d\mathbf{p}$ | differential element of volume in momentum space |
| $d\mathbf{k}$ | differential element of volume in wave number space |
| $d\sigma$ | differential cross section |
| $dW$ | differential element of work |
| $D$ | diffusion coefficient |
| $\mathbf{D}$ | electric displacement |
| $\mathscr{D}f(\xi) = \langle (f - \bar{f})^2 \rangle$ | dispersion in $f$ |
| $e$ | electron charge |
| $\mathbf{E}$ | electric field intensity |
| $\mathscr{E}f(\xi) = \langle f \rangle$ | expectation or average value of $f$ |

[1] Note the difference between lower case italic "vee" ($v$) and lower case greek nu ($\nu$) both in this list of symbols and throughout the text.

| | |
|---|---|
| $\langle \exp iaf \rangle$ | characteristic function of $f$ |
| $f(k, \theta)$ | scattering amplitude |
| $f(\mathbf{v})$ | Boltzmann velocity distribution function |
| $g(\varepsilon)$, $\varrho(E)$ | density of states |
| $g(N_A, N_B, N_{AB})$ | number of arrangements |
| $g(\nu)$ | line shape |
| $G$ | Gibbs function |
| $G(\nu)$ | response function |
| $h$, $\hbar$ | Planck's constant |
| $H$, $H_N$ | Hamiltonian |
| $H$ | enthalpy |
| $\mathbf{H}$ | magnetic field intensity |
| $I(\nu, \theta)$ | intensity in watt/steradian-hertz |
| $I(\mathbf{K})$ | intensity or flux |
| $I$ | intensity in watts or flux of particles |
| $i$ | $\sqrt{-1}$ ; index |
| $j$ | index |
| $\mathbf{J}$ | current |
| $k$ | Boltzmann constant; index |
| $\mathbf{k}$ | wave number; index |
| $K$ | thermal conductivity |
| $\mathbf{K}$ | wave number |
| $K_T$ | isothermal compressibility |
| $l$ | length; index |
| $L$ | length |
| $m$ | mass; quantum number |
| $M$ | mass; quantum number |
| $\mathbf{M}$ | magnetic dipole moment per cubic meter |
| $n$ | index; number per cubic meter |
| $n_k$ | occupation number, particle number, photon number, etc. |
| $N$ | number of particles |
| $O$ | order of |
| $p(n)$ | probability of $n$ |
| $p(x, t)$ | probability of $x$, $t$ |
| $\mathbf{p}$ | momentum |
| $P$ | pressure |
| $P(\xi = n)$ | probability that $\xi = n$ |
| $P(v^2)$ | probability of a Boltzmann velocity distribution |
| $\mathbf{P}$ | electric dipole moment per cubic meter; momentum |
| $P_{xy}$ | stress tensor |
| $q$ | index; coordinate; electric charge; latent heat |
| $\dot{q} = dq/dt$ | velocity, electric current |
| $q_z$ | heat flow |
| $Q$ | heat |
| $Q_N$, $Q$ | canonical generating function or partition function |
| $\mathcal{Q}$ | grand canonical generating or partition function |
| $r$ | index |
| $r$, $\mathbf{r}$ | coordinate |

| | |
|---|---|
| $R$ | gas constant |
| $R(\nu)$ | reflection coefficient |
| $\mathbf{R}$ | coordinate |
| $S$, $s$ | entropy |
| $t$ | time |
| $T$ | temperature |
| $u$ | index |
| $\hat{u}$ | unit vector for electric polarization |
| $U$ | internal energy, interaction energy |
| $v$ | velocity; volume |
| $V$ | volume |
| $\mathbf{V}$ | relative velocity |
| $W$ | work |
| $\mathbf{W}$ | average velocity |
| $W_{ba}$ | time proportional transition rate |
| $W(AB; CD)$ | time proportional transition rate |
| $x,X$ | coordinate |
| $y,Y$ | coordinate |
| $z,Z$ | coordinate |
| $z$ | fugacity |
| $\alpha$ | index |
| $\alpha(\nu)$ | absorption coefficient |
| $\alpha_{xy}$, $\alpha$ | molecular polarizability |
| $\beta$ | index; $(kT)^{-1}$ |
| $\gamma$, $\Gamma$ | damping constant |
| $\Gamma_z$ | particle flow in $z$ direction |
| $\delta$ | Dirac delta function; phase shift |
| $\Delta$ | small increment |
| $\nabla$ | gradient |
| $\nabla^2$ | Laplacian |
| $\varepsilon$ | energy; electric permittivity |
| $\varepsilon_0$ | electric permittivity of free space |
| $\eta$ | coordinate, variable; coefficient of viscosity |
| $\theta$ | polar coordinate |
| $\Theta$ | Debye theta |
| $\varkappa$ | index, dielectric constant |
| $\boldsymbol{\varkappa}$ | wave number |
| $\varkappa_T$ | isothermal compressibility |
| $\lambda$ | wavelength; average path between collisions |
| $\lambda$, $\Lambda$ | rate of occurrence |
| $\mu$ | chemical potential; reduced mass; index of refraction |
| $\mu_0$ | magnetic permeability of free space |
| $\boldsymbol{\mu}$ | magnetic moment |
| $\nu$ | index; frequency |
| $\xi$ | coordinate, variable |
| $\pi$ | 3.1416... |
| $\varrho$ | density matrix; density |
| $\sigma$ | cross section; bilinear operator |

| | |
|---|---|
| $\sigma(\mathbf{K})$, $S(\mathbf{K})$ | elastic cross section or spectral density |
| $\sigma(\boldsymbol{\varkappa}, \omega)$, $S(\boldsymbol{\varkappa}, \omega)$ | inelastic cross section or spectral density |
| $\tau$ | time |
| $\varphi$ | index; polar coordinate; variable |
| $\varphi(a)$ | characteristic function |
| $\Phi$ | wave function; variable |
| $\Phi(a)$ | characteristic function |
| $\chi$ | variable; magnetic susceptibility |
| $\psi$ | variable; wave function |
| $\psi(\nu)$ | spectral density |
| $\varphi(\tau)$ | correlation function |
| $\Psi$ | wave function |
| $\omega$ | angular frequency |
| $\Omega$ | solid angle |
| $\langle f \rangle = \bar{f}$ | average value of $f$ |
| $\hat{f}$ | caret indicates unit vector of any variable |
| $f^+$ | Hermitian conjugate of $f$ |
| $\|a)$, $\|\beta)$, etc. | quantum state |
| $(a \mid a') = \delta_{aa'}$ | orthogonality |
| $(\alpha \mid U \mid \beta)$ | matrix element |
| $\langle x(t)x(t+\tau) \rangle$ | correlation function |

# Mass Point Gas

# I

## 1.1 Introduction

Some important concepts and problems in the kinetic theory of gases can be studied in an elementary manner by examining the average properties of the mass point gas. The gas is regarded as composed of particles of fixed mass and the interaction between particles is regarded as negligible. Since the cross section for collisions is small, the particles are treated as points. The physical aspects of the gas depend on the interaction of the particles with the walls of the containing vessel. Gases at low densities and considerations of their flow are problems of this nature and can be examined within the framework of this model. Before initiating the discussion of this ideal gas of point particles, some elementary properties of space and velocity distributions are introduced.

## 1.2 Uniform Space Distributions

The uniform space distribution provides an excellent introduction to more general distribution functions and also serves as a review of the various coordinate systems. A uniform distribution in space assumes that each region in the space under consideration is equally probable. Thus if a mark is placed at random on a line of length $L$ and each region on the line is regarded as an equally probable location for the mark, the probability of finding the mark in the region between $x$ and $x + dx$ is $dx/L$. Similar arguments can be made for selecting a position on a surface $A$ or in a volume $V$, as shown in Fig. 1.1, and for a position selected at random, the probabilities of it lying in region $dx$, $dA$, or $dV$ are as follows:

$$\begin{aligned}\{\text{probability of lying within } dx\} &= \frac{dx}{L}\\ \{\text{probability of lying within } dA\} &= \frac{dA}{A}\\ \{\text{probability of lying within } dV\} &= \frac{dV}{V}\end{aligned} \tag{1.1}$$

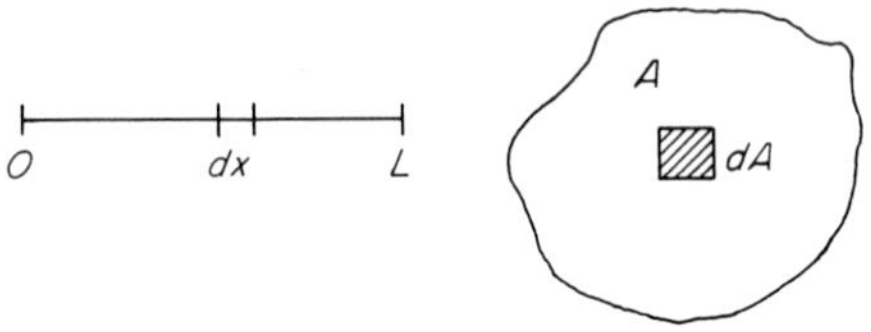


**Fig. 1.1** Probability of given position lying within length $dx$ or area $dA$.

As an example, consider a vector attached at the origin 0 and randomly oriented. A convenient method of visualization is to draw a sphere of radius $a$ about the origin and regard each region on the sphere as equally probable. The probability of the vectors touching a specific area $dA$ or pointing in a direction $d\Omega$ is immediately seen to be $dA/A = a^2\, d\Omega/4\pi a^2$, or

$$\left\{\begin{matrix}\text{probability that a vector}\\ \text{points in direction } d\Omega\end{matrix}\right\} = \frac{d\Omega}{4\pi} = \frac{\sin\theta\, d\varphi\, d\theta}{4\pi} \tag{1.2}$$

The solid angle $d\Omega$ in steradians is a convenient notation and will be used throughout the text. Specific calculations are usually made in polar coordinates with $d\Omega = \sin\theta\, d\varphi\, d\theta$. The direction of emission of a particle from a nucleus for which all directions are equally probable is analogous to the probability that a random vector or ray points in a given direction.

The probability that the tip of a vector ends in a volume $dV$ can be discussed in a similar manner. It is quite common to use the symbolic notation $d\mathbf{r}$ to denote the volume element $dV$. With this notation,

$$\begin{aligned}\left\{\begin{matrix}\text{probability that a vector ends}\\ \text{in volume element } d\mathbf{r}\end{matrix}\right\} &= \frac{d\mathbf{r}}{V} = \frac{dx\, dy\, dz}{V}\\ &= \frac{r^2\, d\Omega\, dr}{V}\end{aligned} \tag{1.3}$$

The various volume elements are shown in Fig. 1.2.

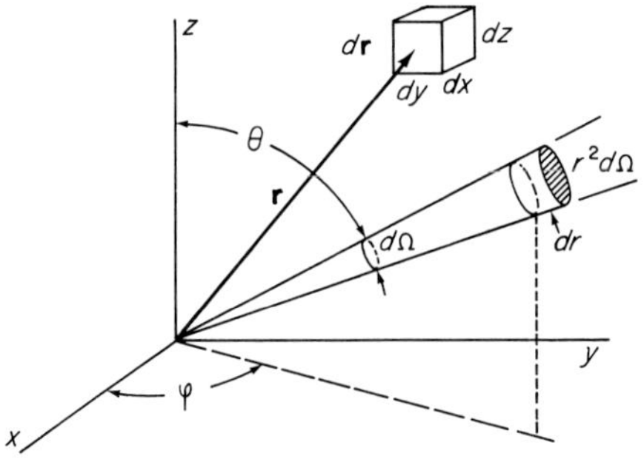


**Fig. 1.2** Probability of a vector pointing in direction $d\Omega$ or ending in volume element $d\mathbf{r} = dx\,dy\,dz = r^2\,d\Omega\,dr = r^2(\sin\theta)\,d\varphi\,d\theta\,dr$.

## 1.3 Nonuniform Space Distributions

For many problems, the physical laws governing the systems cause some regions of space to be more probable than others. A weight function or a distribution function is then introduced to indicate that some regions are more probable than others. Suppose that on a line of length $L$, a mark is placed between 0 and $L$, but now some regions are more probable than others, as indicated by the weight function $W(x)$, such that

$$dP = \{\text{probability of lying within} \quad dx\} = W(x)\frac{dx}{L} = P(x)\,dx \quad (1.4)$$

$P(x) = W(x)/L$ is the probability of lying within the interval $dx$ and includes both the weight function and the aspects of the uniform distribution. Since the mark is found between 0 and $L$ with certainty, the integral

$$\int_0^L dx\,P(x) = 1 \quad (1.5)$$

is often called the normalization integral.

Frequently, a distribution function is introduced rather than a weight function or a probability function. Suppose $N$ particles are distributed over a line of length $L$ according to the distribution law $f(x)$; then, the number of particles in the interval between $x$ and $x + dx$ is

$$\{\text{number of particles in interval} \quad dx\} = f(x)\,dx \quad (1.6)$$

and normalization requires

$$\int_0^L dx\,f(x) = N \quad (1.7)$$

It is immediately apparent that $f(x)$ and $P(x)$ are related in a simple manner by

$$f = NP \tag{1.8}$$

Both the distribution function $f$ and the probability function $P$ will be used throughout the text. The simple arguments given here apply to systems of greater dimensions by changing $x$ to $(x, y)$ and $dx$ to $dx\,dy$, and so on.

As an example, suppose $N$ particles are distributed over the surface of a sphere according to the distribution law $f(\theta) = B\cos^2\theta$. How many particles lie in the area between $\theta$ and $\theta + d\theta$? With the previous definitions,

$$\begin{aligned} \left\{\begin{matrix}\text{number of particles} \\ \text{between } \theta \text{ and } \theta + d\theta\end{matrix}\right\} &= f(\theta)\frac{dA}{A} = f(\theta)\frac{\sin\theta\,d\theta}{2} \\ &= \frac{3}{2}N\cos^2\theta\sin\theta\,d\theta \end{aligned} \tag{1.9}$$

where $B$ is determined from the normalization integral $\int dA\,f = N$.

## 1.4 Average or Expected Value of Observable Quantities

If $A(x)$ is an observable property of a system, then the average or expected value of this quantity for a distribution $f(x)$ is

$$\langle A\rangle = \bar{A} = \frac{\int dx\,f(x)A(x)}{\int dx\,f(x)} \tag{1.10}$$

Both bracket $\langle A\rangle$ and bar $\bar{A}$ notation will be used to denote average values. Since the distribution function $f$ is frequently not normalized, the integral over $f$ is included in the denominator. Spaces of more than one dimension are treated by changing $x$ to $x, y, \ldots$ and $dx$ to the appropriate volume element.

The examples given thus far refer to spaces that are described by continuous variables. Quite often, the system contains $N$ particles and a physical property of each particle is known. In one dimension, this might be the speed $v_\alpha$, which is shown by the vertical lines in Fig. 1.3. The discrete values that the speed can assume are denoted by the index $\alpha$. The distribution law for such a system is expressed by

$$\{\text{probability that } v = v_\alpha\} = P(v = v_\alpha) \tag{1.11}$$

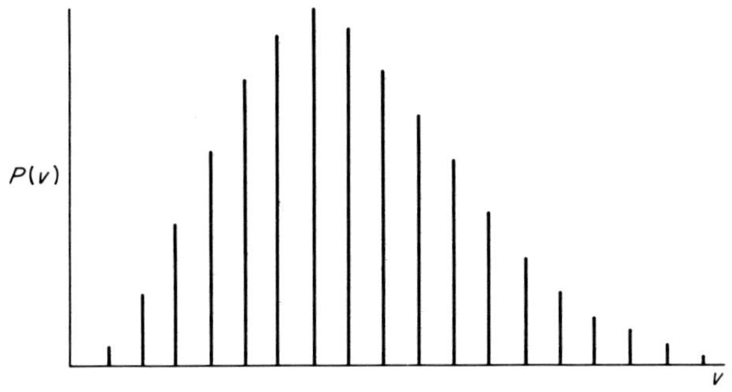


**Fig. 1.3** Distribution law $P(v = v_\alpha)$, or the probability of speed $v_\alpha$ as a function of $v$.

where $P$ is nonzero at $v_\alpha$. Since this is a probability,

$$\sum_\alpha P(v = v_\alpha) = 1 \tag{1.12}$$

and the expectation value of $v$ is

$$\mathscr{E}v = \langle v \rangle = \bar{v} = \sum_\alpha vP(v = v_\alpha) \tag{1.13}$$

In this chapter, discrete variables occur for the velocities of the particles and it is convenient to have a procedure for transforming the sum to an integral. The average value of a physical quantity $A$ is

$$\langle A \rangle = \frac{\sum_\alpha AP(x = x_\alpha)}{\sum_\alpha P(x = x_\alpha)} \to \frac{\sum_a A(x_a) f(x_a)\, \Delta x_a}{\sum_a f(x_a)\, \Delta x_a} \to \frac{\int dx\, Af}{\int dx\, f} \tag{1.14}$$

where $f(x_a)\, \Delta x_a$ is the number of particles in the interval $\Delta x_a$. If $\Delta x_a$ is selected sufficiently large so that many points $x_\alpha$ are included, then $f(x_a)$ has the features of a continuous function and the sum can be transformed to an integral.

**Exercise 1.1** $N$ particles are distributed along a line of length $L$ with a distribution function $f(x) = A[x - (x^2/L)]$. How many particles lie between $x$ and $x + dx$? What is the average distance of a particle from $x = 0$? Give a numerical answer for $N = 1000$, $L = 1$ m, and $dx = 1$ mm.

Ans. $6(x - x^2)$

**Exercise 1.2** $N$ particles are distributed in space according to the distribution function

$$f(r, \theta) = A[\exp(-Br^2)] \cos^2 \theta$$

Show that the number of particles lying in $r^2\,d\Omega\,dr$ is given by

$$3N\left(\frac{B}{\pi}\right)^{3/2}[\exp(-Br^2)](\cos^2\theta)r^2\,d\Omega\,dr$$

Find the average distance of a particle from the origin, that is, $\langle r\rangle$.

Ans. $\langle r\rangle = 2(\pi B)^{-1/2}$

## 1.5 Average or Expected Free Paths

An elementary and useful application of the previous ideas is the calculation of the expected path for a particle in some simple type of enclosure. Let a bug be located at a point $O$ on the inner rim of a circle of radius $a$ as shown in Fig. 1.4. Assume that the bug flies in a straight path toward another part of the circle, and choose the direction in a random manner.

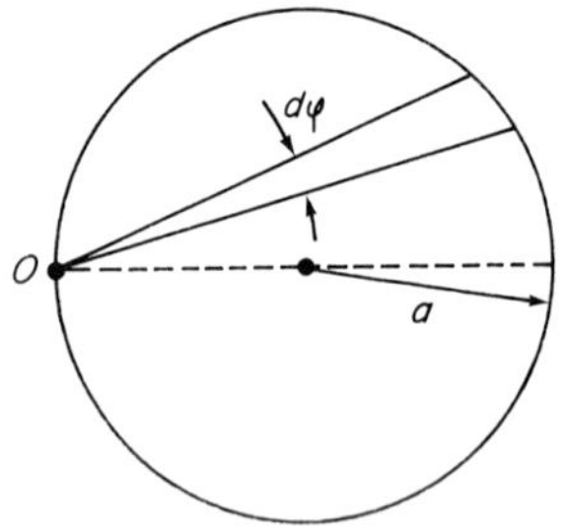


**Fig. 1.4** A bug makes random jump in direction $d\varphi$ from position $O$ inside of circle of radius $a$.

What is the expected length of the path, or $\langle r\rangle$? Since the bug leaves $O$ in a random manner for the accessible region inside the circle, the probability that the path makes an angle between $\varphi$ and $\varphi + d\varphi$ is simply proportional to $d\varphi$. For $\varphi$ between $\pi/2$ and $3\pi/2$, the region is inaccessible and the probability of being in this region is zero. The average value of $r = 2a\cos\varphi$ is

$$\langle r\rangle = \frac{\int_{-\pi/2}^{\pi/2} r\,d\varphi}{\int_{-\pi/2}^{\pi/2} d\varphi} = \frac{4a}{\pi} \tag{1.15}$$

As a second example, consider the expected length for a random jump in an infinitely long tube of radius $a$. Again assume the particle or bug is located at the origin $O$ as shown in Fig. 1.5 and that the particle leaves $O$ in a random manner for the accessible directions. The probability of a path that lies in the solid angle $d\Omega$ is just proportional to $d\Omega$ and the ex-

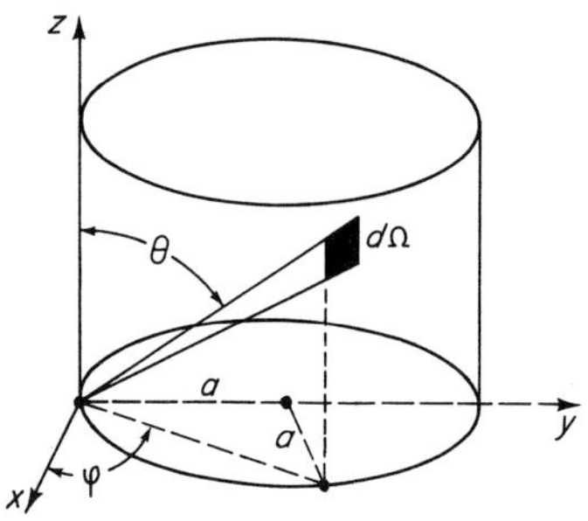


**Fig. 1.5** Path of a bug or molecule making either a random jump or a cosine-law jump in direction $d\Omega$ from position $O$ inside a long tube of radius $a$.

pected path is given by

$$\langle r \rangle = \frac{\int r \, d\Omega}{\int d\Omega} = 2a \tag{1.16a}$$

Integration over the solid angle is limited to the accessible region of $2\pi$ steradians and from Fig. 1.5, $r \sin \theta = 2a \sin \varphi$ is used to evaluate the integral in a convenient manner.

As will be discussed in greater detail in subsequent sections, *particles do not leave surfaces in a random manner* or in the sense that all accessible directions are equally probable. The probability of leaving a surface in the direction $d\Omega$ is proportional to $\cos \alpha \, d\Omega$, where $\alpha$ is the angle between the surface normal and the direction of the particle leaving the surface. The departure of the particles in this manner is often referred to as the *cosine law* or Knudsen's law. Random emission processes are more nearly characteristic of the emission of energetic $\gamma$ rays or nuclear particles from non-oriented atoms. The cosine law is characteristic of the problems encountered in the kinetic theory of gases.

The expected path in a long tube as shown in Fig. 1.5 for a particle leaving the surface according to the cosine distribution law is given by

$$\langle r \rangle = \frac{\int (\cos \alpha \, d\Omega) r}{\int \cos \alpha \, d\Omega} = 2a \tag{1.16b}$$

This integral is readily evaluated by noting in Fig. 1.5 that $\cos \alpha = \sin \theta \times \sin \varphi$. The expected length of a step in the $z$ direction can be evaluated in a similar manner and is

$$\langle | z | \rangle = \langle | r \cos \theta | \rangle = \frac{\int d\Omega \, r \cos \theta \cos \alpha}{\int d\Omega \cos \alpha} = a \tag{1.17}$$

Other interesting cases are examined in the following exercises.

**Exercise 1.3** Show for a long cylinder such as that shown in Fig. 1.5 that $\langle r \rangle = 2a$ for both cosine-law effusion from the surface and for random or diffuse emission from the surface. What is the longest path?

**Exercise 1.4** Repeat Exercise 1.3 for the expected length of a step along the tube, $\langle | z | \rangle$, and show that $\langle | z | \rangle = a$ for cosine-law restitution and $4a/\pi$ for random emission.

**Exercise 1.5** A particle is located on the inside surface of a sphere of radius $a$. Show that a given path has a length $r = 2a \cos \theta$, and then show that the expected path for random emission is

$$\langle r \rangle = a \tag{1.18a}$$

and for cosine-law restitution is

$$\langle r \rangle = \tfrac{4}{3}a \tag{1.18b}$$

## 1.6 Mass Point Gas

As emphasized in the introduction, the properties of the mass point gas are of primary interest in this chapter, and the properties of this gas are now considered in greater detail. The particles composing the gas are placed inside an enclosure of volume $V$ and the coordinate reference frame is attached to this enclosure. For simplicity, external forces acting on the particles are assumed negligible. Interactions between particles are assumed negligible and interaction occurs only at the walls of the enclosure. With this model, the individual particles are unaware of the presence of the other particles when the particles obey the classical equations of motion. The particles are acted upon by no forces and are not accelerated during their motion through $V$. Only at the walls can the energy and momentum of a particle change. Since these ideas are only valid in classical physics, in this chapter, the mass is assumed sufficiently large and the temperature sufficiently high that the de Broglie wavelength is small compared to the interparticle distance and the equation of motion of a particle is almost classical.

In the course of time, the motion of a given particle is such that all elements of volume $\Delta V$ in $V$ are equally probable and the probability of finding a particle in $\Delta V$ is equal to $\Delta V/V$. Also, in the course of time, each collision with the wall changes the velocity and since no direction is preferred, it is expected that all directions of the velocity are equally probable. These two assumptions form the basis for the assumption of molecular chaos. Rather than examine the properties of one particle in the course of time, it is usually more convenient to place $N$ particles in volume $V$ and to discuss the properties of those particles in volume $\Delta V$ at some instant of time. $\Delta V$ is selected sufficiently large to include a reasonably large number of particles. Molecular chaos is assumed and all regions of volume $\Delta V$ are expected to have the same average properties. Thus it is expected that

$$\{\text{average number of particles in } \Delta V\} = N\frac{\Delta V}{V} \tag{1.19}$$

$$\left\{\begin{matrix}\text{average number of particles in } \Delta V \\ \text{traveling in direction } \Delta\Omega\end{matrix}\right\} = \frac{N\,\Delta V}{V}\,\frac{\Delta\Omega}{4\pi} \tag{1.20}$$

Equation (1.20) emphasizes that all regions of $V$ are equally probable and all velocity directions are equally probable; this equation is as yet too restricted. At some instant of time, denote the velocities of the particles by $\mathbf{v}_\alpha$. Let $N_i$ be the number of particles with velocities between $\mathbf{v}_i$ and $\mathbf{v}_i + \Delta\mathbf{v}_i$. Even among this group, the directions are equally probable, or in general, the distribution function in the velocity is spherically symmetric. The number of particles in $\Delta V$ with speeds between $v_i$ and $v_i + \Delta v_i$ and direction $d\Omega$ is

$$\left\{\begin{matrix}\text{number of particles in } \Delta V \text{ with speeds between} \\ v_i \text{ and } v_i + \Delta v_i \text{ and traveling in direction } d\Omega\end{matrix}\right\} = \frac{N_i\,\Delta V}{V}\,\frac{\Delta\Omega}{4\pi} \tag{1.21}$$

A knowledge of the $N_i$ will yield the distribution function for the speeds. Regardless of the detailed form for the $N_i$, the random direction for each velocity group ensures that linear momentum is conserved,

$$\sum_\alpha m\mathbf{v}_\alpha = \sum_i N_i m\mathbf{v}_i = 0 \tag{1.22a}$$

and that angular momentum is conserved,

$$\sum_\alpha m\mathbf{r}_\alpha \times \mathbf{v}_\alpha = \sum_i N_i m(\mathbf{r}_i \times \mathbf{v}_i) = 0 \tag{1.22b}$$

relative to the coordinate frame of the enclosure. Equilibrium or steady state is reached within the enclosure when the number of particles in each velocity group is constant in time, that is,

$$\frac{dN_i}{dt} = \sum_j (N_j P_{ji} - N_i P_{ij}) = 0 \tag{1.23}$$

$P_{ij}$ is the probability of a wall collision changing the speed from $i$ to $j$. For constant $N_i$, the average energy of the particles is a constant,

$$\begin{aligned}\langle E\rangle &= \text{constant} = \sum_{\alpha} \tfrac{1}{2}m{v_\alpha}^2 = \sum_{i} \tfrac{1}{2}N_i m{v_i}^2 \\ &= N\langle\tfrac{1}{2}mv^2\rangle \end{aligned} \tag{1.24}$$

or the average energy $\langle\frac{1}{2}mv^2\rangle$ per particle is a constant. This idea will be developed with greater precision in subsequent chapters.

## 1.7 Effusive Flow of a Mass Point Gas

The number of particles inside volume $V$ crossing a surface $dA$ in direction $d\Omega$ can readily be evaluated using Fig. 1.6 and Eq. (1.21). In order for a particle with speed $v_i$ to cross a surface $dA$ in time $dt$ in direction $d\Omega$, it is nece⸢⸣ary for the particle to lie in the volume $\Delta V = v_i \cos\theta \, dA \, dt$ as

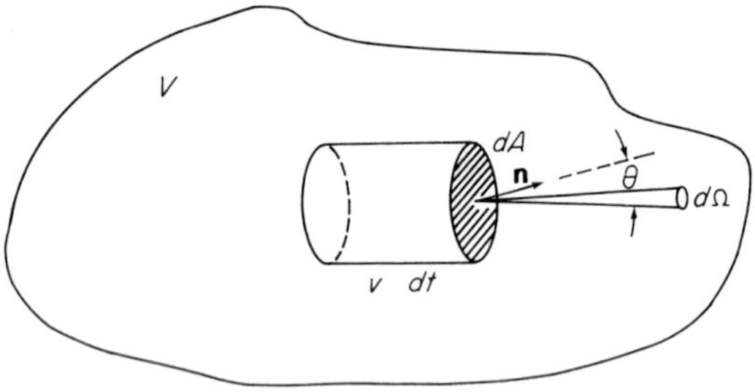


**Fig. 1.6** Particles crossing a surface $dA$ in direction $d\Omega$ in time interval $dt$ with speed $v$ must lie in the volume element $d\mathbf{r} = v(\cos\theta)\, dA\, dt$.

shown in Fig. 1.6. The number of particles that lie in $\Delta V$ and are traveling in direction $d\Omega$ is given by Eq. (1.21) and

$$\begin{aligned}\left\{\begin{matrix}\text{number of particles crossing } dA \\ \text{in interval } dt \text{ in direction } d\Omega\end{matrix}\right\} &= \sum_i \left\{\frac{N_i v_i \cos\theta\, dA\, dt}{V}\right\}\frac{d\Omega}{4\pi} \\ &= n\bar{v}(\cos\theta)\frac{d\Omega}{4\pi}\, dA\, dt \end{aligned} \tag{1.25}$$

In the expression on the right, the average speed is defined as

$$\bar{v} = \frac{\sum N_i v_i}{N} \tag{1.26}$$

and the average number of particles per cubic meter as

$$n = \frac{N}{V} \tag{1.27}$$

Selecting $dA$ as 1 m$^2$, $dt$ as 1 sec, and integrating $(\cos\theta)\,d\Omega/4\pi$ from $\theta = 0$ to $\theta = \pi/2$ yields the number of particles crossing a surface toward one side, that is

$$\Gamma_z = \tfrac{1}{4}n\bar{v} \tag{1.28}$$

The effects of the assumptions in this section are illustrated in Fig. 1.7 by considering the number of particles with speed $v_i$ that cross a surface $dA_1$ and subsequently cross the surface $dA_2$. Since $(\cos\theta_2)\,dA_2 = r^2\,d\Omega_2$, we have

$$\begin{Bmatrix}\text{number of particles per second crossing}\\ dA_1 \text{ and then } dA_2 \text{ with speed } v_i\end{Bmatrix} = \left\{\frac{N_i}{V}\,v_i(\cos\theta_1)\,dA_1\,\frac{d\Omega_2}{4\pi}\right\}$$
$$= \frac{N_i}{V}\,v_i\,\frac{(\cos\theta_1)\,dA_1(\cos\theta_2)\,dA_2}{4\pi r^2} \tag{1.29}$$

following from Eq. (1.25). Previous assumptions of molecular chaos require this equation to be invariant to the location of surfaces 1 and 2 and invariant to the interchange of indices 1 and 2. This is equivalent to a statement of detailed balance, or the number crossing $dA_1$ and then $dA_2$ with speed $v_i$ is equal to the number crossing $dA_2$ and then $dA_1$ with speed $v_i$. If the surface $dA_2$ is adjacent to the wall, the properties of the particles crossing $dA_2$ are the same as those of the particles leaving the wall. From these observations, it follows that the distribution of speeds and the directions,

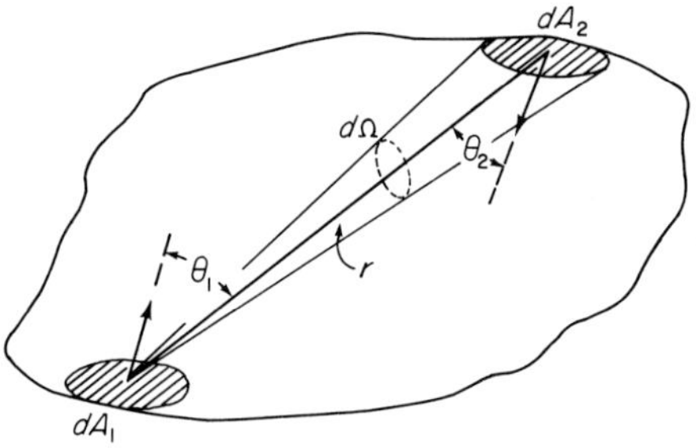


**Fig. 1.7** Particles with speed $v$ which cross surface $dA_1$ and subsequently cross surface $dA_2$ depend on the joint probability of lying in the volume element $d\mathbf{r}_1$ given in Fig. 1.6 and traveling in direction $d\Omega_2 = r^{-2}(\cos\theta_2)\,dA_2$.

that is, the velocity distribution, for particles leaving the wall is the same as for particles crossing any surface $dA$ inside the enclosure. *Particles restituted to the gas from a surface obey the cosine distribution law.*

Equations (1.25) and (1.28) permit the calculation of the effusive flow of atoms from a small hole. As an example, consider a container of volume $V$ which contains $N_0$ particles at time $t = 0$ and has a small opening of area $A$ on the surface. The flow of particles out of the hole is $\frac{1}{4}n\bar{v}A$ and the rate of decrease of particles is given by

$$\frac{dN}{dt} = -\frac{1}{4}n\bar{v}A = -\frac{1}{4}N\bar{v}\frac{A}{V} \tag{1.30a}$$

The solution of this equation,

$$N(t) = N_0 \exp\left(-\frac{1}{4}\bar{v}\frac{A}{V}\right)t \tag{1.30b}$$

is the distribution function for the number of particles in the container as a function of time. The mean time a particle is found in the enclosure is

$$\tau = \frac{\int_0^\infty tN(t)\,dt}{\int_0^\infty N(t)\,dt} = \left(\frac{1}{4}\bar{v}\frac{A}{V}\right)^{-1} \tag{1.30c}$$

## 1.8 Pressure of a Mass Point Gas

Equation (1.25) gives the transport of particles across an area $dA$. Each particle transports momentum $mv_i \cos\theta$ across $dA$ in the direction of the surface normal. The momentum transported is

$$\begin{aligned} dp &= \int \sum_i \left(\frac{N_i}{V} v_i(\cos\theta)\,dA\,dt\,\frac{d\Omega}{4\pi}\right)(mv_i \cos\theta) \\ &= dA\,dt\left[\frac{2}{3}n\left\langle\frac{1}{2}mv^2\right\rangle\right] \end{aligned} \tag{1.31}$$

where $d\Omega$ is integrated over the entire solid angle. This transport of momentum by the gas gives rise to a pressure at the walls of the container. If $dA$ is taken in the interior of the vessel, particles with speed $v_i$ cross from left to right and right to left as shown in Fig. 1.8. At the wall of the vessel, for each particle arriving with speed $v_i$ at angle $\theta$ with the normal, a particle is restituted along the reflected path and the transfer of momentum is

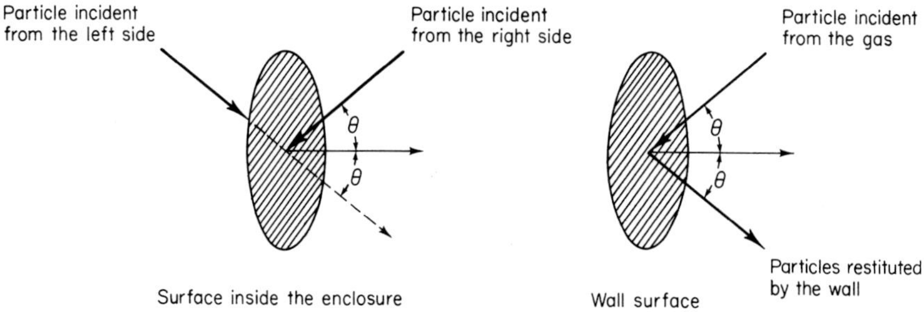


**Fig. 1.8** At the wall, the distribution of velocities of the restituted particles is the same as that for particles crossing a surface $s$ in the interior of the enclosure.

$2mv_i \cos\theta$. Integration is now over the half-space and it is apparent that the net momentum transport across any surface $dA$ is the same as the momentum change at the wall. Thus the pressure exerted by the gas is the force per unit area, and from Eq. (1.31) the pressure is

$$P = \frac{dp}{dt\,dA} = \frac{2}{3}n\left\langle\frac{1}{2}mv^2\right\rangle \tag{1.32}$$

If this expression is compared with the experimentally determined ideal gas laws, the results are consistent if the average kinetic energy per particle is proportional to the temperature $T$. It will be shown in greater detail in subsequent chapters that the average translational energy of a particle is

$$\langle \tfrac{1}{2}mv^2\rangle = \tfrac{3}{2}kT \tag{1.33}$$

and the average energy of a particle in a mass point or ideal gas is a measure of the temperature. The gas law takes on an elementary and useful form,

$$P = nkT \tag{1.34}$$

where $P$ is the pressure in newtons per suare meter (N/m$^2$), $n$ is the number of particles per cubic meter, $k$ is the Boltzmann constant, and $T$ is the absolute temperature in degrees Kelvin (°K). Since Eq. (1.33) is independent of the mass and the species of particles, the pressure exerted by a particle is $kT/V$. The volume can be shown explicitly by writing Eq. (1.34) as

$$PV = NkT \tag{1.35}$$

Either Eq. (1.34) or (1.35) may be compared with the Ideal Gas Laws:

*Boyle's Law*

$$PV = \text{constant} \qquad (\text{for constant } T)$$

*Charles' Law*

$$P = (\text{constant}) \times T \qquad (\text{for constant } V)$$

*Dalton's Law for Partial Pressures*

$$P = nkT = \Big(\sum_{\text{A}} n_{\text{A}}\Big)kT = \sum_{\text{A}} P_{\text{A}}$$

where $n_{\text{A}}$ is the number of particles of species A and $P_{\text{A}}$ is the partial pressure of species A.

*Avogadro's Law*

$$n = \frac{P}{kT}$$

or the number of particles per cubic meter depends on the pressure and temperature and not on the nature of the particle.

Although Eq. (1.34) is extremely simple, many alternate forms of the gas law are used. One of the more common is to define the kilogram-mole as the number of particles in a volume of 22.4136 m³ exerting a pressure of 1 atm at the ice point, or a temperature of 273.15°K, and to denote this number as Avogadro's constant $\mathcal{N}$. Thus Avogadro's constant is

$$\mathcal{N} = \frac{P_{\text{atm}} V_{\text{kg-mole}}}{kT_{\text{ice}}} = 6.0222 \times 10^{26} \tag{1.36}$$

The Boltzmann constant $k$ is

$$k = 1.38062 \times 10^{-23} \quad \text{J/°K}$$

and the gas constant $R$ is then defined as

$$R = \mathcal{N}k = 8314 \quad \text{J/°K-kg-mole} \tag{1.37}$$

Equation (1.35) can now be written as

$$PV = \sum_{\text{A}} \eta_{\text{A}} RT \tag{1.38}$$

where $\eta_{\text{A}}$ is the number of kilogram-moles of species A and use is made of the relationship $Nk = \sum \eta_{\text{A}} \mathcal{N}k = R \sum \eta_{\text{A}}$. The kilogram-mole is not in general use in the mks system of units and Avogadro's constant[1] is usually

[1] A list of fundamental constants can be found in the Appendix.

given as $\mathcal{N} = 6.0222 \times 10^{23}$ and $R$ as 8.314 J/°K-g-mole. Some caution is required in the use of the gas laws as the various values of $R$ are used. Equation (1.34) is insensitive to these choices.

Although the appropriate unit of pressure in the mks system of units is newtons per square meter, this unit is seldom used. The pressure is frequently expressed in terms of the equivalent height of a column of mercury by making use of the expression $P = \varrho g h$, and a pressure of 20 mm Hg means the pressure newtons per square meter exerted by a column of Hg 20 mm in height. In order to avoid ambiguity, a unit of pressure such as the millimeter of Hg requires that a standard acceleration of gravity $g$ and a standard density for mercury be defined. The recommended definitions are

$$g = \{\text{standard acceleration}\} = 9.8062 \quad \text{m/sec}^2$$

$$P_{\text{atm}} = \{\text{standard atmospheric pressure}\} = 101{,}323 \quad \text{N/m}^2$$
$$= 760 \quad \text{Torr}$$

760 mm of Hg pressure is equivalent to one standard atmosphere of pressure. Other units are also used, such as the micron, $10^{-6}$ m Hg, which is used as a unit of pressure in high-vacuum technology. The Torr as a unit of pressure is being used more frequently, and 1 Torr $\approx$ 1 mm Hg.

Exercise 1.6 Show that $n = 9.656 \times 10^{24} P_{\text{Torr}} T^{-1}$. Write a similar expression for $P$ in microns.

Exercise 1.7 Find the number of particles per cubic meter in a vessel at a temperature of 300°K and a pressure of $10^{-7}$ Torr. How many newtons per square meter is equivalent to a pressure of $10^{-7}$ Torr?

Ans. $3.22 \times 10^{15}$; $13.35 \times 10^{-7}$ N/m²

Exercise 1.8 A one-liter vessel contains $10^{16}$ helium atoms and $10^{16}$ argon atoms at 400°K. What is the pressure?

Ans. 0.11 N/m²

## 1.9 Speed Distribution Function

The form of the speed distribution function is contained in a knowledge of the $N_i$, and a theoretical development for the form is given in subsequent chapters. For this theoretical development and for comparison with the experiments discussed in this section, it is convenient to introduce a speed distribution function in terms of continuous variables. The number of particles is always so large that integrals are introduced for the sums. Velocity space is introduced as shown in Fig. 1.9 and the axes are $v_x$, $v_y$, $v_z$. The elements of volume are analogous to those in position space:

$$d\mathbf{v} = dv_x\, dv_y\, dv_z = v^2\, d\Omega\, dv \tag{1.39}$$

$d\mathbf{v}$ is a short notation for the element of volume in velocity space and is analogous to $dV = d\mathbf{r}$ for the element of volume in position space. If $P(v)$ is the probability of a given speed, then Eq. (1.21) can be rewritten as

$$N_i \frac{\Delta V}{V} \frac{\Delta\Omega}{4\pi} = nP(v_i)\, d\mathbf{v}_i\, d\mathbf{r} \tag{1.40}$$

Dropping the subscript $i$, the distribution function is

$$\left\{\begin{matrix}\text{number of particles in volume } d\mathbf{r} \\ \text{with velocity in the range } d\mathbf{v}\end{matrix}\right\} = nP(v^2)\, d\mathbf{v}\, d\mathbf{r} \tag{1.41}$$

Reference to Fig. 1.9 should emphasize the meaning of this equation. The probability that a particle has a velocity vector ending in the element of volume $d\mathbf{v}$ is equal to $P(v^2)\, d\mathbf{v}$. Since the number of particles lying in the element of volume $d\mathbf{r}$ is $n\, d\mathbf{r}$, the product of these two statistically independent quantities results in Eq. (1.41). By our earlier assumptions that all directions are equally probable, the distribution must depend only on the energy.

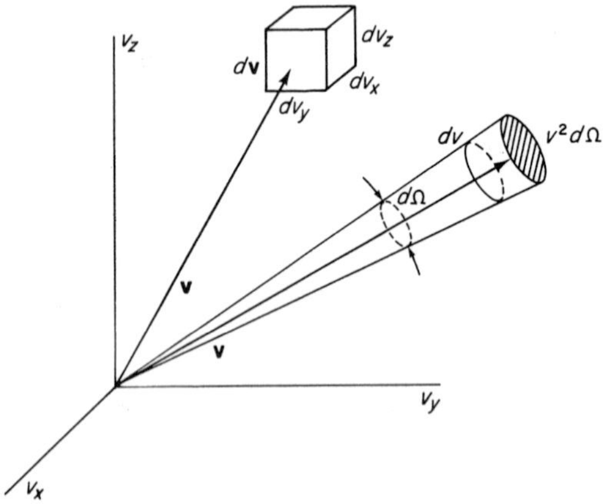


**Fig. 1.9** Elements of volume in velocity space: $d\mathbf{v} = dv_x\, dv_y\, dv_z = v^2\, d\Omega\, dv$.

greement with the Ideal Gas Laws was obtained by assuming an average netic energy of

$$\tfrac{3}{2}kT = \langle \tfrac{1}{2}mv^2 \rangle = \langle \tfrac{1}{2}m(v_x^2 + v_y^2 + v_z^2) \rangle$$

Each direction is equally probable and $\langle \frac{1}{2}mv_z^2 \rangle = \frac{1}{2}kT$. With this assumption, the Doppler effect indicates that the probability that a particle has speed between $v_z$ and $v_z + dv_z$ is given by

$$\left(\frac{m}{2\pi kT}\right)^{1/2} \exp - \frac{mv_z^2}{2kT}\, dv_z \tag{1.44}$$

The Doppler width of a spectral line is usually determined at the half-intensity points, and from the preceding,

$$\frac{\Delta\nu_{\mathrm{D}}}{\nu_0} = \left(\frac{\nu_{+1/2} - \nu_{-1/2}}{\nu_0}\right) = 2\left(\frac{2kT \ln 2}{mc^2}\right)^{1/2} \tag{1.45}$$

Equation (1.43) is written in the typical Gaussian form and is dependent on the dispersion $\langle (\nu - \nu_0)^2 \rangle$. $g$ has its half-value at $\nu_{\pm 1/2}$ and the Doppler width and the dispersion are related by

$$\Delta\nu_{\mathrm{D}} = 2[2\langle (\nu - \nu_0)^2 \rangle \ln 2]^{1/2}$$

Equation (1.43) can be rewritten in terms of $\Delta\nu_D$ and compared with experiment as shown in Fig. 1.10. The agreement is excellent for a Doppler width of 210 MHz and a temperature of 515°K.

A second procedure is to measure the speed distribution of molecules effusing from a small hole. Equation (1.29) for effusive flow can be modified into the continuous distribution form by introducing Eqs. (1.39) and (1.41),

$$\begin{Bmatrix} \text{number of particles per second crossing surface } dA \\ \text{in direction } d\Omega \text{ with speed in interval } dv \end{Bmatrix}$$

$$= nP(v^2)(v^2\, d\Omega\, dv)(v \cos\theta\, dA) = nP(v^2)v_z\, d\mathbf{v}\, dA \tag{1.46}$$

If the surface normal is chosen in the $z$ direction, the flow distribution function may be written in the second form. One of the earliest experiments for determining the speed distribution of molecules was that of Stern [*2*]. A schematic diagram of the apparatus is shown in Fig. 1.11. The source $S_1$, the slit $S_2$, and the drum surface $S_3$, rotate together. The molecules effuse from a small hole in the source $S_1$ and the solid angle $d\Omega$ of effusive flow is limited by $S_2$. When the drum is stationary, the molecules are deposited at $P$, and as the drum rotates, the fast molecules are deposited nearer $P$

One of the simplest experiments for determining the distribution function is to observe the shape of the emis from a hot gas at low densities. If the emitted radiation spectral instrument measuring the radiation traveling in the Doppler shift in frequency and the atomic velocity ar

$$\frac{\nu - \nu_0}{\nu_0} = \frac{v_z}{c}$$

$\nu_0$ is the frequency emitted by atoms at rest, $\nu$ is the measure emitted by atoms moving with velocity $v_z$, and $c$ is the speed of the transverse motion of the atoms causes a much smaller effect, th

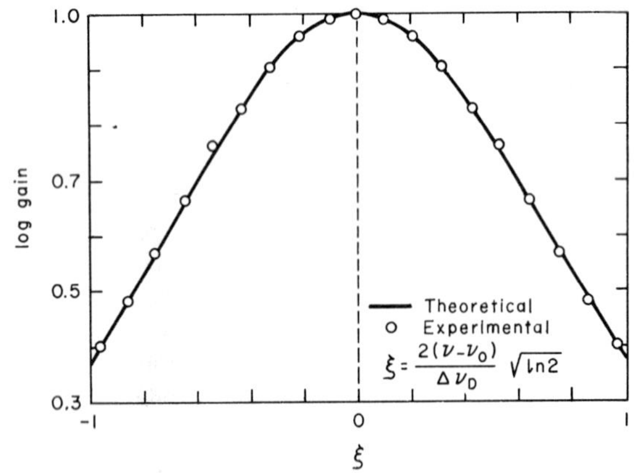


**Fig. 1.10** Doppler motion in a helium–xenon laser amplifier. The gain curve of the amplifier is directly proportional to the Doppler motion of the xenon atoms along the laser axis. $\xi = 2[(\nu - \nu_0)/\Delta\nu_D](\ln 2)^{1/2}$ and $\Delta\nu_D = 210$ MHz. [From C. K. N. Patel, *Phys. Rev.* **131**, 1582 (1963), Fig. 4.]

shift is directly proportional to the velocity in the $z$ direction. Patel [*1*] has measured the Doppler width of the $2.026 \times 10^{-6}$ m line of xenon in a helium–xenon discharge tube by measuring the gain, or optical amplification, as a function of frequency. Typical results are shown in Fig. 1.10, where the gain is plotted as a function of frequency. The spectral line is Gaussian in shape and has the functional form

$$g\left(\frac{\nu - \nu_0}{\nu_0}\right)\frac{d\nu}{\nu_0} = \frac{d\nu}{[2\pi\langle(\nu - \nu_0)^2\rangle]^{1/2}} \exp\left[\frac{-(\nu - \nu_0)^2}{2\langle(\nu - \nu_0)^2\rangle}\right]$$

$$= g\left(\frac{v_z}{c}\right)\frac{dv_z}{c} = \frac{dv_z}{[2\pi\langle v_z^2\rangle]^{1/2}} \exp\left[\frac{-v_z^2}{2\langle v_z^2\rangle}\right] \qquad (1.43)$$

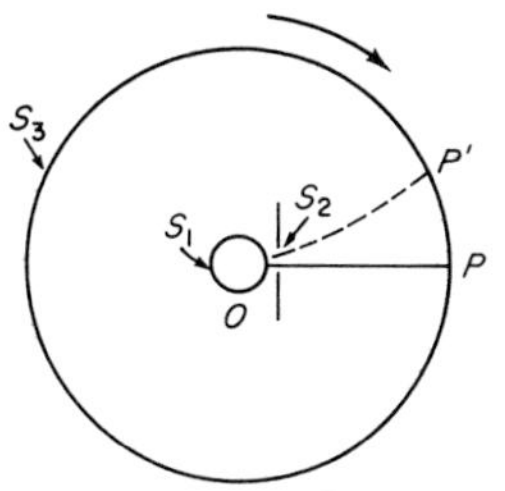


**Fig. 1.11** Early experimental apparatus for determining the speed distribution of molecules.

than the slow molecules. The distance from $P$ can be correlated with the speed and the thickness of the deposited atoms can be used to determine $P(v^2)$.

A more recent experiment of Estermann *et al.* [*3*] determines the speed distribution by defining the beam with horizontal slits and observing the free fall under gravity as shown in Fig. 1.12. Again, the atoms or molecules effuse from a small hole in the oven, and the solid angle $d\Omega$ is formed

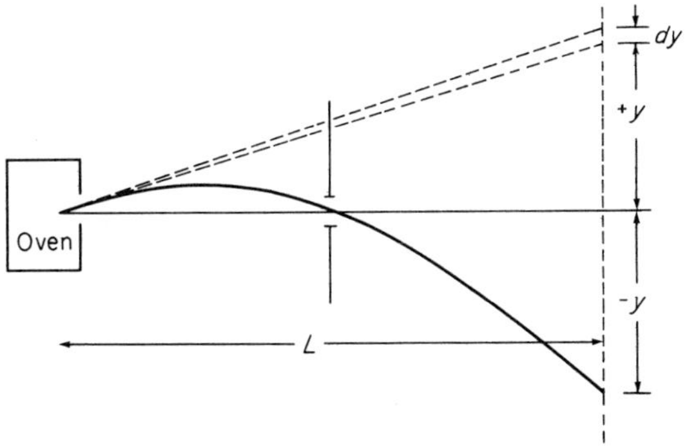


**Fig. 1.12** Free-fall apparatus for measuring the speed distribution of molecules. [From I. Estermann, O. C. Simpson, and O. Stern, *Phys. Rev.* **71**, 238 (1947), Fig. 1.]

into a narrow horizontal strip by the slits. The velocity of a particle is analyzed in the same manner as the trajectory of a bullet. An atom leaving the source toward $+y$ and passing through the middle slit strikes the detector at $-y$. The transit time $t = v/L$, where $L$ is approximately 2 m and the distance fallen is

$$2y = \frac{1}{2} gt^2 = \frac{1}{2} g \frac{v^2}{L^2}$$

All atoms arriving at $-y$ have approximately the same speed $v = 4yL^2/g$ and the number of atoms arriving at $-y$ is directly proportional to the flow distribution given by Eq. (1.46). The intensity as a function of speed for Cs is shown in Fig. 1.13. The experimental points fall near the solid curve which is the value expected with

$$P(v^2) = \left(\frac{m}{2\pi kT}\right)^{3/2} \exp\left(-\frac{mv^2}{2kT}\right)$$

Close examination indicates that there is a deficiency of slow atoms. This is usually attributed to deviations from the model of a mass point gas and is explained by collisions between molecules near the hole for effusive flow in the oven. An extensive list of references is given by Ramsey [*4*] to other experimental techniques used to determine the velocity distribution.

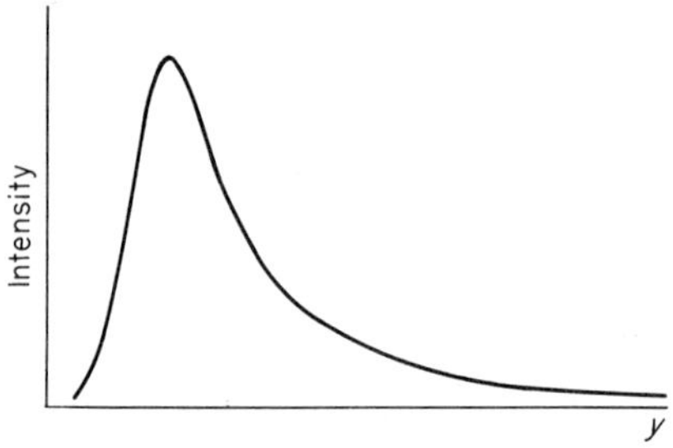


**Fig. 1.13** Intensity as a function of free-fall distance $y$ for apparatus shown in Fig. 1.12. [From I. Estermann, O. C. Simpson, and O. Stern, *Phys. Rev.* **71**, 245 (1947), Fig. 10.]

In summary, the agreement with experiment is excellent if the velocity distribution has the form

$$\left\{\begin{matrix}\text{number of particles per cubic meter}\\ \text{with velocity in the range } d\mathbf{v}\end{matrix}\right\} = n\left(\frac{m}{2\pi kT}\right)^{3/2}\left[\exp\left(-\frac{mv^2}{2kT}\right)\right] d\mathbf{v}$$

$$= nP(v^2)\, d\mathbf{v} = f(v)\, d\mathbf{v} \qquad (1.47)$$

This form will be used rather than the $N_i$ in the subsequent discussions.

The distribution law for velocities which is given by Eq. (1.47) is known as the Maxwellian distribution. Although the validity of the law is based on experimental evidence and requires no further proof, the form of the law was deduced by Maxwell and by Boltzmann on purely theoretical grounds at a time when even the existence of atomic particles was open to question. The development of their ideas will be considered in a systematic manner in later chapters. Historical references are given by Ramsey [*4*].

**Exercise 1.9** Show that the Maxwellian distribution for velocities given by Eq. (1.47) yields the following average values:

(a) $\langle 1 \rangle = 1$ (1.48a)

(b) $$\langle v \rangle = \left(\frac{8kT}{\pi m}\right)^{1/2} \approx 145.5\left(\frac{T}{M}\right)^{1/2} \quad \frac{\text{m}}{\text{sec}} \tag{1.48b}$$

(c) $$\left\langle \frac{1}{2} mv^2 \right\rangle = \frac{3}{2} kT \tag{1.48c}$$

where $M$ is the molecular weight number.

**Exercise 1.10** (a) What is the distribution law for the transport of particles across a surface $dA$ in time $dt$ with velocities in the interval $d\mathbf{v}$ for a Maxwellian gas?

Ans. $nP(v^2)\,d\mathbf{v}\,d\mathbf{r}$ with $d\mathbf{r} = v\,dt \cos\theta\,dA$ (1.49)

(b) Show that the total transport of particles is $\Gamma_z = \frac{1}{4}n\bar{v}$.

(c) Show that the total momentum transported across a surface causes a pressure $P = nkT$.

**Exercise 1.11** (a) Using the transport distribution function in Exercise 1.10, show that the average kinetic energy of a particle transported across a surface or effusing from a small hole is $2kT$.

(b) Show that the average kinetic energy in the direction of the surface normal is $kT$.

**Exercise 1.12** Consider a small, thin disk of area $\Delta A$ placed in a large container of volume $V$ with $n$ particles per cubic meter. Assume that it is possible to construct one surface so that the $\frac{1}{4}n\bar{v}\,\Delta A$ particles that are incident on the surface are returned to the beam with a random orientation of the velocities. The particles leaving the other surface obey the cosine law. Show from previous discussion that the pressure on the cosine-law surface is $P$. Show that on the other surface, the pressure exerted by the incident particles is $\frac{1}{2}P$ and by the particles leaving with random orientation of velocities $\frac{3}{8}P$. Thus a net force of $\frac{1}{8}P\,\Delta A$ acts on the disk and provides an engine which can do work without the usual hot and cold reservoirs. This is a perpetual motion machine of the second kind and violates the Second Law of Thermodynamics [5]. (Note that for random reflection, $d\mathbf{r} = v\,dA\,dt$ and the number arriving is $\frac{1}{4}n\bar{v}$.) Assume the disk does not gain or lose mass or energy. Energy conservation is maintained by choosing the same distribution of speeds for the incident and reflected particles and with this assumption, the problem is independent of this form of the distribution law.

**Exercise 1.13** Show that the number of particles arriving in interval $dy$ in the experiment shown in Fig. 1.12 is proportional to

$$y^{-5/2} \exp\left[-\frac{mg}{4kT}(y + y^{-1})\right]$$

for a Maxwell–Boltzmann velocity distribution for particles effusing from the oven.

## 1.10 Momentum Transfer and Heat Transfer in Free Molecular Flow

The transfer of momentum or energy between surfaces at low pressures or in the free molecular flow region may be discussed in terms of the mass point gas. Consider two large parallel surfaces separated by a distance $D$ as shown in Fig. 1.14a. Gas molecules interact with surfaces $A$ and $B$.

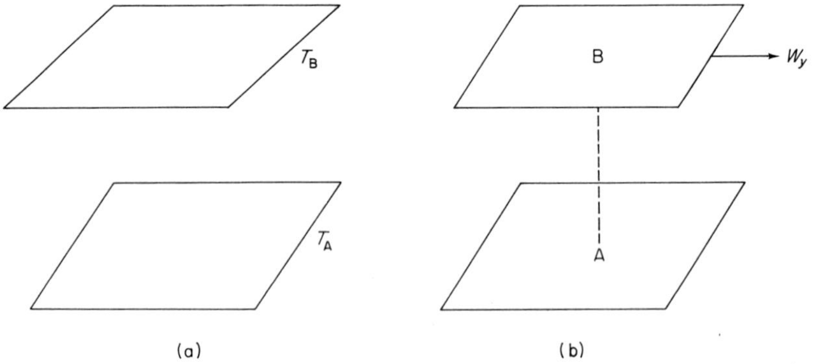


**Fig. 1.14** (a) Heat transfer between surfaces at temperatures $T_A$ and $T_B$. (b) Drag force exerted on surface $A$ by surface $B$ moving with velocity $W_y$ relative to $A$.

During an encounter with the wall, either energy is conserved and the collision is elastic or energy is exchanged with the wall and the collision is inelastic. The wall interactions are discussed in more detail in Chapters IV and IX. In this section, it is assumed that a particle exchanging energy with the wall has a Maxwell–Boltzmann distribution of speeds upon being returned to the gas and this distribution is characteristic of the wall.

Heat transfer between two surfaces provides an elementary example. For this example, let all the particles leaving surface $A$ have properties characteristic of a surface at temperature $T_A$. For a Maxwell–Boltzmann speed distribution, the average kinetic energy transported by this particle is $2kT_A$. Upon arrival at surface $B$, an energy exchange occurs and the particle is restituted to the beam with energy $2kT_B$. The number of particles leaving surface $A$ is $\frac{1}{4}n_A\bar{v}_A$. Since accumulation of molecules at the surface is not desired in this model, the flows are related by $\frac{1}{4}n_A\bar{v}_A = \frac{1}{4}n_B\bar{v}_B = \frac{1}{4}n\bar{v}$, where $n$ and $\bar{v}$ are average numbers appropriate to the gas. The heat transfer is

$$q_z = \left(\frac{1}{4}n\bar{v}\right)[2k(T_B - T_A)] = \frac{1}{2}n\bar{v}k(T_B - T_A) \quad \frac{\mathrm{W}}{\mathrm{m}^2} \tag{1.50}$$

for monatomic molecules.

If surface $B$ has a velocity $W_y$ relative to surface $A$, the molecules restituted from surface $B$ with velocity $v_y$ appear to have velocity $v_y + W_y$ in the inertial frame of surface $A$. A similar argument applies to the molecules leaving surface $A$. A first approximation to the drag force of surface $B$ on surface $A$ is

$$P_{yz} = \{\text{drag force per square meter of } B \text{ on } A\} = \frac{1}{4}n\bar{v}(mW_y) \quad \frac{\mathrm{N}}{\mathrm{m}^2} \tag{1.51a}$$

$P_{yz}$ denotes the momentum in the $y$ direction transported across a surface with a $\hat{z}$ normal. If $\mathbf{C}$ is the velocity distribution observed at surface $B$ and $\mathbf{v}$ the same distribution observed at surface $A$, these two distributions are related by

$$\left[\exp\left(-\frac{mC^2}{2kT}\right)\right] d\mathbf{C} = \left[\exp\left(-\frac{m[v_x^2 + (v_y + W_y)^2 + v_z^2]}{2kT}\right)\right] d\mathbf{v}$$
$$\approx \left[\exp\left(-\frac{mv^2}{2kT}\right)\right] d\mathbf{v}\left(1 - \frac{mv_yW_y}{kT} + \cdots\right)$$

Thus at surface $A$, the molecules appear to arrive with the given distribution and leave with the distribution $\exp(-\frac{1}{2}mv^2/kT)\, d\mathbf{v}$. The momentum transported in the $y$ direction across a surface with a normal in the $z$ direction is

$$P_{yz} = \int d\mathbf{v}\; v_z(mv_y)[f_A(\text{leaving}) - f_A(\text{arriving})]$$
$$= \int d\mathbf{v}\; v_z(mv_y)nP(v^2)\frac{mv_yW_y}{kT}$$
$$= \frac{1}{4}n\bar{v}mW_y \tag{1.51b}$$

where the integral over $d\mathbf{v}$ is for the positive-$z$ side of surface $A$. This technique emphasizes the cancellation occurring in the more elementary development given earlier.

These elementary derivations imply that all the particles are restituted to the beam with a Maxwell–Boltzmann distribution of speeds. A simple modification is to consider a surface for which the probability that an atom is restituted from the surface with a speed distribution characteristic of the surface is $\alpha$ and the probability of an elastic collision is $r$. In a beam particles, $\frac{1}{4}n\bar{v}\alpha$ exchange energy with the wall and $\frac{1}{4}n\bar{v}r$ undergo elastic encounters with the wall. The probability that a molecule leaves surface $A$ with properties characteristic of surface $A$ is $\alpha_A$, and the probability of acquiring the property of wall $B$ is $\alpha_A\alpha_B$ on the first encounter; $\alpha_A r_B r_A \alpha_B$ on the second encounter; and so on. Hence the probability that a molecule leaving surface $A$ with property characteristic of surface $A$ and subsequently taking on the property of surface $B$ is

$$\alpha_A\alpha_B[1 + r_B r_A + (r_B r_A)^2 + \cdots] = \frac{\alpha_A\alpha_B}{1 - r_A r_B} = \frac{\alpha_A\alpha_B}{\alpha_A + \alpha_B - \alpha_A\alpha_B} \tag{1.52}$$

where the last expression follows from $\alpha + r = 1$. If the surfaces are alike, this expression reduces to $\alpha/(2 - \alpha)$. Knudsen refers to $\alpha$ as the accommodation coefficient of a surface. Since the probability of energy exchange with the surface is reduced, it is necessary to *decrease* the heat transfer to the surface or the drag force by quantities of this order of magnitude. Data on the accommodation coefficient are given in Chapter IV. These simple ideas permit an estimate of heat transfer and drag forces in the free molecular flow region and these effects can be reduced by selecting a surface with a small accommodation coefficient.

Exercise 1.14 A bug flies back and forth between pots of red and black paint. Let $\alpha_R$ be the probability that the bug alights in the red paint and $\alpha_B$ that it alights in the black paint. Show that the probability of a bug leaving the red pot covered with red paint and subsequently being covered with black paint is given by $\alpha_R\alpha_B(\alpha_R + \alpha_B - \alpha_R\alpha_B)^{-1}$.

Exercise 1.15 Show that the heat transfer between two surfaces with accommodation coefficients $\alpha_A$ and $\alpha_B$ is given by

$$q_z = (\alpha_A\alpha_B)(\alpha_A + \alpha_B - \alpha_A\alpha_B)^{-1}q_z^{(0)} \tag{1.53}$$

where $q_z^{(0)}$ is given by Eq. (1.50).

**Exercise 1.16** Find the heat transfer per square centimeter between two plates separated by a small distance and at temperatures of 1.0 and 1.1°K at a pressure of $10^{-6}$ Torr helium gas.

Ans. $4.9 \times 10^{-8}$ W/cm$^2$

**Exercise 1.17** Estimate the drag force per centimeter of length between two rotating concentric cylinders of radii 5 and 5.5 cm and which have an angular velocity of rotation of 1 rps. Use helium gas at a pressure of $10^{-6}$ Torr and a temperature of 300°K.

Ans. $P_{r\varphi} = 2.3 \times 10^{-8}$ N/m$^2$
$\{\text{drag force}\} = 6 \times 10^{-11}$ N

**Exercise 1.18** Estimate the drag force on a satellite at a height of 100 miles. Assume a radius of 1 m. Note that the velocity of the satellite is orders of magnitude greater than the average particle velocity.

Ans. $\approx 10^{-2}$ N/m$^2$

**Exercise 1.19** Three surfaces are arranged as shown in Fig. 1.15. Show that

$$\frac{\{q_z \text{ with surface } 3\}}{\{q_z \text{ without surface } 3\}} = \frac{1}{2}$$

What is the effect of an accommodation coefficient $\alpha$? What is the effect of $N$ layers between surfaces 1 and 2? (See the work of Scott [*6*] for insulation materials.)

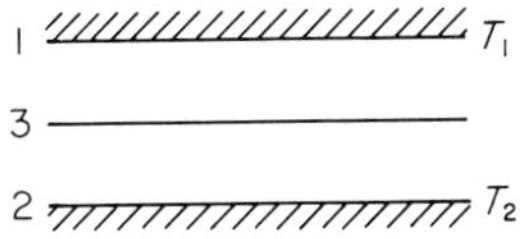


**Fig. 1.15** Diagram for Exercise 1.19.

## 1.11 Free Molecular Flow

The elementary ideas developed for the mass point gas can be used in the study of some simple free molecular flow problems, or flow processes in which wall interactions are dominant. Molecular collisions have a negligible effect. A number of special topics are discussed in this section.

### *1.11.1 Effusive Flow*

The effusive flow from a hole which is small compared to the size of the container is given by Eq. (1.30a) as $\dot{N} = \frac{1}{4}n\bar{v}A$ and since $\bar{v} = (8kT/\pi m)^{1/2}$, isotopes of different mass have different rates of flow out of the hole. Thus if the container shown in Fig. 1.16 contains isotopes of $^3$He and $^4$He,

$$\frac{\dot{N}_3}{\dot{N}_4} = \frac{n_3}{n_4}\left(\frac{m_3}{m_4}\right)^{1/2} \tag{1.54}$$

A suitable arrangement of containers can be used for isotope separation.

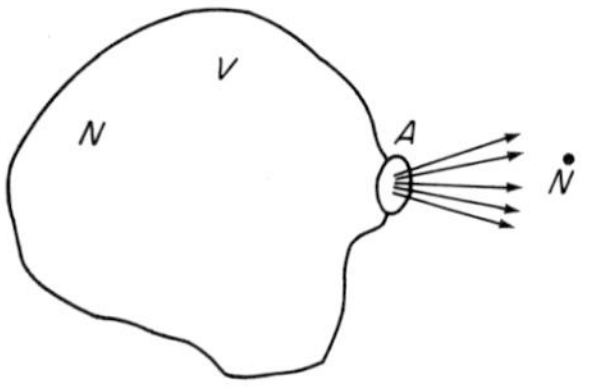


**Fig. 1.16** Effusive flow of molecules from a small hole.

Another interesting example, shown in Fig. 1.17, concerns the rate of evaporation from a surface at low vapor pressures. Since the pressure is low, molecular collisions are not important and the property of the vapor adjacent to the surface is the same as in the bulk gas. Detailed balance requires

$$\begin{Bmatrix}\text{number of molecules evaporating from}\\ \text{1 m}^2\text{ of surface per second}\end{Bmatrix} = \begin{Bmatrix}\text{number striking 1 m}^2\\ \text{of surface per second}\end{Bmatrix}$$

$$= \frac{1}{4}n\bar{v} = \frac{P_v\bar{v}}{4kT} \tag{1.55}$$

and provides a simple method for estimating the rate of evaporation at a given vapor pressure $P_v$.

In high-vacuum technology the speed of an orifice or a hole is referred to as

$$S = \frac{1}{4}\bar{v}A \quad \frac{\text{m}^3}{\text{sec}} \tag{1.56a}$$

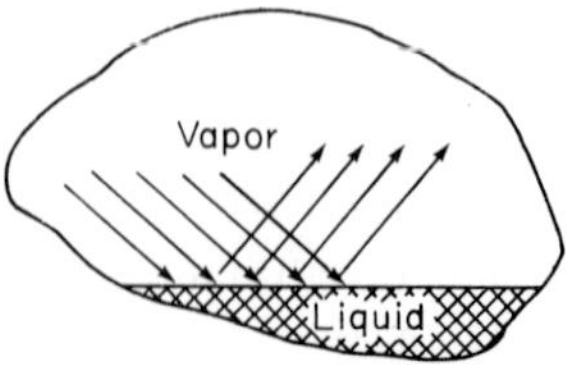


**Fig. 1.17** Evaporation from a liquid surface and sublimation from a solid surface. (1) Vapor; (2) liquid or solid.

Commercial vacuum pumps approach 70% of this ideal speed. Pumps are usually rated in terms of volume flow and, as discussed in Exercise 1.22.

$$\{\text{orifice flow in micron-liters/second}\} = S_\alpha P_\alpha \tag{1.56b}$$

where $S_\alpha$ is the speed and $P_\alpha$ the partial pressure of a given molecular species in microns.

### *1.11.2 Thermal Transpiration and Flow through a Narrow Tube*

Consider the static problem of two vessels $A$ and $B$ which are connected by a small hole. The hole is assumed small, so that the properties of the gas crossing from $A$ to $B$ are characteristic of $A$ and vice versa for those crossing from $B$ to $A$. Usually this is shown as a long, narrow tube as shown in Fig. 1.18, but only the previous assumption is necessary. The steady-state condition is determined by the detailed balance argument that the number of particles leaving $A$ per second must equal the number of particles arriving from $B$ per second and thus maintain the pressure constant in each container. The cross section cancels and

$$\tfrac{1}{4}n_A\bar{v}_A = \tfrac{1}{4}n_B\bar{v}_B \qquad \text{or} \qquad P_A T_A^{-1/2} = P_B T_B^{-1/2} \tag{1.57}$$

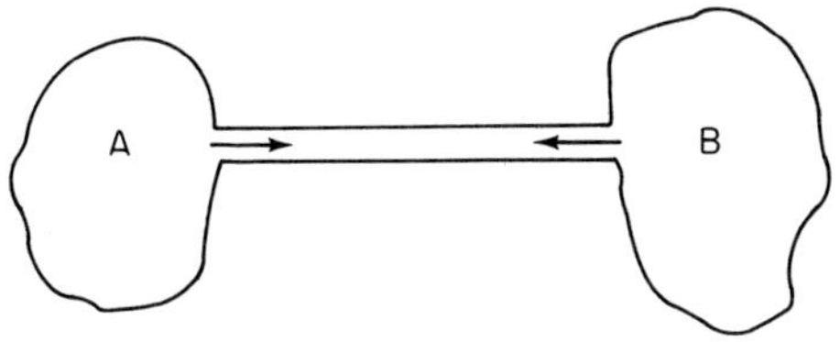


**Fig. 1.18** Pressure $P_A$ and $P_B$ and numbers of molecules $n_A$ and $n_B$ in two vessels at temperatures $T_A$ and $T_B$ and connected by a small-bore tube.

The temperature difference causes a pressure difference between the two bulbs of gas. If the ratio of the temperatures is large, an appreciable pressure difference can occur. Thus gas flows can be caused by a temperature difference and one might consider driving a small turbine through a larger-diameter return tube.

Returning to more conventional problems, Eq. (1.57) indicates that in the pressure region in which wall collisions dominate other collision processes, the pressure along the tube is determined by the wall temperature for static conditions. As a first approximation to the flow through a long, narrow tube, or in general, a tube in which the gas pressure is such that the wall collisions are more important than particle–particle collisions, a flow dependent on the gradient of $n\bar{v}$ or $P/\sqrt{T}$ is expected. In order to obtain the necessary numerical coefficients, consider the tube shown in Fig. 1.19. Using Eq. (1.29) and Fig. 1.8 for cosine-law effusion, one has for the flow from surface $dS'$ on the wall through the surface $dS$ on the tube cross section

$$\begin{Bmatrix}\text{number of particles restituted from}\\ dS' \text{ and crossing } dS \text{ per second}\end{Bmatrix} = \frac{n'\bar{v}' \cos\theta' \, dS' \cos\theta \, dS}{4\pi r^2}$$

$$= n'\bar{v}'(\cos\theta)\, dS \frac{d\Omega}{4\pi}$$

and the flow $\dot{N}$ across the tube cross section is given approximately by

$$\dot{N} = -\frac{1}{8}\frac{d(n\bar{v})}{dz}\left\{\iint_S dS \int_0^{2\pi} s\, d\alpha\right\} \tag{1.58}$$

The general transport equation can be reduced to the approximate form of Eq. (1.58) in the following manner. $n'\bar{v}'$ is used since the *gas properties are characteristic of the wall.* No sources or sinks occur in the tube and the

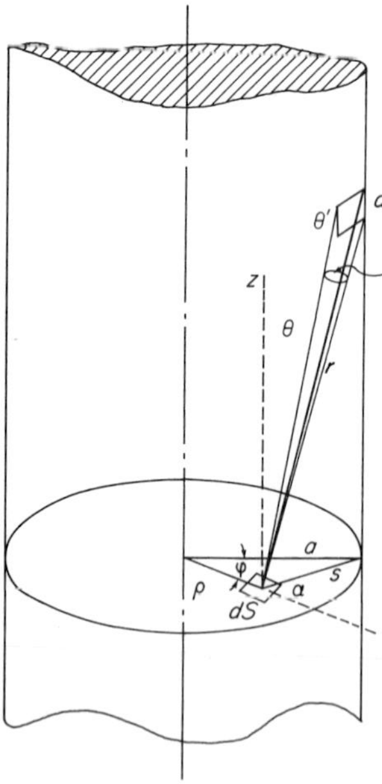


**Fig. 1.19** Flow of molecules through a tube at low gas pressures.

flow is steady, so that $\dot{N}$ is constant and independent of the position of the cross section. Expanding the wall term $n'\bar{v}'$ as a Taylor series expansion about $z = 0$,

$$n'\bar{v}' = (n\bar{v})_{z=0} + \left[\frac{d(n\bar{v})}{dz}\right]_{z=0} z + \cdots$$

and noting that $z = r\cos\theta = s\cot\theta$, the integration over $\theta$ from 0 to $\pi$ gives zero contribution for the first term. The flow is proportional to the gradient for small flows and integration over $\theta$ gives Eq. (1.58). Additional even terms in the expansion are zero and the odd terms are assumed small. The second term in the large parentheses depends on the tube shape. For a cylindrical tube of radius $a$, the expression simplifies and

$$\int dS \int s\,d\alpha = \int d\alpha \int s\,dS = \frac{16\pi a^3}{3}$$

Thus the flow through a circular tube is given by

$$\dot{N} = \{\text{flow through a long cylindrical tube in particles per second}\}$$
$$= -\frac{2}{3}\pi a^3 \frac{d(n\bar{v})}{dz} = -\frac{2}{3}\pi a^3\left(\frac{8}{\pi k m}\right)^{1/2} \frac{d(P/\sqrt{T})}{dz} \tag{1.59}$$

In deriving this expression for flow, the tube is assumed sufficiently long that the end effects can be neglected and the $\theta$ integral is extended from 0 to $\pi$. For a tube at constant temperature, the flow of particles and pressure difference are related by

$$\dot{N} = \frac{2}{3}\pi a^3 L^{-1}\left(\frac{8}{\pi m k T}\right)^{1/2}(P_2 - P_1) \tag{1.60}$$

In high-vacuum flow problems, $\dot{N}$ is often written as an expression in volume flow, and at constant temperature, the flow is written as

$$\{\text{volume flow in (N/m}^2\text{) (m}^3\text{/sec)}\} = \dot{N}kT = \frac{d(PV)}{dt}$$
$$= \left(\frac{2}{3}\frac{\pi a^3 \bar{v}}{L}\right)(P_2 - P_1)$$
$$= C(P_2 - P_1) \tag{1.61}$$

More conventional units are microns for pressure and liters for the volume. Equations (1.60) and (1.61) are often expressed as $C(P_2 - P_1)$, where $C$ is called the conductance of the tube. If tubes of many sizes are connected in series, $\dot{N}$ remains constant throughout, and one can readily show the additive property of conductance. (See Exercise 1.24.)

In all of the following exercises, assume free molecular flow as approximated by the mass point gas.

**Exercise 1.20** In a molecular beam experiment similar to that shown in Fig. 1.7, the source contains 50% monatomic hydrogen and 50% diatomic hydrogen at a pressure of 0.1 Torr. The gas effuses through a $0.1 \times 0.2$ mm slit at the source and then passes through a second slit $0.5 \times 1.0$ cm at a distance of 1 m from the first slit. Find the number of H and $H_2$ particles leaving the first slit and then the number passing through the second slit ($T = 400°K$).

Ans. $\dot{N}_{\mathrm{H}} = 3.5 \times 10^{10} = \sqrt{2}\,\dot{N}_{\mathrm{H}_2}$

**Exercise 1.21** A vacuum gauge at room temperature, or 300°K, is connected to a container at 1°K via a tube. The gauge indicates a pressure of $10^{-6}$ Torr. What is the pressure in the cold container?

Ans. $5.8 \times 10^{-8}$ Torr

**Exercise 1.22** The particle flow through a hole of area $A$ is $\frac{1}{4}n\bar{v}A$; recall that in high-vacuum technology, the speed of an orifice is referred to as $\frac{1}{4}\bar{v}A = S$. Find the speed of a 4-in.-diameter hole for air and helium at 27°C. Give the answer in liters per second. Commercial pumps approach 70% of this ideal speed. Show that the flow through the orifice can be expressed by Eq. (1.56b):

$$\{\text{orifice flow in micron-liters/second}\} = S_\alpha P_\alpha \tag{1.62}$$

where $S_\alpha$ and $P_\alpha$ are the speed and partial pressure of each species. Show that the pressure in newtons per square meter is related to the pressure in microns by $P_{\mu\mathrm{m}} = 0.133\, P_{\mathrm{N/m^2}}$.

Ans. 960 liters/sec and 2570 liters/sec

**Exercise 1.23** A 1000-liter container containing air at $10^{-5}$ Torr and 27°C is evacuated through a 2-m cylindrical tube with a diameter of 10 cm. Find the initial rate of flow in number of particles per second and in terms of micron-liters/second. How long does it take to reduce the pressure to $10^{-8}$ Torr?

Ans. $2 \times 10^{16}$; approximately 112 sec

**Exercise 1.24** Show that in Fig. 1.20,

$$\{\text{flow}\} = C(P_B - P_A), \qquad \frac{1}{C} = \frac{1}{C_1} + \frac{1}{C_2} + \frac{1}{C_3} \tag{1.63}$$

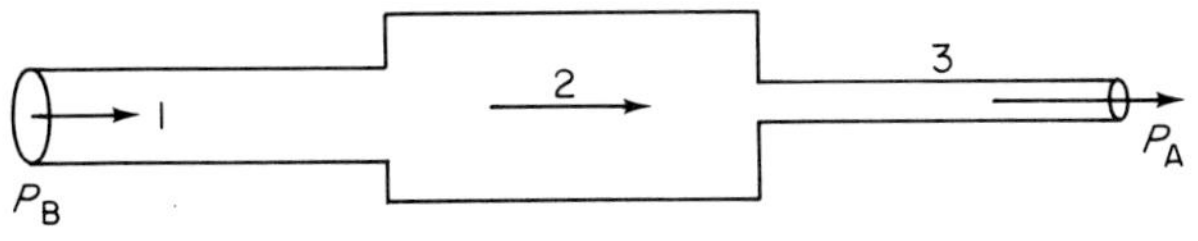


**Fig. 1.20** Diagram for Exercise 1.24.

**Exercise 1.25** Estimate the rate of evaporation of sodium atoms from a surface of sodium at temperature of 50°C.

Ans. $2 \times 10^{16}$ atoms/m²

**Exercise 1.26** Estimate the rates of sublimation of nylon and of evaporation of water from a satellite at a temperature of 0°C.

### *1.11.3 Molecular Pump*

An efficient vacuum pump utilized by Gaede and now commercially available consists of a rotating drum within an outer cylinder. The general features of the pump are shown in Fig. 1.21a and the idealized molecular pump is shown in Fig. 1.21b. Containers $A$ and $B$ are connected by a tube with walls moving with velocity $W$ toward the right. The pressure ratio $P_A/P_B$ can be estimated in the following manner. The flow out of vessel $A$ is $\frac{1}{4}n_A\bar{v}S_A$. Flow into vessel $A$ can be estimated by referring to Fig. 1.19 and Eq. (1.58). The particles restituted from the tube wall flow into vessel $A$, and at steady state,

$$\frac{1}{4} n_A \bar{v} S_A \approx \int n(z)(\bar{v} - W)\, dS \cos\theta_A \frac{d\Omega_A}{4\pi} \tag{1.64a}$$

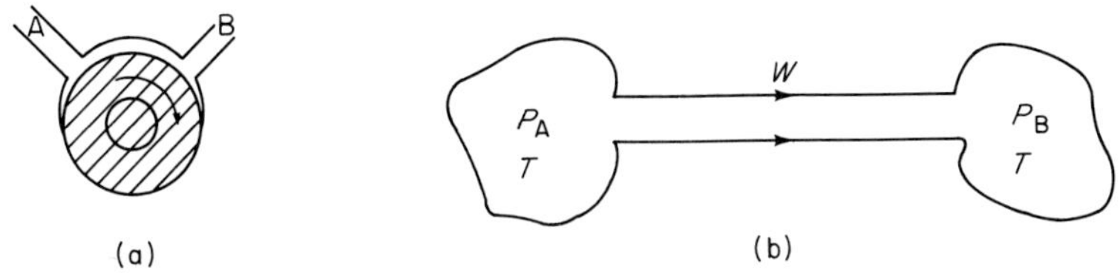


**Fig. 1.21** (a) Schematic of the molecular pump. (b) Pressure difference between two vessels connected by a tube with walls moving with velocity $W$.

A similar equation follows for surface $S_B$ with $A$ changed to $B$ and $W$ to $-W$. The ratio of the pressures is given by

$$\frac{P_A}{P_B} = \frac{n_B}{n_A} \approx \left(1 + \frac{W}{\bar{v}}\right)\left(1 - \frac{W}{\bar{v}}\right)^{-1} \times \text{(constant)} \qquad (1.64b)$$

where the constant depends on the integrals along the tube. The actual system is more complex, but ratios as large as $10^5$ to $10^6$ can be obtained.

### 1.11.4 Knudsen's Absolute Manometer

If two closely spaced surfaces are at temperatures $T_1$ and $T_2$ as shown in Fig. 1.22(a), the momentum transport across any parallel surface $S$ is a

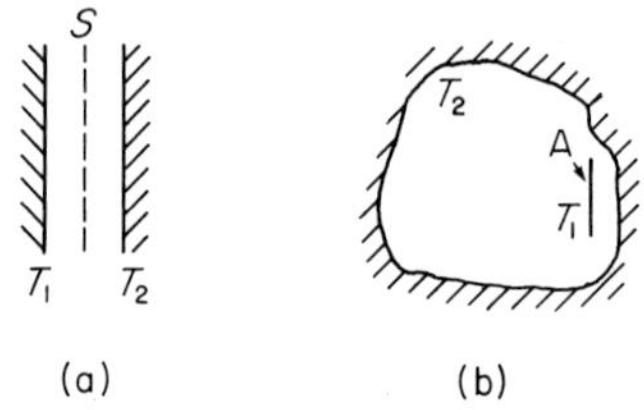


**Fig. 1.22** (a) Pressure between two parallel surfaces at temperatures $T_1$ and $T_2$. (b) Force on surface $A$ at temperature $T_1$ and very close to a portion of the wall of the containing vessel at temperature $T_2$.

constant and defines a pressure

$$P = P_1 = P_2$$

$P_1$ and $P_2$ are the wall pressures. Particles are not permitted to accumulate at the walls and regardless of the character of the walls,

$$\tfrac{1}{4}n_1\bar{v}_1 = \tfrac{1}{4}n_2\bar{v}_2$$

where $\frac{1}{4}n_1\bar{v}_1$ is the number of atoms returned by a surface to a gas that is described by $n_1$ and $T_1$, and so on. For momentum transport across $S$ the corresponding pressure $P$ is given by

$$P = \frac{1}{2}n_1kT_1 + \frac{1}{2}n_2kT_2 = \frac{1}{2}n_2kT_2\left[\left(\frac{T_1}{T_2}\right)^{1/2} + 1\right]^{-1} \qquad (1.65)$$

In Fig. 1.22b, a surface $A$ at temperature $T_1$ is placed near the wall at temperature $T_2$. The pressure between the closely spaced surfaces $T_1$ and $T_2$ is given by Eq. (1.65), and the pressure on the other side of $T_1$ is characteristic of the pressure $P = n_2kT_2$. The net force on surface $A$ and pressure

$P$ are related by

$$P = \left(2\frac{F}{A}\right)\left[\left(\frac{T_1}{T_2}\right)^{1/2} - 1\right] \tag{1.66}$$

Knudsen used a torsion suspension system to measure this small force.

**Exercise 1.27** Show that Eq. (1.66) is consistent with the flow of particles of $\frac{1}{4}n_2\bar{v}_2$ into the small region between the plates and $\frac{1}{2}(\frac{1}{4}n_1\bar{v}_1 + \frac{1}{4}n_2\bar{v}_2)$ out of this same region.

### *1.11.5 Thermal Creep, Radiometer Action, and Thermomolecular Pressure*

It was mentioned in the discussion of Fig. 1.18 that an engine could be driven by the pressure difference created by the temperature difference between the ends of a tube or a set of tubes of small diameter. If the temperature of the hot end is held constant and the flow through a return path reduces the pressure at the hot end, the gas flow through the small-diameter tube is toward the region of higher temperature and higher pressure. This is apparent in Eq. (1.57) or in Eq. (1.59) for the flow through a tube in the free molecular flow region. Equation (1.59) indicates that the flow through the tube at constant temperature is proportional to $-dP/dz$ and the gas flows to the region of lower pressure. Flow at constant pressure is proportional to $+dT/dz$ and is toward the region of higher temperature and pressure and is referred to as "thermal creep."

It is not possible to complete the interesting aspects of the discussion in the free molecular flow region here and some of the results which are derived in Chapters IV and V are used. Figure 1.23 shows a tube of large diameter with a temperature difference $T_2 - T_1$ between the ends of the

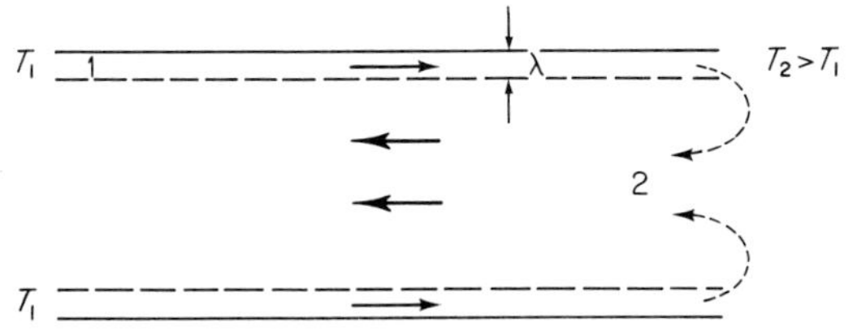


**Fig. 1.23** Thermomolecular pressure caused by a temperature difference between the ends of the tube. Thermal creep near the tube walls causes free molecular flow (1) toward the hot end of the tube and the return flow through the central part of the tube is by normal viscous flow (2).

tube. Collisions between gas particles can occur and a particle restituted from the wall has its first collision at an average distance $\lambda$ from the wall. The gas in this region of thickness $\lambda$ has the characteristics of the free molecular flow region and thermal creep occurs in this thin layer toward the hot end. Return flow occurs in the large central region and $P_2 = P_1$. This real example has an almost constant pressure throughout the tube. The flow in the boundary layer is proportional to $d(P/\sqrt{T})/dz$ or $+dT/dz$ and toward the hot end, and the viscous return flow is proportional to $-dP/dz$ and toward the low-pressure end. The pressure difference caused by thermal creep is referred to as the "thermomolecular pressure difference." Watkins *et al.* [*7*] discuss the measurement of thermomolecular pressure differences for $^3$He and $^4$He and give a list of references pertaining to the completion of the problem discussed here. When a movable disk is placed adjacent and parallel to a wall with a temperature gradient along the wall, the thermal creep of the gas toward the higher-temperature region causes a drag force on the movable disk. This tangential force has been measured [*8*].

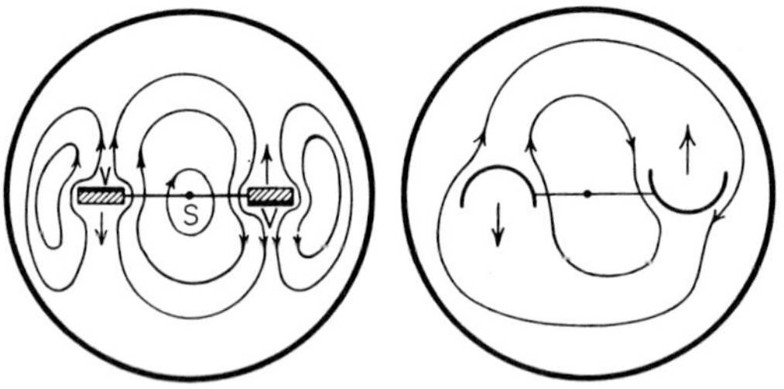


**Fig. 1.24** Thermal creep and return gas flow, causing the radiometer effect. The source of radiation is at the bottom of the figure. (From "Kinetic Theory of Gases," p. 334, Fig. 78, by E. H. Kennard. Copyright 1938, McGraw-Hill, New York. Used with permission of McGraw-Hill Book Company.)

The radiometer effect is also due to thermal creep, and the elementary explanation which is usually given is wrong. If the vane in the radiometer is a poor thermal conductor and is blackened on one side and acts as a reflector on the other as is shown in Fig. 1.24, then radiation incident from one direction causes a thermal gradient across the vane. Thermal creep occurs around the edge of the vane from the cold side to the hot side, causes a higher pressure on the hot side, and the gas then returns by a viscous flow path in the manner shown [*9*]. The temperature difference across the vane gives rise to a small pressure difference which causes the black side of the vane to move away from the light source. The radiometer effect is greatest at a pressure of the order of 0.05 Torr and needs both viscous and free molecular flow aspects for its explanation.

Exercise 1.28 Cup-shaped radiometer vanes are often used and it is observed that the open side of the cup moves toward the light. Explain, by noting that the bottom of the cup has a higher temperature, and the temperature gradient is between the bottom and the lip of the cup.

## References

1. C. K. N. Patel, *Phys. Rev.* **131**, 1582 (1963).
2. O. Stern, *Z. Phys.* **3**, 418 (1920).
3. I. Estermann, O. C. Simpson, and O. Stern, *Phys. Rev.* **71**, 238 (1947).
4. N. F. Ramsey, "Molecular Beams." Oxford Univ. Press, London and New York, 1956.
5. P. S. Epstein, *Phys. Rev.* **23**, 710 (1924).
6. R. B. Scott, "Cryogenic Engineering." Van Nostrand-Reinhold, Princeton, New Jersey, 1959.
7. R. A. Watkins, W. L. Taylor, and W. J. Hauback, *J. Chem. Phys.* **46**, 1007 (1967).
8. M. Czerny and G. Hettner, *Z. Phys.* **30**, 258 (1924).
9. P. S. Epstein, *Z. Phys.* **54**, 537 (1929).

# Radiation

# II

## 2.1 Introduction

Planck's introduction of the concept of the quantization of electromagnetic radiation to provide a satisfactory explanation of thermal or "blackbody" radiation has led to the description of thermal radiation as a "gas" composed of photons. The linear form of Maxwell's equations for electromagnetic phenomena indicates that photons do not interact with each other and the photon gas is an ideal gas. In many respects, the discussion given for particles in Chapter I applies to photons. In Chapter I, the particles were treated in a classical manner, and quantum mechanics is needed to explain binary encounters and molecular forces. For photons, a quantum mechanical explanation is required at the beginning of the development.

Although the historical approach to problems of radiant energy are of considerable interest, the study is greatly facilitated by the use of more recent developments. This latter approach is used here; therefore the elegant deductions made from experimental studies prior to 1900 will often appear as almost trivial. This is indeed not the case; the author does not wish to retrace this tortuous path of discovery, but wishes to devote the present chapter to problems to which this work is applicable. An excellent review is given in Planck's "Warmestrahlung" [*1*].

The next section discusses some aspects of electromagnetic radiation that follow directly from Maxwell's equations. Measurable aspects of radiation from independent sources and the concept of the modes of the radiation field are introduced. This is then combined with the basic postulate of Planck to form the basis for a consideration of thermal radiation. Section 2.4 starts the discussion of the transfer of radiant energy and includes most of the simple and useful concepts in the transfer of thermal radiation. The next two sections can be omitted for the reader primarily interested in the transfer of radiant energy.

## 2.2 Electromagnetic Radiation

An understanding of electromagnetic radiation starts with Maxwell's equations for free space or the vacuum. A simple solution to these equations is the electric vector (the caret denotes a unit vector)

$$\mathbf{E}_k(\mathbf{r}, t) = E_k \hat{u}_k \cos(\mathbf{k} \cdot \mathbf{r} - \omega t) \tag{2.1a}$$

This is a traveling wave in direction $\hat{k}$ with amplitude $E_k$ in volts/meter, with linear polarization $\hat{u}_k$, and with frequency $\nu = \omega/2\pi$. It is a solution to Maxwell's equations if the wave number $k$, the velocity of light $c$, and angular frequency $\omega$ are related by

$$k^2c^2 = \omega^2 \qquad \text{or} \qquad \nu\lambda = c \tag{2.1b}$$

and $\hat{k} \cdot \hat{u}_k = 0$. The wavelength $\lambda$ and $k$ are related by $k = 2\pi/\lambda$. The average energy crossing a square meter per second for this plane wave follows from the Poynting vector $\mathbf{E} \times \mathbf{H}$ and is

$$\hat{k}\left(\frac{1}{2}\,\varepsilon_0 E_k{}^2\right)c \quad \frac{\mathrm{W}}{\mathrm{m}^2} \tag{2.2}$$

The plane wave given in Eq. (2.1) could be used to approximate the spherical wave emitted by a distant source as shown in Fig. 2.1a. Since Maxwell's equations are linear, the electric field at some point in space is the sum of the electric fields generated by different sources and this sum is referred to as the principle of superposition. If the sources are distant, then a good approximation to the electric field is a set of plane waves,

$$\mathbf{E}(\mathbf{r}, t) = \sum_k \mathbf{E}_k(\mathbf{r}, t)$$

where each $\hat{k}$ is in the direction of the source. The energy flow measured by a detector located in the region near $r$ is usually of primary interest

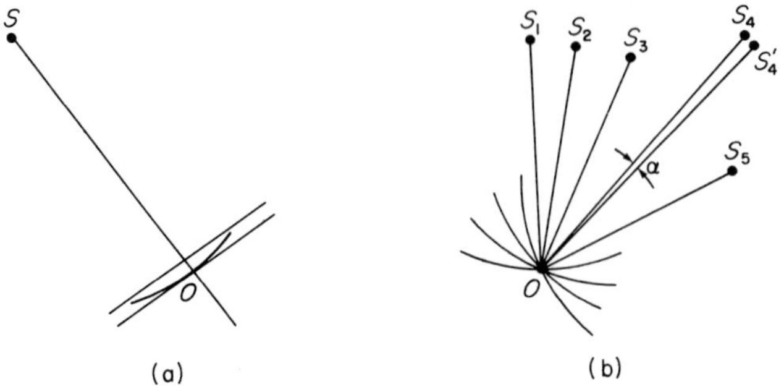


**Fig. 2.1** (a) Spherical wave emitted at $S$ and observed at $O$. (b) Spherical waves emitted by sources $S_1, \ldots, S_N$ and observed at $O$. $\alpha$ is the angle between $S_4$ and $S_4'$.

and some care is necessary in forming $\mathbf{E} \times \mathbf{H}$ for a set of plane waves. Direct superposition yields the double sum

$$\mathbf{E} \times \mathbf{H} = \sum_k \sum_{k'} \mathbf{E}_k \times \mathbf{H}_{k'}$$

If the sources are independent or incoherent, waves with different $k$ do not interfere and the double sum reduces to a single sum,

$$\mathbf{E} \times \mathbf{H} \approx \sum_k (\mathbf{E}_k \times \mathbf{H}_k)$$

This seems like an easily measured quantity, but at a particular point in space, an observer can measure only the power crossing a surface within a given solid angle $d\Omega$ and frequency range $d\nu$. The size of the detector will limit the solid angle and the response time of the detector will limit the frequency range. This is quite apparent from elementary considerations. If the measuring interval is $\Delta t$, the frequency cannot be measured more accurately than

$$\Delta\nu\, \Delta t \approx 1 \tag{2.3a}$$

Elementary considerations of physical optics indicate that the angular spread of radiation from a slit, or the angular resolution of a lens, of size $D$ is of the order of $\alpha \approx \lambda/D$. Thus the direction cannot be determined more accurately than

$$\Delta\Omega \approx \frac{\lambda^2}{\Delta S} \tag{2.3b}$$

where $\Delta S$ is the area of the detector. If, in Fig. 2.1b, the two distant sources are separated by an angle of less than $\lambda/D$, a detector of size $D$ will observe only one source and interpret this as an energy flow $\frac{1}{2}c\varepsilon_0 E_k^2$. It is in this sense that the intensity is defined for independent sources by the expression

$$I(\nu, \theta, \varphi, p)\, d\Omega\, d\nu = \frac{1}{2}\, c\varepsilon_0 E_k^2 \quad \frac{\mathrm{W}}{\mathrm{m}^2} \tag{2.4}$$

$I(\nu, \theta, \varphi, p)$ is the intensity for the radiation in approximate direction $\hat{k}$ or $(\theta, \varphi)$, at frequency $\nu$, and with polarization $p$. The unit vector $\hat{k}$ lies in range $d\Omega$ and the frequency $\nu$ lies in the range $d\nu$. The unit of intensity $I(\nu, \theta, \varphi, p)$ is watts per square meter per hertz per steradian, and not watts per square meter.

The above discussion is typical for wave phenomena, where the method of observation influences the result. A small detector measures the two sources as a single source, while a large detector can resolve the angular separation of the two sources so that two independent sources are observed. Similar arguments can be made in the time domain, but these are discussed in greater detail in Chapter X. It is sufficient to note here that if we wish to measure with great accuracy a plane wave of the type given by Eq. (2.1), a large detector surface and a long time constant for the detector are needed.

Time and solid angle play important roles in the discussion of plane waves. Limitations are introduced by Maxwell's equations which are not always obvious in the analysis. Some of these features are made more apparent by considering electromagnetic waves confined to finite enclosures with perfectly conducting walls. The boundaries limit the allowed electric field configurations within the enclosure and these allowed configurations are usually referred to as the normal modes of the radiation field.

Consider first the rectangular enclosure shown in Fig. 2.2 and assume that the walls are perfectly conducting or are perfect reflectors. The electric vector must be zero at the walls and a transverse electric vector which

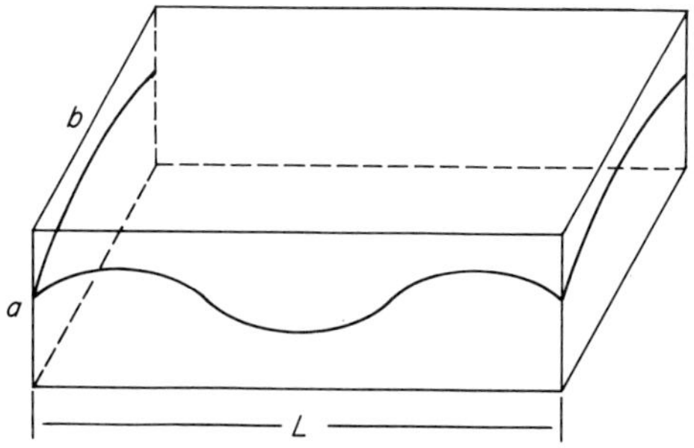


**Fig. 2.2** Standing waves in a rectangular cavity.

meets this condition is

$$\mathbf{E}_{mnq}(\mathbf{r}) = E_{mnq}\left\{\text{grad}\left[\cos\left(\frac{m\pi x}{a}\right)\cos\left(\frac{n\pi y}{b}\right)\right]\right\}\sin\left(\frac{q\pi z}{L}\right) \tag{2.5}$$

where $m$, $n$, and $q$ are integers. Note that integral values of $m$, $n$, and $q$ are needed to make the electric field zero at the walls. This is the $\text{TE}_{mn}$ mode for a waveguide, and, with closed ends, is a resonant cavity with allowed frequencies $\nu_{mnq}$. The standing-wave pattern for the low-frequency modes is sketched in Fig. 2.2 and at low frequencies, the spatial dependence of these modes can be important. A similar set of transverse magnetic, or TM, modes occurs for this resonant cavity. In order to simplify the notation, let the triple index $mnq$ be represented by $\alpha$. The stored energy in the cavity is given by the integral

$$\int \frac{1}{2}\left(\varepsilon_0 E^2 + \frac{B^2}{\mu_0}\right) d\mathbf{r}$$

Since the normal modes are orthogonal, the integral over the volume of terms like $\sum E_\alpha E_{\alpha'}$ reduce to a single sum over $\alpha$, and the stored energy becomes the sum of the stored energy in the individual modes,

$$\varepsilon = \sum_\alpha \varepsilon_\alpha \tag{2.6}$$

There is no exchange of energy between normal modes, and the energy in each normal mode is a constant of the motion. Thus if a particular mode $mnq$ is excited in the lossless rectangular cavity, the energy in this mode remains constant and is not transferred to other modes. The electric field $\mathbf{E}_\alpha(\mathbf{r})$ changes in amplitude throughout the entire cavity at frequency $\nu_\alpha$.

### 2.2.1 Quanta and Photons

In his study of thermal radiation, Planck made a discovery that has had a profound influence on physics. He realized that an explanation of thermal radiation required that the energy $\varepsilon_\alpha$ in a normal mode must be composed of a discrete number of quanta and that the energy $\varepsilon_\alpha$, number of quanta $n_\alpha$, and frequency of oscillation $\nu_\alpha$ must be related by

$$\varepsilon_\alpha = n_\alpha h\nu_\alpha \tag{2.7}$$

where the constant of proportionality $h$ is now known as Planck's constant.

The rectangular cavity discussed earlier is not unique and $\alpha$ can be used to denote the modes of a cavity of arbitrary shape and $n_\alpha$ the number of quanta associated with each mode.

### 2.2.2 Box Modes

As the size of the cavity becomes large or $L \gg \lambda$, the boundaries become of less importance. One could let $a = b = L \gg \lambda$ in Eq. (2.5) and discuss the resulting problem. It is more convenient to use a traveling-wave analysis in which the plane-wave mode is given by

$$\mathbf{E}_k(\mathbf{r}, t) = c_k \hat{u}_k[\exp i(\mathbf{k} \cdot \mathbf{r} - \omega_k t)] + \text{complex conjugate} \tag{2.8}$$

where

$$k_x = \frac{2\pi n_x}{L}, \qquad k_y = \frac{2\pi n_y}{L}, \qquad k_z = \frac{2\pi n_z}{L} \tag{2.9}$$

and $n_x$, $n_y$, and $n_z$ are integers. This periodic boundary condition

$$e^{ik_x(L+x)} = e^{ik_x x} \qquad \text{or} \qquad e^{ik_x L} = 1 \tag{2.10}$$

maintains the wave-packet nature of the normal modes of a real cavity, but is not sensitive to the shape or character of the boundaries for $L \gg \lambda$. Since it is also a plane wave, it includes some of the features of Eq. (2.1). In the limit of $L \to \infty$, they become identical. For these reasons, box modes are used in almost all problems that obey a wave equation. Maxwell's equations in free space are obeyed with

$$\omega_k^2 = k^2 c^2$$

and only discrete values of frequency $\nu_k$ are allowed. Maxwell's equations also require $\hat{k} \cdot \hat{u}_k = 0$ and there are always two polarization vectors $\hat{u}_1$ and $\hat{u}_2$ which are perpendicular to $\hat{k}$. The most common choice is the unit vectors $\hat{\theta}$ and $\hat{\varphi}$ or the two states of linear polarization. Other polarization states are given by Eq. (2.51). In Eq. (2.8), $k$ implies four indices, that is, $n_x n_y n_z$ and the state of polarization, and each index $k$ describes a mode. According to the requirement that each normal mode contain only a discrete number of quanta,

$$\varepsilon_k = n_k h \nu_k \tag{2.11}$$

where $n_k$ is the number of quanta or photons. The momentum of the photon with energy $h\nu$ is

$$\mathbf{p} = \left(\frac{h\nu}{c}\right)\hat{k} \tag{2.12}$$

The radiation energy in $L^3$ is

$$\varepsilon = \sum_k n_k h\nu_k \tag{2.13}$$

For large $L^3$, it is convenient to transform the sums to integrals. This is accomplished with

$$\sum_{\Delta k} \to \sum_p L^3 \frac{\nu^2 \, d\Omega \, d\nu}{c^3} \tag{2.14}$$

where $p$ indicates the sum over the two states of polarization. This can be shown by noting in Fig. 2.3 that within a sphere of a radius $k$ in $k_x$, $k_y$, $k_z$ space, the number of space modes with values less than $k$ is

$$N(k) = \frac{\frac{4}{3}\pi k^3}{(2\pi/L)^3}$$

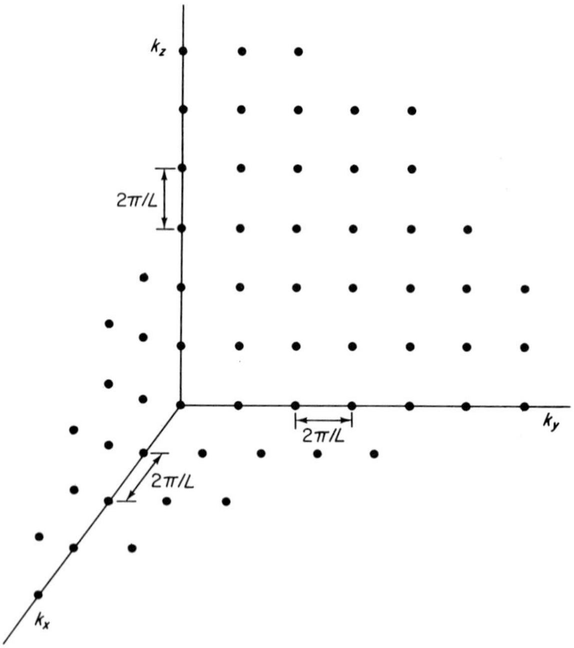


**Fig. 2.3** Counting of modes in a plane wave or **k**-space for cubic boxes of volume $L^3$.

The distance between modes in **k**-space follows from Eq. (2.9) and is $2\pi/L$ and a volume $(2\pi/L)^3$ can be associated with each mode. The number of modes between $k$ and $k + dk$ is given by $N(k + dk) - N(k) = dN/dk$, or between $\nu$ and $\nu + d\nu$ by

$$N(\nu + d\nu) - N(\nu) = \frac{dN}{dk}\frac{dk}{d\nu}$$

Then,

$$\begin{Bmatrix}\text{number of space modes} \\ \text{in range } d\mathbf{k} \text{ or } \nu^2\, d\Omega\, d\nu\end{Bmatrix} = \frac{L^3\, dk_x\, dk_y\, dk_z}{(2\pi)^3} = \frac{L^3\nu^2\, d\Omega\, d\nu}{c^3} \tag{2.15}$$

and the frequency expression follows from $(2\pi\nu)^2 = k^2c^2$.

The intensity from independent sources is given by

$$\sum_{\Delta k} \frac{n_k}{L^3} h\nu c = I(\nu, \theta, \varphi, p)\, d\Omega\, d\nu \tag{2.16}$$

where the sum is over the modes in range $\Delta k$ or $\nu^2\, d\Omega\, d\nu/c^3$. This is a natural expression in this type of expansion since $n_k/L^3$ is the number of photons per cubic meter and $(n_k/L^3)h\nu c$ is the flux of photons per second across 1 $m^2$ in direction $\hat{k}$. The photon number, or the number of photons in interval $d\Omega\, d\nu$, is defined by the relationship

$$\sum_{\Delta k} \frac{n_k}{L^3} = n(\nu, \theta, \varphi, p) \frac{\nu^2\, d\Omega\, d\nu}{c^3} \tag{2.17}$$

Intensity and photon number are related by

$$I(\nu, \theta, \varphi, p) = \frac{n(\nu, \theta, \varphi, p) h\nu^3}{c^2} \tag{2.18}$$

It will be shown in Section 2.11 that invariance under a Lorentz transformation implies this functional relationship. Note that

$$n_k \rightarrow n(\nu, \theta, \varphi, p)$$

and is not the number of photons per unit volume.

## 2.3 Thermal Radiation

Thermal radiation inside an enclosure is now discussed in terms of the basic postulate of Planck. Each mode contains a discrete number of quanta, and if this mode can exchange energy with a thermal reservoir at temperature $T$, the average number of thermal quanta associated with the mode is

$$\bar{n}_\alpha = \frac{1}{\exp(h\nu_\alpha/kT) - 1} \tag{2.19}$$

The derivation of this expression is given in Chapter VII.

In general, the average number of thermal quanta or photons associated with a mode is dependent only on the frequency of the mode and is given by

$$\bar{n}_\nu = \frac{1}{\exp(h\nu/kT) - 1} \tag{2.20}$$

At any location inside $L^3$, the thermal energy crossing a surface $dS$ in direction $d\Omega$ with frequency in the range $d\nu$ and polarization $p$ follows from Eq. (2.16) and (2.17) and the Planck Distribution Law:

$$\{\text{thermal energy crossing } dS \text{ per second in } d\Omega\, d\nu \text{ with polarization } p\}$$

$$= \bar{n}_\nu \frac{h\nu^3}{c^2}\, d\Omega\, d\nu(\cos\psi)\, dS = I(\nu, T, p)\, d\Omega\, d\nu(\cos\psi)\, dS \tag{2.21}$$

where $\psi$ is the angle between the normal $\hat{n}$ of $dS$ and the direction $\hat{k}$ of $d\Omega$. Since the distribution in the enclosure $\bar{n}_\nu$ is a function of the frequency only, and since the distribution of normal modes is isotropic, the radiation inside the enclosure is also isotropic. The angular dependence $\theta$, $\varphi$ does not occur and the thermal intensity and average number of photons are related by

$$I(\nu, T, p) = \bar{n}_\nu \frac{h\nu^3}{c^2} \tag{2.22}$$

The thermal power crossing per second per square meter is given by the integral of Eq. (2.21) over frequency and solid angle $d\Omega = 2\pi \sin\psi\, d\psi$, to yield

$$P_T = 2\int \bar{n}_\nu \frac{h\nu^3}{c^2} (\cos\psi)\, d\Omega\, d\nu = \sigma T^4 \quad \frac{\text{W}}{\text{m}^2} \tag{2.23a}$$

where $\sigma$ is the Stefan–Boltzmann constant,

$$\sigma = \frac{2\pi^5 k^4}{15h^3c^2} = 5.6697\times 10^{-8} \quad \frac{\text{W}}{\text{m}^2\ {}^\circ\text{K}^{-4}} \tag{2.23b}$$

This is the expression used most frequently in elementary physics.

It is interesting to note that the average number of thermal quanta or photons in volume $V$ is finite and is given by

$$N = 2V \int_0^\infty \bar{n}_\nu \nu^2 \frac{d\Omega\, d\nu}{c^3} = 0.244V\left(\frac{2\pi kT}{hc}\right)^3 \tag{2.24}$$

Exercise 2.1 (a) Use the definite integral

$$\int_0^\infty \frac{x^3\, dx}{e^x - 1} = \frac{\pi^4}{15} \tag{2.25a}$$

to show that $\sigma T^4$ follows from Eq. (2.23a).

(b) Use the definite integral

$$\int_0^\infty \frac{x^2\, dx}{e^x - 1} = 2\zeta(3) \tag{2.25b}$$

where $\zeta(3) = 1.202$ to show that Eq. (2.24) follows.

Exercise 2.2 Show that the thermal energy stored in a cavity of volume $V$ in thermal equilibrium with a thermal reservoir at temperature $T$ is given by

$$\frac{U}{V} = 4\left(\frac{\sigma}{c}\right)T^4 \tag{2.26}$$

Exercise 2.3 Sketch some of the allowed modes for the $TE_{10q}$ mode of a rectangular cavity.

## 2.4 Surface Absorption and Emission of Thermal Radiation

In the previous section, it was assumed that each mode was in equilibrium with a heat reservoir at temperature $T$. Since photons do not interact, the exchange of energy between the mode and the heat reservoir is due to the interaction of radiation with matter. This exchange can occur either at the walls or at matter which is contained in the cavity. At a surface, the incident thermal radiation can be absorbed, reflected, or scattered. Absorption and reflection are the only two processes that need to be considered for wavelengths that are short compared to the object under consideration. Scattering with or without a change in frequency is sensitive to size and the more general normal-mode analysis for the entire cavity might

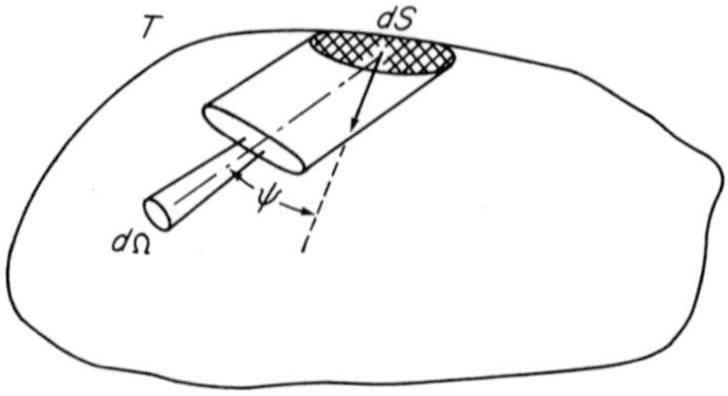


**Fig. 2.4a** Energy absorbed and radiated by a surface element $dS$ of an enclosure at temperature $T$.

be needed. At a surface $dS$, the probability of annihilation of a photon of frequency $\nu$, at an angle of incidence $\psi$ with the surface normal, and of polarization $p$ is defined as

$$A(\nu, \psi, p)$$

Since either absorption or reflection can occur at the surface, the sum of their probabilities is unity, or

$$A(\nu, \psi, p) + R(\nu, \psi, p) = 1 \tag{2.27}$$

If the surface shown in Fig. 2.4a is considered and the surface is in equilibrium with the thermal radiation, detailed balance requires that

$$\{\text{radiation emitted}\} = \{\text{radiation absorbed}\}$$

or for each interval $d\Omega\, d\nu$,

$$J_S(\nu, \psi, p)\, d\Omega\, d\nu\, dS = A(\nu, \psi, p) I(\nu, T, p)(\cos \psi)\, d\Omega\, d\nu\, dS \tag{2.28}$$

The term on the right-hand side is the radiation incident on $dS$ multiplied by the probability of absorption $A(\nu, \psi, p)$. The term $J_S$ describes the emission properties of the surface and is expected to differ for various surfaces.

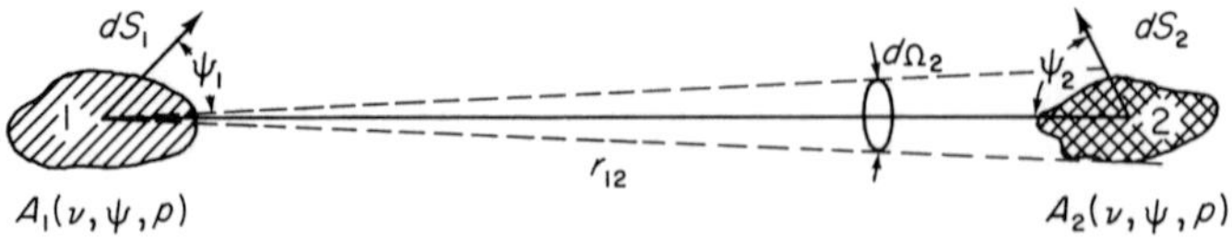


**Fig. 2.4b** Energy exchange between two surface elements $dS_1$ and $dS_2$ separated by a distance $r_{12}$ and with absorption coefficients $A_1(\nu, \psi_1, p)$ and $A_2(\nu, \psi_2, p)$. $\psi_1$ and $\psi_2$ are the angles between the surface normals and $r_{12}$. Equation (2.31) gives the thermal power emitted by $dS_1$ and absorbed by $dS_2$.

The ratio

$$\frac{J_S(\nu, \psi, p)}{A(\nu, \psi, p)} = I(\nu T p) \cos \psi = \bar{n}_\nu \left(\frac{h\nu^3}{c^2}\right) \cos \psi \tag{2.29}$$

depends only on the property of thermal radiation in the enclosure and is independent of the surface properties. Equation (2.29) is equivalent to a statement of Kirchhoff's Law for Surfaces:

*Kirchhoff's Law for Surfaces*

At any temperature, the ratio of the emissive power to the absorptive power is constant and equal to the emissive power of a perfect blackbody.

Defining a perfect blackbody as a surface for which $A(\nu, \psi, p) = 1$, the energy emitted by a surface can be given in terms of the absorption coefficient $A(\nu, \psi, p)$,

$$\begin{Bmatrix}\text{thermal power emitted by surface } dS \text{ in direction } d\Omega, \\ \text{frequency interval } d\nu, \text{ and polarization } p\end{Bmatrix}$$

$$= A(\nu, \psi, p) I(\nu, T, p)(\cos \psi)\, d\Omega\, d\nu\, dS$$

$$= A(\nu, \psi, p) \frac{(h\nu^3/c^2) \cos \psi}{\exp(h\nu/kT) - 1}\, d\Omega\, d\nu\, dS \tag{2.30}$$

This is probably the most useful single equation for thermal radiation. It gives the thermal energy per second, or thermal power emitted by a source of area $dS$ into solid angle $d\Omega$ making an angle $\psi$ with the surface normal, in the frequency range $d\nu$, and of given polarization $p$.

This may be immediately extended to the exchange of energy between two surfaces, as shown in Fig. 2.4b:

$$\{\text{Thermal power emitted by } dS_1 \text{ and absorbed by } dS_2\}$$

$$= \{\text{thermal power emitted by } dS_1 \text{ into } d\Omega_2\} \times \{\text{probability of absorption by } dS_2\}$$

$$= [A_1(\nu, \psi_1, p) I(\nu, T, p)(\cos \psi_1)\, d\Omega_2\, d\nu\, dS_1][A_2(\nu, \psi_2, p)]$$

$$= A_1(\nu, \psi_1, p) A_2(\nu, \psi_2, p) I(\nu, T_1, p)\, d\nu \frac{(\cos \psi_1)\, dS_1 (\cos \psi_2)\, dS_2}{r_{12}^2} \tag{2.31}$$

Equation (2.30) clearly indicates the cosine law for thermal radiation emitted by a surface and Eq. (2.31) emphasizes the inverse square law for the transport of energy between two surfaces. Some surface absorption coefficients are shown in Fig. 2.5 and some average absorption coefficients

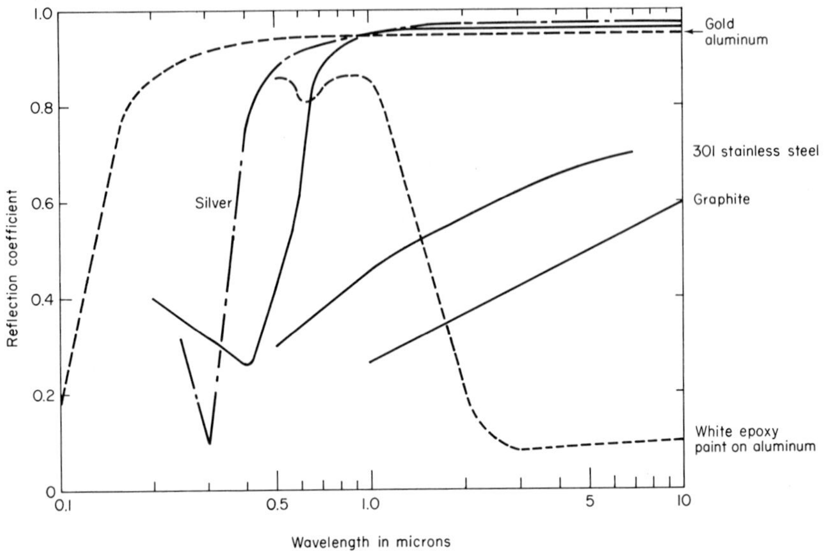


**Fig. 2.5** Reflection coefficient for frequently used materials as a function of the wavelength. (Data is taken from G. G. Gubareff, J. E. Janssen, and R. H. Torborg, "Thermal Radiation Properties Survey," Honeywell Research Center, Minneapolis, Minnesota, 1960; D. K. Edwards, K. E. Nelson, R. D. Roddick, and J. T. Gier, "Basic Studies on the Use and Control of Solar Energy," Univ. of California Dept. of Engineering Rep. 60-93, 1960; A. J. Chapman, "Heat Transfer," Macmillan, New York, 1967; T. J. Love, "Radiative Heat Transfer," Merrill Books, Columbus, Ohio, 1968.)

are given in Table 2.1.[1] Since the condition of the surface is of considerable importance, these average coefficients must be regarded as very crude values.

## 2.5 Radiation Pressure

Photons carry momentum $h\nu/c$ in the direction $\hat{k}$ in the plane-wave analysis and hence transport momentum. Pressure is the momentum transported per second along the direction of the surface normal $\hat{n}$ per square meter of surface; that is,

$$P = \sum_k \left(\frac{n_k}{L^3}\right)\left[\left(\frac{h\nu}{c}\right)\hat{k} \cdot \hat{n}\right](\hat{k} \cdot \hat{n})\, c \tag{2.32a}$$

[1] See Hottell and Sarofim [2] for additional coefficients, references, and an extended discussion of applied problems in radiation transfer.

**Table 2.1**

Absorption Coefficients[a]

| Material | Temperature (°C) | $A$ |
|---|---|---|
| Al, polished | 100 | 0.095 |
| Al, rough plate | 40 | 0.06 |
| Al, oxidized | 100–400 | 0.35 |
| Brass, polished | 30 | 0.04 |
| Cu, polished | 100 | 0.02 |
| Au, polished | 200–600 | 0.02 |
| Stainless steel, polished | 100 | 0.07 |
| W filament | 2500 | 0.4 |
| Red brick | 40 | 0.93 |
| Glass | 30 | 0.94 |
| White paint | 30 | 0.9 |
| Water | 30 | 0.95 |

[a] An average absorption coefficient $A$ is defined in terms of the total thermal power radiated by a surface,

$$P_T = \sum_p \int \{\text{Eq. (2.30)}\} = A\sigma T^4$$

Some common values of $A$ are given in this table and a more extensive list is given in the "Handbook of Chemistry and Physics" [*2a*]. Total emissivity and absorption are regarded as equivalent in this table.

where $(h\nu/c)\hat{k} \cdot \hat{n}$ is the momentum of the photon along the normal $\hat{n}$ and $\hat{k} \cdot \hat{n} = \cos \psi$ is the area perpendicular to $\hat{k}$. This can be related to the total energy density or intensity and upon integration yields

$$P = \frac{U}{3V} \tag{2.32b}$$

This is the same relationship between energy density and pressure as that obtained in classical electromagnetic theory, in which the photon concept is not used.

For some calculations, the flux of photons is of interest and for thermal radiation, the photons leaving a surface per second is

$$\begin{Bmatrix}\text{number of thermal photons emitted per second by}\\ \text{surface } dS \text{ into } d\Omega\, d\nu \text{ with polarization } p\end{Bmatrix}$$
$$= A(\nu, \psi, p)\left[\bar{n}_\nu \nu^2\, d\nu\, (\cos\psi)\frac{d\Omega\, dS}{c^2}\right] \tag{2.33}$$

The number of photons crossing a surface inside the cavity is given by Eq. (2.33) with $A = 1$. It is interesting to note that the pressure per photon is one-third of the photon energy and the pressure per particle in Chapter I was two-thirds of the kinetic energy of the particle. This is a direct consequence of the linear relationship between the energy and momentum of the photon, $E = cp = \hbar kc$.

Radiation pressure is quite significant in astrophysical problems, but is quite difficult to observe in the laboratory. Heating caused by the radiation introduces thermal creep or radiometer pressures, which were discussed in Section 1.11.5, and are more than an order of magnitude greater than radiation pressure.

The radiation pressure of a 1-W laser is appreciable [3] and can be used to accelerate small particles. In order to avoid the radiometric effect, the particles, of the order of 1 μm in diameter, were placed in a liquid. Not only were the particles accelerated by the radiation pressure in accordance with the theory, but particles with an index of refraction greater than that of the liquid were drawn into the laser beam and bubbles with an index of refraction less than that of the liquid were pushed out of the laser beam. This observation is in accord with the forces on a dielectric sphere in an inhomogeneous electric field and in the opposite sense to the radiometric force.

A laser source can be used to deflect atoms or molecules. Absorption of the laser radiation is directional and emission is isotropic in the rest frame of the atom. Thus a momentum $h\nu/c$ is acquired for each absorption–emission event.

Exercise 2.4 (a) Find the approximate force exerted by a 1-W argon laser beam at $\lambda = 0.5145$ μm focused to a cross section of $\lambda^2$.

(b) Show that the acceleration of an average-density particle of diameter $\lambda$ which reflects approximately 10% of the light backward is of the order of $10^5 g$, where $g$ is the acceleration of gravity. Note that the extent of the focal region is $\lambda$.

(c) Repeat parts (a) and (b) for a beam cross section of $1 \times 1$ mm. Over what length does the beam retain this approximate cross section?

## 2.6 Experimental Verification of the Frequency Distribution of Thermal Radiation

In previous sections, the radiation was regarded to be isotropic within an enclosure and to have a frequency distribution given by Eq. (2.21) The spectral distribution of thermal radiation emanating from a small hole of area $\Delta S$ in the wall of the enclosure at temperature $T$ is characteristic of the radiation arriving at the wall and is given by (see Fig. 2.6)

$$I(\nu, T)\, d\nu (\cos\psi)\, d\Omega\, \Delta S = \frac{2h(\nu^3/c^2)\, d\nu\, (\cos\psi)\, d\Omega\, \Delta S}{e^{h\nu/kT} - 1} \tag{2.34}$$

$\Delta S$ must be large compared to the wavelength of most of the photons so that the laws of geometric optics apply. This avoids the diffraction effects of physical optics. $\Delta S$ should be small compared to the total surface area of the enclosure so that the energy loss is small compared to the stored

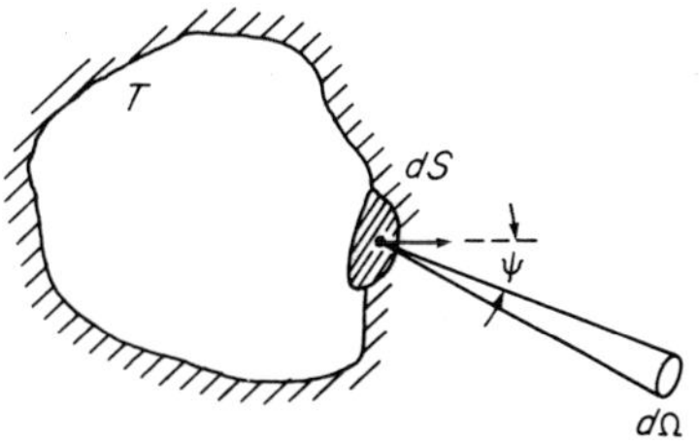


**Fig. 2.6** Thermal power emanating from a small hole in an enclosure with walls at temperature $T$. [See Eq. (2.34).]

thermal energy. Since $(\cos\psi)\, d\Omega\, \Delta S$ is a geometric factor depending on the arrangement of the detector and the hole, this term can be omitted in the consideration of the spectral distribution. A plot of $I(\lambda)$ as a function $\lambda$ is shown in Fig. 2.7, where $I(\lambda)$ is given by

$$I(\lambda)\, d\lambda = \frac{hc^2}{\lambda^5} \frac{d\lambda}{e^{hc/\lambda kT} - 1} \tag{2.35}$$

and is usually referred to as the spectral distribution of blackbody radiation as a function of the wavelength. The maximum of the curve occurs at a wavelength $\lambda_{max}$, that is, $dI/d\lambda = 0$ yields

$$\lambda_{max} T = hc(4.9651k)^{-1} \tag{2.36a}$$

Wien [*4*] showed in 1893 using classical thermodynamics and electromagnetic theory of radiation that $I(\lambda)\, d\lambda = \text{constant} \times \lfloor f(\lambda T)/\lambda^5 \rfloor\, d\lambda$ and then that $\lambda_{max} T = \text{constant}$. Paschen and Lummer *et al.* subjected the

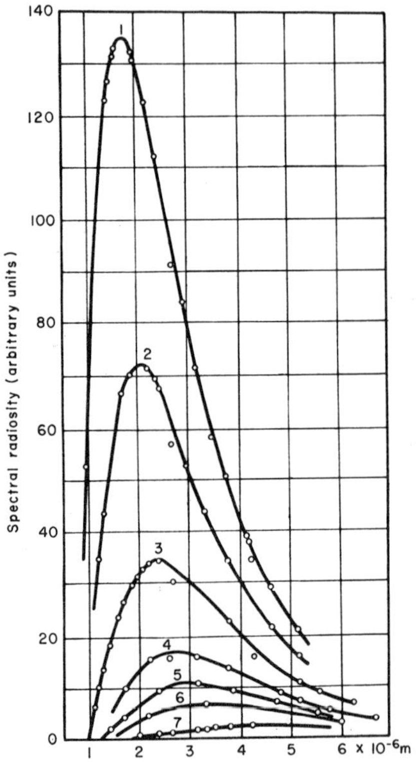


**Fig. 2.7** Radiation from a blackbody for various temperatures (°K) as a function of wavelength. 1, 1646 °K; 2, 1449 °K; 3, 1259 °K; 4, 1095 °K; 5, 998 °K; 6, 904 °K; 7, 723 °K. (From P. Moon, "Scientific Basis of Illuminating Engineering," p. 111, Fig. 5.04. Dover, New York, 1936.)

deductions of Wien's Displacement Law to experimental test [*5*] and obtained the set of curves shown in Fig. 2.7. They indeed observed that Eq. (2.36a) gave a suitable description of the relationship between the wavelength at maximum intensity and the absolute temperature and they also determined the constant. This relationship was exceedingly useful in the following years since it allowed the temperature of a blackbody to be determined from the spectral distribution. The shape of the curve could not be explained in terms of the classical statistics used for gas particles, which led to the Rayleigh–Jeans Law for the spectral distribution. The explanation of the shape of the distribution curve was first given by Planck and required the introduction of discrete levels for the harmonic oscillators, or quantization of the system. His conclusions are expressed by Eq. (2.35)

Due to the great importance of Planck's formula for modern developments in physics, it has been subjected to numerous experimental tests [*6*]. The functional form of Eqs. (2.34) and (2.35) is such that this almost resolves itself into the determination of the term $hc/k$ from radiation experiments and comparing this value with that determined from a knowledge of the fundamental constants from other experiments [7]. The experimental values are not very accurate and the first radiation constant $8\pi hc$, the second radiation constant $hc/k$, and the Wien displacement constant $h\nu_{\max}/kT$ are taken from other experiments [8] and theory.

Some care is needed in discussing the position of the maximum intensity. If intensity is plotted as a function of frequency, the maximum occurs at $dI(\nu)/d\nu = 0$ and

$$\frac{h\nu_{\max}}{kT} = 2.898 \tag{2.36b}$$

This may be compared with the plot of intensity as a function of wavelength, which by Eq. (2.36a) is

$$\frac{hc}{\lambda_{\max}kT} = 4.9651$$

**Exercise 2.5** Find the total radiant energy per cubic meter inside an enclosure at 3000°K; inside an enclosure at $10^7$ °K.

Ans. 0.06 and $7.56\times 10^{12}$ J/m$^3$

Find the pressure exerted by the thermal radiation and the number of photons per cubic meter.

Ans. $2.5\times 10^{12}$ N/m$^2$; $2\times 10^{28}$ (at $10^7$ °K)

**Exercise 2.6** Write a general expression for the net heat transfer between surfaces $dS_1$ and $dS_2$ shown in Fig. 2.5.

**Exercise 2.7** Two plane parallel surfaces with absorption coefficients $A_1(\nu)$ and $A_2(\nu)$ are at temperatures $T_1$ and $T_2$ and are separated by a distance $d$. Prove the following:

(a) The energy emitted per second by surface 1 of both polarizations is

$$\pi A_1(\nu)I(\nu, T_1)\, d\nu \tag{2.37a}$$

(b) The energy emitted by surface 1 and absorbed by surface 2 is

$$\frac{A_1(\nu)A_2(\nu)}{1-R_1(\nu)R_2(\nu)}\,\pi I(\nu, T_1)\, d\nu \tag{2.37b}$$

(c) The net heat transfer is

$$\pi\frac{A_1A_2}{A_1+A_2-A_1A_2}\,[I(\nu, T_1)-I(\nu, T_2)]\, d\nu \tag{2.37c}$$

Exercise 2.8 (a) Show the thermal power emitted by a sphere into solid angle $d\Omega$ is

$$\pi R^2 I(\nu, T)\, d\Omega\, d\nu$$

for $A = 1$.

(b) Show from Eq. (2.31) that in Fig. 2.5,

$$\{\text{thermal power emitted by } dS_1 \text{ and absorbed by } dS_2\}$$
$$= A_1(\nu, \psi, p) A_2(\nu, \psi_2, p) I(\nu, T_1, p)\, d\nu\, d\Omega_1 (\cos\psi_2)\, dS_2 \qquad (2.38)$$

can be used also. The solid angle $d\Omega_1$ is measured from surface 2.

Exercise 2.9 (a) Estimate, for a surface temperature of the sun of $T_s = 5732°\text{K}$ and for $A_s \approx 1$, the number of photons received per square meter at an earth–sun distance of $1.49 \times 10^{11}$ m for a solar diameter of $1.39 \times 10^9$ m.

(b) Estimate the intensity $I(\nu, T)$ and integrated intensity $P_T$ at the same distance from the sun. Compare with the solar constant of 1390 W/m$^2$.

(c) Estimate the pressure exerted by the solar radiation at the earth–sun distance.

Exercise 2.10 Estimate the temperature of a penny at the earth–sun distance from the sun. Assume the penny is perpendicular to the sun's rays and $A \approx 1$ for both surfaces.

Ans. $\approx 326°\text{K}$

Estimate for $A$ toward the sun of 0.1 and $A$ away from the sun of 1.

Exercise 2.11 Show that the thermal power radiated by a sphere with absorption coefficient $A(\nu)$ is

$$(4\pi R^2)\pi A(\nu) I(\nu, T)\, d\nu \qquad (2.39)$$

Exercise 2.12 A spherical satellite of 1 m diameter receives energy on one side from the sun. The absorption coefficient is given by

$$A(\nu) = \frac{0.4a}{a + [(\nu - \nu_1)/\nu_1]^2} + \frac{0.1\text{a}}{a + [(\nu - \nu_2)/\nu_1]^2}$$

$$\nu_1 = 0.2 \times 10^{14}, \qquad \nu_2 = 6 \times 10^{14}, \qquad a = 0.04$$

The satellite is the same distance from the sun as the earth. Find the maximum rate of heat generation within the satellite if the temperature is to remain at 400°K. Show that

$$\{\text{total energy lost per second}\} = \pi^2 R_{\text{sat}}^2 \int_0^\infty d\nu\, A(\nu) \times \left[4I(\nu, T_{\text{sat}}) - \left(\frac{R_{\text{sun}}}{R_{\text{ss}}}\right)^2 I(\nu, T_{\text{sun}})\right] \quad (2.40)$$

where $R_{\text{sat}}$ and $R_{\text{sun}}$ are the satellite and sun radii, respectively, and $R_{\text{ss}}$ is the distance between the sun and the satellite. Evaluate the integral by making a plot of $I(\nu)$ for the sun and for the satellite. Superimpose $A(\nu)$ on these curves and approximate by numerical methods.

Ans. 30–50 W

**Exercise 2.13** The results of early experiments on radiation are stated below. Show that these results are consistent with developments in this chapter.

(a) Prevost's theory of exchanges states that the rise or fall of temperature which is observed in a body is due to exchange of energy with surrounding bodies.

(b) The radiation inside a hollow enclosure is independent of the nature or geometric shape of the walls of the enclosure or of any body placed inside it.

(c) Discuss the character of the radiation at the focal point of the mirror in Fig. 2.8.

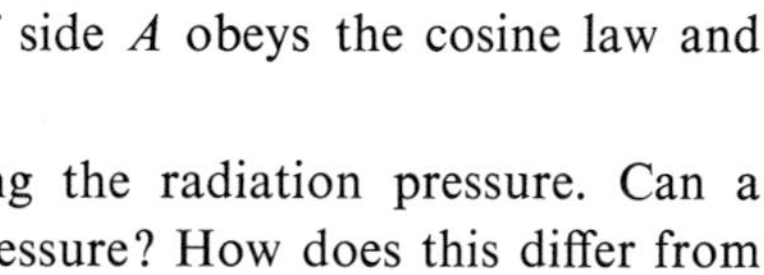

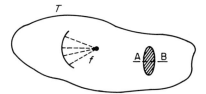

**Fig. 2.8** Radiation at the focal point of a mirror and the pressure on a flat disk if surface $B$ does not obey the cosine law for thermal emission.

(d) What happens to the flat disk if side $A$ obeys the cosine law and side $B$ does not?

(e) Discuss an experiment measuring the radiation pressure. Can a sail be used in space to use radiation pressure? How does this differ from part (d)?

(f) Rayleigh argued that each normal mode has energy $kT$. Show that the Rayleigh–Jeans law is appropriate for the long-wavelength radiation of bodies at high temperatures.

## 2.7 Radiation Temperature

Equation (2.30) gives the thermal radiation emitted by a surface. This radiation is independent of the surroundings and propagates through space in accord with Eq. (2.30). Consider a detector $\Delta S$ which accepts radiation in a small solid angle $\Delta\Omega$ and frequency range $\Delta\nu$, of polarization $p$. If the sources are separated by angles greater than $\Delta\Omega$, then for known $A_i$, the temperature of $S_i$ can be measured from a plot of $I(\nu, T_i)$ as a function of $\nu$. If the maximum intensity can be determined, then the temperature of the source follows from Eq. (2.36). Of course, if more than one source is observed in $\Delta\Omega$, some average spectral distribution is measured.

The energy received per second by an isolated detector of area $\Delta S$ which measures the radiation from a source which fills the entire solid angle $\Delta\Omega$ and frequency range $\Delta\nu$ is given by $I(\nu, \theta, \varphi, p)\,\Delta\Omega\,\Delta\nu$. Assuming such a device can be constructed, then the intensity $I(\theta, \varphi, \nu, p)$ can be measured, where $\theta$ and $\varphi$ are the directions of the surface normal of $\Delta S$ with respect to space-fixed axes. Thus the energy crossing $\Delta S$ along the surface normal $\hat{k}$ which makes angles $\theta$ and $\varphi$ with respect to space-fixed directions is, by Eq. (2.18),

$$I(\theta\varphi\nu p)\,\Delta\Omega\,\Delta\nu\,\Delta S = n(\theta, \varphi, \nu, p)\frac{h\nu^3}{c^2}\,\Delta\Omega\,\Delta\nu\,\Delta S \tag{2.41a}$$

An anisotropic radiation temperature $T(\theta, \varphi, \nu, p)$ can be introduced by defining

$$n(\theta, \varphi, \nu, p) = \frac{1}{\exp[h\nu/kT(\theta, \varphi, \nu, p)] - 1} \tag{2.41b}$$

or

$$T(\theta, \varphi, \nu, p) = \frac{h\nu/k}{\ln\{1 + [1/n(\theta, \varphi, \nu, p)]\}} \tag{2.41c}$$

If one is located at a particular point in space and examines ones surroundings by measuring the amount of radiation received from each direction and frequency interval, one can define a "brightness temperature" as follows:

$$\{\text{brightness temperature}\} = T(\theta, \varphi, \nu, p)$$

If the detector system can radiate only into the same solid angle $\Delta\Omega$ and frequency range $\Delta\nu$, then it will come to the same temperature as the distant source which emitted the radiation. The source must fill the entire solid angle for this measurement. If the source does not fill the solid angle, the detector measures the power incident on the detector. Thus $T(\theta, \varphi, \nu, p)$ can be measured for the sun or a galactic radiofrequency or microwave source and only the radiated power can be measured for a star, which appears as a point source to the detector.

As an interesting example, consider the lens–detector system shown in Fig. 2.9. The lens of area $S$ has a short focal length. The detector is highly reflecting on the side opposite the lens and has an absorption coefficient $A = 1$ for the side toward the lens. The detector is placed at the focus of the short-focal-length lens. Power is incident over a large solid angle $\Omega = S/f^2$ and is absorbed according to the cos $\psi$ law with respect to the surface normal. The power received by the lens–detector system in the diffraction-limited angle $1.22\lambda/D$ or $\Delta\Omega\ S/\lambda^2 \approx 1$ is focused to an area of the order of $\lambda^2$. Energy received at other angles is focused on adjacent areas. If the radiation from a given source completely covers the area of the detector and the detector absorbs almost all the energy in the solid angle

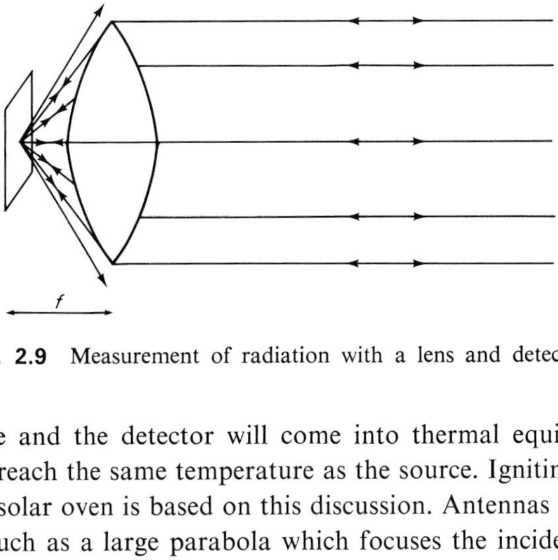


**Fig. 2.9** Measurement of radiation with a lens and detector.

$\Omega$, the source and the detector will come into thermal equilibrium. The detector will reach the same temperature as the source. Igniting paper with a lens or the solar oven is based on this discussion. Antennas used in radio astronomy, such as a large parabola which focuses the incident power on a detector, or a horn in which the detector is coupled so as to absorb power over the entire sensitive surface, are examples of devices which measure the temperature by coming into equilibrium with the source. Detection systems which come into equilibrium with the source must receive and emit into the same solid angle. Such detectors can measure $T(\theta, \varphi, \nu, p)$ by orienting the detector at various angles $(\theta, \varphi)$ and using a filter for $\nu$ and $p$.

Exercise 2.14 A long-focal-length lens is used to focus the energy from the sun on a detector. Why is the temperature of the detector much less than for the short-focal-length lens? Which is more important in the lens shown in Fig. 2.9, the short focal length or the quality of the lens? Can a poor-quality, short-focal-length reflector be used for a solar oven?

## 2.8 Interaction of Radiation with an Atom

In Chapter I and in this chapter, the kinetic energy of the particle or the energy of the photon has been determined at the wall of the container. In subsequent chapters, the interactions of two or more particles will be considered. Photons do not interact with each other, but they interact with matter. The simplest problem of this type is now considered.

Assume a single atom is placed in the cavity shown in Fig. 2.2. Assume further that the atomic states are labeled $a$, $b$, $c$, and so forth as shown in Fig. 2.10, where $| a)$ is the lowest state. $| a)$ is a short-hand notation for

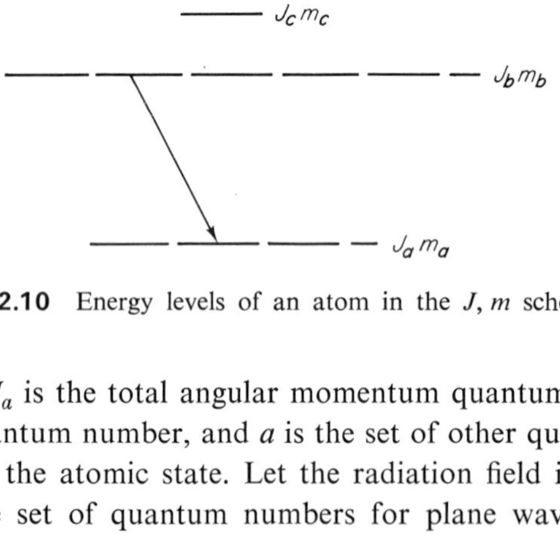


**Fig. 2.10** Energy levels of an atom in the $J$, $m$ scheme.

$| aJ_a m_a)$, where $J_a$ is the total angular momentum quantum number, $m_a$ is the magnetic quantum number, and $a$ is the set of other quantum numbers used to describe the atomic state. Let the radiation field in the cavity be described by the set of quantum numbers for plane waves, that is, $n_k$. Again, the index $k$ indicates the photons with momentum $\hbar\mathbf{k}$, energy $h\nu$, and polarization $p$. The energy of the radiation field is $E_r = \sum_k n_k h\nu_k$. The interaction between the atom and the radiation field is given by the quantum mechanical operators $-\mathbf{P} \cdot \mathbf{E}$ for electric dipole transitions, $-\mathbf{M} \cdot \mathbf{B}$ for magnetic dipole transitions, and so on. This interaction causes a transition of the type

$$E_b + E_r \rightarrow E_a + E_r + h\nu_k$$

where the atom changes from state $b$ to state $a$ and adds a photon to the radiation field. Absorption is the inverse process. Standard first-order perturbation calculations yield

$$W(ba;k) = \frac{\omega_k}{2\varepsilon_0 \hbar L^3} \,|\, (a \,|\, \mathbf{P} \cdot \hat{u}_k \,|\, b) \,|^2 (n_k + 1_k) g_k \tag{2.42a}$$

where

$$g_k = 2\Gamma_{ba}[(\omega_{ab} - \omega_k)^2 + \Gamma_{ba}]^{-1} \tag{2.42b}$$

$W(ba; k)$ is the transition rate for the process and is the probable rate that a photon of index $k$ is added to the radiation field and the atomic state changes from $b$ to $a$. This transition rate is discussed in greater detail in Chapter XI.

If the inverse process, or absorption, is denoted as the transition rate $W(ab; k)$, then the ratio of the transition rates for absorption and emission is

$$\frac{W(ba;k)}{W(ab;k)} = \frac{n_k + 1_k}{n_k} \tag{2.43}$$

The transition rates differ by the spontaneous emission term $1_k$. If all $n_k = 0$, or the radiation field is the vacuum state, and the atom is in state $|\, b)$, a sum over all $k$ and over all final states $|\, a)$ yields the rate of spontaneous emission $\Gamma_{bb}$, or the Einstein $A$ coefficient:

$$A_b = \Gamma_{bb} = \sum_k W(ba;k) = \frac{16\pi^3}{3(2J_b + 1)h\lambda^3\varepsilon_0} \,|\, (J_b \,\|\, P \,\|\, J_a) \,|^2 \tag{2.44}$$

Equation (2.14) can be used to change the $\sum_k$ to an integral. $\Gamma_{ab}$ in Eq. (2.42) is now defined as

$$\Gamma_{ba} = \tfrac{1}{2}(\Gamma_{bb} + \Gamma_{aa}) \tag{2.45}$$

where $\Gamma_{aa}$ is the spontaneous emission rate out of state $|\, a)$. If $|\, a)$ is the lowest-energy state, $\Gamma_{aa} = 0$.

Most recent discussions use the expansion in terms of spherical basis vectors,

$$\mathbf{P} \cdot \hat{u}_k = \sum_M P_M(\hat{e}_M{}^* \cdot \hat{u}_k) \tag{2.46}$$

where

$$\hat{e}_\pm = \mp 2^{-1/2}(\hat{x} \pm i\hat{y}) \qquad \text{and} \qquad \hat{e}_0 = \hat{z} \tag{2.47}$$

and the values of $M$ are equal to 1 and 0. Unit vectors $\hat{x}$, $\hat{y}$, and $\hat{z}$ refer to space-fixed axes. In this notation, the Wigner–Eckart theorem can be used to write the matrix elements for atomic states as

$$(b \mid \mathbf{P} \cdot \hat{u}_k \mid a) = (2J_b+1)^{-1/2}(J_b \,\|\, P \,\|\, J_a)(J_a 1 m_a M \mid J_b m_b)(\hat{e}_M{}^* \cdot \hat{u}_k) \quad (2.48)$$

$(J_a 1 m_a M \mid J_b m_b)$ is a Clebsch–Gordon coefficient and can be found from tables. $(J_b \,\|\, P \,\|\, J_a)$ is the reduced matrix element and is related to the line strength by

$$\mid (J_b \,\|\, P \,\|\, J_a) \mid^2 = (ea_0)^2 S(J_b J_a) = 0.72 \times 10^{-58} S(J_b J_a) \qquad (2.49)$$

The polarization operator $\mathbf{P} = \sum e\mathbf{r}_n$, where $\mathbf{r}_n$ is the electron coordinate and $e$ the charge of the electron. The line strength depends on $ea_0$, where $a_0$ is the Bohr radius. A detailed knowledge of the atomic wave functions is required for the line strength. If either $S(J_b J_a)$ or the spontaneous emission lifetime is known, the transition rate can be written as

$$W(ba; k) = \frac{3\lambda^2 c}{8\pi} A_b L^{-3} (J_b 1 m_b M \mid J_b m_b)^2 \mid \hat{e}_M{}^* \cdot \hat{u}_k \mid^2 (n_k + 1_k) g_k \quad (2.50)$$

States of polarization have been discussed in vague terms in the earlier part of this chapter. Two states of polarization are associated with photons. In the plane-wave expansion, the only requirement is that $\hat{k} \cdot \hat{u}_k = 0$, and there are two such vectors. Right and left circular polarization are quite convenient, and for a wave with $\hat{k}$ in directions $(\theta, \varphi)$ with the space-fixed axes,

$$\hat{u}_{r,l} = \tfrac{1}{2}[\hat{e}_+(1 \mp \cos\theta)e^{-i\varphi} + \hat{e}_-(1 \pm \cos\theta)e^{i\varphi}] \mp 2^{-1/2}\hat{e}_0 \sin\theta \qquad (2.51)$$

and $\hat{u}_l$ has the lower sign. $\hat{u}_r$ is defined so that the transition $m_b \rightarrow m_a = m_b + 1$ with the emitted radiation along $+\hat{z}$ is right circular and agrees with the classical convention that the electric vector for polarization rotates in the clockwise sense for the observer. Photons carry angular momentum and a left circularly polarized photon has angular momentum $+\hbar$. Linear polarization vectors are

$$\hat{\theta} = 2^{-1/2}(-\hat{e}_+ e^{-i\varphi} + \hat{e}_- e^{+i\varphi}) \cos\theta - \hat{e}_0 \sin\theta \qquad (2.52)$$

$$\hat{\varphi} = -i2^{-1/2}(\hat{e}_+ e^{-i\varphi} + \hat{e}_- e^{i\varphi}) \qquad (2.53)$$

where $\hat{\theta}$ and $\hat{\varphi}$ are in the directions of increasing $\theta$ and $\varphi$, respectively. With this notation, $\hat{u}_k$ or $\hat{u}(\theta, \varphi, \nu, p)$ denotes two distinct sets of polarization vectors. Either set may be used, but the sets cannot be mixed. Experimental arrangements only permit observing one set at a given position in space at a given time.

### 2.8.1 Atomic Motion

If the atom has a velocity $\mathbf{v}$ with respect to the cavity, then the Lorentz transformation of the radiation from the cavity frame of reference to the frame of reference of the atom requires changing

$$\omega_k \rightarrow \omega_k(1 + \mathbf{v} \cdot \hat{k}/c) \tag{2.54}$$

in the line shape $g_k$ in Eq. (2.42). If the atomic velocities are thermal,

$$\langle g_k \rangle = \int_{-\infty}^{\infty} P(v)\, dv \frac{2\Gamma_{ab}}{\{\omega_k[1 + (v/c)] - \omega_{ab}\}^2 + \Gamma_{ab}^2} \tag{2.55}$$

where

$$P(v) = \left(\frac{m}{2\pi kT}\right)^{1/2} \exp\left(-\frac{mv^2}{2kT}\right)$$

for $\mathbf{v}$ along $\hat{k}$. This integral can be given in terms of the plasma dispersion functions [*9*], but for small $\Gamma_{ab}$, this becomes the Doppler line shape

$$\langle g_k \rangle \approx (\Delta\nu_D \pi^{1/2})^{-1/2} \exp\left(-\frac{(\nu - \nu_{ab})^2}{\Delta\nu_D{}^2}\right) \tag{2.56}$$

where $\Delta\nu_D$ is given by Eq. (1.45): $\Delta\nu_D{}^2 = \nu^2(8kT/mc^2) \ln 2$. Spontaneously emitting atoms with a thermal velocity distribution have this line shape and an experimental example is shown in Fig. 1.10.

### 2.8.2 Absorption Cross Section

The cross section for stimulated absorption by the atom in volume $L^3$ is defined as

$$\{\text{cross section}\} = \frac{\{\text{transition rate}\}}{\{\text{flux of photons}\}}$$

In a large cavity with well-defined $\hat{k}$, the flux of photons is $L^{-3}n_k c$. The cross section is

$$\sigma(ab; k) = \frac{W(ab; k)}{L^{-3}n_k c} \quad \text{m}^2 \tag{2.57}$$

For radiation of index $k$, the atom appears to have a size of $\sigma(ab; k)$ for the absorption of this radiation. This cross section can be written as

$$\sigma(ab; k) = \frac{\omega_k}{2\varepsilon_0 \hbar c} |(a| \mathbf{P} \cdot \hat{u}_k | b)|^2 g(\nu) \tag{2.58}$$

If $| a)$ is the ground state and $g(\nu)$ is the normalized line shape, this equation can be placed in a convenient form for estimating the cross section. Using Eq. (2.50),

$$\sigma(J_a m_a, J_b m_b; k) = \frac{\lambda^2}{8\pi} [3(J_a 1 m_a M \mid J_b m_b)^2 \mid \hat{e}_M{}^* \cdot \hat{u}_k \mid^2][Ag(\nu)] \quad (2.59)$$

The term in the first square brackets is of the order of unity and the second is of the order of unity for $\omega_k \approx \omega_{ab}$, or resonance. Thus the cross section for $\omega_k = \omega_{ab}$ is of the order of $\lambda^2/4\pi$ and this can be regarded as the maximum size of an atom interacting with resonant photons. This statement is of rather general validity and suggests that the apparent size of a highly resonant antenna, or in this case an atom, can approach the square of the wavelength of the photons being observed. The size for other processes can be much smaller. For convenience, the atom is assumed at rest, but for moving atoms, $\omega_k \to \omega_k[1 + (\mathbf{v} \cdot \hat{k}/c)]$ in $g(\nu)$ in the given cross section.

### 2.8.2.1 Absorption and Emission Coefficients

If $N$ atoms are randomly distributed in a cavity of volume $L^3$ having $\eta(a, v)$ atoms with velocity $\mathbf{v}$ and in state $| a)$, then the rate of change of the number of photons in the cavity mode of index $k$ is

$$\frac{dn_k}{dt} = \left\langle \sum_{a,b} [W(b, a; k, v)\eta(b, v) - W(a, b; k, v)\eta(a, v)]L^3 \right\rangle \quad (2.60)$$

If the spontaneous emission term is omitted, the stimulated effect is given by the equation

$$\frac{dn_k}{dt} = \gamma_k n_k \quad (2.61)$$

For many problems, the change of $n_k$ with position is of greater interest than the change in time, and

$$\frac{dn_k}{dz} = \alpha_k n_k \quad (2.62)$$

may be inferred from the above by replacing $dt$ with $c^{-1}\, dz$ or $\gamma_k = c\alpha_k$. With this notation, the growth in time coefficient for a cavity, the growth in space coefficient, and the cross section are related by

$$\alpha_k = c^{-1}\gamma_k = \left\langle \sum_{a,b} \sigma(a, b; k, v)[\eta(b, v) - \eta(a, v)] \right\rangle \quad (2.63)$$

Stimulated emission occurs for $\eta(b, \nu) > \eta(a, \nu)$ and stimulated absorption for $\eta(b, \nu) < \eta(a, \nu)$.

As mentioned earlier, the concept of modes is closely related to the expansion in terms of a traveling wave of the type $\exp[i(kz - \omega t)]$ for the electric vector. Maxwell's equations are consistent with a complex $\omega$ or a complex $k$ and these correspond to a growth of the electric vector in time by $e^{\gamma t/2}$ or in space by $e^{\alpha z/2}$. The term $L^{-3} n_k c$ is proportional to the square of the electric vector and the exponential growth of $n_k$ as $e^{\gamma t}$ or as $e^{\alpha z}$ is consistent with Maxwell's equations.

A more conventional approach is illustrated in Fig. 2.11. The cylinder of length $\Delta\zeta$ and cross section $\Delta S$ with the surface normal to $\hat{\xi}$ along $\hat{k}$ has energy flowing across the surface at $\zeta$ which stimulates emission or absorption in $\Delta\zeta$ and changes the energy crossing $\zeta + \Delta\zeta$,

$$[I(\theta, \varphi, \nu, p)\, d\Omega\, d\nu]_{\zeta+\Delta\zeta}\, \Delta S = [I(\theta, \varphi, \nu, p)\, d\Omega\, d\nu]_{\zeta}\, \Delta S$$
$$+ \{\text{stimulated change in } \Delta S\, \Delta\zeta\}$$

The stimulated change in $\Delta S\, \Delta\zeta$ is $\alpha_k I(\theta, \varphi, \nu, p)\, d\Omega\, d\nu\, \Delta S\, \Delta\zeta$ and the change in intensity in distance $\Delta\zeta$ becomes

$$\frac{dI(\theta, \varphi, \nu, p)}{d\zeta} = \alpha_k I(\theta, \varphi, \nu, p) \tag{2.64a}$$

where $\alpha_k$ is given by Eq. (2.63). This is the most frequently used form for the change in intensity along a path.

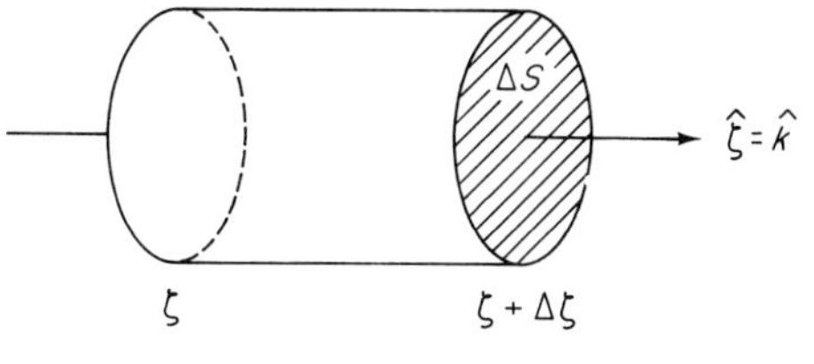


**Fig. 2.11** Attenuation coefficient. [See Eq. (2.64).]

The absorption coefficient $\alpha_k$ is not necessarily isotropic. Equation (2.63), or Eq. (2.50), illustrates that either an anisotropy in the population of the atomic states, $\eta(a) \to \eta(J_a m_a)$, or a magnetic or electric field can cause $\alpha_k$ to be dependent on direction. Thus the subscript $k$ referring to the direction $\hat{k}$ in space is kept, as is polarization. If the static magnetic and electric fields are not important and $\eta$ is a function of $J$ and not $m$, then $\alpha_k$ is a function of frequency only, $\alpha(\nu)$. Under these isotropic conditions,

Eq. (2.64) takes on its most usual form:

$$\frac{dI(\nu)}{dz} = \alpha(\nu)I(\nu) \tag{2.64b}$$

where $z$ is along the ray. In the absence of an electric or magnetic field or other orienting aspects, the axis of quantization $\hat{z}$ can have an arbitrary direction $(\theta, \varphi)$ with respect to a space-fixed axis $\hat{Z}$. An equivalent statement is that different $\hat{z}$ axes can be used for different atoms or molecules. An average of $|\hat{e}_M{}^* \cdot \hat{u}_k|^2$ over $\theta$ and $\varphi$ yields an average value of $\frac{1}{3}$ for each value of $M$ if $\hat{u}_k = \hat{u}_r$. A sum over $M$ yields all possibilities for a given value of $m_a$ as the initial state, and the sum over the Clebsch–Gordon coefficient is 1. The cross section can be estimated as

$$\sigma(\nu) \approx \frac{\lambda^2}{8\pi} Ag(\nu)$$

For an isotropic response of the atom, the stimulated absorption coefficient becomes, from Eqs. (2.63) and (2.59),

$$\alpha(\nu) \approx \left(\frac{\lambda^2}{8\pi}\right) g(\nu) A_b \left[\eta_b - \frac{(2J_b + 1)}{(2J_a + 1)} \eta_a\right] \tag{2.64c}$$

and $\langle g(\nu) \rangle$ is used for thermal radiation.

## 2.9 Thermal Radiation in Enclosures Containing Matter

If the material medium in the enclosure is described by an index of refraction $\mu = c/v$ as in Fig. 2.12, the exact solution can always be found for the cavity modes. The short-wavelength approximation is much simpler and one need only note that in the interior of the medium of index of refraction $\mu$, $d\mathbf{k}$ remains unchanged, but Maxwell's equations require

$$k^2 v^2 = \frac{k^2 c^2}{\mu^2} = \omega^2 \qquad \text{or} \qquad \lambda \nu = v = \frac{c}{\mu}$$

and in Eq. (2.5), $dk_x\, dk_y\, dk_z$ becomes

$$\frac{dk_x\, dk_y\, dk_z}{(2\pi)^3} = \mu^3 \nu^2 \frac{d\Omega\, d\nu}{c^3} \tag{2.65}$$

$\bar{n}_\nu$ remains the same in the medium, but the energy crossing a surface $dS$

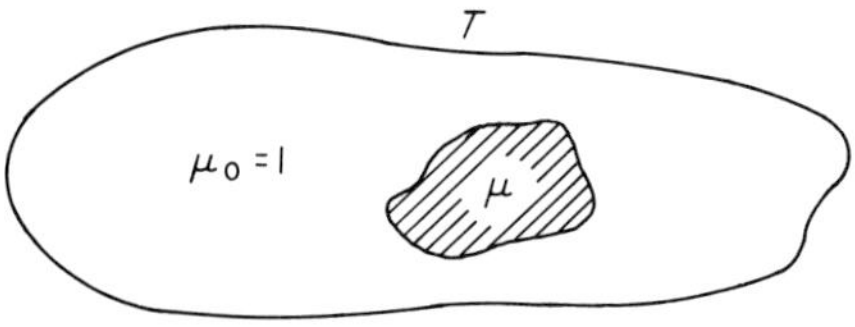


**Fig. 2.12** Thermal radiation in a medium with an index of refraction $\mu$. [See Eq. (2.66).]

depends on $v(\cos \psi)\, dS$ and the expression for the intensity $I(\nu, T, p)$ given by Eq. (2.18) has $c^2$ replaced by $v^2$. It immediately follows that $I/\mu^2$ is a constant, that is,

$$\frac{I(\nu, T, p)}{\mu(\nu)^2} = \frac{\bar{n}_\nu h\nu^3}{c^2} \tag{2.66}$$

is a universal function of frequency and temperature and is independent of the index of refraction of the medium. This conclusion may also be reached by applying the laws of geometric optics and detailed balance for the energy crossing an interface between media of indices of refraction $\mu = 1$ and $\mu$, respectively [*10*].

## 2.10 Absorption and Emission of Radiation by Matter

The absorption and emission by bulk matter [*11*] is treated in a manner similar to that for surfaces. Radiation incident on matter is regarded as either transmitted, absorbed, or reflected in the usual elementary discussions. Scattering is discussed in a later section. These coefficients are dependent on the frequency, the angle of incidence of the radiation, and on the polarization. Then,

$$a(\nu, p) + t(\nu, p) + r(\nu, p) = 1 \tag{2.67}$$

where $a$ is the probability of absorption, $t$ the probability of transmission, and $r$ the probability of reflection of a photon. The exact dependence of these quantities requires either experimental data or a theory of the optical properties of the system. Each quantity in Eq. (2.67) depends on the state of polarization of the radiation, and this is readily seen from the optical absorption properties of, say, a Polaroid film.

If a piece of matter is placed inside the thermal enclosure shown in Fig. 2.13, and hence comes into thermal equilibrium with thermal radiation, the principle of detailed balance applies for each direction $d\Omega$, frequency interval $d\nu$, and polarization, that is,

$$\{\text{energy emitted}\} = \{\text{energy absorbed}\}$$

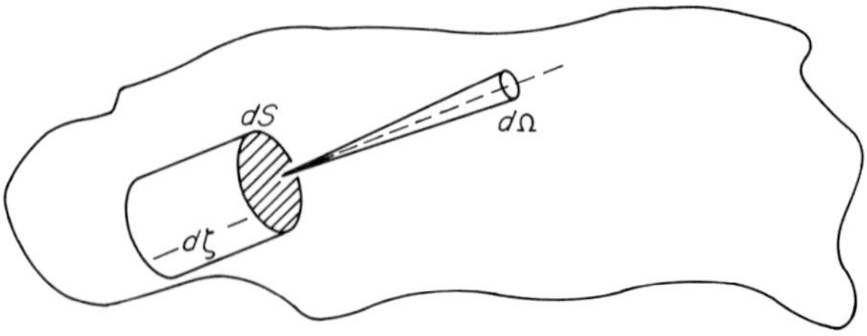


**Fig. 2.13** Emission properties of a partially transparent medium at temperature $T$. [See Eqs. (2.68) and (2.69).]

The matter in volume element $dS\, d\zeta$ emits power $j(\nu, T, p)\, d\Omega\, d\nu\, dS\, d\zeta$ into $d\Omega\, d\nu$ and absorbs power from the surroundings according to

$$j(\nu, T, p)\, d\Omega\, d\nu\, dS\, d\zeta = a(\nu, p) I(\nu, T, p)\, d\Omega\, d\nu\, dS\, d\zeta$$

Canceling similar terms and replacing $I(\nu, T, p)$ by Eq. (2.67) yields

$$\frac{j(\nu, T, p)}{\mu^2(\nu) a(\nu, p)} = \frac{\bar{n}_\nu h\nu^3}{c^2} \tag{2.68}$$

This is Kirchhoff's Law for matter in a thermal enclosure and the quantity on the left is a constant throughout the enclosure, is the same constant for any two enclosures at the same temperature $T$, and is a universal function of frequency and temperature. The thermal energy emitted by matter at temperature $T$ and in an enclosure can be inferred from Eq. (2.68), and

$$\left\{\begin{matrix}\text{thermal power emitted by matter in} \\ dS\, d\zeta \text{ into } d\Omega\, d\nu \text{ of polarization } p\end{matrix}\right\} = a(\nu, p) I(\nu, T, p)\, d\Omega\, d\nu\, dS\, d\zeta$$

$$= \frac{\mu^2(\nu) a(\nu, p)(h\nu^3/c^2)\, d\Omega\, d\nu\, dS\, d\zeta}{e^{h\nu/kT} - 1} \tag{2.69}$$

Note that in these equations $d\Omega$ is along $\zeta$, which is the surface normal of $dS$. Some care is needed if the index of refraction changes continuously, and the complete cavity modes may be needed.

### 2.10.1 *Thermal Power Emitted by an Infinite Slab*

Consider the semiinfinite slab shown in Fig. 2.14. The energy emitted by the cross-hatched area a distance $z$ from the surface crosses the surface $dS$ in direction $d\Omega$ making angle $\psi$ with $z$. The thermal power emitted by the matter in this volume element $(z \sec \psi)^2 \, d\Omega \sec \psi \, dz$ into solid angle $d\Omega' = (\cos \psi)\, dS/(z \sec \psi)^2$ is, by Eq. (2.69),

$$a(\nu)I(\nu, T)\, d\nu\, d\Omega\, dS\, dz$$

Part of this energy is absorbed in traversing the path $z \sec \psi$, and the thermal power crossing $dS$ into $d\Omega\, d\nu$ is

$$[a(\nu)I(\nu, T)\, d\Omega\, d\nu\, dS\, dz]\left[\exp - \int_0^z a(\nu) \sec \psi\, dz\right]$$

Let

$$\tau_\psi = (\sec \psi) \int_0^z a(\nu)\, dz \tag{2.70}$$

and $d\tau_\psi = a(\nu)\, dz \sec \psi$. Equation (2.70) can be written in terms of $\tau_\psi$ as

$$[I(\nu, T)\, d\nu\, d\Omega\, dS\, (\cos \psi)\, d\tau_\psi]e^{-\tau_\psi}$$

Integration over $d\tau_\psi$ yields for the thermal power crossing a square meter into $d\nu\, d\Omega$

$$\left\{\begin{matrix}\text{thermal power per } m^2 \text{ emitted into} \\ d\Omega\, d\nu \text{ by a slab at temperature } T\end{matrix}\right\} = (1 - e^{-\tau_\psi})I(\nu, T)\, d\Omega\, d\nu \cos \psi$$
$$= (1 - e^{-\tau_\psi})\, \frac{2(h\nu^3/c^2)\, d\Omega\, d\nu \cos \psi}{e^{h\nu/kT} - 1} \tag{2.71}$$

Equation (2.70) determines the optical thickness of the slab and is a function of the angle $\psi$ with respect to the surface normal and the absorption coefficient $a(\nu)$. For a thick sample, $\tau_\psi \to \infty$ for all frequencies and the infinite slab emits a continuous spectrum of frequencies with the same distribution as a blackbody surface or as the thermal power crossing a surface inside a blackbody enclosure. A thin slab can emit line spectra along the surface normal, but as the angle $\psi$ is increased toward 90°, or

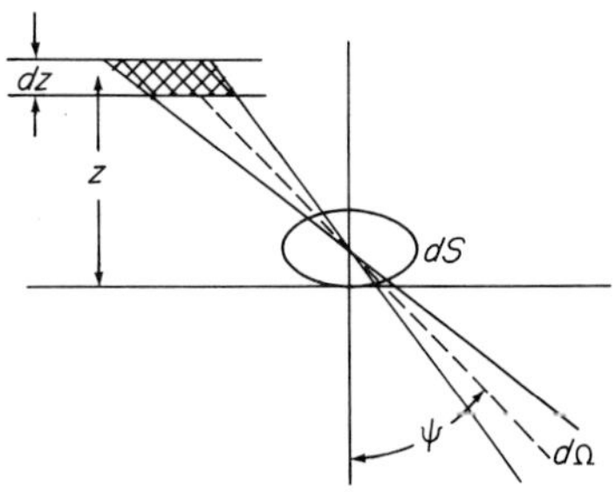


**Fig. 2.14** Thermal power emitted by a semi-infinite slab. [See Eq. (2.71).]

the surface is viewed tangentially, $\sec \psi \to \infty$, the thin slab becomes a blackbody emitter. A thin layer is brighter at glancing incidence than at normal incidence. A thin sheet of glowing metal appears as equally bright in all directions and its optical thickness $\tau_\psi$ is effectively infinite.

For some systems, $a(\nu)$ is so small that even for large path lengths $\tau_\psi$ remains small for the viewing angle. Thus gaseous nebulas show bright lines without continuous spectra.

It should again be noted that a sufficient thickness of material at a given temperature will emit a continuous blackbody spectrum. No line structure will be observed. Line structure requires an optically thin layer at the frequency of interest.

Again it must be emphasized that this section has not included scattered radiation.

### 2.10.2 Scattering at a Surface

The earlier sections have assumed that a photon is either absorbed or reflected at a surface. Scattering of a photon can also occur at a surface and two types of scattering are important. If the index of the incident photon is $k$ and the index of the scattered photon is $k'$, then two cases are possible: (a) $k \to k'$ for $\nu = \nu'$; (b) $k \to k'$ for $\nu \neq \nu'$ (fluorescence). Scattering without change in frequency requires that all directions be included in the detailed balance argument at a surface $dS$, that is,

$$\int J_S(\nu, \psi, p)\, d\Omega = I(\nu, T, p) \int A(\nu, \psi, p) \cos \psi \, d\Omega \tag{2.72}$$

Scattering with change in frequency, or fluorescence, requires an integral over $d\nu\, d\Omega$, or at equilibrium, the radiation from surface $dS$ is given by

$$\int J_S(\nu, \psi, p)\, d\Omega\, d\nu = \int A(\nu, \psi, p) I(\nu T p) \cos \psi \, d\Omega \, d\nu \tag{2.73}$$

Kirchhoff's Law for a surface can quite generally be stated as

$$\{\text{thermal power emitted by a surface}\} = \begin{cases} [\text{Eq. (2.30)}] & (2.74\text{a}) \\ \int d\Omega \;\; [\text{Eq. (2.30)}] & (2.74\text{b}) \\ \int d\Omega \, d\nu \;\; [\text{Eq. (2.30)}] & (2.74\text{c}) \end{cases}$$

for no scattering, scattering without a change in frequency, and scattering with a change in frequency, respectively.

Extensive discussions of the transfer of radiation are given by Rutgers [*12*].[2]

## 2.11 Motion through Thermal Radiation

It is interesting to use the concepts introduced and discussed earlier in this chapter to explore the consequences of motion through thermal radiation. The temperature measured by a moving observer is velocity-dependent and permits a measurement of the velocity of the observer.

It has been suggested that there is a residue of 3°K cosmic radiation which is the remnant of a primeval fireball, which would survive to the present time in the form of a universal thermal radiation [*13*]. A thermal or cosmic radiation at a temperature of 3°K has been confirmed by a number of observers [*14*]. The existence of such a background immediately leads to the suggestion that the earth's motion can be measured with respect to the rest frame of the 3°K cosmic background.[3]

Equation (2.4) relates the measurable intensity and the amplitude of a plane wave in a given frame of reference

$$I(\nu, \theta, \varphi, p)\, d\Omega\, d\nu = (\text{constant}) \times A^2_{\nu\theta\varphi p}$$

In order to discuss the intensity in another frame of reference, consider a source of electromagnetic radiation in reference frame $S'$ very far from the observer in $S$. Let $S$ be moving with velocity $\mathbf{v}$ along the $+\hat{z}$ direction away from $S'$ as shown in Fig. 2.15 and let $(\theta, \varphi)$ be the direction of the radiation. A Lorentz transformation between the two frames of reference yields the frequency relationship

$$\nu = \nu' \frac{1 - \beta \cos \theta'}{(1 - \beta^2)^{1/2}} = \nu' \frac{(1 - \beta^2)^{1/2}}{1 + \beta \cos \theta} \tag{2.75}$$

where $\beta = v/c$, $\theta$ is the angle in reference frame $S$, and $\theta'$ is the angle in $S'$. The wave amplitude transforms as

$$\left(\frac{A'}{\nu'}\right)^2 = \left(\frac{A}{\nu}\right)^2 \tag{2.76}$$

[2] Thermal emission from cavities of various shapes is also discussed by Rutgers [*12*].
[3] This measurement is discussed by Heer and others [*15*].

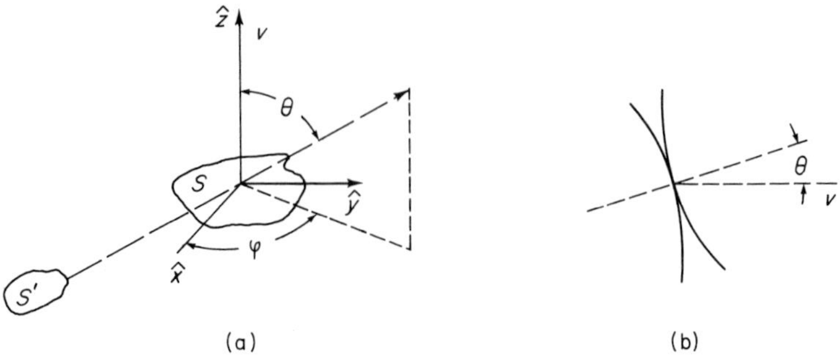


**Fig. 2.15** (a) Reference frame $S$ is moving relative to reference frame $S'$ with velocity $\mathbf{v}$ along $+z$, and radiation from $S'$ makes angle $\theta$ with the $z$-axis. (b) Two temperature measuring parabolic antennas are used to measure the absolute velocity through thermal radiation. The temperature difference measured by the two antennas is given by $\Delta T(\theta)/T \approx (v/c) \cos \theta$.

since $A/\nu$ is an invariant for a plane wave. A transformation law for intensities follows immediately from Eq. (2.76), and

$$\frac{I(\nu, \theta, \varphi, p)\, d\Omega\, d\nu}{\nu^2} = \frac{I'(\nu', \theta', \varphi', p')\, d\Omega'\, d\nu'}{(\nu')^2} \tag{2.77}$$

Since the solid angles are related by the invariant

$$(\nu')^2\, d\Omega' = \nu^2\, d\Omega \tag{2.78}$$

and since the intensity can be written in the alternate form given by Eq. (2.41a), the photon number is an invariant under a Lorentz t·ansformation:

$$n(\nu, \theta, \varphi, p) = n'(\nu', \theta', \varphi', p') \tag{2.79}$$

If the alternate form given by Eq. (2.41b) is used for the photon number, the radiation fields in $S$ and $S'$ are related by

$$\frac{\nu}{T(\nu, \theta, \varphi, p)} = \frac{\nu'}{T'(\nu', \theta', \varphi', p')} \tag{2.80}$$

If, in some reference frame $S'$, the radiation is thermal, then

$$T'(\nu', \theta', \varphi', p') = T'$$

and the temperature is independent of frequency and is isotropic. In reference frame $S$ moving in the $+\hat{z}$ direction, the radiation is described by an anisotropic temperature $T(\theta) = (\nu/\nu')T'$, or

$$T(\theta) = \frac{T}{(1 + \beta \cos \theta)} \tag{2.81}$$

where $T = (1 - \beta^2)^{1/2}T'$. $T(\theta)$ completely describes the anisotropy of the radiation, and the spectral distribution in $S$ is given by

$$I(\nu, \theta)\, d\Omega\, d\nu = \frac{(2h/c^2)\nu^3\, d\Omega\, d\nu}{\exp[h\nu/kT(\theta)] - 1} \tag{2.82}$$

If two identical antennas point in opposite directions and are temperature-measuring devices as described in Section 2.7, the temperature difference they measure is

$$\frac{T(\pi + \theta) - T(\theta)}{T(\pi + \theta) + T(\theta)} = \frac{v}{c} \cos \theta \tag{2.83}$$

This is a true velocity-measuring device and can in principle measure the velocity through space or through the thermal background. At a velocity of 100 km/sec, a temperature difference of 0.002° occurs for a background temperature of 3°K. The earth's velocity is estimated to be of this order of magnitude. Heer and Kohl [*15*] have suggested a novel speedometer for measuring the speed through space. If two sheets of material are parallel to and thermally isolated from each other, and are thermally black on the outward surfaces, a temperature difference $\Delta T$ occurs between the two surfaces,

$$\Delta T \approx \frac{4vT}{3c} \tag{2.84}$$

Again, at a 100 km/sec, the temperature difference is of the order of a millidegree.

There is some experimental evidence that the motion of the earth through this thermal radiation has been measured [*16*]. Since the antennas are located on the earth and temperature changes of the order of a few millidegrees are observed, the thermal noise from other causes makes the measurements quite difficult.

**Exercise 2.15** Show that Eq. (2.78) follows from

$$\frac{d\Omega}{d\Omega'} = \frac{d(\cos \theta)}{d(\cos \theta')}$$

## References


1. M. Planck, "Warmesstrahlung." Barth, Leipzig, 1913.
2. H. C. Hottell and A. F. Sarofim, "Radiative Transfer." McGraw-Hill, New York, 1967.

2a. "Handbook of Chemistry and Physics," 52nd ed. Chemical Rubber Co., Cleveland, Ohio, 1971.

3. A. Askin, *Phys. Rev. Lett.* **24**, 156 (1970); **25**, 1321 (1970).
4. W. Wien, Berl Ber. p. 55 (1893); *Wied. Ann.* **52** (1894).
5. B. F. Paschen, *Wied. Ann.* **60**, 662 (1897); *Ann. Phys.* **4**, 277 (1901); O. Lummer, E. Pringsheim, H. Rubens, and F. Kurlbaum, *Verh. Deut. Phys. Ges.* **1**, 213 (1899).
6. H. T. Wensel, (*U.S.*) *Bull. Bur. Stand.* **22**, 375 (1939).
7. M. N. Saha and B. N. Srivastava, "Theory of Heat," p. 675. The Indian Press, Allahabad, 1958.
8. B. N. Taylor, W. H. Parker, and D. N. Langenberg, *Rev. Mod. Phys.* **41**, 375 (1969).
9. B. D. Fried and S. D. Conte, "The Plasma Dispersion Function-The Hilbert Transform of a Gaussian." Academic Press, New York, 1961.
10. M. Planck, Theory of heat. "Introduction to Theoretical Physics," Vol. 5. Macmillan, New York, 1932.
11. A. Schuster, K. Schwarzschild, A. S. Eddington, S. Rosseland, and E. A. Milne, "Selected Papers on the Transfer of Radiation." Dover, New York, 1966; R. W. Preisendorfer, "Radiative Transfer on Discrete Spaces. Pergamon, Oxford, 1965; V. Kourganoff, "Basic Methods in Transfer Problems." Oxford Univ. Press (Clarendon), London and New York, 1952; H. Thomas, "Some Aspects of Non-Equilibrium Thermodynamics in the Presence of a Radiation Field." Univ. of Colorado, Boulder, Colorado, 1965.
12. G. A. Rutgers, *in* "Handbuch der Physik" (S. Flügge, ed.), Vol. 26, p. 138. Springer-Verlag, Berlin and New York.
13. G. Gamow, "Vistas in Astronomy" (A. Beer, ed.), Vol. 2, p. 1726. Pergamon, Oxford, 1956.
14. A. A. Penzias, R. W. Wilson, *Astrophys. J.* **142**, 419 (1965); R. B. Partridge and D. T. Wilkinson, *Phys. Rev. Lett.* **18**, 557 (1967); R. H. Dicke, P. J. E. Peebles, P. G. Roll, and D. T. Wilkinson, *Astrophys. J.* **142**, 414 (1965); P. G. Roll and D. T. Wilkinson, *Phys. Rev. Lett.* **16**, 405 (1966); T. F. Howell and J. R. Shakeshaft, *Nature* (*London*) **210**, 1318 (1966); G. B. Field and J. L. Hitchcock, *Phys. Rev. Lett.* **16**, 817 (1966); P. Thadeus and J. F. Clauser, *Phys. Rev. Lett.* **16**, 819 (1966); T. F. Howell and J. R. Shakeshaft, *Nature* (*London*) **216**, 753 (1967); E. K. Conklin and R. N. Bracewell, *Nature* (*London*) **216**, 777 (1967); R. K. Sachs and A. M. Wolfe, *Astrophys. J.* **147**, 73 (1967).
15. C. V. Heer and R. H. Kohl, *Phys. Rev.* **174**, 1611 (1968); G. R. Henry, R. B. Feduniak, J. E. Silver, and M. A. Peterson, *Phys. Rev.* **176**, 1431 (1968); P. J. E. Peebles and D. T. Wilkinson, *Phys. Rev.* **174**, 2168 (1968).
16. E. K. Conklin, *Nature* (*London*) **222**, 971 (1969).

# Probability, Statistics, and Conditional Probability

# III

## 3.1 Introduction

Earlier chapters have made use of some elementary concepts in the theory of probability and their introduction was based on intuitive physical arguments. A somewhat more rigorous discussion of probability is given in this chapter, but again the emphasis is placed on physical examples of interest in a select group of physical problems. These problems are discussed in terms of a very powerful technique introduced into physics by Khinchin [*1*] in recent years. The integral-valued random variable and the associated characteristic function find a very natural place in the techniques used by present-day physicists. These techniques are applied to problems in statistics and systems with a single constraint. Fluctuations and noise are considered in greater detail in Chapter X.

## 3.2 Integral-Valued Random Variables

### *3.2.1 Distribution Law of Integral-Valued Random Variables*

Some elementary concepts in probability are introduced by considering the integral-valued random variable. Figure 3.1 shows a line with a series of dots which are given integer labels. The physical process under consideration will correspond to assigning a probability to the occupancy of each dot. Thus the distribution law of an integral-valued random variable $\xi$ is completely determined by assigning to each integer $n$ the probability that $\xi$ takes on the value $n$. It is convenient to let the notation $P(\xi = n)$ denote the probability that $\xi$ takes on the value $n$ and this notation is used

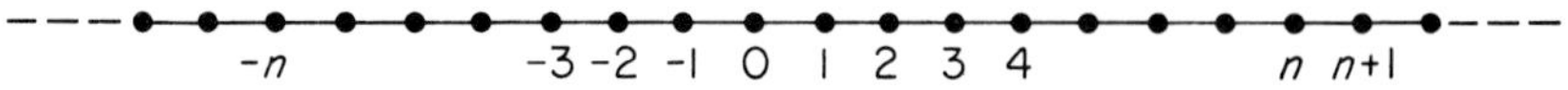


**Fig. 3.1** Labeling of points along a line for a discussion of integral-valued random variables.

throughout this chapter. $\xi$ is an integral-valued random variable and whenever $\xi$ is equal to a *particular* or *selected* value of $n$, the probability of this value of $n$ is given by $p(n)$. With this definition,

$$P(\xi = n) = p(n) \tag{3.1}$$

is a way of stating that an integer-valued random variable $\xi$ is distributed according to the law $p(n)$. All integers between $-\infty$ and $+\infty$ are included and since the event occurs with certainty,

$$\sum_{n=-\infty}^{\infty} p(n) = 1 \tag{3.2}$$

The mathematical expectation value of a physical quantity $f(\xi)$ is expressed as

$$\mathscr{E}f(\xi) = \sum_{n} f(n)p(n) \tag{3.3}$$

Two common quantities of interest are the expectation of $\xi$, $\mathscr{E}\xi$, and the dispersion of $\xi$, $\mathscr{D}\xi$. Thus the expectation value of the variable $\xi$ is

$$\mathscr{E}\xi = \sum np(n) = \bar{n} \tag{3.4}$$

and the dispersion or variance of the variable $\xi$ is

$$\mathscr{D}\xi = \mathscr{E}(\xi - \mathscr{E}\xi)^2 = \sum (n - \bar{n})^2 p(n) \tag{3.5}$$

As in earlier chapters, the notation

$$\mathscr{E}\xi = \bar{\xi} = \langle\xi\rangle \tag{3.6}$$

is frequently used. The $k$th moment of the variable $\xi$ is defined as

$$\mathscr{E}\xi^k = \sum n^k p(n) \tag{3.7}$$

In the discussion here, it has been assumed that the expansions form convergent series.

### 3.2.2 *Characteristic Function of an Integral-Valued Random Variable*

If $f(\xi) = e^{ia\xi}$, the expectation value is denoted as the characteristic function, and in this text, the characteristic function of an integral-valued random variable with a distribution law $p(n)$ is

$$\varphi(a) = \mathscr{E} e^{ia\xi} = \langle e^{ia\xi} \rangle = \sum p(n) e^{ian} \tag{3.8}$$

The Fourier series transform of $\varphi(a)$ is obtained by multiplying by $e^{-iam}$ and integrating from $-\pi$ to $+\pi$,

$$p(m) = \frac{1}{2\pi} \int_{-\pi}^{\pi} da\, \varphi(a) e^{-iam} \tag{3.9}$$

This inversion formula shows that the distribution law of an integral-valued random variable is completely determined by its characteristic function $\varphi(a)$. This simple dependence and its natural place among the physicist's analytical techniques comprise one of the reasons for its introduction in this text. Some other properties of the characteristic function are

$$\varphi(0) = 1 \tag{3.10a}$$

$$|\, \varphi(a) \,| \leq \sum p(n) = 1 \tag{3.10b}$$

$$\mathscr{E}\xi = -i\left(\frac{d\varphi}{da}\right)_{a=0} = -i\left[\frac{d(\ln \varphi)}{da}\right]_{a=0} \tag{3.11a}$$

$$\mathscr{D}\xi = -\left[\frac{d^2(\ln \varphi)}{da^2}\right]_{a=0} \tag{3.11b}$$

**Exercise 3.1** Show that Eqs. (3.11a) and (3.11b) follow from Eqs. (3.4), (3.5), and (3.8).

### 3.2.3 *Rule of Composition of Distributions*

Let $\xi'$ and $\xi''$ be integral-valued random variables obeying the distribution laws $p'(n)$ and $p''(n)$. The distribution law for the random variable $\xi$ which is the sum of the mutually independent random variables $\xi'$ and $\xi''$

is given by

$$\xi = \xi' + \xi''$$
$$P(\xi = n) = p(n) = \sum_{k=-\infty}^{\infty} P(\xi' = k, \xi'' = n - k) = \sum_{k=-\infty}^{\infty} p'(k)p''(n - k) \tag{3.12}$$

The characteristic function of the distribution law for $p(n)$, or the rule of composition of characteristic functions, is

$$\begin{aligned}\varphi(a) &= \mathscr{E} \exp ia\xi = \mathscr{E} \exp ia(\xi' + \xi'') = \mathscr{E}[(\exp ia\xi') \exp ia\xi''] \\ &= \mathscr{E}(\exp ia\xi')\mathscr{E} \exp ia\xi'' = \varphi'(a)\varphi''(a)\end{aligned} \tag{3.13a}$$

This rule, that the characteristic function of the sum of mutually independent random variables is equal to the product of the individual characteristic functions, is the second property of characteristic functions that makes them extremely useful. This rule of composition may be shown by direct substitution in Eq. (3.12) or may be regarded as the definition of mutual independence for random variables $\xi'$ and $\xi''$. Equation (3.13) can be written in the alternate form

$$\begin{aligned}\varphi(a) &= \langle \exp ia\xi \rangle = \langle \exp ia(\xi' + \xi'') \rangle = \langle \exp ia\xi' \rangle \langle \exp ia\xi'' \rangle \\ &= \varphi'(a)\varphi''(a)\end{aligned} \tag{3.13b}$$

Consider as a simple example the throw of an ordinary six-sided die and that of a four-sided tetrahedral die. Let the sides of the six-sided die be labeled 1–6 and those of the four-sided die 0–3. If the dice are "honest" and each side equally probable, then $p(n) = \frac{1}{6}, \frac{1}{4}$, or 0. The characteristic functions are

$$\varphi' = \varphi_6(a) = \sum p(n)e^{ina} = \tfrac{1}{6}(e^{ia} + e^{i2a} + e^{i3a} + e^{i4a} + e^{i5a} + e^{i6a})$$

and

$$\varphi'' = \varphi_4(a) = \tfrac{1}{4}(1 + e^{ia} + e^{i2a} + e^{i3a})$$

A question of interest is, "What is the probability in the throw of the two dice that the sum of the numbers on, say, the bottom of the two dice is 8? Equation (3.12) yields

$$p(8) = \sum_{k=-\infty}^{\infty} p'(k)p''(8 - k) = \tfrac{1}{6}\,\tfrac{1}{4} + \tfrac{1}{6}\,\tfrac{1}{4} = \tfrac{1}{12}$$

In this simple example, the characteristic function is more tedious, but it is more straightforward. Thus

$$\varphi(a) = \varphi'(a)\varphi''(a)$$

and

$$p(8) = \frac{1}{2\pi}\int_{-\pi}^{\pi} da\, \varphi(a)e^{i8a} = \frac{1}{12}$$

If the calculation is extended to $s$ mutually independent random variables, the random variable $\xi$,

$$\xi = \sum_{r=1}^{s} \xi_r \qquad \text{or} \qquad n = \sum_{r=1}^{s} m_r$$

has the distribution law

$$P(\xi = n) = p(n) = \sum_{\text{all } m_r} \left[ \prod_{i=1}^{s} p(m_r)\, \delta\left(n - \sum_{r=1}^{s} m_r\right)\right] \tag{3.14}$$

where $p(m_r)$ is the distribution law for the random variable $\xi_r$ and the Kronecker $\delta$ function is zero unless $n = \sum m_r$. Normally, the Kronecker $\delta$ function is defined as $\delta_{mn}$ and is zero for $m \neq n$ and one for $m = n$ but it is more convenient to use the notation

$$\delta(m - n) = \delta_{mn} = \begin{cases} 0, & m \neq n \\ 1, & m = n \end{cases}$$

for the more involved discrete indices. The sum over "all $m_r$" is over all values permitted by the constraint $n$, and the enumeration of all the permitted possibilities can become rather involved. This can be avoided by using the characteristic function for the sum of $s$ mutually independent random variables. Following the procedure used for the development of Eq. (3.13),

$$\varphi(a) = \langle e^{ia\xi}\rangle = \left\langle \prod_{r=1}^{s} e^{ia\xi_r}\right\rangle = \prod_{r=1}^{s} \langle e^{ia\xi_r}\rangle$$
$$= \prod_{r=1}^{s} \varphi_r(a) \tag{3.15}$$

the characteristic function of the sum of $s$ mutually independent random variables is the product of the individual characteristic functions. $p(n)$ follows from $\varphi(a)$ and the inversion formula (3.9).

As an illustrative example, consider a sequence of 500 throws of a die. What is the probability that the total is 1500? Direct enumeration of all possible sequences of 500 throws yielding a sum of 1500 is quite tedious, but the characteristic function for the sequence of 500 throws is

$$\varphi(a) = [\varphi_6(a)]^{500} \tag{3.16a}$$

$\varphi_6(a)$ is the characteristic function given earlier for the six-sided die. Using the inversion formula (3.9),

$$p(1500) = \frac{1}{2\pi} \int_{-\pi}^{\pi} da\, e^{-ia1500} \left( \frac{1}{6} \sum_{m=1}^{6} e^{ima} \right)^{500} \tag{3.16b}$$

the appropriate terms can be selected by expanding the term inside the large parentheses with the multinomial expansion. If one performs 300 throws of a six-sided die and 200 throws of a four-sided die, the characteristic function is readily obtained as

$$\varphi(a) = [\varphi_6(a)]^{300} [\varphi_4(a)]^{200} \tag{3.17}$$

and the inversion formula can be used to obtain $p(n)$.

## 3.3 Useful Mathematical Formulas

*Binomial Expansion*

$$(a+b)^N = \sum_{m_a m_b} \left[ \frac{N!}{m_a!\, m_b!} (a)^{m_a} (b)^{m_b} \delta(m_a + m_b - N) \right]$$

$$= \sum_m \left[ \frac{N!}{m!(N-m)!} a^m b^{N-m} \right] = \sum_m \binom{N}{m} a^m b^{N-m} \tag{3.18a}$$

where

$$\binom{N}{m} = \frac{N(N-1), \ldots, (N-m+1)}{m!} \tag{3.18b}$$

*Multinomial Expansion*

$$(a_1 + a_2 + a_3 + \cdots + a_s)^N = \sum_{\text{all}} \left[ N! \prod_{r=1}^{s} \left( \frac{a_r^{m_r}}{m_r!} \right) \delta\left( N - \sum_{r=1}^{s} m_r \right) \right] \tag{3.19}$$

*Expansions for Large Numbers*

$$N! \cong N^N e^{-N} (2\pi N)^{1/2}$$

$$\ln N! \cong N \ln N - N + \tfrac{1}{2} \ln(2\pi N) \qquad \text{(Stirling's expansion)} \tag{3.20}$$

$$\lim_{N\to\infty}\left(1-\frac{a}{N}\right)^N = e^{-a} \tag{3.21}$$

$$\lim_{N\to\infty}\cos^N\left(\frac{a}{\sqrt{N}}\right) = \exp\left(-\frac{a^2}{2}\right) \tag{3.22}$$

$$\frac{(A+a)^N}{A^N} \xrightarrow[\text{large } N]{} \left(\exp\frac{Na}{A}\right)\exp\left(-\frac{Na^2}{2A^2}\right) \tag{3.23}$$

$$\left(\sum_{m=0}^{M} e^{iam}\right)^N = \left(\frac{1-e^{ia(M+1)}}{1-e^{ia}}\right)^N = (e^{iaM/2})^N\left\{\frac{\sin[(2M+1)a/2]}{\sin(a/2)}\right\}^N \tag{3.24}$$

## 3.4 Random Walk

The random walk is a problem of considerable importance in physics and provides an excellent example of the use of the previous techniques for integral-valued random variables. In the random walk, each step is an independent event, or random variable. The probability of taking a step to the right is often denoted as $p$ and the probability of a step to the left as $q$ In terms of the earlier notation for the dots on a line shown in Fig. 3.1, the distribution law of a step is

$$\begin{aligned} P(\xi_r = +1) &= p(+1) = p \\ P(\xi_r = -1) &= p(-1) = q \\ \{\text{all other} \quad p(n)\} &= 0 \end{aligned} \tag{3.25}$$

The characteristic function of this distribution law is

$$\varphi_r = qe^{-ia} + pe^{+ia} \tag{3.26}$$

and the expectation value and dispersion are

$$\begin{aligned} \mathscr{E}\xi_r &= -i\left(\frac{d(\ln\varphi_r)}{da}\right)_{a=0} = p-q \\ \mathscr{D}\xi_r &= -\left(\frac{d^2(\ln\varphi_r)}{da^2}\right)_{a=0} = 4pq \end{aligned} \tag{3.27}$$

for an individual step.

Now, consider the probability of taking $m$ steps to the right after taking a total of $N$ steps. The sum of $N$ steps is a mutually independent random variable $\xi$,

$$\xi = \sum_{r=1}^{N} \xi_r$$

with a distribution law

$$p(m; N) = \sum_{\text{all}} \left[ \prod_r p(m_r)\, \delta\left(m - \sum_{r=1}^{N} m_r\right) \right] \tag{3.28}$$

and a characteristic function

$$\Phi_N(a) = [\varphi(a)]^N = (qe^{-ia} + pe^{+ia})^N \tag{3.29}$$

$p(m; N)$ can be evaluated by a direct enumeration of Eq. (3.28) or by the use of the inversion formula with Eq. (3.29). Using the inversion formula and the binomial expansion for $\varphi^N$,

$$\begin{aligned} p(m; N) &= \frac{1}{2\pi} \int_{-\pi}^{\pi} da\, e^{-ima} \Phi_N(a) = \frac{1}{2\pi} \int_{-\pi/2}^{\pi/2} da\, e^{-ima} \Phi_N [1 + (-)^{N+m}] \\ &= \frac{N!}{[\frac{1}{2}(N+m)]!\,[\frac{1}{2}(N-m)]!}\, p^{(N+m)/2} q^{(N-m)/2} \end{aligned} \tag{3.30}$$

Note that $p(m; N)$ is zero for $m$ even and $N$ odd, or for $m$ odd and $N$ even, and that $\frac{1}{2}(N+m)$ and $\frac{1}{2}(N-m)$ are integers. This expresses the fact that one cannot move an even number of steps to the right while taking an odd number of steps, and vice versa. In the binomial expansion $\varphi^N$, only the term $e^{+ima}$ contributes to the integral.

Another procedure is often used to obtain $p(m; N)$. The number of ways of taking $m$ steps to the right while taking a total of $N$ steps is considered, and is the coefficient of $p$ and $q$ in Eq. (3.30). The probability of a particular "way" is given by the $pq$ term.

If $N$ is large, Eq. (3.30) can be evaluated in an approximate manner. Stirling's expansion for $N!$ is often used. It is more convenient to make an expansion of the characteristic function. A Taylor series expansion of $\ln \Phi_N$ about $a = 0$ is chosen:

$$\begin{aligned} \ln \Phi_N(a) &= \ln 1 + \left[\frac{d(\ln \Phi)}{da}\right]_{a=0} a + \frac{1}{2}\left[\frac{d^2(\ln \Phi)}{da^2}\right]_{a=0} a^2 + \cdots + \theta(a^3) \\ &= ia\mathscr{E}\,\xi_N - \frac{1}{2} a^2 \mathscr{D}\,\xi_N + \cdots \end{aligned} \tag{3.31}$$

Equation (3.30) takes the form

$$p(m; N) = \frac{1}{\pi} \int_{-\pi/2}^{\pi/2} da\, [\exp(-ima)] \exp[ia\mathscr{E}\,\xi_N - \tfrac{1}{2}a^2\mathscr{D}\,\xi_N + \cdots] \tag{3.32}$$

Since $\ln \Phi_N(a) = N \ln \varphi(a)$, the expectation value and dispersion

$$\mathscr{E}\,\xi_N = N\mathscr{E}\,\xi = N(p - q) \tag{3.33a}$$

$$\mathscr{D}\,\xi_N = N\mathscr{D}\,\xi = N(4pq) \tag{3.33b}$$

increase linearly with $N$. Thus the coefficient of $a^2$ in the exponential increases linearly with $N$ and for large $N$, the integrand is appreciable only near $a = 0$. The limits of the integral can, to a good approximation, be changed from $\pm\pi/2$ to $\pm\infty$ and the resulting definite integral yields

$$p(m; N) \approx \frac{2}{(2\pi\mathscr{D}\,\xi_N)^{1/2}} \exp\left[\frac{-(m - \mathscr{E}\,\xi_N)^2}{2\mathscr{D}\,\xi_N}\right] \tag{3.34}$$

This expansion illustrates the usefulness of the characteristic function. The validity of the expansion is considered in greater detail during the discussion of the central limit theorem.

Some care is required in the use of Eq. (3.34) in that $m$ changes by two. If $N$ is even, then $m$ takes on only even values. The sum of the probabilities

$$1 = \sum_{\substack{m \text{ even} \\ \text{or odd} \\ \text{as } N}} p(m; N) \approx \frac{1}{2} \sum_{\text{all } m} p(m; N) \tag{3.35}$$

The approximate procedure used to evaluate Eq. (3.30) is shown in Fig. 3.2 for $p = q = \frac{1}{2}$. For this example,

$$\varphi(a) = \tfrac{1}{2}(e^{-ia} + e^{ia}) = \cos a$$

and

$$\Phi_N(a) = \varphi^N = \cos^N a \approx \exp -\tfrac{1}{2}Na^2$$

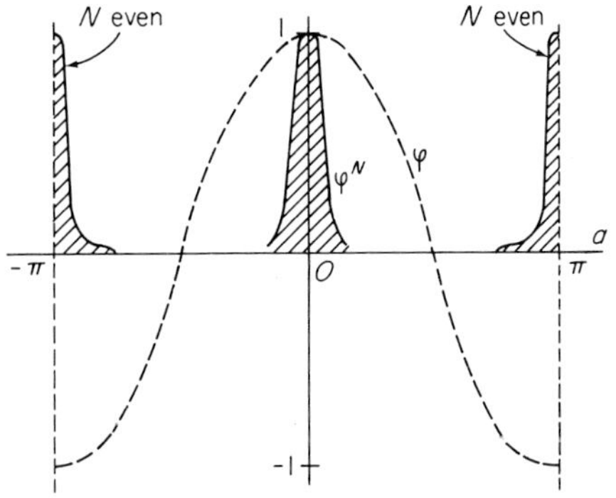


**Fig. 3.2** Characteristic function for the random walk problem for $\varphi(a)$ and $[\varphi(a)]^N$. Note that the cosine function to the $N$th power becomes a very sharp function.

As $N$ increases, $\cos^N a$ becomes a sharp function near $a = 0$ and $a = \pm\pi$. The sign at $\pm\pi$ depends on $N$ being odd or even. If $m$ is even and $N$ even, the integral is twice the value between $\pm\pi/2$. If $N$ is odd and $m$ even, the integral is zero. This is easily seen by sketching $\cos a$ on the figure. With this in mind, the primary contribution to the integral occurs near $a = 0$ for large $N$ and with negligible error, the integral may be extended from $\pm\pi/2$ to $\pm\infty$:

$$p(m; N) = \frac{1}{2\pi}\int_{-\pi}^{\pi} da\,[\exp(-ima)]\Phi_N = \frac{1}{\pi}\int_{-\pi/2}^{\pi/2} da\,[\exp(-ima)]\Phi_N$$

$$\approx \frac{1}{\pi}\int_{-\infty}^{\infty} da\,[\exp(-ima)]\exp\left(-\frac{Na^2}{2}\right)$$

$$\approx \frac{2}{(2\pi N)^{1/2}}\exp\left(-\frac{m^2}{2N}\right)$$

Exercise 3.2 Show that Eqs. (3.18)–(3.20) are correct.

Exercise 3.3 A container contains 300 six-sided dice and 200 four-sided dice. The dice are noninteracting and the container is shaken in a random manner. What is the probability that the sum of the numbers facing down is 500? 1500? 2600? Why is this problem similar to that of independent throws? Develop an approximate procedure for evaluation. A procedure for the expansion of $\ln \varphi$ as indicated in Eq. (3.31) is suggested for $p(1500)$, and $\ln \varphi \approx 1550ia - 562a^2$.

Exercise 3.4 A man has been carousing for the evening, and upon starting home, his power of decision becomes completely random. He is 10 steps from his door. What is the probability that he arrives home in 10 steps? 100 steps? 200 steps? What is the most probable number of steps required? If the experiment is repeated many times, what is the average number of steps required to reach the door?

### *3.4.1 Difference Equation for the Random Walk*

The random walk can also be discussed as random steps on the set of integers between $-\infty$ and $+\infty$; the steps are governed by the difference equation

$$p(m; N+1) = qp(m+1; N) + pp(m-1; N) \tag{3.36}$$

$$p(0; 0) = 1, \qquad p(m; 0) = 0 \tag{3.37}$$

$p(m; N+1)$ is the probability of being at position $m$ on the $(N+1)$th step and depends only on being at either position $m-1$ or position $m+1$ on the $N$th step. The initial condition indicates that $m = 0$ when $N = 0$. This equation can be readily solved by multiplying by $e^{ima}$ and summing over $m$ to yield an equation in terms of the characteristic functions,

$$\Phi_{N+1}(a) = (qe^{-ia} + pe^{ia})\Phi_N(a) = \varphi(a)\Phi_N(a) \tag{3.38}$$

This equation has the obvious solution

$$\Phi_N(a) = \varphi^N(a) \tag{3.39}$$

and all of the previous discussion applies to this problem. It should be noted that

$$\sum_m e^{ima}p(m; N) = \Phi_N(a)$$

Exercise 3.5 A student learning statistics is told that the probability of a passenger carrying a bomb aboard an airplane is $10^{-6}$ and the probability of two passengers carrying bombs aboard the same plane is $10^{-12}$. Being very worried about explosions, he now carries his own bomb. Explain the error in his reasoning.

Exercise 3.6 Show that Eq. (3.38) describes the difference equation (3.36). Show that Eq. (3.39) is a solution of Eq. (3.38).

Exercise 3.7 If the average length of a step in the random walk problem is $\lambda$, what is an estimate of the distance traveled to the right in $N$ steps? ($p = q = \frac{1}{2}$.)

Ans. $\lambda\sqrt{N}$

Exercise 3.8 What is the average value of $m^2$ for large $N$, for $p \neq q$ and $p = q$? Find for fixed $m$ the value of $N$ for which $p(m; N)$ has a maximum.

Exercise 3.9 Show for $p = q$ and steps of length $\lambda$ that $\bar{x} = 0$:

$$\langle(x - \bar{x})^2\rangle = \langle x^2\rangle = N\lambda^2 \tag{3.40}$$

Exercise 3.10 A particle passes through a slab of material and is deflected by a small angle $\pm\theta_0$ on each collision. Show that the angular scattering after $N$ scattering events is

$$\bar{\theta} = 0, \qquad \langle\theta^2\rangle = N{\theta_0}^2 \tag{3.41}$$

**Exercise 3.11** Show in Exercise 3.10 that a particle undergoing random small-angle scattering will exhibit a radius of curvature for its path of the order of

$$R \approx \sqrt{N}\left(\frac{\lambda}{\theta_0}\right) \tag{3.42}$$

where $N\lambda$ is the distance traveled while the $N$ small-angle scattering events occur.

## 3.5 Binomial and Poisson Distributions

Some physical problems are analogous to the consideration of the occupancy of a cell as shown in Fig. 3.3. In order to formulate the problem, a system is divided into $g$ cells. If a particle is placed in the system in a

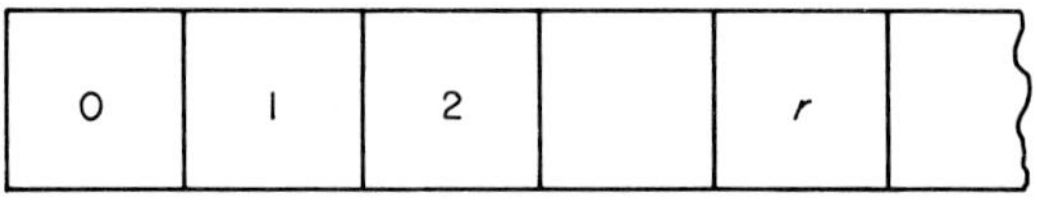


**Fig. 3.3** Labeling of boxes or cells for a discussion of placing $N$ particles in $g$ cells.

random manner, each cell is regarded as equally probable. The only cell in which measurements are made is the $r$th cell, and the distribution law for the $r$th cell is

$$P(\xi_r = 1) = \frac{1}{g}, \qquad P(\xi_r = 0) = 1 - \frac{1}{g} \tag{3.43}$$

This distribution law emphasizes that for the $r$th cell, only two possibilities can occur, either the particle is found in it or it is not. The characteristic function for the $r$th cell is

$$\varphi(a) = \left(1 - \frac{1}{g}\right) + \frac{1}{g}\, e^{ia} \tag{3.44}$$

with expectation value and dispersion

$$\mathscr{E}\xi_r = \frac{1}{g} \qquad \text{and} \qquad \mathscr{D}\xi_r = \frac{1}{g}\left(1 - \frac{1}{g}\right) \tag{3.45}$$

If $N$ particles are placed in the system, the random variable

$$X_r = \sum_{s=1}^{N} \xi_{r;s}$$

and the distribution law $P(X_r = m)$ correspond to the physical quantities of interest:

$$\Phi_N(a) = \varphi^N(a) = \left[\left(1 - \frac{1}{g}\right) + \frac{1}{g} e^{ia}\right]^N \tag{3.46}$$

Using the inversion formula, the exact form of the distribution law is

$$p(m; N) = \frac{1}{2\pi} \int_{-\pi}^{\pi} da\, e^{-ima} \Phi_N(a) = \frac{N!}{m!(N-m)!} \left(\frac{1}{g}\right)^m \left(1 - \frac{1}{g}\right)^{N-m} \tag{3.47}$$

If $N$ and $g$ take on large values such that

$$\frac{N}{g} \rightarrow \lambda \tag{3.48}$$

then the characteristic function is approximately

$$\ln \Phi_N(a) = N \ln \varphi(a) \approx -\lambda(1 - e^{ia})$$

This is usually referred to as the Poisson distribution, and the Poisson distribution is described by the characteristic function

$$\Phi(a) = \exp[-\lambda(1 - \exp ia)] \tag{3.49}$$

and the distribution law

$$p(m) = \frac{1}{2\pi} \int_{-\pi}^{\pi} da\, e^{-ima} \Phi(a) = e^{-\lambda} \frac{\lambda^m}{m!} \tag{3.50}$$

The Poisson distribution will be used frequently in Chapters X and XI in the discussion of random pulses and scattering by fluctuations.

If $\lambda$ is also a large number, the Poisson distribution tends to a Gaussian or normal distribution. This can be shown by a direct expansion of $e^{ia}$, and

$$\Phi(a) \xrightarrow[\text{large } \lambda]{} \exp \lambda(ia - \tfrac{1}{2}a^2) \tag{3.51}$$

Expansions of this type illustrate the convenience of the characteristic function. In evaluating $p(m)$ with the inversion formula, for large $\lambda$, the integral can be taken from $-\infty$ to $+\infty$ with negligible error. Equation (3.53) can be derived from $\sum p(m) = 1$ by taking the first and second derivatives with respect to $\lambda$. Thus

$$\frac{\partial}{\partial \lambda} \sum_m \frac{e^{-\lambda} \lambda^m}{m!} = 0 \qquad \text{yields} \quad \overline{m} = \lambda \tag{3.52}$$

It is shown in Exercise 3.13 that the average value of $m$ is $\bar{m} = \lambda$ and the dispersion in $m$ is $\langle(m - \bar{m})^2\rangle = \lambda$ for the Poisson distribution. Thus the average number of particles in a cell and the fluctuation in the number of particles in the cell are equal.

Exercise 3.12 Show that Eq. (3.47) follows from a direct evaluation of the number of arrangements that place $m$ particles in the $r$th box and the probability of each arrangement.

Exercise 3.13 Using the approximation $\ln(1 + x) \approx x$, show that $\Phi_N(a)$ has the form given by Eq. (3.49). Using an expansion for $\Phi(a)$, show that Eq. (3.50) follows. Equations (3.11a) and (3.11b) can be used to obtain the expectation value and dispersion for the Poisson distribution:

$$\mathscr{E}X_r = \lambda = \bar{m}, \qquad \mathscr{D}X_r = \lambda = \langle(m - \bar{m})\rangle^2 \tag{3.53}$$

Exercise 3.14 What is the average number of particles in volume $\delta V$ in a vessel which contains $N$ particles in volume $V$? What is the probability of finding one, two, or three particles in volume $\delta V = \Delta = (V/N)$? Repeat for $2\Delta$ and $100\Delta$. What is the fluctuation in the number of particles in $\delta V$?

Exercise 3.15 Current $I = \dot{N}e$ is flowing across a vacuum tube or in a diode. The detecting instrument has a response time $\tau$. What is the probability of receiving $m$ electrons in the time interval $\tau$? Show $\langle(m - \bar{m})^2\rangle = \dot{N}\tau$ if the expected number is large in time $\tau$. If $\tau^{-1} = \dot{N}$, what is the probability of receiving one, two, or three electrons during this interval?

Exercise 3.16 Radiation of 0.5 μm passes through a container filled with $N_2$ gas at 400°K and a pressure of 1 atm. What is the average square fluctuation of the index of refraction in a cell of the order of the (wavelength)$^3$?

### 3.5.1 Difference Equation for the Binomial Distribution

The Poisson distribution can also be expressed as the difference equation

$$p(m; N + 1) = \left(1 - \frac{1}{g}\right)p(m; N) + \frac{1}{g}\,p(m - 1; N) \tag{3.54}$$

where $p(m; N + 1)$ is the probability of having $m$ particles in the selected, or $r$th, cell after distributing $N + 1$ particles in the $g$ cells. The subsidiary conditions for the initial condition and the first event are, respectively,

$$p(m; 0) = \delta(m - 0) \tag{3.55a}$$

and

$$p(0; 1) = \left(1 - \frac{1}{g}\right)p(0; 0) \tag{3.55b}$$

If the characteristic function is assumed of the form

$$\Phi_N(a) = \sum_m p(m; N)e^{ima}$$

and

$$p(m; N) = \frac{1}{2\pi}\int_{-\pi}^{\pi} da\, e^{-ima}\Phi_N(a)$$

then direct substitution into Eq. (3.54) yields

$$\Phi_{N+1}(a) = \left(1 - \frac{1}{g}\right)\Phi_N(a) + \frac{1}{g}\, e^{ia}\Phi_N(a) = \varphi(a)\Phi_N(a) \tag{3.56}$$

in terms of the characteristic functions. This has the obvious solution

$$\Phi_N(a) = \varphi(a)^N \tag{3.57a}$$

where

$$\varphi(a) = \left(1 - \frac{1}{g}\right) + \frac{1}{g}\, e^{ia} \tag{3.57b}$$

Since $\varphi(a)$ is the same characteristic function as that used in Eq. (3.46), the previous discussion applies.

## 3.6 Central Limit Theorem

The discussion of the random walk and the Poisson distribution indicated that under certain conditions, the distributions tended toward the Gaussian or normal distributions. This is true for a large class of problems. The characteristic function for the sum of $N$ mutually independent random variables is

$$\xi = \sum_{r=1}^{N} \xi_r \tag{3.58a}$$

$$\Phi_N(a) = \prod_{r=1}^{N} \varphi_r(a) \tag{3.58b}$$

where $\varphi_r(a)$ is the characteristic function that describes the distribution law $P(\xi_r = n) = p(n)$. Expanding $\ln \Phi_N(a)$ about $a = 0$ yields the expansion

$$\ln \Phi_N(a) = ia\mathscr{E}\xi_N - \tfrac{1}{2}a^2\mathscr{D}\xi_N + \cdots$$

where

$$\mathscr{E}\xi_N = -i\left[\frac{d(\ln \Phi_N)}{da}\right]_{a=0} = \sum_{r=1}^{N} -i\left[\frac{d(\ln \varphi_r)}{da}\right]_{a=0} = \sum_{r=1}^{N} \mathscr{E}\xi_r \tag{3.59a}$$

$$\mathscr{D}\xi_N = -\left[\frac{d^2(\ln \Phi_N)}{da^2}\right]_{a=0} = \sum_{r=1}^{N} -\left[\frac{d^2(\ln \varphi_r)}{da^2}\right]_{a=0} = \sum_{r=1}^{N} \mathscr{D}\xi_r \tag{3.59b}$$

The distribution law $P(x = m) = p(m; N)$ is given by the Fourier inversion formula

$$\begin{aligned} p(m; N) &= \frac{1}{2\pi}\int_{-\pi}^{\pi} da\,(\exp -ima)\Phi_N(a) \\ &\approx \frac{1}{2\pi}\int_{-\infty}^{\infty} da\,(\exp -ima)\exp\left(ia\mathscr{E}\xi_N - \frac{1}{2}a^2\mathscr{D}\xi_N\right) \\ &\approx \frac{1}{(2\pi\mathscr{D}\xi_N)^{1/2}}\exp\left[-\frac{(m - \mathscr{E}\xi_N)^2}{2\mathscr{D}\xi_N}\right] \end{aligned} \tag{3.60a}$$

Since the dispersion $\mathscr{D}\xi_N$ increases linearly with $N$, the integral has an appreciable value only when $a$ is small. Higher-order terms $a^3, a^4, \ldots$ can be omitted and the limits of integration can be changed from $\pm\pi$ to $\pm\infty$ with negligible error. Thus the distribution law $p(m; N)$ tends toward a Gaussian or normal distribution for large $N$. This is not true for all distributions described by $\varphi_r(a)$. The Poisson distribution does not become Gaussian for large $N$, but it does become Gaussian for large $\lambda$, that is, for a large number of events per interval or of particles per cell.

The criterion for the validity of Eq. (3.60a) is known as the central limit theorem. A rigorous proof is given by Khinchin [*1*] and the proof is not given here. His proof indicates that if $p(m_r)$ has moments up to the fifth-order inclusive, that is, $\mathscr{E}\xi_r^5$ is finite, then

$$\begin{aligned} p(m; N) = {} & \frac{1}{(2\pi\mathscr{D}\xi_N)^{1/2}}\exp\left(-\frac{(m - \mathscr{E}\xi_N)^2}{2\mathscr{D}\xi_N}\right) \\ & + O[N^{-3/2}(1 + |\, m - \mathscr{E}\xi_N \,|)] \end{aligned} \tag{3.60b}$$

## 3.7 Conditional Probability

A physical system subject to a constraint is now considered. Suppose an ensemble of one million dice are placed in a container and shaken in such a fashion that the sum of the numbers facing upward is two million after every shake. This is a constraint on this system. Two quantities are of interest. How can the total number of arrangements of the dice subject to this constraint be computed, and what is the probability that the number on a given die is, say, three? Khinchin [*1*] has introduced a very simple and elegant method for treating this problem. It forms the basis for all discussions in statistical mechanics in this text and is used in the discussion in Chapter VI.

### *3.7.1 An Approximate Method for Counting Arrangements*

Before considering the specific problem proposed in the preceding paragraph or similar problems, consider the general physical system that obeys the distribution law

$$u(m) = g(m)\frac{e^{-\beta m}}{z(\beta)} \tag{3.61a}$$

where $u(m)$ is the probability of occupancy of any dot $m$ on the line shown in Fig. 3.1. $g(m)$ is the *a priori* weight, or number of arrangements associated with this dot, $e^{-\beta m}$ is another weighting factor, and $z(\beta)$ is selected so that

$$\sum_m u(m) = 1 \tag{3.61b}$$

or

$$z(\beta) = \sum_m g(m)e^{-\beta m} \tag{3.61c}$$

If $N$ such systems are considered, the distribution law for the mutually independent random variable $M$,

$$M = \sum_{r=1}^{N} m_r \tag{3.62}$$

is

$$\begin{aligned} P(X_N = M) = U(M; N) &= \sum_{\text{all}} \left[\prod_{r=1}^{N} u(m_r)\, \delta\left(M - \sum_{r=1}^{N} m_r\right)\right] \\ &= \left\{\sum_{\text{all}} \left[\prod_{r=1}^{N} g(m_r)\, \delta\left(M - \sum_{r=1}^{N} m_r\right)\right]\right\} \frac{e^{-\beta M}}{z^N(\beta)} \\ &= G(M; N)\frac{e^{-\beta M}}{Z_N(\beta)} \end{aligned} \tag{3.63}$$

where $Z_N(\beta) = z^N(\beta)$. $G(M; N)$ is the number of arrangements of $N$ systems subject to the constraint $M$ and is given by the quantity inside the curly brackets. For a large number of systems, direct enumeration is very tedious and perhaps almost impossible in a reasonable amount of time. In order to find an approximate method for evaluating $G(M; N)$, the characteristic function is introduced.

The characteristic function of the distribution is

$$\varphi(a) = \sum u(m)e^{iam} = z^{-1}\sum g(m)e^{-\beta m}e^{iam} = \frac{z(\beta - ia)}{z(\beta)} \tag{3.64a}$$

and the characteristic function for the distribution $U(M; N)$ is

$$\Phi_N(a) = \frac{Z_N(\beta - ia)}{Z_N(\beta)} = \left[\frac{z(\beta - ia)}{z(\beta)}\right]^N \tag{3.64b}$$

$\varphi(a)$ or $\Phi_N(a)$ is frequently expressed in terms of the normalization factor $z$ or $Z_N$ as indicated earlier. Expanding $\ln \Phi_N$,

$$\ln \Phi_N(a) = [\ln \Phi_N(a)]_{a=0} + \left\{\frac{d[\ln \Phi_N(a)]}{da}\right\}_{a=0} a + \cdots = ia\bar{M} - \frac{1}{2} D_N a^2 + O(a^3) \tag{3.65}$$

where

$$\bar{M} = \left[\frac{-i\, d(\ln \Phi_N)}{da}\right]_{a=0} = -\frac{d(\ln Z_N)}{d\beta} = -N\frac{d(\ln z)}{d\beta} = N\bar{m} \tag{3.66a}$$

and

$$D_N = \left[-\frac{d^2(\ln \Phi_N)}{da^2}\right]_{a=0} = \frac{d^2(\ln Z_N)}{d\beta^2} = N\frac{d^2(\ln z)}{d\beta^2} = N\langle(m-\bar{m})^2\rangle \tag{3.66b}$$

Since $\langle(m - \bar{m})^2\rangle$, or the dispersion of the distribution described by $\varphi(a)$, can never be negative, the dispersion of the distribution described by $\Phi_N$ increases linearly with $N$. For large $N$, the limits on the inversion formula may be changed to $\pm\infty$ with negligible error and

$$U(M; N) = \frac{1}{2\pi}\int_{-\pi}^{\pi} da\, e^{-iam}\Phi_N(a) \approx \frac{1}{(2\pi D_N)^{1/2}} \exp\left[-\frac{(M - \bar{M})^2}{2D_N}\right] \tag{3.67}$$

Combining Eqs. (3.63) and (3.67) yields an approximate expression for the number of arrangements subject to the constraint $M$,

$$G(M; N) \approx \frac{Z_N(\beta)e^{\beta M}}{(2\pi D_N)^{1/2}} \exp\left[-\frac{(M-\bar{M})^2}{2D_N}\right] \tag{3.68}$$

The requirement that $M = \bar{M}$ determines $\beta$. $\bar{M}$ is given by Eq. (3.66a) and

$$\frac{d(\ln Z_N)}{d\beta} + M = 0 \tag{3.69}$$

must be solved for $\beta$. For $D_N \neq 0$, this equation has a single root $\beta$. Equation (3.66b) for

$$\frac{d^2(\ln Z_N)}{d\beta^2} = D_N > 0$$

indicates that $d(\ln Z_N)/d\beta$ is always convex. It can be shown that Eq. (3.69) has unique value of $\beta$ for a given $M$. Therefore a constraint $M$ can be introduced by the proper choice of $\beta$, and

$$G(\bar{M}; N) \approx \frac{Z_N(\beta)\exp(\beta\bar{M})}{(2\pi D_N)^{1/2}} \tag{3.70}$$

where $\beta$ follows from a solution of Eq. (3.69). Knowledge of $z(\beta)$ and $\beta$ is adequate to determine the number of arrangements to a high degree of accuracy.

**Exercise 3.17** Suppose Eq. (3.61) applies to a four-sided die labeled 0, 1, 1, 2. What does $g(m)$ equal? If $10^6$ dice are placed in a container and $\beta = 1$, estimate the number of permitted configurations. What average value of $M$ corresponds to $\beta = 1$?

**Exercise 3.18** Show that in a problem of limited complexity, if $\bar{m}$ is given, $\beta$ can be evaluated by numerical methods and $z(\beta)$ determined.

**Exercise 3.19** Use a table of definite integrals and show that Eq. (3.67) follows from Eq. (3.65) for $\Phi_N(a)$.

### *3.7.2 Distribution Law for a System Subject to a Constraint*

Suppose $N_0$ systems are placed in an isolated enclosure and subject to the single constraint

$$M_0 = \sum_{r=1}^{N_0} m_r \tag{3.71}$$

In the absence of the constraint, the number of arrangements associated with each value of $m$ for the $r$th system is $g(m_r)$. Assume that each configuration or arrangement of the $m_r$ permitted by the constraint is equally probable and the number of such arrangements is

$$G(M_0; N_0) = \{\text{number of arrangements with constraint} \quad M_0\} \tag{3.72}$$

If $u(m_r)$ is written in the form of Eq. (3.61), $G(M_0; N_0)$ can be determined approximately by Eq. (3.70) and by Eq. (3.69) for $\beta$.

The probability of finding the $r$th system in the state with label $m_r$ is proportional to the number of arrangements for this configuration,

$$\begin{aligned} p(m) &= \{\text{probability of finding a system in state} \quad m\} \\ &= \frac{g(m)G(M_0 - m; N_0 - 1)}{G(M_0; N_0)} \end{aligned} \tag{3.73}$$

where $g(m)$ is the number of arrangements with label $m_r$ and $G(M_0 - m; N_0 - 1)$ is the number of arrangements of $N_0 - 1$ systems with $\sum m_r = M_0 - m$. Using the value of $\beta$ determined from Eq. (3.69) and $M_0$ from Eq. (3.70) for $\bar{M}$, Eq. (3.73) is approximately given by

$$\begin{aligned} p(m) &\approx \frac{g(m)G(\bar{M} - m; N_0 - 1)}{G(\bar{M}; N_0)} \\ &\approx \frac{g(m)Z_{N-1} \exp[\beta(\bar{M} - m)]}{Z_N \exp(\beta\bar{M})} \left(\frac{D_N}{D_{N-1}}\right)^{1/2} \end{aligned}$$

and

$$p(m) \approx \frac{g(m)e^{-\beta m}}{z(\beta)} \tag{3.74}$$

Since $D_N$ increases linearly with $N$, $D_{N-1} \approx D_N$, and this approximation has been introduced here. $p(m)$ is the conditional probability for state $m$. Equation (3.75) expresses the rule of composition for the function $G(M; N)$. Khinchin refers to $G(M; N)$ as the structure function. If $N - 1$ systems are combined with a single system, then the number of arrangements for a particular value of $m$ is the product of the arrangements of the mutually independent systems, that is,

$$G(m; 1)G(M; N - 1)\, \delta[M - (M_0 - m)] = g(m)G(M_0 - m; N - 1)$$

subject to the constraint of the $\delta$ function. The sum over $m$ yields all the possible arrangements of the two systems subject to the constraint $M = M_0 - m$.

Exercise 3.20 Show that

$$\sum_m g(m)G(M_0 - m; N_0 - 1) = G(M_0; N_0) \tag{3.75}$$

and therefore $\sum p(m) = 1$ in Eq. (3.73).

### *3.7.3 Entropy and the Number of Configurations with a Constraint*

In the previous discussion, each dot on the line had an *a priori* weight $g(m)$. If $g(m) \neq 1$, then the $m$th dot is degenerate and a second label $g(m\mu)$ is needed. The *a priori* weight of $g(m\mu) = 1$ and

$$g(m) = \sum_\mu g(m\mu) \tag{3.76}$$

Thus in the example of a six-sided die with five faces labeled $m = 0$ and with colors $\mu =$ red, black, orange, yellow, and green on the faces, and the sixth side labeled $m = 1$, it may be noted that each state $g(m\mu)$ has weight 1. The constraint $\sum m_r = M$ does not involve $\mu$ or the color index in the example. The generating function for the two indices is

$$z(\beta) = \sum_m \sum_\mu g(m\mu)e^{-\beta m} = \sum_m g(m)e^{-\beta m} \tag{3.77}$$

and is the same as discussed in the previous section. The probability of a particular value of $m$ is

$$p(m) = z^{-1}g(m)e^{-\beta m} \tag{3.78}$$

and of a particular state $(m\mu)$,

$$p(m\mu) = z^{-1}e^{-\beta m} \tag{3.79}$$

In many problems, it is desirable to label each state of the system and this labeling scheme is used in the following discussion.

If an ensemble of $N$ systems has a constraint

$$\sum_{r=1}^{N} m_r = M \tag{3.80}$$

on the $m$ index, the number of configurations is given by

$$G(M; N) = \sum_{\text{all}} \prod_{r=1}^{N} g(m_r)\, \delta\left(M - \sum_{r=1}^{N} m_r\right) \tag{3.81}$$

This rather tedious calculation in combinatorial analysis can be approximated by the use of Eq. (3.70), or

$$G(M;N) \approx Z_N(\beta)e^{\beta M}$$

The entropy of this ensemble of $N$ systems is defined as

$$S(M;N) = \ln G(M;N) \tag{3.82a}$$

and

$$S(\bar{M};N) \approx N[\ln z(\beta) + \beta\bar{m}] \tag{3.82b}$$

If $\sigma$ is defined in terms of the properties of the system as

$$\sigma = -\sum_{m\mu} p(m\mu) \ln p(m\mu) \tag{3.83a}$$

then with Eq. (3.79) for $p(m\mu)$, it follows that

$$\sigma = \ln z(\beta) + \beta\bar{m} \tag{3.83b}$$

and for $N$ systems,

$$S(\bar{M};N) \approx N\sigma \tag{3.84}$$

The error is of the order of $\ln(2\pi D_N)^{-1/2}$ and can be neglected for large $N$. It can be noted that $\sigma$ follows from a knowledge of $\beta$ and the generating function $z$ for a single system, and that $S \approx N\sigma$. The number of configurations $G \approx e^S \approx e^{N\sigma}$ can be estimated from $\sigma$. The entropy $S(M;N)$ has its maximum value at $M = \bar{M}$.

As an example of estimating the number of configurations in an ensemble of systems, consider the problem of $10^6$ dice placed in a container and shaken in such a manner that the numbers facing upward total $3\times 10^6$. The number of configurations is approximately given by Eq. (3.70); $M = 3\times 10^6$ and from Eq. (3.69),

$$N\frac{d(\ln z)}{d\beta} + M = 0$$

or

$$\frac{\sum me^{-\beta m}}{\sum e^{-\beta m}} = 3$$

This problem can be simplified by noting that

$$\sum_{r=0}^{R} x^r = \frac{1 - x^{R+1}}{1 - x}$$

and

$$z = \sum_{m=1}^{6} e^{-\beta m} = e^{-\beta} \frac{1 - e^{-6\beta}}{1 - e^{-\beta}}$$

Then,

$$\frac{d(\ln z)}{d\beta} = \frac{6}{e^{6\beta} - 1} - \frac{1}{1 - e^{-\beta}} = -3$$

and this equation can be solved for $\beta$ to yield

$$\beta \approx 0.20$$

Then, $z(0.20) = 3.17$ and

$$S = 10^6(1.15 + 0.6) = 1.75 \times 10^6 = \ln G$$

The number of configurations is the very large number

$$G \approx \exp(1.75 \times 10^6)$$

The probability of throwing a six is now given by Eq. (3.78) with the given values for $\beta$ and $z$, or

$$p(6) = 0.095$$

and can be compared with the no-constraint value of $\frac{1}{6}$.

Exercise 3.21 Show that $\sigma = \ln z + \beta \bar{m}$ and $d\sigma/d\bar{m} = \beta$.

Exercise 3.22 (a) Consider a set of $10^6$ six-sided dice labeled 1–6. Show that $\beta = 0$ corresponds to an average value of $m = 21/6$. Show also that as $\beta \to +\infty$, the average value of $\bar{m} \to 1$. What is $\bar{m}$ as $\beta \to -\infty$?

(b) Suppose in the example just given that the number of die is $10^6$ and the box is shaken in such a manner that the sum of numbers facing up is $10^6 + 1$. Compare the number of permitted configurations obtained by the approximate method with the exact answer. What value of $\beta$ is needed?

Ans. $e^{-\beta} \approx 10^{-6}$

Exercise 3.23 Suppose five sides of the die are labeled 0 and one side is labeled 1. If $10^6$ dice are placed in a box and shaken in a constrained manner such that the sum of the numbers facing up is 100,000, what is the probability of a particular die having a 1 facing up? Of a particular face facing up?

**Exercise 3.24** Suppose $10^6$ four-sided dice (labeled 1–4) and $10^6$ six-sided dice (labeled 1–6) are placed in a box and the constraint limits the sum facing down after a shake to $4 \times 10^6$. What is the probability of a four facing down on a six-sided die? On a four-sided die? Is a single value of $\beta$ adequate for this problem? How is $\beta$ determined?

### 3.7.4 *Fluctuations*

If in the ensemble of $N$ systems subject to a single constraint the value of $m$ is measured for a group of systems, the values of $m$ will take on a distribution about the average value of $\bar{m}$ that is in accord with the probability law $p(m\mu)$. The dispersion about the average value and the average value follow from $z(\beta)$ in the usual manner, and

$$\frac{d(\ln z)}{d\beta} = -\bar{m}$$
$$\frac{d^2(\ln z)}{d\beta^2} = \langle (m - \bar{m})^2 \rangle = -\frac{d\bar{m}}{d\beta} \tag{3.85a}$$

If a sequence of measurements is made on a particular system and if in the course of time this system takes on the values suggested by $p(m\mu)$, then $\langle (m - \bar{m})^2 \rangle$ can be interpreted as the fluctuation in the value of $m$ in a sequence of measurements.

The constraint $\bar{M} = M$ fixes $\beta$ at $\beta_0$ for the ensemble of $N$ systems, there is no dispersion in the values of $M$, and $S(\bar{M}; N)$ takes a fixed value. This is not true for a group of single systems and the quantity $\sigma$ which is defined by Eq. (3.83) can take on a distribution of values about $\sigma(\beta_0)$. Again, this can be regarded as a fluctuation of $\sigma$ in a sequence of measurements on a particular system. Expanding $\ln z$ about $\beta_0$ yields

$$\ln z(\beta) = \ln z(\beta_0) - \bar{m}(\beta - \beta_0) + \tfrac{1}{2}(\beta - \beta_0)^2 \langle (m - \bar{m})^2 \rangle + \cdots \tag{3.85b}$$

With these same definitions, $p(m\beta)$ and $p(m\beta_0)$ are related by

$$p(m\beta) \approx p(m\beta_0)\{\exp[-(\beta - \beta_0)(m - \bar{m})]\} \exp[-\tfrac{1}{2}(\beta - \beta_0)^2 \langle (m - \bar{m})^2 \rangle] \tag{3.85c}$$

In order to avoid confusion, it is again emphasized that $\langle (m - \bar{m})^2 \rangle$ is a function of $\beta$, but not of $m$, and is a number for a given value of $\beta$. Using the notation

$$\Delta\beta = \beta - \beta_0 \qquad \text{and} \qquad \Delta m = m - \bar{m}$$

the probability law for $p(\Delta\beta) = p(\bar{m}, \beta)$ is given by

$$p(\Delta\beta) = \text{constant} \times \exp[-\tfrac{1}{2}(\Delta\beta)^2\langle(\Delta m)^2\rangle] \tag{3.85d}$$

It is interesting to note that the fluctuations in $\Delta\beta$ and $\Delta m$ are related by

$$\overline{(\Delta\beta)^2}\,\overline{(\Delta m)^2} = 1 \tag{3.85e}$$

and this follows from the form of the distribution law for $p(\Delta\beta)$. Another form is given with $\Delta\beta = \Delta(d\sigma/d\bar{m})$.

In later calculations in this text, the generating function $z(\beta)$ will be referred to as the generating function for a canonical ensemble, the quantity $\sigma$ as the entropy, $\bar{m}$ as the energy, $d\bar{m}/d\beta$ as the heat capacity, and $p(\Delta\beta)$ as the distribution law for fluctuations in the canonical ensemble.

### 3.7.5 *Detailed Balance and a Constraint*

Again consider a system with a constraint

$$M_0 = \sum_{r=1}^{N} m_r$$

given by Eq. (3.71) and assume that the probability of finding a system in state $m'$ or $m$ is given by

$$\frac{p(m')}{p(m)} = \frac{g(m')G(M_0 - m'; N - 1)}{g(m)G(M_0 - m; N - 1)} \approx \frac{g(m')}{g(m)} \exp[-\beta(m' - m)] \tag{3.86}$$

Again, $\beta$ replaces the constraint in the estimate of the number of configurations and is given by Eq. (3.69).

If, on each shake of this fictitious system, the probability of going from $m$ to $m'$ is denoted by $W(mm')$ and the inverse process by $W(m'm)$, the steady state, or the state in which the average values of the $p(m)$ do not change on each shake, requires

$$p(m)W(mm') = p(m')W(m'm) \tag{3.87a}$$

Although we might not know any detailed aspects of the cause of a particular die changing the number facing down from a 1 to a 6, if the entire ensemble of dice has a single constraint on the total facing down, we do know from the detailed balance equation and our earlier considerations that

$$\frac{W(mm')}{W(m'm)} = \frac{p(m')}{p(m)} \approx \frac{g(m')}{g(m)} \exp[-\beta(m' - m)] \tag{3.87b}$$

This equation is frequently used to infer the ratio of the transition rates in a system. It can also be noted that the number of accessible configurations and rates are related by

$$W(mm')g(m)G(M_0 - m; N - 1) = W(m'm)g(m')G(M_0 - m'; N - 1) \tag{3.88}$$

This equation implies that the probability of changing from $m$ to $m'$, multiplied by the number of configurations for $m$, is equal to the rate from $m'$ to $m$, multiplied by the number of configurations for $m'$. Again, this statement does not require a knowledge of the mechanism causing the system to change from $m$ to $m'$ on a given "shake"; only the existence of a constraint and a knowledge of the number of accessible configurations are needed.

Many of the results of equilibrium and nonequilibrium statistical mechanics result from a study of this simple shaking of a set of dice with a constraint. The single constraint implies that the transition probability $W(m\mu \rightarrow m'\mu')$ of making a transition from the nondegenerate states $m\mu$ to the state $m'\mu'$ on a given shake and the inverse process are related by $W(m\mu \rightarrow m'\mu')/W(m'\mu' \rightarrow m\mu) = \exp[-\beta(m' - m)]$. If $\beta$ is a positive number, then the transition probability for going from a state with a larger value of $m$ to a smaller value of $m'$ is greater than for the inverse process.

## 3.8 Physical Processes Continuous in Space and Time

Many physical processes are more conveniently discussed as processes which are continuous in time or are continuous in space and time. In this section, the Poisson process is considered first and then the random walk or diffusion problem is discussed.

### *3.8.1 Poisson Distribution of a Physical Process Continuous in Time*

Suppose that the expected rate at which discrete events occur is given by $\dot{N} = \alpha$. Then, the probability that an event occurs during the interval

$\Delta t$ is proportional to $\dot{N}\,\Delta t$ or $\alpha\,\Delta t$ and the probability of no occurrence in $\Delta t$ is $1 - \alpha\,\Delta t$. The probability that $m$ events have occurred in the time interval between $t = 0$ and $t = t$ is described by the difference equation

$$p(m; t + \Delta t) = (1 - \alpha\,\Delta t)p(m; t) + \alpha\,\Delta t\, p(m - 1; t) + O(\Delta t) \quad (3.89)$$

Expansion of the difference equation as a Taylor series expansion in time yields, in the limit of small $\Delta t$, the differential equation

$$\frac{dp(m; t)}{dt} = -\alpha p(m; t) + \alpha p(m - 1; t) \quad (3.90)$$

for the $m$th step. The initial condition

$$p(m; 0) = \delta(m - 0)$$

and the equation for the zeroth event is

$$\frac{dp(0; t)}{dt} = -\alpha p(0; t)$$

If the characteristic function is assumed to be of the form

$$\Phi(a; t) = \sum p(n; t)e^{ian}$$

and

$$p(m; t) = \frac{1}{2\pi}\int_{-\pi}^{\pi} da\, e^{-iam}\Phi(a; t)$$

direct substitution into the difference equation yields an equation for the characteristic function

$$\frac{d\Phi(a; t)}{dt} = -\alpha(1 - e^{ia})\Phi(a; t) \quad (3.91)$$

Integration of this equation yields the characteristic function of the Poisson process continuous in time

$$\Phi(a; t) = \exp\{-[1 - (\exp ia)]\alpha t\} \quad (3.92)$$

and substitution in the inversion formula yields the distribution law for the Poisson process continuous in time:

$$p(m; t) = \frac{e^{-\alpha t}(\alpha t)^m}{m!} \quad (3.93)$$

The similarity with the earlier development, which is given in Eq. (3.51), is apparent, with

$$\dot{N}t = \alpha t = \lambda \tag{3.94}$$

Equation (3.93) gives the probability of $m$ events occurring between $t = 0$ and $t = t$ if the events occur at the expected rate of $\alpha$ events per second.

Exercise 3.25 (a) Show that $p(m; t)$ follows from the characteristic function.

(b) Show that successive integration of the differential equation for $p(m; t)$ yields

$$p(m;\ t) = e^{-\alpha t}\left[\alpha \int_0^t p(m-1;\ t)e^{\alpha y}\,dy\right] = \frac{e^{-\alpha t}(\alpha t)^m}{m!} \tag{3.95}$$

(c) Use the Laplace transform to solve the differential equation. If the Laplace transform is defined as

$$g(m;\ u) = \int_0^\infty e^{-ut}p(m;\ t)\,dt$$

show that the difference equation has the form

$$g(m;\ u) = \frac{\alpha}{\alpha + s}\, g(m-1;\ u)$$

Find $g(m; u)$ and $p(m; t)$.

Ans. $g(0;\ u) = (\alpha + u)^{-1}$;
$g(m;\ u) = \alpha^m/(\alpha + u)^{m+1}$;
$p(m;\ t) = e^{-\alpha t}(\alpha t)^m/m!$

### 3.8.2 *Random Walk Continuous in Space and Time*

The random walk discussed in an earlier section in terms of the difference equation (3.36) or the characteristic function $\Phi_N(a)$ given by Eq. (3.38) can be transformed to continuous variables. Let $\Delta t$ be an increment of time and $\Delta x$ an increment of length such that

$$N = \frac{t}{\Delta t}, \qquad m = \frac{x}{\Delta x}, \qquad a = u\,\Delta x \tag{3.96a}$$

For small increments of $\Delta t$ and $\Delta x$, $N$ and $m$ are large numbers, but the limiting process must be taken in a manner such that $m$ does not exceed $N$.

This is accomplished most easily with the characteristic function. For large $N$, Eq. (3.31) can be used and only $\mathscr{E}x$ and $\mathscr{D}x$ are needed. Using the transformation just given,

$$a\mathscr{E}x_N = aN(p-q) \to u\,\Delta x \frac{t}{\Delta t}(p-q) \to v_d t \tag{3.96b}$$

$$a^2\mathscr{D}x_N = a^2N4pq \to u^2(\Delta x)^2 \frac{t}{\Delta t}(4pq) \to 2u^2Dt \tag{3.96c}$$

where the drift velocity $v_d$ is defined as

$$v_d = (p-q)\frac{\Delta x}{\Delta t} \tag{3.97}$$

and the diffusion coefficient as

$$D = \frac{1}{2}\frac{(\Delta x)^2}{\Delta t} \tag{3.98}$$

In the limit of small $\Delta x$ and $\Delta t$, the product $4pq \to 1$. With this notation, the characteristic function $\Phi_N(a)$ becomes

$$\Phi(u;t) = \exp(iuv_d t - u^2Dt) \tag{3.99}$$

and the probability distribution is given by the Fourier integral transform,

$$\begin{aligned} p(x;t)\,dx &= \frac{dx}{2\pi}\int_{-\infty}^{\infty} du\,\Phi(u;t)\exp(-iux) \\ &= \frac{dx}{(4\pi Dt)^{1/2}}\exp\left[-\frac{(x-\bar{v}_d t)^2}{4Dt}\right] \end{aligned} \tag{3.100}$$

One can show by direct differentiation that $p(x;t)$ is a solution to the differential equation

$$\frac{\partial p}{\partial t} = -v_d\frac{\partial p}{\partial x} + D\frac{\partial^2 p}{\partial x^2} \tag{3.101}$$

This differential equation is known as the Fokker–Planck equation for diffusion with drift and will be discussed in greater detail in Chapter V and in the discussion of Brownian motion in Chapter X. It is often solved by the use of the Fourier transform of $p(x;t)$, and this transform is the function denoted as the characteristic function in the earlier discussion.

Although the limiting process in transforming to continuous variables must be done with some care, it illustrates the importance of discreteness in many problems in which the Fokker–Planck equation is used. As an illustration, consider the random walk of a particle down a long tube. The expected length of a step in the $x$ direction along the axis of the tube is $\Delta x = a$ and the expected time between steps is $\Delta t = 2a/v$. The diffusion coefficient is $D = \frac{1}{4}a\bar{v}$. The drift velocity is zero. Equation (3.100) gives the probability of finding the particle at position $x$ at time $t$. In this example, $p = q = \frac{1}{2}$ and $4pq = 1$. If drift occurs, the discrete development permits any value for $p$ between 0 and 1. If $N$ becomes large, this implies large drift to the right if $p \approx 1$, but such a problem has a definite answer in the discrete case. The limit of small $\Delta t$, or $N \to \infty$, requires $p \approx q \approx \frac{1}{2}$ for the continuous case, i.e., $p = \frac{1}{2}v_d(\Delta t/2D)^{1/2}$. For those physical processes for which the Fokker–Planck equation applies, the probability of an event to the right must be almost equal to that to the left. The large number of steps helps to make the process correspond to the discrete solution in that both have about the same expected value $\mathscr{E}x$ and dispersion $\mathscr{D}x$.

**Exercise 3.26** Show that the Fokker–Planck equation can be derived from the difference equation (3.36). First show that the difference equation has the functional form

$$p(x, t + \Delta t) = qp(x + \Delta x; t) + pp(x - \Delta x; t) \tag{3.102}$$

and then expand for small $\Delta t$ and $\Delta x$.

**Exercise 3.27** Show that $\langle x^2 \rangle = 2Dt$ and compare with $\langle m^2 \rangle \lambda^2 = N\lambda^2$. If $\bar{v}$ is the average speed, show that

$$\langle x^2 \rangle = 2Dt \tag{3.103a}$$

and

$$D = \tfrac{1}{2}N\lambda^2/t = \tfrac{1}{2}\lambda\bar{v} \tag{3.103b}$$

**Exercise 3.28** In a one-dimensional diffusion problem, the flow is assumed proportional to the concentration gradient, that is,

$$\Gamma_z = -D\frac{\partial n}{\partial z} \tag{3.104a}$$

Show that

$$\frac{\partial n}{\partial t} = D\frac{\partial^2 n}{\partial z^2} \tag{3.104b}$$

and compare with the random walk development.

**Exercise 3.29** Estimate the flow through a long, narrow tube as a random walk process. Assume $\langle x^2 \rangle/\tau = D$ is the average flow to the right. Show that $D = \frac{1}{2}a\bar{v}$. Compare the flow with the results in Chapter I. If the tube has a 1-cm radius, what is the probable number of steps required for a molecule to travel a distance of 2 m to the right?

**Exercise 3.30** Show that Eq. (3.100) is a solution of Eq. (3.101).

An atom is released at $x = 0$ at time $t = 0$ and there is an absorbing barrier at $x = x_0$; a question of interest is determining the probability of finding the particle at $x$ at time $t$. Equation (3.101) with $v_d = 0$ gives the probability distribution. A boundary condition

$$p(x_0, t; x_0) = 0 \qquad \text{(absorbing surface)}$$

must now be included and the solution is of the form

$$p(x, t; x_0) = \frac{1}{(4\pi Dt)^{1/2}} \left\{ \exp\left(\frac{-x^2}{4Dt}\right) \mp \exp\left[\frac{-(2x_0 - x)^2}{4Dt}\right] \right\} \tag{3.105}$$

with the minus sign for the absorbing surface. The boundary condition for a reflecting surface at $x = x_0$ is

$$\frac{\partial p(x_0, t; x_0)}{\partial x} = 0$$

where the plus sign is used in Eq. (3.105).

**Exercise 3.31** A particle is placed between parallel absorbing plates at $x = 0$ and $x = x_0$. Show that the solution to Eq. (3.101) with $v_d = 0$ is of the form

$$p(x, t) = \sum_n A_n \left(\sin \frac{n\pi x}{x_0}\right) \exp\left[-\left(\frac{n\pi}{x_0}\right)^2 Dt\right] \tag{3.106}$$

Show that the $n = 1$ mode has the longest average lifetime and for many purposes can be regarded as the dominant mode.

**Exercise 3.32** If $p(x, 0) = 1/\Delta$ is the probability of finding the particle between $x = \frac{1}{2}(x_0 \pm \Delta)$, find the amplitudes $A_0$ and $A_1$ in Eq. (3.106) and the decay of these modes. What is the probability of finding the particle in the modes with $n > 1$?

## 3.9 Random Walk with Absorbing Boundaries

The random walk with absorbing boundaries is often discussed as the "gambler's ruin" problem, extinction of a population after $n$ generations, diffusion with absorbing boundaries, first passage, and so on. In order to formulate the basic problem, consider the set of integers between 0 and $L$ shown in Fig. 3.4. "Gambler's ruin" occurs when a gambler with an initial

**Fig. 3.4** Random walk with boundaries at 0 and $L$.

stake of $r$ dollars loses his last dollar. First passage occurs when one initially $r$ steps from his goal, crosses the goal for the first time. These ideas are summarized in the following definition. If the initial position is $r$, the probability of reaching 0 on the $n$th step is of primary interest. Let $u(r; n)$ describe the following probability:

$$u(r;n) = \left\{\begin{matrix}\text{probability that from initial position } r \\ \text{the process ends on the } n\text{th step}\end{matrix}\right\}$$

$$= \left\{\begin{matrix}\text{probability that a gambler with } r \text{ dollars} \\ \text{is ruined on the } n\text{th trial}\end{matrix}\right\}$$

$$= \left\{\begin{matrix}\text{probability that a particle starting from } r \text{ is} \\ \text{absorbed at position } 0 \text{ on the } n\text{th step}\end{matrix}\right\} \tag{3.107a}$$

The difference equation describing the process places the position at $r + 1$ or $r - 1$ after the first step, and

$$u(r; n + 1) = pu(r + 1; n) + qu(r - 1; n) \tag{3.107b}$$

Again, $p$ is the probability of a step to the right or gain, and $q$ a step to the left or loss. Thus having started at $r$, the probability of ruin on the $(n + 1)$th step, $u(r; n + 1)$, is equal to the probability of making a step to the right multiplied by the probability of ruin from the $r + 1$ position on the $n$th step, $pu(r + 1; n)$, plus the probability of making a step to the left multiplied by the probability of ruin from the $r - 1$ position on the $n$th step, $qu(r - 1; n)$. Equation (3.107) describes the process for $1 < r < L - 1$ and $n \geq 1$. In order that Eq. (3.107) be valid at $r = 1$, $r = L - 1$, and $n = 0$, it is necessary for

$$u(0; n) = u(L; n) = 0, \qquad n \geq 1$$

and

$$u(0;0) = 1 \quad \text{and} \quad u(r;0) = 0, \quad r > 0$$

Equation (3.107) is now valid in the range $0 < r < L$ and $n \geq 0$.

In order to introduce an alternate method of solution for integers ranging from zero to infinity, the generating function is used for this problem. The generating function for this probability distribution is

$$U(r, s) = \sum_{n=0}^{\infty} u(r; n)s^n \tag{3.108a}$$

where $s$ replaces $e^{ia}$ in our previous considerations. Differentiation with respect to $s$ replaces integration over $da$ for selecting the coefficient $u(r; n)$. Thus

$$u(r; n) = \frac{1}{n!}\left[\frac{d^n U(r; s)}{ds^n}\right]_{s=0} \tag{3.108b}$$

Multiplication of Eq. (3.107) by $s^{n+1}$ and summation over $n$ yields

$$U(r; s) = s[pU(r+1; s) + qU(r-1; s)] \tag{3.109a}$$

with conditions

$$U(0; s) = 1 \quad \text{and} \quad U(L; s) = 0 \tag{3.109b}$$

Assume a solution of the form

$$U(r; s) = g^r(s) \tag{3.110a}$$

and substitute into Eq. (3.109a) to yield

$$g = s(q + pg^2) \tag{3.110b}$$

This equation has two roots,

$$g_{\pm} = \frac{1 \pm (1 - 4pqs^2)^{1/2}}{2ps} \tag{3.111}$$

The generating function is a linear combination of the two particular solutions, and

$$U(r; s) = c_+ g_+{}^r + c_- g_-{}^r$$

The $c$'s are selected to fit the boundary conditions given by Eq. (3.109b). The generating function for absorption or ruin after having started at the

$r$th position is

$$U(r; s) = \left(\frac{q}{p}\right)^r \frac{g_+^{L-r} - g_-^{L-r}}{g_+^L - g_-^L} \tag{3.112}$$

### 3.9.1 Gambler's Ruin, Victory, and Duration of Game

Starting from the $r$th position (or with $r$ dollars), the generating function for absorption at 0 (or ruin) is given by Eq. (3.112). A considerable amount of cleverness [2] permits $u(r; n)$ to be written as

$$u(r; n) = L^{-1}2^n p^{(n-r)/2} q^{(n+r)/2} \sum_{b=1}^{L-1} \left(\cos^{n-1} \frac{\pi b}{L} \sin \frac{\pi b}{L} \sin \frac{\pi r b}{L}\right) \tag{3.113}$$

Starting from the $r$th position, the generating function for absorption at $L$ (or the victory of a gambler over his opponent) is

$$U(r; s) = \left(\frac{p}{q}\right)^{L-r} \frac{g_+{}^r - g_-{}^r}{g_+{}^L - g_-{}^L} \tag{3.114}$$

that is, $p \to q$, $q \to p$, and $r \to L - r$ in Eq. (3.112).

The generating function for the duration of the game is the sum of the generating functions given by Eqs. (3.112) and (3.114).

### 3.9.2 First Passage

The generating function for first passage follows by allowing $L \to \infty$ in Eq. (3.112). Starting from position $r$, the first passage of position 0 follows from the generating function

$$U(r; s) = g_-{}^r(s) \tag{3.115}$$

This is also the generating function for absorption at the origin, and for gambler's ruin when playing against an infinitely rich adversary. The coefficient of $s^n$ in Eq. (3.115) indicates that the probability of ruin, first crossing, and so forth on the $n$th trial is

$$u(r; n) = \frac{r}{n} \binom{n}{\frac{1}{2}(n-r)} p^{(n-r)/2} q^{(n+r)/2} \tag{3.116}$$

Maximum probability for first passage occurs for

$$\frac{d[\ln u(r; n)]}{dn} = 0$$

Direct evaluation of Eq. (3.116) for $p = q = \frac{1}{2}$ yields

$$n \approx r^2 \tag{3.117}$$

for the most probable value. This can be compared with $\langle m^2 \rangle$ of Eq. (3.33b) for the random walk without barriers.

It is common practice in physics to estimate the probability of first crossing by using $\langle m^2 \rangle \approx N$ for the unrestricted random walk. The above development illustrates the correctness of these crude estimates.

**Exercise 3.33** Show, for the random walk with absorbing barriers at 0 and $L$, that the probability of starting at position $r$ and being at $r'$ on the $n$th step is given by the difference equation

$$v(r, r'; n+1) = pv(r+1, r'; n) + qv(r-1, r'; n) \tag{3.118}$$

with boundary conditions

$$v(0, r'; n) = v(L, r'; n) = 0, \qquad n \geq 1 \tag{3.119a}$$

$$v(r, r'; 0) = 0, \quad r \neq r'; \qquad v(r', r'; 0) = 1 \tag{3.119b}$$

**Exercise 3.34** Show, for reflecting boundaries at 0 and $L$, that the difference equation is the same as Eq. (3.118), but the boundary conditions are

$$\begin{aligned} v(0, r'; n) &= v(1, r'; n) \\ v(L, r'; n) &= v(L-1, r'; n) \end{aligned} \tag{3.120}$$

and Eq. (3.119b).

### 3.9.2.1 Alternate Method for First Passage

In the earlier section on the random walk, the characteristic function for the net gain in $N$ steps was developed. Rather than use the notation $m = \sum_{n=1}^{N} x_n$, this section denotes the net gain as

$$S_N = \sum_{n=1}^{N} x_n$$

First passage is denoted as the first time the net gain exceeds zero, and the sequence

$$S_1 < 0, \quad S_2 < 0, \quad \ldots, \quad S_N = 1$$

is considered. A possible path is shown in Fig. 3.5 and direct enumeration

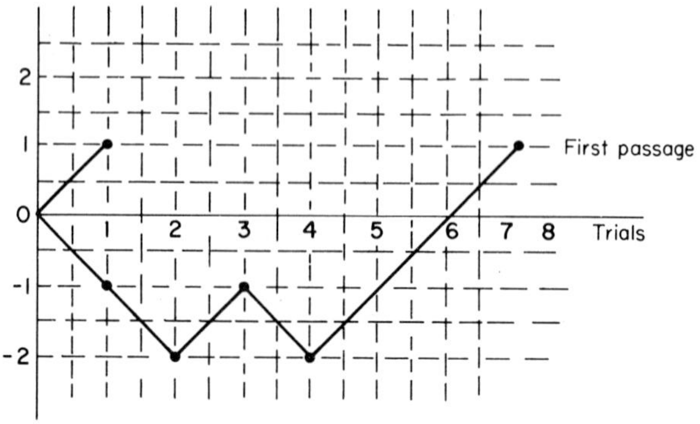


**Fig. 3.5** Two routes for first passage to position $+1$ as a function of the number of trials.

indicates first passage will occur on trials 1, 3, 5, ... with probabilities $p, qp^2, 2q^2p^3, \ldots$. Let $g(s)$ denote the generating function for first passage through 1 starting from 0. The generating function for first passage through 2 is the product of one-step generating functions. The two sequences are independent. The generating function for first passage through $x = m$ is

$$g(m; s) = g^m(s) \tag{3.121}$$

Only the one-step generating function is needed.

In order to find $g(s)$, consider the two possibilities: (1) $x_1 = 1$ and first passage occurs on the first trial; (2) $x_1 = -1$ and the net gain must increase by 2 on subsequent trials. Let

$$w(m; n) = \{\text{probability of net gain of } m \text{ on } n\text{th trial}\} \tag{3.122}$$

The one-step generating function is

$$\begin{aligned} g(s) &= \sum_{n=1}^{\infty} w(1; n)s^n \\ &= ps + q \sum_{n=2}^{\infty} w(2; n-1)s^n \\ &= ps + qsg^2(s) \end{aligned} \tag{3.123}$$

The last step follows from

$$\sum_{n=2} w(2; n-1)s^n = s \sum_{k=1} w(2; k)s^k = sg(2; s)$$

and $g(2; s) = g^2(s)$. The solution of the quadratic yields

$$g(s) = \frac{1 - (1 - 4pqs^2)^{1/2}}{2qs} \tag{3.124}$$

A binomial expansion of $g(s)$ yields, for the probability of first passage on $n = 2k - 1$,

$$\begin{aligned} w(1; 2k - 1) &= \frac{(-1)^{k-1}}{2q} \binom{\frac{1}{2}}{k} (4pq)^k \\ w(1; 2k) &= 0 \end{aligned} \tag{3.125a}$$

First passage through $m$ on the $n$th trial is given by

$$w(m; n) = \frac{1}{m!} \left[ \frac{d^n g^m}{ds^n} \right]_{s=0} \tag{3.125b}$$

If $p$ and $q$ are interchanged, the generating function for first passage is the same as that given by Eq. (3.115) in the previous section.

Chandrasekhar [*3*] has considered the discrete random walk with either an absorbing or reflecting barrier at position $m_0$. The problem is similar to the continuous distribution considered earlier. For the discrete random walk, the probability of being at position $m$ on the $N$th step when there is either an absorbing or reflecting barrier at $m = m_0$ is,

$$p(m, N; m_0) \approx \left( \frac{2}{\pi N} \right)^{1/2} \left\{ \exp\left( \frac{-m^2}{2N} \right) \mp \exp\left[ \frac{-(2m_0 - m)^2}{2N} \right] \right\} \tag{3.126}$$

for large $N$. The minus sign is for absorbing barriers and the plus sign for reflecting barriers. This can be compared with Eq. (3.105), which is a solution to the diffusion equation and which is discussed in Exercise 3.30.

## 3.10 Some Probability Laws Frequently Encountered in Physics

This section summarizes some of the probability laws used in physics. Examples of discrete and continuous probability laws, their characteristic functions, averages, and so on, are given.

### *3.10.1 Discrete Probability Laws*

(a) *Random walk*: Equations (3.26), (3.29), (3.30), (3.36).
(b) *Binomial*: Equations (3.44)–(3.46), (3.54). The binomial distribu-

tion is also used to describe "success" and "failure" for the occurrence of an event: $1/g$ is success and $1 - (1/g)$ is failure. The probability of $m$ successes in $N$ trials is given by Eq. (3.49).

(c) *Poisson*: Equations (3.50), (3.51), and (3.54) are based on an average $\lambda = N/g$. Section 3.8 discusses discrete processes that occur at a rate of $\alpha$ per second.

(d) *Geometric*: The probability that exactly $k$ failures precede the first success, or the probability that the first success occurs on the $(k + 1)$th trial, is the geometric distribution. Denote this probability by $u(k + 1; 1)$. Let $p = 1/g$ be the probability of success and $q = 1 - (1/g)$ be the probability of failure; then the characteristic function describing the process is

$$\varphi(a) = \frac{pe^{ia}}{1 - qe^{ia}} \tag{3.127a}$$

$$\bar{k} = \frac{1}{p}; \qquad \langle (k - \bar{k})^2 \rangle = \frac{q}{p^2} \tag{3.127b}$$

$$u(k + 1; 1) = \begin{cases} pq^k & \text{for } k = 0, 1, 2, \ldots \\ 0 & \text{otherwise} \end{cases} \tag{3.127c}$$

Equation (3.127a) is readily shown by noting that success on the first trial is equal to $p$, to $pq$ on the second trial, $pq^2$ on the third, and so on; and the characteristic function is the sum of terms like $pe^{ia} + pqe^{i2a} + \cdots$. $k$ is the average number of trials required to obtain the first success. Thus, as $q \to 0$, success occurs on the first trial.

(e) *Negative binomial*: Denote the probability that exactly $k$ failures precede the $r$th success, or the probability that the $r$th success occurs at the trial number $r + k$, by $u(k; r)$. The characteristic function is

$$\Phi_r(a) = \left( \frac{p}{1 - qe^{ia}} \right)^r = [\varphi(a)]^r \tag{3.128a}$$

and the probability is given by the Fourier transform, or

$$u(k; r) = \begin{cases} \dbinom{r + k - 1}{k} p^r q^k & \text{for } r = 1, 2, \ldots; \ k = 0, 1, 2, \ldots \\ 0 & \text{otherwise} \end{cases} \tag{3.128b}$$

Then,

$$\bar{k} = r\left(\frac{q}{p}\right); \qquad \langle (k - \bar{k})^2 \rangle = r\left(\frac{q}{p^2}\right) \tag{3.128c}$$

$k$ is the running index in the formation of the characteristic function

$$\Phi_r(a) = \sum_k u(k; r)\, e^{ika}$$

Since Eq. (3.128a) can be written as $[\varphi(a)]^r$, it is clear that the process is described by $\varphi(a)$ or $u(k; 1)$. $u(k; 1)$ is the probability that $k$ failures precede the first success; $u(0; 1) = p$, $u(1; 1) = pqe^{ia}$, ..., to yield $\varphi(a)$. $k$ is the average number of failures before $r$ successes occur.

### 3.10.2 Continuous Probability Laws

(f) *Uniform*: All points between $A < x < B$ are equally probable, and the probability of striking position $x$ is given by

$$\varphi(a) = \frac{e^{iaB} - e^{iaA}}{ia(B - A)} \tag{3.129a}$$

$$p(x) = \frac{1}{B - A} \tag{3.129b}$$

$$\bar{x} = \frac{A + B}{2}; \qquad \langle(x - \bar{x})^2\rangle = \frac{(B - A)^2}{12} \tag{3.129c}$$

(g) *Normal or Gaussian*: Equations (3.99)–(3.101):

$$\Phi(u; t) = \exp(iuv_{\mathrm{d}}t - u^2Dt) \tag{3.130a}$$

$$p(x; t)\, dx = \frac{dx}{(4\pi Dt)^{1/2}} \exp\left[-\frac{(x - v_{\mathrm{d}}t)^2}{4Dt}\right] \tag{3.130b}$$

$$\bar{x} = v_{\mathrm{d}}t \tag{3.130c}$$

$$\langle(x - \bar{x})^2\rangle = 2Dt \tag{3.130d}$$

(h) *Exponential*: Events occur at a rate of $\alpha$ per second; the waiting time required to observe the first event $u(1, t)$ is given by

$$\varphi(a) = \frac{1}{1 - (ia/\alpha)} \tag{3.131a}$$

$$u(1, t) = \alpha e^{-\alpha t}, \qquad t > 0 \tag{3.131b}$$

$$\bar{t} = \frac{1}{\alpha}; \qquad \langle(t - \bar{t})^2\rangle = \frac{1}{\alpha^2} \tag{3.131c}$$

(i) *Gamma distribution*: The waiting time required to observe the $r$th occurrence of events that occur at the rate of $\alpha$ per second is given by

$$\Phi_r(a) = \frac{1}{[1 - (ia/\alpha)]^r} \tag{3.132a}$$

$$u(r; t) = \frac{\alpha}{\Gamma(r)} (\alpha t)^{r-1} e^{-\alpha t} \tag{3.132b}$$

$$\bar{t} = \frac{r}{\alpha}; \qquad \langle (t - \bar{t})^2 \rangle = \frac{r}{\alpha^2} \tag{3.132c}$$

where $\Gamma(r)$ is the $\Gamma$ function.

## References

1. Q. Y. Khinchin, "Mathematical Foundations of Quantum Statistics." Graylock Press, Albany, New York, 1960.
2. W. Feller, "Introduction to Probability Theory and Its Applications," p. 322. Wiley, New York, 1957.
3. S. Chandrasekhar, *Rev. Mod. Phys.* **15**, 1 (1943).

## General References

Bartlett, M. S., "An Introduction to Stochastic Processes." Cambridge Univ. Press, London and New York, 1965.

Beckmann, P., "Elements of Applied Probability Theory." Harcourt, New York, 1967.

Bharucha-Reid, A. T., "Elements of the Theory of Markov Processes and Their Applications." McGraw-Hill, New York, 1960.

Cramer, H., "The Elements of Probability Theory." Wiley, New York, 1955.

Feller, W., "An Introduction to Probability Theory and Its Applications," 2nd ed. Wiley, New York, 1957.

Gnedenko, B. V., "The Theory of Probability." Chelsea, Bronx, New York, 1967.

Parzen, E., "Stochastic Processes." Holden-Day, San Francisco, California, 1962.

# Intermolecular Interactions IV

## 4.1 Introduction

In Chapter I, the properties of a mass point gas were studied in some detail. This chapter is primarily concerned with the interactions that take place between atoms or molecules. Such interactions give rise to terms in the potential energy of the gas and to an apparent size for the interacting molecules. The term "apparent size" is used since the size of an atom will depend on the experiment. Experiments with energetic nuclear particles indicate a diameter of the order of $10^{-15}$ m, atom–atom scattering of the order of $10^{-10}$, the disorientation of magnetic atomic dipoles of the order of $10^{-8}$ m, and so on. The effect of the encounter or interaction between two particles is usually described in terms of a collision cross section. If two particles approach each other so that at some instant of time they can be regarded as randomly located on 1 m$^2$, the probability of an observable effect is denoted as the cross section $\sigma$. Nuclear cross sections are of the order of magnitude of $10^{-30}$ m$^2$, molecular cross sections of the order of $10^{-20}$ m$^2$, spin disorienting cross sections for atoms of the order of $10^{-16}$ m$^2$, and so on. In this chapter, attention is directed toward atom–atom or molecule–molecule interactions at kinetic energies which are appropriate for conventional thermal energies. Diffusion and viscosity cross sections, which are important for the transport properties discussed in Chapter V, are developed. Binary reaction rates of interest in nuclear, chemical, and laser physics are also discussed. A brief discussion of the interaction of an atom or molecule with a surface is given. Although the student might expect to approach the question of intermolecular forces from first principles by a consideration of the electromagnetic forces between orbital electrons of the molecules, this fundamental approach is in fact too complex for the present problem.

Three simple categories of intermolecular forces are as follows.

(a) *Interaction between ions.* If $U(r)$ is the potential energy of two ions at a distance $r$ apart, then the repulsive, long-range Coulomb interaction is

$$U(r) = \frac{Z_1 Z_2 e^2}{4\pi\varepsilon_0 r_{12}} \tag{4.1}$$

for ions with charge $Z_1 e$ and $Z_2 e$, respectively, and separation $r_{12}$.

(b) *van der Waals Interaction.* The attractive interaction between two neutral atoms in their ground state is often referred to as the van der Waals interaction, since it is similar to the attractive force introduced by van der Waals in his discussion of the equation of state of gases. London [*1*] gave a quantum explanation of these forces, and due to their relationship to atomic polarizability and optical dispersion, they are also called dispersion forces. The effect is roughly described by the interaction between two well-separated hydrogen atoms. The wave functions are spherically symmetric for the ground state of each atom and give rise to no first-order interactions. In second order, the Coulomb interaction between the electrons perturbs the wave functions and gives rise to a dynamic polarization of one atom by the other in which the electrons adjust their relative phase so that their average separation is as large as possible. This results in an attractive energy, and the main part of the dispersion energy varies as

$$U(r) = -\frac{C}{r^6} \tag{4.2}$$

(c) *Repulsive forces.* At distances of close approach, repulsive interaction occurs between atoms. This is due in part to the repulsive forces between like ionic cores and to the impenetrability of electron clouds. The latter is a typical quantum effect and arises as the electron clouds of the two atoms overlap. For inert gases, a solution of the Schrödinger equation yields a repulsive energy of the form $f(r)e^{-\alpha r}$. A suitable first approximation is

$$U(r) \approx Ae^{-\alpha r} \tag{4.3}$$

From these very elementary discussions, the first approximation for the total potential energy between two neutral atoms is of the form

$$U(r) = Ae^{-\alpha r} - \frac{C}{r^6} \tag{4.4}$$

and in many of the following sections, we assume that the interaction between complex neutral particles can be expressed in terms of $U(r)$. The detailed structure of the particles or molecules need not be considered.

Also in subsequent sections, the determination of the form of $U(r)$ and of the appropriate values for $C$, and so forth by experimental methods is considered. Atomic and molecular beams and diffusion are used to study the cross sections of unlike atoms. Viscosity and second virial coefficient measurements are used to obtain information about the interaction between like atoms. The student may be surprised at the crudeness of the methods and the answers obtained for a subject in which the answers are available in principle from purely theoretical considerations.

## 4.2 Classical Elastic Collisions

Consider two particles of masses $m_1$ and $m_2$ located at distances $r_1$ and $r_2$ from an origin in the laboratory frame of reference. $\dot{\mathbf{r}}_1$ and $\dot{\mathbf{r}}_2$ are the velocities relative to this origin. The interaction between the two particles is most conveniently studied in an inertial frame moving with the center of mass. Let $\mathbf{R}$ denote the distance to the center of mass and $\mathbf{r}$ the distance between the particles, with these quantities are defined by

$$(m_1 + m_2)\mathbf{R} = m_1\mathbf{r}_1 + m_2\mathbf{r}_2, \qquad \mathbf{r} = \mathbf{r}_2 - \mathbf{r}_1 \tag{4.5}$$

$\mathbf{G} = \dot{\mathbf{R}}$ is the velocity of the center of mass and $\dot{\mathbf{r}}$ is the relative velocity. Since the potential energy of the interaction decreases rapidly with distance, at large distances of separation, the energy of the particles is primarily kinetic and $\dot{\mathbf{r}}_1$ and $\dot{\mathbf{r}}_2$ tend to $\mathbf{v}_1$ and $\mathbf{v}_2$ prior to the encounter and $\mathbf{v}_1'$ and $\mathbf{v}_2'$ after the encounter.

### 4.2.1 *Dynamic Aspects of Elastic Collisions*

The dynamic aspects of the encounter follow from a knowledge of the velocities at large separation and the conservation laws for energy and momentum.

*Conservation of Linear Momentum*

$$m_1\dot{\mathbf{r}}_1 + m_2\dot{\mathbf{r}}_2 = (m_1 + m_2)\mathbf{G} = m_1\mathbf{v}_1 + m_2\mathbf{v}_2 = m_1\mathbf{v}_1' + m_2\mathbf{v}_2' \tag{4.6}$$

*Conservation of Energy for an Elastic Encounter*

$$\tfrac{1}{2}m_1\dot{\mathbf{r}}_1^2 + \tfrac{1}{2}m_2\dot{\mathbf{r}}_2^2 + U(r) = \tfrac{1}{2}m_1\mathbf{v}_1^2 + \tfrac{1}{2}m_1\mathbf{v}_2^2 = \tfrac{1}{2}(m_1 + m_2)\mathbf{G}^2 + \tfrac{1}{2}\mu\mathbf{V}^2$$
$$= \tfrac{1}{2}m_1(\mathbf{v}_1')^2 + \tfrac{1}{2}m_2(\mathbf{v}_2')^2 = \tfrac{1}{2}(m_1+m_2)\mathbf{G}^2 + \tfrac{1}{2}\mu(\mathbf{V}')^2 \tag{4.7}$$

At large separation, the relative velocity is defined as

$$\mathbf{V} = \mathbf{v}_2 - \mathbf{v}_1 \qquad \text{and} \qquad \mathbf{V}' = \mathbf{v}_2' - \mathbf{v}_1' \tag{4.8}$$

where $1/\mu = (1/m_1) + (1/m_2)$ is the reduced mass. Linear momentum remains constant throughout the collision and is the momentum of the center of mass. The energy of the center of mass remains constant and the only effect of an elastic collision is to change the direction of the relative velocity. It follows from Eq. (4.7) that

$$|\mathbf{V}| = |\mathbf{V}'| = V \tag{4.9}$$

and the magnitude of the relative velocity is not changed by an elastic collision. These ideas are summarized in Fig. 4.1, in which $\mathbf{v}_1$, $\mathbf{v}_2$, $\mathbf{G}$, and so on are shown in velocity space. The most general effect is the rotation of the relative velocity through an angle $\theta$ about the tip of the vector $\mathbf{G}$.

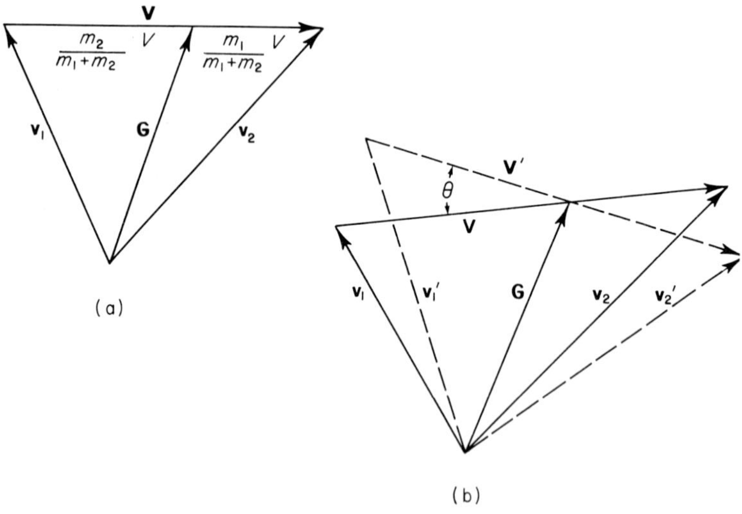


**Fig. 4.1** (a) Relative velocity diagram. $\mathbf{v}_1$ and $\mathbf{v}_2$ are the particle velocities, $\mathbf{V}$ the relative velocity, and $\mathbf{G}$ the velocity of the center of mass. (b) Change in relative velocity from $\mathbf{V}$ to $\mathbf{V}'$ or by angle $\theta$ which is caused by an elastic collision.

As is readily apparent in Fig. 4.1, and for future reference, the velocities are related by

$$\mathbf{v}_1 = \mathbf{G} - \left(\frac{m_2}{m_1 + m_2}\right)\mathbf{V}, \qquad \mathbf{v}_1' = \mathbf{G} - \left(\frac{m_2}{m_1 + m_2}\right)\mathbf{V}'$$
$$\mathbf{v}_2 = \mathbf{G} + \left(\frac{m_1}{m_1 + m_2}\right)\mathbf{V}, \qquad \mathbf{v}_2' = \mathbf{G} + \left(\frac{m_1}{m_1 + m_2}\right)\mathbf{V}' \tag{4.10}$$

The change in linear momentum of each particle is

$$\Delta \mathbf{p}_1 = m_1(\mathbf{v}_1' - \mathbf{v}_1) = \mu(\mathbf{V} - \mathbf{V}') = -\Delta \mathbf{p}_2 \tag{4.11}$$

### *4.2.2 Geometric Aspects of a Collision*

A determination of the angle $\theta$ between $\mathbf{V}$ and $\mathbf{V}'$ requires that the geometric aspect of the collision be considered. If the force of interaction is along the line of centers, or the interaction potential energy $U(r)$ depends only on the distance between the two particles, the angular momentum of the two-particle system is conserved during the encounter. Referring to Fig. 4.3, the angular momentum with the origin at the center of mass is

$$L = \mu V b \tag{4.12}$$

and directed perpendicular to the plane formed by $b$ and $V$. $b$ is often called the "impact parameter." The plane in which the motion occurs is defined by the impact parameter $b$ and the relative velocity $\mathbf{V}$. $\mathbf{V}'$ lies in this same plane.

An elementary and illustrative elastic collision is that between two hard spheres of mass $m$ and radius $a$. The collision of two billiard balls is an example of such a collision. Assume two billiard balls approach with relative velocity $\mathbf{V}$ and a separation between the line of centers of $b$ as shown in Fig. 4.2. At the instant of impact, the force between the balls is along the line of centers. Only the momentum along the line of centers reverses and the component perpendicular to the line of centers is unchanged. Thus the angle $\theta$ between $\mathbf{V}$ and $\mathbf{V}'$ is given by $b = 2a\cos(\theta/2)$. The probability that $A$ is scattered into solid angle $d\Omega$ by the collision is proportional to the probability that the path $PP'$ lies between $b$ and $b + db$, or denoting the probability for the scattering of an incident particle per

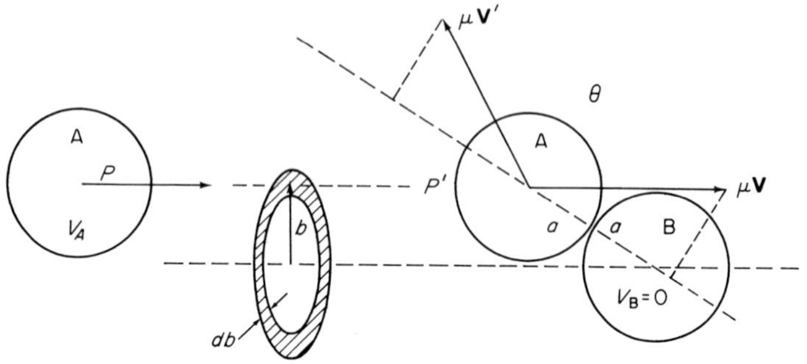


**Fig. 4.2** Collision of two hard spheres. The relative velocity along the line of centers reverses upon impact. $b$ is the "impact parameter."

square meter by the differential cross section $d\sigma = 2\pi b\, db$ per square meter

$$\begin{aligned} d\sigma &= \{\text{probability that } A \text{ is scattered into } d\Omega \text{ by } B\} \\ &= \sigma_{AB}(\Omega)\, d\Omega = 2\pi b\, db = a^2\, d\Omega \end{aligned} \tag{4.13}$$

Integration over the solid angle $d\Omega$ yields a total cross section of $\pi(2a)^2$.

For the more general case of a potential energy $U(r)$, it is convenient to use an inertial frame attached to the center of mass. The motion occurs in the plane that contains $\mathbf{V}$ and $b$. Equations (4.7) and (4.12), which govern the motion, can be rewritten as

$$\tfrac{1}{2}\mu V^2 = \tfrac{1}{2}\mu(\dot{r}^2 + r^2\dot{\alpha}^2) + U(r) \tag{4.14a}$$

$$L = \mu V b = \mu r^2 \dot{\alpha} \tag{4.14b}$$

where $\dot{\alpha}$ is the angular rate of change of $\mathbf{r}$. Equations (4.14a) and (4.14b) are the same as for the scattering of a particle with mass $\mu$, relative velocity $V$, and angular momentum $L$ by a scattering potential $U(r)$. This is shown in Fig. 4.3(b) and is somewhat simpler than the equivalent formulation shown in Fig. 4.3(a). Since the angular momentum $L$ is a constant, $\dot{\alpha}$ can be eliminated from Eq. (4.14a), and $\dot{r}$ can be rewritten as

$$\dot{r} = \frac{dr}{d\alpha}\frac{d\alpha}{dt} = \frac{Vb}{r^2}\frac{dr}{d\alpha} \tag{4.15}$$

Substitution into Eqs. (4.14) for $\dot{r}$ and $\dot{\alpha}$ yields an equation for the trajectory of the form

$$\frac{dr}{d\alpha} = \pm\frac{r^2}{b}\left[1 - \frac{b^2}{r^2} - \frac{U(r)}{\frac{1}{2}\mu V^2}\right]^{1/2} \tag{4.16}$$

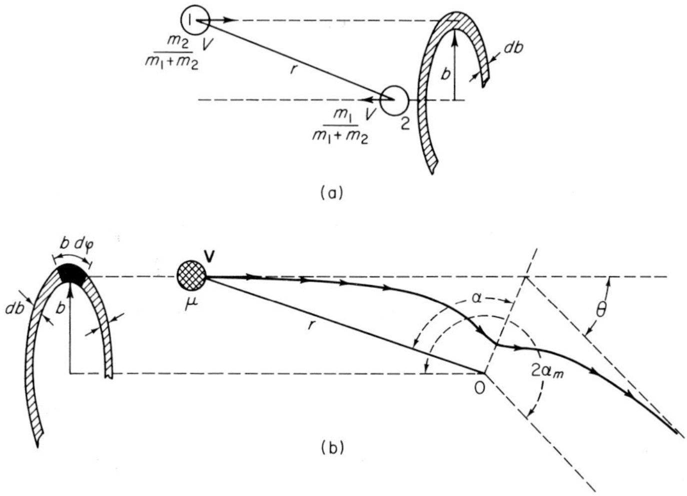


**Fig. 4.3** (a) Definition of the impact parameter for two colliding particles. (b) Trajectory of a particle of mass $\mu$ which is scattered by an attractive central potential. $\theta$ is the change in relative velocity and $2\alpha_m$ is the angular change of the asymptotes of the path "in" and the path "out."

The trajectory is symmetric in the angle $\alpha$ about the line for which the term in square brackets is zero or $dr/d\alpha = 0$. This line is chosen as the origin for $\alpha$. The distance of closest approach is also along this line, since $\dot{r} = 0$ occurs when $dr/d\alpha = 0$. If $\alpha$ is measured from the symmetry line and the asymptotic value at large $r$ is denoted by $\alpha_m$, the scattering angle between $V$ and $V'$ is given by

$$\theta = \pi - 2\alpha_\mathrm{m} = \pi - 2b \int_{r_m}^{\infty} \frac{dr}{r^2} \left[ 1 - \frac{b^2}{r^2} - \frac{U(r)}{\frac{1}{2}\mu V^2} \right]^{-1/2} \tag{4.17}$$

where the substitution

$$\alpha_\mathrm{m} = \int_0^{\alpha_m} d\alpha \, [\ ] = \int_{r_m}^{\infty} dr \, [\ ]$$

is used. An examination of Fig. 4.3(b) indicates that the impact parameter and the probability of scattering into solid angle $d\Omega$ are related by

$$\begin{aligned} d\sigma &= \{\text{probability that } V \text{ changes to } V' \text{ in direction } d\Omega\} \\ &= \sigma(V, \theta)\, d\Omega = b\, d\varphi\, db \end{aligned} \tag{4.18a}$$

The interval $db$ defines the angular change $\theta$ in $V$, and $d\varphi$ defines the plane of the encounter. If $V$ of the approaching particle passes through the element of area $b\,d\varphi\,db$, $V'$ will pass through an area of the unit sphere defined by $(\sin\theta)\,d\varphi\,d\theta = d\Omega$. A differential classical cross section follows from Eq. (4.18a) and is given by

$$\sigma(V,\theta) = \frac{b}{\sin\theta}\left|\frac{db}{d\theta}\right| \tag{4.18b}$$

At small angles, the classical scattering is extremely large and the total cross section is infinite for a scattering potential of the type to be given by Eq. (4.19a). This is of course expected, since during the encounter, the particles are always deflected through some small angle.

Exercise 4.1 Let particles 1 and 2 be hard or elastic spheres of radii $a_1$ and $a_2$, respectively. Show by elementary means that the total cross section $\sigma_0 = \pi r_0^2$, where $r_0 = a_1 + a_2$. Why is $\sigma_0$ the same as the probability that particles 1 and 2 are in contact if they are randomly placed on an area of 1 $m^2$?

Exercise 4.2 Refer to Fig. 4.2 and show for hard spheres of radii $a_1$ and $a_2$ that the probability that the relative velocity shifts by angle $\theta$ is $\sigma_0(\theta) = \frac{1}{4}r_0^2$. By examination of the figure, deduce the relationship between $b$ and $\theta$. Show that this is the same as that deduced for a hard-sphere potential energy, that is, $U(r) = 0$ for $r > r_0$ and $U(r) \to \infty$ for $r \leq r_0$.

Exercise 4.3 Using direct integration of Eq. (4.17), show that, for Coulomb scattering,

$$\tan\frac{\theta}{2} = \frac{Z_1 Z_2 e^2}{4\pi\varepsilon_0}\,\frac{1}{\mu V^2 b} \tag{4.19a}$$

and show that the differential cross section for Coulomb scattering is

$$d\sigma = \sigma(V,\theta)\,d\Omega = \left(\frac{Z_1 Z_2 e^2}{4\pi\varepsilon_0}\,\frac{1}{2\mu V^2}\right)^2 \frac{1}{\sin^4(\theta/2)}\,d\Omega \tag{4.19b}$$

## 4.3 Quantum Aspects of Elastic Collisions

### 4.3.1 *Two-Particle Problem*

The motion of two particles interacting with a potential energy function of the form $U(|\mathbf{r}_2 - \mathbf{r}_1|)$ is described by the Hamiltonian

$$H = \frac{p_1^2}{2m_1} + \frac{p_2^2}{2m_2} + U(|\mathbf{r}_2 - \mathbf{r}_1|) = \frac{P^2}{2(m_1 + m_2)} + \frac{p^2}{2\mu} + U(r) = E' \tag{4.20}$$

**P** and **p** are the momentum for the center of mass and the relative momentum, respectively, and follow from the coordinate transformation given by Eq. (4.5). Since the Hamiltonian is invariant under a space translation and a time translation, linear momentum and energy are conserved. These dynamic aspects are summarized for an elastic collision by Eq. (4.9), which states that only the direction of the asymptotic value of the relative velocity **V** can change. An equivalent statement is that only the direction of **p** can change during an elastic encounter, that is,

$$|\mathbf{p}| = |\mathbf{p}'| \tag{4.21}$$

In the quantum problem, the quantities in Eq. (4.20) are regarded as operators and the invariance of the Hamiltonian under space and time translation requires only that the direction of the relative momentum change during the encounter. In the classical problem, the asymptotic values of $\mathbf{r}_1$ and $\mathbf{p}_1$ before the ẹncounter can be specified with great precision. Asymptotic values are more difficult in the quantum problem. The wave nature of the quantum problem requires that the uncertainity in position and momentum be of the order of $\Delta\mathbf{r}\,\Delta\mathbf{p} = h^3$. Not only is a wave equation necessary for a discussion of the geometric aspects of the collision, but some care is necessary in specifying the source of particles and the detector of the scattered particles.

For a study of the quantum aspects of a two-particle collision, the operator form of the Hamiltonian given by Eq. (4.20) is written as the time-independent Schrödinger equation in terms of the six coordinates $\mathbf{r}_1$ and $\mathbf{r}_2$:

$$\begin{aligned} H\Psi &= \left[-\left(\frac{\hbar^2}{2m_1}\right)\nabla_1{}^2 - \left(\frac{\hbar^2}{2m_2}\right)\nabla_2{}^2 + U(|\mathbf{r}_2 - \mathbf{r}_1|)\right]\Psi \\ &= \left\{-\left[\frac{\hbar^2}{2(m_1+m_2)}\right]\nabla_R{}^2 - \left(\frac{\hbar^2}{2\mu}\right)\nabla^2 + U(r)\right\}\Psi = E'\Psi \end{aligned} \tag{4.22}$$

and then transformed by Eq. (4.5) to the center-of-mass coordinates **R** and **r**. This differential equation can be separated into a part that depends on the center-of-mass motion,

$$-\left[\frac{\hbar^2}{2(m_1+m_2)}\right]\nabla_R{}^2\psi(\mathbf{R}) = E_R\psi(\mathbf{R}) \tag{4.23}$$

and a part that depends on the relative coordinates,

$$\left[-\left(\frac{\hbar^2}{2\mu}\right)\nabla^2 + U(r)\right]\psi(\mathbf{r}) = E\psi(\mathbf{r}) \tag{4.24}$$

where $\Psi = \psi(\mathbf{R})\psi(\mathbf{r})$ and $E' = E_R + E$. Equation (4.24) is the Schrödinger equation for the motion of a particle of mass $\mu$ in a central potential $U(r)$ and can be compared with Eq. (4.14a) for the classical motion. A single-valued solution of Eq. (4.24) is of the form

$$\psi_{lm}(\mathbf{r}) = r^{-1}u_l(r)Y_l^m(\theta\varphi) \tag{4.25}$$

where $Y_l^m(\theta\varphi)$ denotes a spherical harmonic. Direct substitution into Eq. (4.24) yields a radial equation

$$\left[-\left(\frac{\hbar^2}{2\mu}\right)\frac{d^2}{dr^2} + \frac{l(l+1)(\hbar^2/2\mu)}{r^2} + U(r) - E\right]u_l(r) = 0 \tag{4.26a}$$

The supplemental condition

$$u_l(0) = 0 \tag{4.26b}$$

is needed at the origin and the problem is similar to the one-dimensional problem of a particle of mass $\mu$ in a potential

$$\frac{l(l+1)(\hbar^2/2\mu)}{\xi^2} + U(\xi)$$

for $\xi$ in the region $(0, +\infty)$. This problem and much of the subsequent discussion are given in considerable detail by Messiah [*1a*, Chapters IX and X]. If the potential $U(r)$ approaches zero more rapidly than $r^{-1}$,

$$\lim_{r\to\infty} rU(r) = 0$$

the energy spectrum contains two parts. For $E < 0$, only discrete solutions are permitted and the wave functions for these discrete energies decrease as $\exp[-(-2\mu E/\hbar^2)^{1/2}r]$. For $E > 0$, any value of $E$ is an allowed solution and the wave function $u_l$ oscillates as

$$u_l \underset{r\to\infty}{\sim} a_l \sin(kr - \tfrac{1}{2}l\pi + \delta_l) \tag{4.27a}$$

where

$$k^2 = \frac{2\mu E}{\hbar^2} \tag{4.27b}$$

$\delta_l$ is the phase shift and $-\frac{1}{2}l\pi$ has been added so that $\delta_l = 0$ when $U(r) = 0$. In principle, the detailed form of $u_l(r)$ can be found by integrating Eq. (4.26a). Equation (4.24) has an integral equation solution of the form

$$\psi_k(\mathbf{r}) = e^{i\mathbf{k}\cdot\mathbf{r}} - \frac{(2\mu/\hbar^2)}{4\pi}\int \frac{e^{ik|\mathbf{r}-\mathbf{r}'|}}{|\mathbf{r}-\mathbf{r}'|} U(r')\psi_k(\mathbf{r}')\, d\mathbf{r}' \tag{4.28}$$

For large $r$, the asymptotic behavior of $\psi_k$ is

$$\psi_k(r) \approx e^{i\mathbf{k}\cdot\mathbf{r}} + f(k, \theta)\frac{e^{ikr}}{r} \tag{4.29}$$

where $\theta$ is the angle between $\hat{k}$ and $\hat{r}$. The amplitude $f(\theta)$ is given by

$$f(k, \theta) = f(\mathbf{k}, \hat{r}) = -\frac{(2\mu/\hbar^2)}{4\pi}\int (\exp -ik\hat{r}\cdot\mathbf{r}')U(r')\psi_k(\mathbf{r}')\,d\mathbf{r}' \tag{4.30}$$

The line passing through the center of force and parallel to $\hat{k}$ is the symmetry axis.

Although Eqs. (4.28) and (4.29) are not properly normalized, the relative importance of the two terms is fixed by this solution. In the presence of a center of force, a proper solution must contain both terms and in the asymptotic form, the term $r^{-1}f(\theta)e^{ikr}$ must occur with the term $e^{i\mathbf{k}\cdot\mathbf{r}}$. $f(\theta)$ is a measure of the strength of the interaction and has the dimensions of length. $|f(\theta)|^2$ has the dimensions of area, and this quantity can be interpreted as the size of the center of force, or the cross section $\sigma(k, \theta)$:

$$\sigma(k, \theta) = |f(k, \theta)|^2 \tag{4.31}$$

A physical interpretation is given in the next section. The fixing of the relative size of these two terms can be compared with a barrier in the one-dimensional problem; the ratio of the coefficients of $e^{ikz}$ and $e^{-ikz}$ is fixed by the barrier $U(z)$.

### 4.3.2 Scattering by a Central Potential

It is common practice to regard $e^{i\mathbf{k}\cdot\mathbf{r}}$ as an incoming plane wave with momentum $\mathbf{p} = \hbar\mathbf{k}$ in direction $\hat{k}$ and $r^{-1}f(\theta)e^{ikr}$ as the scattered wave in direction $d\Omega$ or $\hat{r}$. An experimental arrangement must include a source and a detector as shown in Fig. 4.4 and these are perturbations which are not

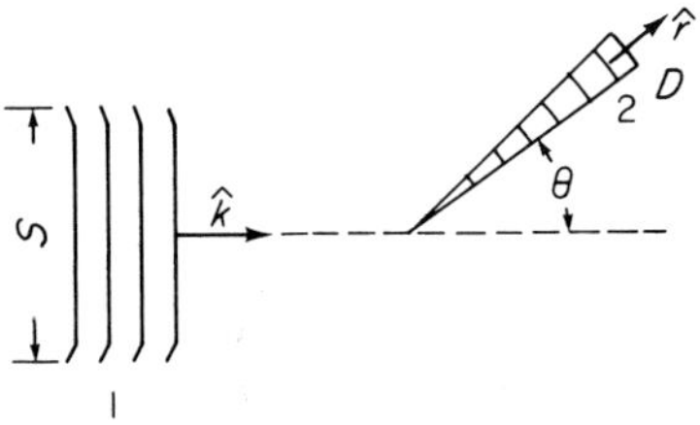


**Fig. 4.4** Scattering of an almost plane wave by a central potential. $S$: Source. $D$: Detector. (1) $\exp i\mathbf{k}\cdot\mathbf{r}$; (2) $f(\theta)e^{ikr}/r$.

normally included in the potential $U(r)$. The flux of particles from the source is confined to a finite cross section and, on the average, a current of particles crossing 1 $m^2$ per second is $V = \hbar k/\mu$ and if the center of force is located somewhere on this square meter, the current of particles crossing a surface $dS$ with a normal along $\hat{r}$ and in direction $d\Omega$ along $\hat{r}$ is given by $V\,|\,f(k, \theta)\,|^2\, d\Omega$. The ratio of these two fluxes is the probability of the scattering of the beam through angle $\theta$ in direction $d\Omega$ and leads to the differential cross section for scattering,

$$d\sigma = \sigma(k, \theta)\, d\Omega = |\,f(k, \theta)\,|^2\, d\Omega \tag{4.32}$$

$\sigma(k, \theta)$ is the probability that a pair of particles which approach each other with relative velocity $V = \hbar k/\mu$ and which pass through the same square meter change $\mathbf{k}$ to $\mathbf{k}'$, where $\theta$ is the angle between $\mathbf{k}$ and $\mathbf{k}'$ and $\hat{k}'$ is in direction $d\Omega$. Messiah [*1a*] discusses in considerable detail the scattering of a wave packet by a potential and gives a formal discussion of incoming and outgoing waves, the scattering matrix, and the $S$ matrix.

### *4.3.3 Phase-Shift Method*

In the phase-shift method, it is noted that the asymptotic solutions for $r^{-1}u_l$ and Eq. (4.29) must be equal since all regular solutions of Eq. (4.26a) yield the same phase shift $\delta_l$ and differ by the normalization $a_l$. A plane wave can be expanded as

$$e^{i\mathbf{k}\cdot\mathbf{r}} = 4\pi \sum_{l=0} \sum_{m=-l}^{l} i^l [Y_l^m(\hat{k})]^* Y_l^m(\hat{r}) j_l(kr) \tag{4.33}$$

and can be simplified with the addition theorem for spherical harmonics,

$$P_l(\cos\theta) = \frac{4\pi}{(2l+1)} \sum_{m=-l}^{l} [Y_l^m(\hat{k})]^* Y_l^m(\hat{r}) \tag{4.34}$$

where $\cos\theta = \hat{k} \cdot \hat{r}$. For large $r$, the spherical Bessel function

$$j_l(kr) \underset{r\to\infty}{\longrightarrow} \frac{\sin(kr - \frac{1}{2}l\pi)}{kr} \tag{4.35}$$

and Eq. (4.29) can be rewritten for large $r$ as

$$\frac{1}{r}\sum_l \left[(-)^{l+1}\frac{2l+1}{2ik}\, e^{-ikr} + \left(\frac{2l+1}{2ik} + f_l\right)e^{ikr}\right]P_l(\cos\theta)$$

$f(\theta)$ is replaced by its Fourier expansion in Legendre polynomials $f(\theta) = \sum f_l P_l(\cos\theta)$. Equating this expression to $r^{-1}u_l$, which is given by Eq. (4.27a), yields $a_l$ and $f_l$ in terms of the phase shift $\delta_l$. From this, the amplitude $f(k, \theta)$ and the phase shifts are related by

$$f(k, \theta) = k^{-1} \sum_{l=0} (2l + 1)e^{i\delta_l}(\sin\delta_l)P_l(\cos\theta) \tag{4.36}$$

The total cross section is obtained by integrating over the solid angle $d\Omega$ and by using the orthogonality of the Legendre polynomials.

$$\sigma(k) = \int d\Omega \,|\, f(k, \theta)\,|^2 = \frac{4\pi}{k^2} \sum_{l=0} (2l + 1)\sin^2\delta_l \tag{4.37}$$

The phase shift $\delta_l$ follows from a solution of Eq. (4.26a) for $u_l(r)$.

### 4.3.4 Born Approximation

A first approximation to $\psi_k$ can be obtained from Eq. (4.28) by iteration and in this approximation, $\psi_k(\mathbf{r}')$ is replaced by $e^{i\mathbf{k}\cdot\mathbf{r}}$. Equation (4.30) for $f(k, \theta)$ is

$$f^0 = -\frac{(2\mu/\hbar^2)}{4\pi} \int [\exp -i(\mathbf{k} - \mathbf{k}') \cdot \mathbf{r}']U(r')\, d\mathbf{r}' \tag{4.38}$$

where the notation is changed so that $\mathbf{k}' = k\hat{r}$. Since $U(r)$ is independent of the angles, the angle integration can be carried out and

$$f^0(k, \theta) = -\left(\frac{2\mu}{\hbar^2}\right) \int_0^\infty U(r) \frac{\sin Kr}{Kr} r^2\, dr \tag{4.39}$$

where

$$K = 2k \sin\frac{\theta}{2} \tag{4.40}$$

As the first example of the Born approximation, consider a finite square well

$$U(r) = \begin{cases} -U_0, & r < r_0 \\ 0, & r > r_0 \end{cases} \tag{4.41}$$

Direct substitution into Eq. (4.39) yields

$$f^0(k, \theta) = \frac{2\mu U_0}{\hbar^2 K^3} [(\sin Kr_0) - Kr_0(\cos Kr_0)] \tag{4.42}$$

for the scattering amplitude. The differential scattering cross section is

$$\sigma^0(k, \theta) = |f^0(k, \theta)|^2 \tag{4.43a}$$

and the total cross section is

$$\sigma^0(k) = \frac{2\pi\mu^2 U_0^2 r_0^6}{\hbar^4} g(\xi) \tag{4.43b}$$

where $\xi = 2kr_0$. $g(\xi)$ is the integral formed from Eq. (4.43a): $g(0) = 1/18$ and $g(\xi) = 1/4\xi^2$ for $\xi > 1$.

As a second example, consider the screened Coulomb field,

$$U(r) = -\frac{Ze^2}{r} e^{-r/r_0} \tag{4.44a}$$

The Thomas–Fermi statistical theory of an atom suggests that $r_0 = \hbar^2/\mu e^2 Z$ for the elastic scattering of an electron by a neutral atom. $\mu$ is now the electron mass. Direct substitution into Eq. (4.39) yields a scattering amplitude of

$$f^0(k, \theta) = \frac{2\mu Ze^2}{\hbar^2[K^2 + (1/r_0^2)]} \tag{4.44b}$$

When $r_0$ can be neglected, this yields the classical differential cross section for Coulomb scattering. The exponential cutoff in $U(r)$ yields a finite total cross section of

$$\sigma^0(k) = \frac{16\pi\mu^2 Z^2 e^4 r_0^4}{\hbar^4(4k^2 r_0^2 + 1)} \tag{4.44c}$$

For $U = Ze^2/r_0$, the results for the square-well and screened Coulomb potentials are quite similar.

The Born approximation is valid under the conditions

$$\frac{\mu U_0 r_0^2}{\hbar^2} \ll 1 \qquad \text{for} \quad kr_0 \ll 1$$

and

$$\frac{\mu U_0 r_0}{\hbar^2 k} \ll 1 \qquad \text{for} \quad kr_0 \gg 1$$

With $U_0 = Ze^2/r_0$, it is apparent that the Born approximation cannot be used for the scattering of slow electrons by neutral atoms and the electron energy should exceed 100 eV. The same arguments apply to molecule–molecule collisions, and the Born approximation is of limited usefulness in problems in this text.

### 4.3.5 Scattering of Hard Spheres

The classical scattering cross section for hard spheres of radius $a$ was discussed in Section 4.2.2; the differential cross section was given by Eq. (4.13), and the total cross section was $\pi(2a)^2$. An equivalent quantum mechanical problem occurs with the potential

$$U(r) = 0, \qquad r > r_0 = 2a$$
$$U(r) \to \infty, \qquad r < r_0$$

The wave nature is most clearly seen by choosing a small relative energy $E = \frac{1}{2}\mu V^2 = \hbar k^2/2\mu$, or a large value for the de Broglie wavelength

$$\lambda = \frac{h}{p} = \frac{2\pi}{k} \gg r_0$$

The radial equation (4.26a) is readily solved for $l = 0$ and the wave function is

$$u_0(r) = \begin{cases} a_0 \sin[k(r - r_0)], & r > r_0 \\ 0, & r < r_0 \end{cases}$$

Comparison with Eq. (4.27a) indicates that the phase shift

$$\delta_0 = -kr_0$$

For $kr_0 \ll 1$, or large $\lambda$, the phase shift $\delta_0$ is small and the phase shifts $\delta_l$ are much smaller. Since $\lambda$ is very much larger than $r_0$, it is expected that the scattered wave is spherically symmetric and is referred to as s-wave scattering. Equation (4.36) reduces to the $l = 0$ term for s-waves or

$$f_0(k, \theta) = k^{-1}\delta_0 \tag{4.45a}$$

and the differential scattering cross section is

$$\sigma_0(k, \theta) = r_0^2 \tag{4.45b}$$

The total scattering cross section is

$$\sigma_0(k) = 4\pi r_0^2 \qquad (k \to 0 \quad \text{or} \quad \lambda \to \infty) \tag{4.45c}$$

and is four times larger than the classical cross section.

At short wavelengths, as $\lambda \to 0$ and $k \to \infty$, classical paths are expected for the particles. Even so, the cross section is not classical. For $l(l + 1)\hbar^2$

$< (\hbar k r_0)^2$, or angular momentum

$$l\hbar < \hbar k r_0$$

the phase shift $\delta_l$ is almost a random variable between 0 and $2\pi$ and $\sin^2 \delta_l$ has a probable value of $\frac{1}{2}$. For values of angular momentum $l\hbar > \hbar k r_0$, the phase shift tends rapidly to zero and the cross section is approximately given by

$$\sigma(k) = \frac{4\pi}{k^2} \sum_{l=0}^{kr_0} (2l+1)\left(\frac{1}{2}\right)$$
$$\approx 2\pi r_0^2 \qquad (kr_0 \gg 1) \tag{4.45d}$$

This remains valid as $k \to \infty$ or $\lambda \to 0$ for a fixed value of $r_0$ and is twice the classical cross section. Usually, in the short-wavelength limit for wave phenomena, the paths of ray optics can be used. Since all rays in direction $\hat{k}$ and within a bundle of area $\pi r_0^2$ are reflected or are deflected in accord with the classical change of $\mathbf{p} \to \mathbf{p}'$ by $b = r_0 \cos(\theta/2)$ for the collision of elastic spheres (see Section 4.2), the particle aspect leads to a cross section of $\pi r_0^2$. In physical optics, the shadow behind the sphere disappears at sufficiently large distances and this is the diffraction or wave aspect of the problem. The diffraction angle is of the order of $\theta \approx \lambda/r_0$ and this small-angle diffraction introduces an additional $\pi r_0^2$ to the cross section. This effect persists in the short-wavelength limit. For angles $\theta > \lambda r_0 \ll 1$, the differential cross section for hard-sphere scattering is approximately given by the classical equation (4.13). An experiment which excludes angles less than $\lambda/r_0$ measures an almost classical value. It is shown in Section 4.4 and Chapter V that these small angles are excluded in most viscosity and diffusion experiments.

A general solution of the radial equation (4.26a) can be given in terms of spherical Bessel and Neumann functions and is of the form

$$u_l(r) = a_l[\,j_l(kr) \cos \delta_l + y_l(kr) \sin \delta_l] \tag{4.46a}$$

The boundary condition, or infinite barrier at $r_0$, requires $u_l(r_0) = 0$ and yields a phase shift of

$$\tan \delta_l = -\frac{j_l(kr_0)}{y_l(kr_0)} \tag{4.46b}$$

In the use of Eq. (4.46b), it seems to have been overlooked in the literature that

$$\delta_l = -\theta_{l+1/2} + \tfrac{1}{2}\pi \tag{4.47a}$$

where $\theta_{l+1/2}$ is defined in the "Handbook of Mathematical Functions" [2, Eq. (9.2.29)]. (Note that $y_l$ has the opposite sign of that used in this text.) An asymptotic expansion for $\delta_l$ is

$$\delta_l \approx -kr_0 + \frac{1}{2} l\pi - \frac{4(l+\frac{1}{2})^2 - 1}{8kr_0} - \frac{l^4}{24(kr_0)^3} + \cdots \tag{4.47b}$$

for $|kr_0| > l$. With this notation, Eq. (4.37) yields

$$\sigma(k) \approx \frac{4\pi}{2k^2} \int_0^{kr_0} 2l\, dl \approx 2\pi r_0^2 \tag{4.45e}$$

and is in agreement with Eq. (4.45d). This can be shown by writing $\sin^2 \delta_l = \frac{1}{2}(1 - \cos 2\delta_l)$ and observing that $\cos(kr_0 - l\pi)$ is as often positive as negative. For large $l$, the phase shift tends toward the values

$$\delta_l(l) \approx -\frac{1}{\sqrt{3}}, \qquad kr_0 = l \tag{4.47c}$$

and

$$\delta_l(kr_0) \approx -\frac{1}{2}\left(\frac{ekr_0}{2l+1}\right)^{2l+1}, \qquad kr_0 < l \tag{4.47d}$$

This exact expression [Eq. (4.46b)] for the phase shift can be used to check the qualitative discussion.

The result of a calculation [3] for an intermediate value of $kr_0 = 20$ is shown in Fig. 4.5. It should be noted that $kr_0 = 20$ corresponds to $\lambda/r_0$

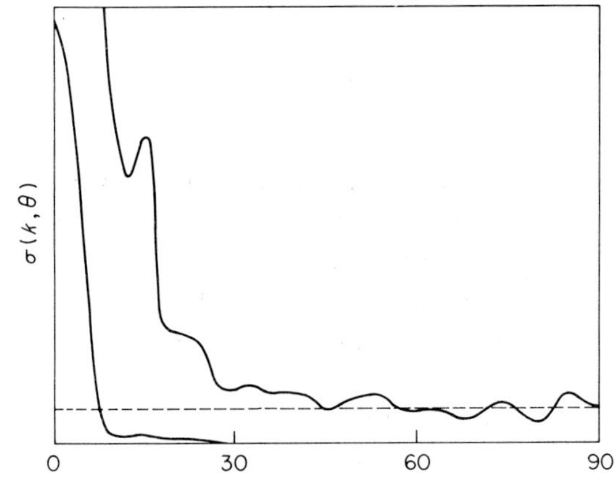


**Fig. 4.5** Angular distribution for the scattering of a rigid sphere for $kr_0 = 20$. Scattering near $\theta = 0$ is reduced by 1/40. The dashed line is the classical value. [From H. S. W. Massey and C. B. O. Mohr, *Proc. Roy. Soc. Ser. A* **141**, 434 (1933), Fig. 3A.]

$\approx 18^\circ$ and this is comparable to the half-width of the strong peak in the forward direction. At larger angles, the scattering tends toward the classical differential cross section. The area under the dashed line, or the classical value, is $\pi r_0^2$ and the area under the peak above the classical value is greater than $\pi r_0^2$.

### *4.3.6 Optical Theorem*

A comparison of Eq. (4.36) for the scattering amplitude $f(k, \theta)$ and Eq. (4.37) for the total scattering cross section $\sigma(k)$ shows that

$$\sigma(k) = \frac{2\pi}{ik} [f(k, 0) - f^*(k, 0)] = \frac{4\pi}{k} \operatorname{Im} f(k, 0) \tag{4.48}$$

This is a rather general feature of scattering and is referred to as the "optical theorem." In this particular example, the total cross section is directly related to the scattering amplitude for $\theta = 0$, or in the forward direction, and follows from the conservation of the probability within a sphere surrounding the scattering center. The current of particles $\mathbf{J}(\mathbf{r})$ is

$$\mathbf{J}(\mathbf{r}) = \frac{\hbar}{2\mu i} (\psi^* \operatorname{grad} \psi - \psi \operatorname{grad} \psi^*)$$

and the surface integral over a sphere surrounding the origin yields this optical theorem:

$$0 = \int d\mathbf{r} \operatorname{div} \mathbf{J} = \int \mathbf{J} \cdot d\mathbf{S}$$

The term $f(k, 0)$ interferes with the first term in Eq. (4.29) and is a measure of the depletion of the incident wave by the scattering center. It is therefore directly related to the total probability of scattering into other states.

The scattering amplitude in the forward direction, $f(k, 0)$, must be complex and in a certain sense implies the attenuation of the incident wave. Since the amount of attenuation is the same as the probability $\sigma(k)$ of being observed in the other states with $k'$ in any direction $\theta$, it is apparent that the optical theorem is related to the completeness theorem. The transition from the wave function of a particle in a central potential to the more general problem, definition of "before" and "after" or "incident" and "observed," and the relationship of completeness and the forward wave are more properly studied in a formal treatment of scattering and are not pursued here.

Exercise 4.4 Use Eq. (4.48) to find $\sigma(k)$ for a hard sphere for $kr_0 \ll 1$ and show the answer is the same as Eq. (4.45c), $\sigma = 4\pi r_0^2$.

### 4.3.7 *The Green Function for Elastic Scattering*

The basic differential equation for scattering, or describing the relative motion of two particles, is of the form

$$(\nabla^2 + k^2)\psi(r) = \frac{2\mu}{\hbar^2} U(r)\psi(r) \tag{4.49}$$

A general solution is

$$\psi_{\mathbf{k}}(r) = g_{\mathbf{k}}(r) - \frac{2\mu/\hbar^2}{4\pi} \int \frac{\exp ik \mid \mathbf{r} - \mathbf{r}' \mid}{\mid \mathbf{r} - \mathbf{r}' \mid} U(\mathbf{r}')\psi(\mathbf{r}')\, d\mathbf{r}' \tag{4.50a}$$

where $g_{\mathbf{k}}$ is a solution of $(\Delta^2 + k^2)g_k = 0$ and

$$G(\mathbf{rr}') = -\frac{1}{4\pi} \frac{\exp ik \mid \mathbf{r} - \mathbf{r}' \mid}{\mathbf{r} - \mathbf{r}'} \tag{4.50b}$$

is the Green function for $(\Delta^2 + k^2)\psi = 0$. This is similar to Eq. (4.28) used in the earlier discussion. Since $\mid \mathbf{r} - \mathbf{r}' \mid = r - \hat{r} \cdot \mathbf{r}' + r^{-1}$, for large $\mid \mathbf{r} - \mathbf{r}' \mid$,

$$\frac{\exp ik \mid \mathbf{r} - \mathbf{r}' \mid}{\mid \mathbf{r} - \mathbf{r}' \mid} \to \frac{\exp ikr}{r} \exp -ik\hat{r} \cdot \mathbf{r}' \tag{4.51}$$

and this is the approximation used in Eq. (4.30).

## 4.4 Cross Sections for Diffusion and Viscosity

Finite cross sections for small-angle scattering for molecular beam experiments verify the correctness of the quantum approach. At larger angles of scattering, the predicted scattering is in accord with either classical or quantum considerations. As shown in Chapter V, the cross sections for diffusion and for viscosity are related to the interaction cross section by

$$\sigma_d(V) = \int \sigma(V, \theta)(1 - \cos\theta)\, d\Omega \qquad \text{(diffusion)} \tag{4.52a}$$

and

$$\sigma_v(V) = \int \sigma(V, \theta)(1 - \cos^2\theta)\, d\Omega \qquad \text{(viscosity)} \tag{4.52b}$$

These equations give additional weight to scattering at large angles and the *classical approximation is adequate over the entire range* for almost all gases above 100°K. These cross sections measure the change in energy and

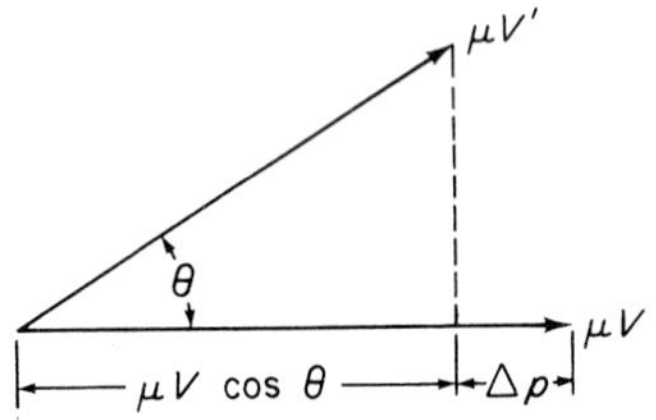


**Fig. 4.6** Change in momentum as the relative velocity changes by angle $\theta$.

momentum with respect to the initial relative velocity direction. Referring to the relative velocity diagram in Fig. 4.6 and to Eq. (4.11), it is apparent that

$$d\sigma_d = \begin{Bmatrix}\text{probability that the momentum changes by } \Delta p \text{ in the} \\ \text{direction of the initial relative motion per collision}\end{Bmatrix}$$

$$= \sigma(V, \theta)\left(\frac{\Delta p}{p}\right) d\Omega = \sigma(V, \theta)(1 - \cos\theta)\, d\Omega \tag{4.53a}$$

and, in a similar manner,

$$d\sigma_v = \begin{Bmatrix}\text{probability that the energy changes by } \Delta E \text{ in the} \\ \text{direction of the initial relative motion per collision}\end{Bmatrix}$$

$$= \sigma(V, \theta)(1 - \cos^2\theta)\, d\Omega \tag{4.53b}$$

In a quantum mechanical analysis, the relative velocity $V$ is replaced by the relative momentum $\hbar k$ or wave vector $\mathbf{k}$. The cross sections for diffusion and viscosity are given by

$$\sigma_d = \int \sigma(k, \theta)(1 - \cos\theta)\, d\Omega \tag{4.54a}$$

$$\sigma_v = \int \sigma(k, \theta)(1 - \cos^2\theta)\, d\Omega \tag{4.54b}$$

Equations (4.31) and (4.36) can be introduced for $\sigma(k, \theta)$. Since $\cos\theta = P_1$

and $\frac{1}{2}(3\cos^2\theta - 1) = P_2$, the integral of the product of three spherical harmonics or Legendre polynomials (Messiah [*1a*, Vol. II, Appendix C]) can be used to give an expression for the diffusion and viscosity cross sections in terms of the phase shifts. The total quantum mechanical and classical cross sections of primary interest are as follows:

*Scattering Cross Section*

$$\sigma = \frac{4\pi}{k^2} \sum_l{}' (2l+1) \sin^2 \delta_l \tag{4.55a}$$

$$\sigma = \int 2\pi b \, db \tag{4.55b}$$

*Diffusion Cross Section*

$$\sigma_d = \frac{4\pi}{k^2} \sum_l{}' (l+1) \sin^2(\delta_{l+1} - \delta_l) \tag{4.56a}$$

$$\sigma_d = 4\pi \int b \, db \sin^2 \tfrac{1}{2}\theta \tag{4.56b}$$

*Viscosity Cross Section*

$$\sigma_v = \frac{4\pi}{k^2} \sum_l{}' \frac{(l+2)(l+1)}{(2l+3)} \sin^2(\delta_{l+2} - \delta_l) \tag{4.57a}$$

$$\sigma_v = 2\pi \int b \, db \sin^2 \theta \tag{4.57b}$$

Quantum cross sections are given by the sums and the classical cross sections by the integral expressions. A detailed knowledge of these cross sections requires a numerical analysis for the phase shifts. The prime is defined by Eq. (4.58).

For unlike colliding particles, the sum in Eq. (4.55a) is over all $l$. For like particles, the wave functions must be symmetric or antisymmetric under exchange of the like particles, and this limits the sum for Bose–Einstein or for Fermi–Dirac particles. This aspect is discussed in greater detail in Chapter VIII. Denote the particle spin by $s$, then the following rule can be used:

| spin $s$ | even | odd |
|---|---|---|
| half-integer | $2\left(\frac{s}{2s+1}\right)$ | $2\left(\frac{s+1}{2s+1}\right)$ |
| integer | $2\left(\frac{s+1}{2s+1}\right)$ | $2\left(\frac{s}{2s+1}\right)$ |

(4.58)

For half-integer $s$, multiply the sum over even $l$ by $2s/(2s+1)$ and the sum over odd $l$ by $2(s+1)(2s+1)$. For integer spins, even and odd are interchanged. Bose–Einstein particles, with $s=0$, are summed over even $l$ and this sum is multiplied by 2. Fermi–Dirac particles with $s=\frac{1}{2}$, are summed over both even and odd $l$, and the even sum is multiplied by $\frac{1}{2}$ and the odd sum by $\frac{3}{2}$.

If one assumes that the cross section for viscosity is the same at large $k$ for the quantum and classical calculations, then Eqs. (4.57a) and (4.57b) are approximately the same with the classical impact parameter $b = l/k$ or with the classical angular momentum equal to $l\hbar$,

$$\mu V b \approx l\hbar$$

Then, the phase shift and the classical angular deflection are related by

$$\theta \approx \delta_{l+2} - \delta_l \tag{4.59}$$

Higher-order corrections can be developed by the WKB method [*1b*] and are given in great detail by Hirschfelder *et al.* [*4*].

Classical hard-sphere cross sections follow from Exercise 4.2, in which $\sigma(V, \theta) = \frac{1}{4}r_0^2$ is used. The three classical hard-sphere cross sections are

$$\begin{aligned} \sigma(V) &= \pi r_0^2 \\ \sigma_{\mathrm{d}}(V) &= \pi r_0^2 \\ \sigma_{\mathrm{v}}(V) &= \tfrac{2}{3}\pi r_0^2 \end{aligned} \tag{4.60}$$

Quantum mechanical expressions can be estimated by the asymptotic expression for the phase, Eq. (4.47b). Then, the general cross section is

$$\sigma(k) \approx 2\pi r_0^2 \tag{4.61a}$$

A phase difference is needed for diffusion and $\delta_l - \delta_{l+1} = \pi/2$. Integration over $l$ between 0 and $kr_0$ yields

$$\sigma_{\mathrm{d}}(k) \approx \pi r_0^2 \tag{4.61b}$$

Viscosity is interesting in that the first two phase terms cancel or are a multiple of $\pi$ and the third term in the asymptotic expansion is needed, $\delta_l - \delta_{l+2} = -\pi + (2l/kr_0)$ and this can be related by Eq. (4.59) to the classical angular deflection

$$\theta \approx \delta_{l+2} - \delta_l = \pi - \frac{2l}{kr_0} \approx \pi - \frac{2b}{r_0} \tag{4.62}$$

The term on the right is the term that follows from the relationship $b = r_0 \times \cos(\theta/2)$ for hard spheres for $\theta > \frac{3}{4}\pi$, or collisions in which the relative velocity reverses.

Exercise 4.5 Find $\sigma_d$, $\sigma_v$ and $\sigma$ for small $k$, that is, $\delta_0 = -kr_0$ and $\delta_l = 0$ for $l > 0$.

## 4.5 Elastic and Inelastic Binary Collisions

### 4.5.1 Classical Discussion of Binary Collisions

Binary collisions are considered from the point of view of nearly ideal gas and in terms of the velocities and impact parameter at large separation. Assume two particles of masses $m_1$ and $m_2$ are placed in a volume of 1 m$^3$. The probability of finding particle $m_1$ in volume element $d\mathbf{r}_1$ with velocity in interval $d\mathbf{v}_1$ is $P(v_1^2)\,d\mathbf{v}_1\,d\mathbf{r}_1$, and the probability of finding particle $m_2$ in volume element $d\mathbf{r}_2$ with velocity in interval $d\mathbf{v}_2$ is $P(v_2^2)\,d\mathbf{v}_2\,d\mathbf{r}_2$. The joint probability of finding both conditions at the same time is the product of the independent probabilities

$$[P(v_1^2)\,d\mathbf{v}_1\,d\mathbf{r}_1][P(v_2^2)\,d\mathbf{v}_2\,d\mathbf{r}_2] \tag{4.63}$$

The probability of a 1–2 collision in time $dt$ requires that the centers of both particles lie in the same element of volume $V\,dt\,2\pi b\,db$ shown in

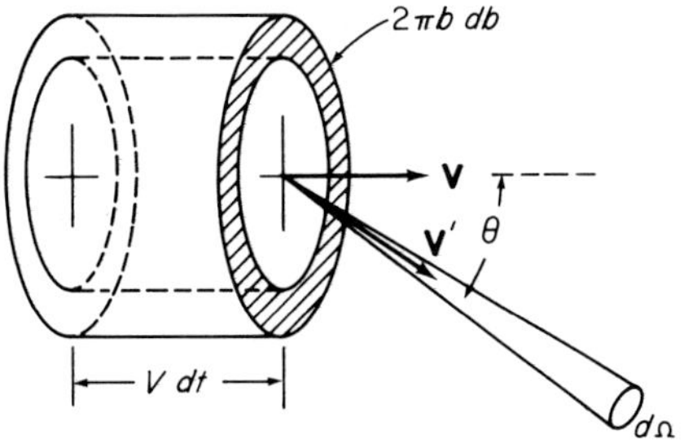


**Fig. 4.7** Important element of volume for the occurrence of a binary encounter.

Fig. 4.7. If the origin of coordinates is selected at particle 2, then the center of particle 1 must lie in this element of volume for a collision to occur, and $d\mathbf{r}_1 = V\,dt\,2\pi b\,db$. The impact parameter is between $b$ and $b + db$

and the relative velocity $\mathbf{V}$ changes its direction to $\mathbf{V}'$ in direction $d\Omega$. If the numbers of particles are increased to $n_1$ and $n_2$ of each species, then, since the particles are again treated as independent, the number of binary encounters in interval $dt$ in volume $d\mathbf{r}_2$ that change $\mathbf{V}$ into direction $d\Omega$ is

$$[n_1 P(v_1^2)\, d\mathbf{v}_1][n_2 P(v_2^2)\, d\mathbf{v}_2](V\, dt\, 2\pi b\, db)\, d\mathbf{r}_2 \tag{4.64}$$

The choice of $d\mathbf{r}_2$ is arbitrary and for convenience, $d\mathbf{r}_2$ is chosen as 1 m³ and $dt$ as 1 sec. Thus the basic relationship for binary encounters is

$$\begin{aligned}&\{\text{number of 1–2 encounters per m}^3 \text{ per sec that change } \mathbf{V} \text{ into direction } d\Omega\}\\ &\quad = [n_1 P(v_1^2)\, d\mathbf{v}_1][n_2 P(v_2^2)\, d\mathbf{v}_2](V 2\pi b\, db)\\ &\quad = [n_1 P(v_1^2)\, d\mathbf{v}_1][n_2 P(v_2^2)\, d\mathbf{v}_2][V\sigma(V, \Omega)\, d\Omega]\end{aligned} \tag{4.65}$$

where the relationship between the differential cross section and the impact parameter is the same as given earlier. Again, $\sigma$ can be interpreted in terms of the probability of an event occurring and Eq. (4.65) is valid for elastic and inelastic collisions.

The Maxwell–Boltzmann distribution of speeds applies to most of the problems of interest in kinetic theory, and it is convenient to transform the product $P(v_1^2)P(v_2^2)$ to the center-of-mass system. Using the transformation given by Eq. (4.5), the product becomes

$$\begin{aligned}P(v_1^2)P(v_2^2) &= \left(\frac{m_1}{2\pi kT}\right)^{3/2}\left(\frac{m_2}{2\pi kT}\right)^{3/2} \exp\left(-\frac{\frac{1}{2}m_1 v_1^2 + \frac{1}{2}m_2 v_2^2}{kT}\right)\\ &= \left(\frac{m_1 + m_2}{2\pi kT}\right)^{3/2}\left(\frac{\mu}{2\pi kT}\right)^{3/2} \exp\left[-\frac{\frac{1}{2}(m_1 + m_2)G^2 + \frac{1}{2}\mu V^2}{kT}\right]\end{aligned} \tag{4.66}$$

The volume of integration in velocity space can be transformed by the use of Eq. (4.10), and

$$d\mathbf{v}_1\, d\mathbf{v}_2 = d\mathbf{G}\, d\mathbf{V} \tag{4.67}$$

In texts on integral calculus, it is shown that the generalized volume elements in two different coordinate systems are related by

$$dx_1\, dx_2 \cdots dx_N = |J|\, dy_1\, dy_2 \cdots dy_N \tag{4.68}$$

where $J$ denotes the absolute value of the Jacobian determinant,

$$J = \begin{vmatrix} \dfrac{\partial x_1}{\partial y_1} & \dfrac{\partial x_2}{\partial y_1} & \cdots & \dfrac{\partial x_N}{\partial x_1} \\ \dfrac{\partial x_1}{\partial y_2} & \dfrac{\partial x_2}{\partial y_2} & \cdots & \dfrac{\partial x_N}{\partial y_2} \\ \vdots & \vdots & \vdots & \vdots \\ \dfrac{\partial x_1}{\partial y_N} & \dfrac{\partial x_2}{\partial y_N} & \cdots & \dfrac{\partial x_N}{\partial y_N} \end{vmatrix} \tag{4.69}$$

In the transformation $d\mathbf{v}_1\, d\mathbf{v}_2 = J\, d\mathbf{G}\, d\mathbf{V}$, $J$ is a six–six determinant. Direct evaluation with Eq. (4.10) yields a value of $|J| = 1$.

These two simplifying substitutions permit integration over the center-of-mass coordinates $d\mathbf{G}$. Thus the number of 1–2 encounters per second in a volume of 1 m³ becomes

$$\begin{aligned} \dot{n}_{12} &= \{\text{number of 1–2 encounters per m}^3 \text{ per sec}\} \\ &= \int [n_1 P(v_1^2)\, d\mathbf{v}_1][n_2 P(v_2^2)\, d\mathbf{v}_2]\left[V \int \sigma(V, \Omega)\, d\Omega\right] \\ &= n_1 n_2 \left(\frac{\mu}{2\pi kT}\right)^{3/2} \int \exp\left(-\frac{\frac{1}{2}\mu V^2}{kT}\right) 4\pi V^2\, dV \left[\int V\sigma(V, \Omega)\, d\Omega\right] \end{aligned} \tag{4.70a}$$

or in symbolic notation,

$$\dot{n}_{12} = n_1 n_2 \langle V \sigma_{12}(V) \rangle \tag{4.70b}$$

The $\langle\ \rangle$ indicate a Maxwell–Boltzmann average with the reduced mass $\mu$ and relative velocity $V$, and

$$\sigma_{12}(V) = \int \sigma(V, \Omega)\, d\Omega \tag{4.71}$$

If the collision occurs between particles with the same mass, the velocities still carry labels 1 and 2 and the development for 1–1 or 2–2 collisions follows in the same manner. Equation (4.70b) becomes

$$\dot{n}_{11} = n_1^2 \langle V_{11} \sigma_{11}(V) \rangle \tag{4.72}$$

where $V_{11}$ is used to emphasize that the reduced mass is for a 1–1 collision. Interchanging 1 and 2 yields the appropriate expressions for 2–2 collisions.

If a given atom is selected, the expected time $\tau_{12}$ between collisions for this *particular atom* and those atoms labeled 2 is

$$\frac{1}{\tau_{12}} = \frac{\dot{n}_{12}}{n_1} \tag{4.73}$$

A similar expression follows for $\tau_{11}$ and $\tau_{22}$. The expected distance between collisions is defined as

$$\lambda_{11} = \bar{v}_1 \tau_{11}, \qquad \lambda_{12} = \bar{v}_1 \tau_{12} \tag{4.74}$$

$\lambda_{12}$ is usually called the mean free path or the average free path between collisions of a 1 particle with a 2 particle. If the particles can be treated as hard spheres, $\sigma(V, \Omega)$ is independent of the relative velocity $V$, and with $\sigma_T = \sigma_{12}$,

$$\dot{n}_{12} = n_1 n_2 \sigma_{12} \bar{V}_{12} \tag{4.75}$$

where $\bar{V}_{12} = (8kT/\pi\mu_{12})^{1/2}$. Collisions also occur between particles of the same species but with different velocities $\mathbf{v}_1$ and $\mathbf{v}_2$. The above discussion remains valid and the number of 1–1 collisions is

$$\dot{n}_{11} = n_1^2 \sigma_{11} \bar{V}_{11} \tag{4.76}$$

where $\bar{V}_{11} = (8kT/\pi\mu_{11})^{1/2} = \sqrt{2}\,\bar{v}_1$. A similar expression is valid for 2–2 collisions. The use of $V_{11} = \sqrt{2}\,v_1$ often leads to a $\sqrt{2}$ in many expressions involving collisions. Thus in the expression for the mean free path for similar particles, $\lambda_{11}$ is

$$\lambda_{11} = \bar{v}_1 \tau_{11} = \frac{1}{\sqrt{2}\,n_1 \sigma_{11}} \tag{4.77}$$

Exercise 4.6 Assume two particles of masses $m_1$ and $m_2$ collide "head on," or that the collision occurs in one dimension. Show that $dv_{1z}\,dv_{2z} = dG_z\,dV_z$.

Exercise 4.7 In Eq. (4.66), show that

$$\int d\mathbf{G} \exp\left[-\frac{\frac{1}{2}(m_1 + m_2)G^2}{kT}\right] = \left(\frac{2\pi kT}{m_1 + m_2}\right)^{3/2}$$

and

$$\int V\, d\mathbf{V} \exp\left(-\frac{\frac{1}{2}\mu V^2}{kT}\right) = 2\pi\left(\frac{2kT}{\mu}\right)^2$$

**Exercise 4.8** Compute the number of Li–He collisions per cubic meter in a gas composed of 10% Li and 90% He at a pressure of $10^{-3}$ Torr and a temperature of 600°K. $\sigma_{\text{Li–He}} = 106 \times 10^{-20}\ \text{m}^2$. Find the average time between collisions for a given Li atom with the He atoms and that for a given He atom with the Li atoms.

Ans. $5.5 \times 10^{22}/\text{sec-m}^3$

**Exercise 4.9** Show that the integration of Eq. (4.70a) for an arbitrary cross section $\sigma(V)\, d\Omega$ which does not depend on $\theta$ gives Eq. (4.70b).

**Exercise 4.10** Show that the average value of the kinetic energy of two colliding hard sphere atoms is $\frac{3}{2}kT + 2kT$, where $\frac{3}{2}kT$ is associated with the center of mass and $2kT$ is associated with the relative motion. Explain why the integrand of Eq. (4.70a) is the appropriate distribution function for finding the averages for collision processes.

**Exercise 4.11** Show in detail that Eq. (4.70b) follows from Eq. (4.66).

### *4.5.2 Quantum Aspects of Binary Collisions*

Some general aspects of collisions between particles are now considered. The particles can be atoms or molecules and it is necessary to specify the internal states of these particles as well as their momenta. A very general binary encounter can be written as

$$\text{A} + \text{B} \rightarrow \text{C} + \text{D} \tag{4.78}$$

where A and B describe the particles before collision and C and D describe the particles after collision.

Let the two particles A and B be placed in a cubic vessel of volume $L^3$ and ignore the interaction between the particles. The spatial aspects of the particle wave functions are similar to those used in Chapter II and have the general form

$$\psi_{\mathbf{k}} = L^{-3/2} e^{i\mathbf{k}\cdot\mathbf{r}}$$

where $k_x = 2\pi n_x/L, \ldots$. The physical aspects of these waves follow from the Schrödinger equation, and the momentum and energy have average values of

$$\mathbf{p}_\alpha = \hbar \mathbf{k}_\alpha, \qquad E_\alpha = \hbar^2 k_\alpha{}^2/2m_\alpha \tag{4.79}$$

where $\alpha$ refers to A, B, C, or D. Since only two particles are present in such a large box, it is assumed that even in the presence of the two-particle interaction the particle states well before and well after the collisions are described by the asymptotic states in the accompanying tabulation. The

| Before | | After | |
|---|---|---|---|
| $\mathbf{p}_A$ | $\mathbf{p}_B$ | $\mathbf{p}_C$ | $\mathbf{p}_D$ |
| $\mathbf{k}_A$ | $\mathbf{k}_B$ | $\mathbf{k}_C$ | $\mathbf{k}_D$ |
| $E_A$ | $E_B$ | $E_C$ | $E_D$ |
| $a$ | $b$ | $c$ | $d$ |

quantum numbers $n_x$, $n_y$, and $n_z$ can be used to specify the momentum or wave number, but it is more convenient to use the index $\mathbf{k}_\alpha$ to denote the wave packet that is implied by these three indices. $a$, $b$, $c$, and $d$ are used to denote the internal states of the particles.

The question of primary interest is the following: If two particles A and B are placed in volume $L^3 = 1\ \text{m}^3$, what is the probability that during the next second the particles C and D are produced? Let this time-proportional transition rate be designated by $W(\text{AB; CD})$. Increase the number of type A to $n_A$ and those of type B to $n_B$. If it is further assumed that even with this increased number the probability of the events are independent, then the number of transitions per second in $L^3 = 1\ \text{m}^3$ is

$$n_A n_B W(\text{AB; CD}) \tag{4.80}$$

The problem is seldom approached in this manner, and a cross section for the process is introduced. Let $\hbar k^2/2\mu$ be the relative kinetic energy and note that this average energy is approximately equal to the classical value $\frac{1}{2}\mu V^2$. The A–B current of particles is the probability of finding A and B in $1\ \text{m}^3$ times their relative velocity $V_{AB}$, and the cross section is the transition rate divided by the flux, that is,

$$\sigma(\text{AB; CD}) = \frac{W(\text{AB; CD})}{V_{AB}} \tag{4.81}$$

With this definition, the number of transitions per second is

$$n_A n_B V_{AB} \sigma(\text{AB; CD}) \tag{4.82}$$

Exercise 4.12 Show that the number of A–B pairs in a gas of $n_A$ and $n_B$ species per cubic meter is $n_A n_B$ and, therefore, that the number of transitions is $n_A n_B W(\mathrm{AB}; \mathrm{CD})$.

### 4.5.3 Elastic Collisions

Internal states of the particles do not change during an elastic collision and these collisions are characterized by the conservation of energy and momentum. The factor $\hbar$ is frequently omitted in the expression for the average momentum and energy, that is, $\mathbf{p}_\alpha = \hbar\mathbf{k}_\alpha$; omitting $\hbar$, the conservation laws are

$$\mathbf{k}_A + \mathbf{k}_B = \mathbf{k}_C + \mathbf{k}_D = \mathbf{K} \tag{4.83a}$$

$$\frac{k_A{}^2}{2m_A} + \frac{k_B{}^2}{2m_B} = \frac{k_C{}^2}{2m_A} + \frac{k_D{}^2}{2m_B} \tag{4.83b}$$

or

$$\frac{K^2}{2(m_A + m_B)} + \frac{k^2}{2\mu} = \frac{K^2}{2(m_A + m_B)} + \frac{(k')^2}{2\mu} \tag{4.83c}$$

and

$$k^2 = (k')^2 \tag{4.83d}$$

Again the coordinate transformation given in Eq. (4.5) is used to transform to the center-of-mass system and the most general aspect of an elastic collision is to change the direction of $\hat{k}$ to $\hat{k}'$. This differential cross section was given in Section 4.3 and was denoted by

$$\sigma(\mathbf{k}, \mathbf{k}') = \sigma(k, \theta) \tag{4.84}$$

In a system in which binary transitions are very much greater than three-particle and higher-order aspects, the number of transitions per second is

$$n(k_A)n(k_B)\left(\frac{\hbar k}{\mu}\right)\sigma(k, \theta) \tag{4.85}$$

where $\mu V = \hbar k$. For a physical system, this must be averaged over the number of particles in states $k_A$ and $k_B$. The integers $n_x$, $n_y$, and $n_z$ can be used to count the allowed states, but it is more convenient to transform the sum to an integral,

$$\sum_{n}^{n+\Delta n} \to \frac{L^3\, d\mathbf{k}}{(2\pi)^3} = \frac{L^3 k^2\, d\Omega\, dk}{(2\pi)^3} \tag{4.86}$$

and this is the number of allowed states in volume $L^3$ with $\mathbf{k}$ in the range $d\mathbf{k}$ or momenta in the range $d\mathbf{p}/h^3$.

In a thermal system, the probability of finding a particle in state $k_A$ is proportional to $\exp(-E_A/2kT) = \exp(-\hbar^2 k_A{}^2/2m_A \varkappa T)$ if the probability of finding a particle in this state is much less than 1. This is discussed in greater detail in Chapter VII. The $\varkappa$ is used here for the Boltzmann constant to avoid confusion with the relative wave number $k$. Thus the average number of particles in range $d\mathbf{k}_A$ or with momenta in the range $d\mathbf{p}_A$ is given by

$$n(k_A) \to n_A\left(\frac{2\pi\hbar^2}{m\varkappa T}\right)^{3/2} \exp\left(-\frac{\hbar^2 k_A{}^2}{2m_A\varkappa T}\right)\frac{d\mathbf{k}_A}{(2\pi)^3} \tag{4.87}$$

In forming the average over $n(k_A)n(k_B)(\hbar k/\mu)\sigma(k, \theta)$, the kinetic energy in the product of exponentials can be written in terms of the energy of the center of mass and the relative kinetic energy. The same coordinate transformation which was used for Eq. (4.67) is used to show

$$d\mathbf{k}_A\, d\mathbf{k}_B = d\mathbf{k}\, d\mathbf{K}$$

and then the integration over $d\mathbf{K}$ is taken. The average number of thermal encounters that change the states is given by

$$\begin{aligned}\left\langle n_A n_B\left(\frac{\hbar k}{\mu}\right)\sigma(k)\right\rangle \\ &= \langle n_A n_B V\sigma(k)\rangle \\ &= n_A n_B\left(\frac{\hbar^2}{2\pi\mu\varkappa T}\right)^{3/2}\int_0^\infty 4\pi k^2\, dk \exp\left(-\frac{\hbar^2 k^2}{2\mu\varkappa T}\right)\left(\frac{\hbar k}{\mu}\right)\sigma(k)\end{aligned} \tag{4.88}$$

and is quite similar to Eq. (4.70a). The total cross section

$$\sigma(k) = \int d\Omega\, \sigma(k, \theta)$$

is used since the direction can no longer be measured in the center-of-mass system and is no longer of interest.

Exercise 4.13 Use Eqs. (4.88) and (4.47c) to estimate the collision rate for hard spheres and compare the result with Eq. (4.76).

### 4.5.4 *Inelastic Collisions*

Kinetic energy is not conserved in an inelastic collision, but the total energy is conserved. Thus for a binary collision

$$A + B \rightarrow C + D$$

we have

$$E_A + E_B = E_C + E_D + Q \tag{4.89}$$

Linear momentum is conserved for the system and the kinetic energy of the center of mass does not change. Only the relative kinetic energy can change and

$$\frac{\hbar^2 k^2}{2\mu} = \frac{\hbar^2 (k')^2}{2\mu'} + Q \tag{4.90}$$

It is common practice to denote this as a channel for a binary collision. It can happen that other final states (C′, D′), (C″, D″), (C‴ and no D), . . . occur, and each is referred to as a possible channel for the inelastic collision. $Q$ will be different for each channel. A channel is regarded as open as soon as $\hbar^2(k')^2/2\mu' > 0$, or as soon as the relative kinetic energy of the final state is greater than zero. A channel is open when

$$E = \frac{\hbar^2 k^2}{2\mu} > Q \tag{4.91}$$

The reaction rate for a thermal system is given by an expression similar to Eq. (4.88). The number of encounters is the same for both the elastic and inelastic processes. They differ in the cross section, and $\sigma(k)$ is changed to $\sigma(\mathrm{AB; CD}; k)$. It is convenient to replace $k$ with the relative kinetic energy $E = \hbar^2 k^2/2\mu = \frac{1}{2}\mu V^2$:

$$\begin{aligned}
&\{\text{number of reactions per m}^3 \text{ per sec} \quad A + B \rightarrow C + D\} \\
&= n_A n_B [2^{3/2}\pi^{-1/2}\mu^{-1/2}(kT)^{-3/2}] \int_0^\infty \exp\left(-\frac{E}{kT}\right) E\sigma(\mathrm{AB; CD}; E)\, dE \\
&= n_A n_B \left(\frac{\mu}{2\pi kT}\right)^{3/2} \int_0^\infty \left[4\pi V^2 \exp\left(-\frac{\mu V^2}{2kT}\right)\right] V\sigma(\mathrm{AB; CD}; E)\, dV \\
&= n_A n_B \langle V\sigma(\mathrm{AB; CD}; E)\rangle
\end{aligned} \tag{4.92}$$

Collision data are usually given in terms of the relative kinetic energy $E$ and this is noted in the cross section $\sigma$. Since $\sigma = 0$ for

$$Q < E = \frac{\hbar^2 k^2}{2\mu} = \frac{1}{2}\mu V^2$$

the integral is zero until $Q$ is reached and then changes in accord with the energy dependence of the cross section $\sigma(E)$.

### *4.5.5 Reaction Kinetics*

As indicated in the previous section, the reaction rate between two constituents of a gas can often be considered in terms of the number of binary collisions that occur per second between the two species. Three examples are now considered.

#### 4.5.5.1 Nuclear Kinetics

A very interesting problem of this type is the fusion rate of deuterium [5]. The nuclear interaction is

$$\mathrm{D} + \mathrm{D} \to {}^3\mathrm{He} + \mathrm{n} + 3.25 \quad \mathrm{MeV}$$

The cross section for this reaction is

$$\sigma_{\mathrm{DD}} = \frac{288 \times 10^{-28}}{E} \exp\left(-\frac{45.8}{\sqrt{E}}\right) \quad \mathrm{m}^2 \tag{4.93}$$

where $E$ is the relative kinetic energy of the colliding nuclei in kilo-electron volts (keV). Using Eq. (4.92), the number of encounters producing a $^3$He nucleus, a neutron, and 3.25 MeV of relative kinetic energy for these particles is

$$\begin{aligned} \dot{n}_{\mathrm{DD}} &= \{\text{number of fusion processes per m}^3 \text{ per sec}\} \\ &= n_{\mathrm{D}}{}^2 \langle V\sigma_{\mathrm{DD}}\rangle \end{aligned} \tag{4.94a}$$

The power, or energy released per cubic meter per second, is

$$\{\text{energy released per m}^3 \text{ per sec}\} = n_{\mathrm{D}}{}^2 \langle V\sigma_{\mathrm{DD}}\rangle Q_{\mathrm{DD}} \tag{4.94b}$$

where $Q_{\mathrm{DD}} = 3.25$ MeV. In this field of research, the temperature is often measured in electron volts or, in this problem, in kilo-electron volts and the appropriate conversion is made by observing that

$$kT = eT_{\mathrm{e}} \times 10^3 \qquad \text{or} \qquad T_{\mathrm{e}} = \frac{kT}{1000e}$$

$e$ is the charge of the electron in coulombs and $T_{\mathrm{e}}$ the temperature in kilo-electron volts.

Substituting the cross section into Eq. (4.94a) and converting to the appropriate units yield

$$\langle V\sigma_{\mathrm{DD}}\rangle = 10^{-20} \int_0^\infty dE \, \frac{1}{T_{\mathrm{e}}^{3/2}} \exp\left(-\frac{45.8}{\sqrt{E}} - \frac{E}{T_{\mathrm{e}}}\right) \tag{4.94c}$$

and for $T_{\mathrm{e}} < 50$ keV,

$$\langle V\sigma_{\mathrm{DD}}\rangle \approx 2.6 \times 10^{-20} \frac{1}{T_{\mathrm{e}}^{2/3}} \exp\left(-18.76 \frac{1}{T_{\mathrm{e}}^{1/3}}\right)$$

The units are cubic meters per second. Differentiation of the integrand with respect to $T$ indicates a maximum at $8.1/T_{\mathrm{e}}^{1/3}$. A large part of value of the integral comes in the region of this maximum. For $T_{\mathrm{e}} = 1$ keV, most of the reactions occur for particles with eight times the thermal temperature. Substituting $\langle V\sigma_{\mathrm{DD}}\rangle$ into Eq. (4.94b) allows the power release to be estimated. At thermal energies of the order of 100 keV, or approximately $10^9$ °K, the energy released per cubic meter is of the order of 100 W at a deuterium pressure of the order of $10^4$ atm.

Cross sections for ($^2$H–$^2$H), $^2$H–$^3$H, and $^2$H–$^4$He fusion reactions are shown in Fig. 4.8. The power density that follows from Eq. (4.94b) at

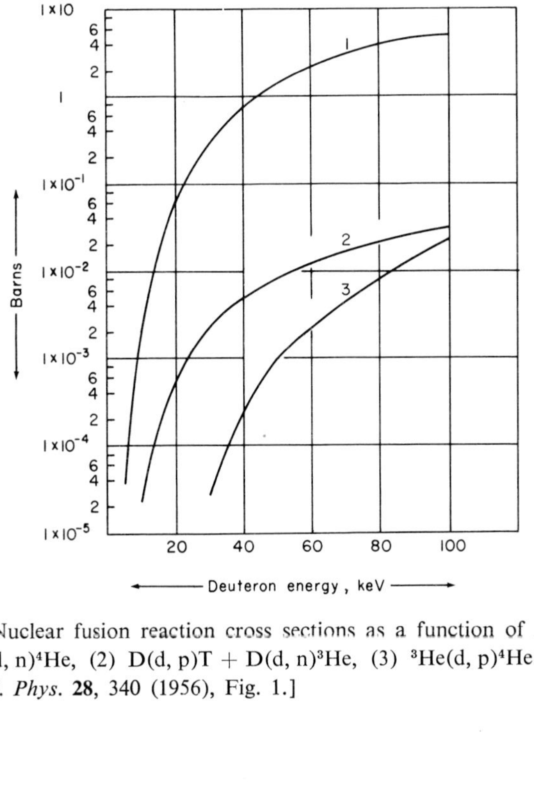


**Fig. 4.8** Nuclear fusion reaction cross sections as a function of relative particle energy. (1) T(d, n)$^4$He, (2) D(d, p)T + D(d, n)$^3$He, (3) $^3$He(d, p)$^4$He. [From R. F. Post, *Rev. Mod. Phys.* **28**, 340 (1956), Fig. 1.]

temperatures of 100 keV is shown in Fig. 4.9. These temperatures of $10^9$ °K seem rather large at the present time and temperatures of the order of $10^7$ °K, or 1 keV, and plasma densities of the order of $10^{14}/cm^3$ seem realistic. These values are reported in the current literature with a confinement time of the order of 0.02 sec.

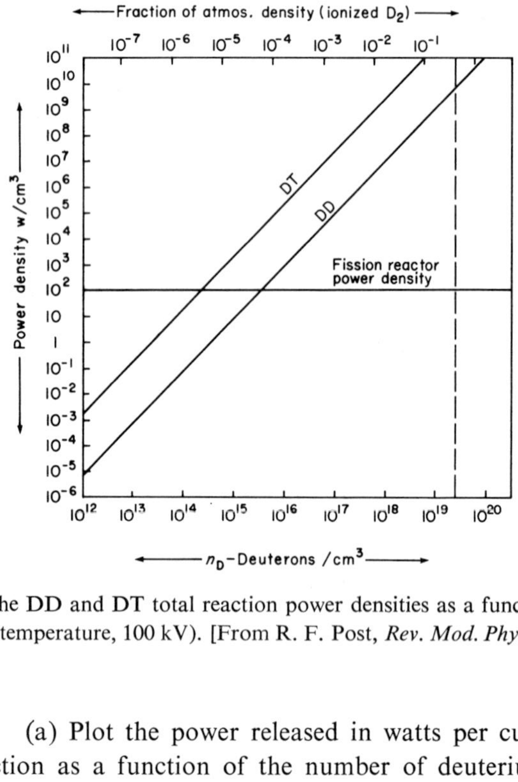


**Fig. 4.9** The DD and DT total reaction power densities as a function of deuteron particle density (temperature, 100 kV). [From R. F. Post, *Rev. Mod. Phys.* **38**, 342 (1956), Fig. 3.]

Exercise 4.14 (a) Plot the power released in watts per cubic meter for the D–D reaction as a function of the number of deuterium nuclei per cubic meter at temperatures of 10, 50, and 100 keV. Use a log–log plot. Show that 1 keV $= 1.16 \times 10^7$ °K. Estimate the mean path between collisions for a deuterium nucleus at an energy of 100 keV and a gas density of $5 \times 10^{15}$ nuclei/cm³.

Ans. $5 \times 10^9$ cm

(b) What is the mean lifetime of a D nucleus for this gas?

Exercise 4.15 Estimate the average path a $^2H$ nucleus travels before the occurrence of a fusion process in $^2H$. Assume the process is thermal, the particle density of $^2H$ is $10^{14}/cm^3$, and $T = 1, 10, 100$ keV. Repeat for $^3H$

in the same $^2$H gas. Find the average lifetime the $^2$H and $^3$H nuclei have. Compare these distances with the earth's diameter.

Ans. $\lambda = 3 \times 10^{10}$ and $2 \times 10^9$ m for $^2$H at 10 and 100 keV; $2 \times 10^8$ and $5 \times 10^7$ m for $^3$H at 10 and 100 keV, respectively; $\tau \approx 2 \times 10^4$, $5 \times 10^2$, $10^2$, 20 sec.

#### 4.5.5.2 Chemical Kinetics

An elementary theory of chemical reaction rates assumes that two approaching atoms colliding with sufficient energy to surmount the repulsive potential barrier combine to form an activated complex, that is,

$$\mathrm{A} + \mathrm{B} \rightarrow \mathrm{AB} \tag{4.95a}$$

The number of such encounters leading to reactions is given by Eq. (4.92), and

$$\dot{n}_{\mathrm{AB}} = \{\text{number of AB complexes formed per m}^3 \text{ per sec}\} = n_{\mathrm{A}} n_{\mathrm{B}} \langle V \sigma_{\mathrm{AB}} \rangle \tag{4.95b}$$

Fowler and Guggenheim [6] treat the collisions as hard-sphere collisions and assume an activated complex is formed when the kinetic energy along the line of centers, $\frac{1}{2}\mu(V\cos\theta)^2$, is greater than $E^*$. Present [7] uses an equivalent definition by introducing a reaction cross section

$$\sigma_{\mathrm{AB}} = \begin{cases} \sigma_0\left(1 - \dfrac{E^*}{E}\right), & E > E^* \\ \sigma(E) = 0, & E < E^* \end{cases} \tag{4.96}$$

Either procedure yields

$$\dot{n}_{\mathrm{AB}} = n_{\mathrm{A}} n_{\mathrm{B}} \sigma_0 \left(\frac{8kT}{\pi\mu}\right)^{1/2} \exp\left(-\frac{E^*}{kT}\right) \tag{4.97}$$

The rate of a second-order bimolecular reaction of the type given by Eq. (4.95a) is often expressed by

$$\frac{\partial n_{\mathrm{A}}}{\partial t} = -k_{\mathrm{AB}}^* n_{\mathrm{A}} n_{\mathrm{B}} \tag{4.98}$$

From Eq. (4.95a), since $\dot{n}_{\mathrm{A}} = \dot{n}_{\mathrm{B}} = \dot{n}_{\mathrm{AB}}$, $k_{\mathrm{AB}}^*$ is given by

$$k_{\mathrm{AB}}^* = \sigma_0 \left(\frac{8kT}{\pi\mu}\right)^{1/2} \exp\left(-\frac{E^*}{kT}\right) \tag{4.99}$$

In the often used example of the $2HI \rightleftarrows H_2 + I_2$ reaction, the reaction rate is well represented by

$$k^* \approx 3 \times 10^{-18} \sqrt{T} e^{-22,000/T}$$

The extreme sensitivity of the rates of reaction to temperature is well known and is indicated in these equations by the exponential term.

**Exercise 4.16** Show that Eq. (4.97) follows from Eqs. (4.92) and (4.96).

**Exercise 4.17** Show that $\sigma$ can be estimated from the reaction rate $k^*$ and that this cross section is of the same order as that deduced from viscosity experiments.

#### 4.5.5.3 Laser Kinetics

Atomic and molecular cross sections and reaction rates play an important role in lasers. The He–Ne and $CO_2$–$N_2$ lasers form illustrative examples and are considered in this section. Equation (2.63) indicates that a gaseous medium will have either a gain or a loss depending on whether the number of atoms or molecules in the upper state is greater than or less than the number in the lower state. Thus gain occurs for $\eta(b) > \eta(a)$ and with a suitable resonant cavity, oscillation can occur for the optical transition from $b$ to $a$. The frequency of oscillation is approximately $\nu_{\mathrm{ba}} = (E_{\mathrm{b}} - E_{\mathrm{a}})/h$. Some dynamic processes for creating population inversion, or for causing $\eta(b) > \eta(a)$, in the gaseous system are now considered.

A He–Ne gas mixture was used for the first gas laser in the optical spectral region [8]. The He atoms in the gas mixture were excited in an electrical discharge by collisions with electrons:

$$\mathrm{He}(1s^2\ {}^1S) + e \rightarrow \mathrm{He}^*(1s2s\ {}^3S) + e$$

The cross section for inelastic collisions with electrons is considered in greater detail in a subsequent section. A very interesting feature of this first optical gas laser was the transfer of excitation from an excited helium atom to a neon atom, or the reaction

$$\mathrm{He}^*({}^3S) + \mathrm{Ne} \rightarrow \mathrm{He}({}^1S) + \mathrm{Ne}^*$$

An energy level diagram showing some of the states of interest in He and Ne is shown in Fig. 4.10. Since the energy levels of the excited states of He and Ne are almost coincident in energy, the cross section for energy ex-

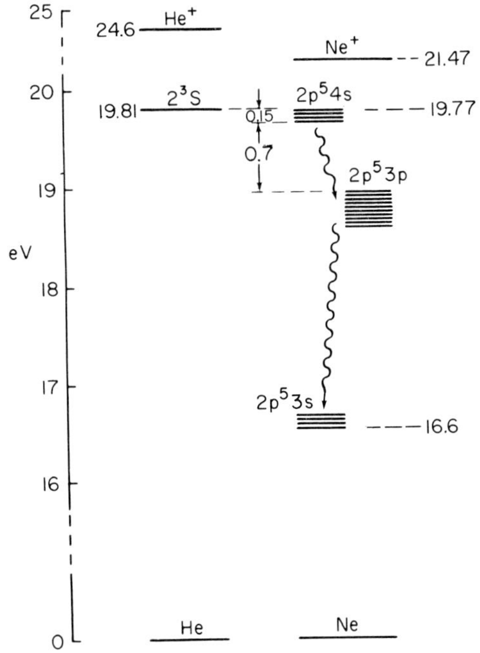


**Fig. 4.10** Energy level diagrams of He and Ne atoms. [From A. Javan, W. R. Bennett, Jr., and D. R. Herriott, *Phys. Rev. Sett.* **6**, 106 (1961).]

change is quite large. An estimate of the excitation transfer cross section is [*8*]

$$\sigma[\text{He}^*(^3\text{S}), \text{Ne}; \text{He}, \text{Ne}^*] \approx 3.7\times 10^{-21} \quad \text{m}^2$$

It was not possible to determine the cross section to particular energy levels of the $2p^54s$ configuration of Ne* and this cross section is an average. Theoretical considerations suggest that excitation cross sections are quite large at resonance, are of the order of $\pi a_0^2$ within a few $kT$ of resonance, and fall rapidly toward zero for larger energy differences [*9*]. The $2p^54s$ configuration is split into four energy levels, but a notation introduced by Racah is necessary to give a spectroscopic description of the levels and this description is not obvious. Since the excitation transfer process populates the $2p^54s$ configuration and the cross section for excitation of the $2p^53p$ levels is quite small, the population of these two levels is inverted, or $\eta(b) > \eta(a)$. Electric dipole optical transitions are permitted between these two levels and in the He–Ne laser, oscillation occurs at 1.15 μm. The specific levels involved are, in Racah notation [*10*],

$$2\text{p}^5 4\text{s}'[\tfrac{1}{2}]_1^{\circ} \rightarrow 2\text{p}^5 3\text{p}'[1\tfrac{1}{2}]_2 + h\nu(1.15\ \mu\text{m})$$

where the subscript on the square bracket is the value of the total angular momentum $J$. The $2^1S_0$ excited state of He is almost coincident with some of the energy levels of the $2p^55s$ configuration of Ne and excitation transfer also occurs for these levels. Population inversion occurs and the laser transition that is characteristic of so many small laboratory lasers is the red transition at 0.6328 μm, or

$$2p^55s'[\tfrac{1}{2}]_1^{\circ} \rightarrow 2p^53p'[1\tfrac{1}{2}]_2 + h\nu(0.6328\ \mu m)$$

Other excitation mechanisms are also important in the operation of the He–Ne laser. An inelastic collision can cause a direct excitation to the excited states of Ne,

$$Ne(2p^6) + e \rightarrow Ne^*(2p^5ns) + e$$

Since a $p \rightarrow s$ transition is an optically allowed electric dipole transition, the cross section for this process is quite large. Excitation to the $2p^5np$ levels is not optically allowed and the inelastic cross section is much smaller for this process. Thus the entrance rate to the $ns$ levels exceeds the entrance rate to the $np$ levels and inversion and gain can occur. A third process is also effective. The spontaneous emission lifetime of the $2p^5ns$ states is less than $10^{-8}$ sec and an ultraviolet photon is emitted. The number of ground-state atoms is usually quite large, that is, pressures of the order of a few Torr, and the ultraviolet photon is absorbed before it can escape from the tube. Absorption and reemission [*11*] occur and the phenomenon is referred to as "trapped resonance radiation." This trapping enhances the number of Ne atoms in the $2p^5ns$ states.

The $CO_2$–$N_2$ gas laser system [*12*] provides a high-power laser in the kilowatt power range and as such has a wide range of usefulness in studies of nonlinear optics, molecular energy transfer, communication systems, precision machining, plasma generation, and nuclear fusion. Its importance has encouraged the study of many rate processes and much data are currently available. Vibrational energy levels of interest in $CO_2$, $N_2$, $O_2$, and $H_2O$ are shown in Fig. 4.11, where some 14 transitions are shown [*13*]. The basic inelastic collision for laser action is labeled (5) and is the reaction

$$CO_2 + N_2^* + Q \rightarrow CO_2^*(\nu_3) + N_2$$

where $N_2^*$ denotes a nitrogen molecule excited to the $v = 1$ vibrational state. $CO_2^*(\nu_3)$ or (001) denotes the excitation of the asymmetric stretching mode $\nu_3$ to the $v_3 = 1$ vibrational energy level and this energy level lies 18 cm$^{-1}$ above the vibrational level of $N_2^*$. Since $kT \approx 300$ cm$^{-1}$, the cross

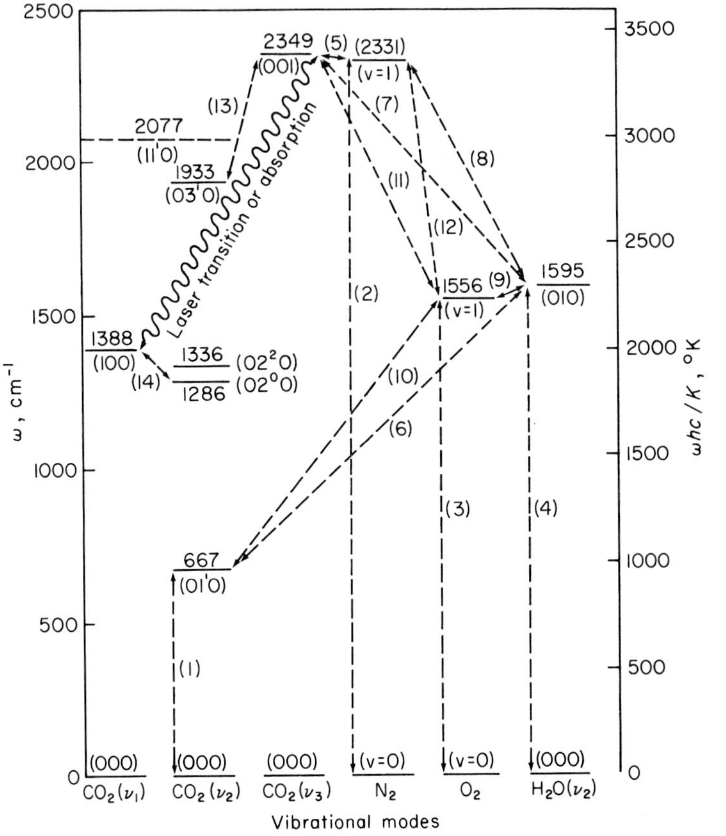


**Fig. 4.11** Vibrational energy level diagram for $CO_2$–$N_2$–$O_2$–$H_2O$ molecular system. Shown are a few of the lowest vibrational levels of the $\nu_1$, $\nu_2$, $\nu_3$ modes and the combination level $\nu_1 + \nu_2$ of $CO_2$, levels of $N_2$ and $O_2$, and of the $\nu_2$ mode of $H_2O$. The arrows indicate the assumed paths for vibrational energy transfer within this system of molecules. The wavy arrow indicates the $CO_2$–$N_2$ laser position at 10.6 $\mu$m. [From R. L. Taylor and S. Bitterman, *Rev. Mod. Phys.* **41**, 27 (1969), Fig. 1.]

section for energy transfer is quite large and quite rapid. The reaction rate [*14*] is shown in Fig. 4.12, and in the notation of Eq. (4.80),

$$W(CO_2, N_2^*; CO_2^*, N_2; T) = K \tag{4.100a}$$

and is in the units of $CO_2$–$N_2^*$ particle pairs per cubic meter per second. It implies that the rate of increase in the number of $CO_2^*$ per cubic meter is given by

$$\frac{\partial n(CO_2^*)}{\partial t} = Kn(N_2^*)n(CO_2) \tag{4.100b}$$

when the inverse reaction can be neglected. A crude estimate of the cross section can be made by assuming $V\sigma \approx K \approx 4\times 10^{-19}$ and then $\sigma \approx 4\times 10^{-22}$ m$^2$ is the approximate cross section. This is a factor of $10^{-2}$ smaller than the cross section for viscosity. As the temperature increases, the cross section becomes larger and approaches the viscosity cross section. It is suggested that at low temperatures, long-range forces are dominant and the rate is proportional to $T^{-1/2}$. At high temperatures, short-range forces become dominant and the rate is proportional to $T^{3/2}$.

A high-power $CO_2$–$N_2$ laser requires some of the other 14 processes indicated in Fig. 4.11 and is limited by some of the processes. These are discussed in detail in a recent review article [*13*], which gives references to the original papers.

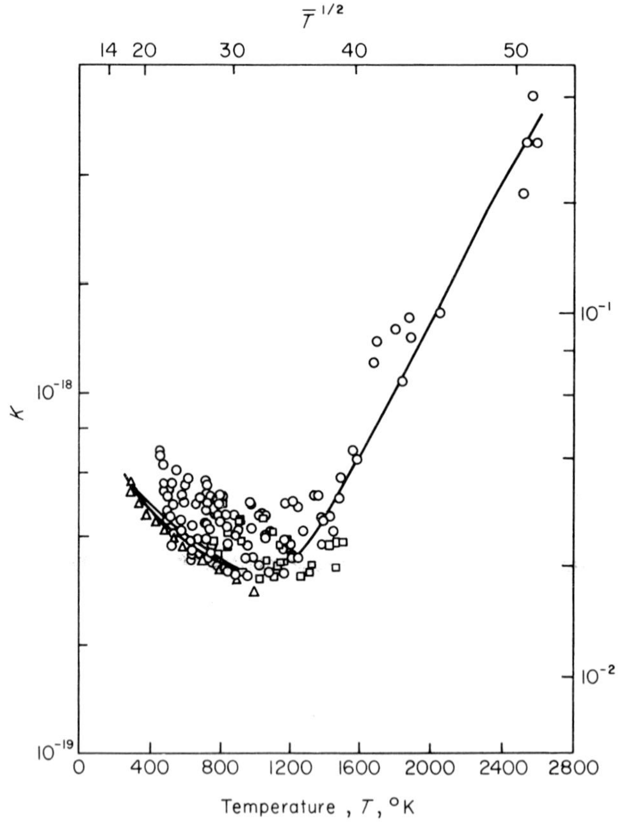


**Fig. 4.12** Experimental data and some theoretical curves for the reaction rate [*13*] $W(CO_2, N_2^*; CO_2^*, N_2; T) = K$. [See Eq. (4.100b).] [From R. L. Taylor and S. Bitterman, *Rev. Mod. Phys.* **41**, 38 (1969), Fig. 11.]

## 4.6 Inelastic Scattering Cross Section

The previous sections have included some aspects of inelastic cross sections. This section pursues the subject somewhat further and gives some typical inelastic cross sections. One of the primary requirements for an inelastic collision is that *at least one of the colliding particles must have internal degrees of freedom.* The collision of an electron with a hydrogen atom is probably the most elementary inelastic collision, and can occur in the 0–15-eV range for the relative kinetic energy. The electron has only spin as an internal degree of freedom and the hydrogen atom has the usual atomic energy levels. The simplest problem for atom–atom collisions is the collision of two hydrogen atoms. Both of these problems are considered in this section and other, more complex problems are discussed briefly.

### *4.6.1 Collisions of Electrons with Atoms or Molecules*

The simplest electron–atom collision is the collision of an electron with the hydrogen atom. The time-independent Schrödinger equation can be written, with an origin of coordinates at the nucleus, as

$$\left[-\frac{\hbar^2}{2m}(\nabla_1{}^2+\nabla_2{}^2)+\frac{e^2}{4\pi\varepsilon_0}\left(-\frac{1}{r_1}-\frac{1}{r_2}+\frac{1}{r_{12}}\right)-E\right]\psi(\mathbf{r}_1\mathbf{r}_2)=0 \tag{4.101}$$

where 1 and 2 are the labels for the electrons. The motion of the center of mass has been omitted from the equation, and spin–orbit coupling and hyperfine interactions are neglected. No attempt is made to solve this scattering problem. Many excellent and extensive review articles are given in the literature [*15–17*] and a considerable degree of sophistication is needed for these problems. In these collision problems, the initial state is an atom in a given state and a distant incoming electron, and the final state is the atom in a certain state and an outgoing electron. Since three or more particles are involved, a many-body problem results and in general an exact solution is not possible. The incident and atomic electrons are identical and the overall wave function must obey the Pauli exclusion principle.

If the relative kinetic energy of the incident electron is not sufficient to excite the atom, then only elastic collisions can occur and are given by the collision equation

$$\mathrm{H} + \mathrm{e} \rightarrow \mathrm{H} + \mathrm{e}$$

$$\mathbf{k}_\mathrm{A} + \mathbf{k}_\mathrm{B} \rightarrow \mathbf{k}_\mathrm{C} + \mathbf{k}_\mathrm{D}$$

The cross section for the elastic collision of an electron with ground state hydrogen atoms is given in Fig. 4.13 for energies from 0 to 12 eV.[1] Experimental cross-section measurements were made by employing a chopped atomic hydrogen beam which is crossed by a dc electron beam. It should be noted that this cross section is quite large at low energies. The cross section for elastic scattering by helium is comparable to that for hydrogen. Cross sections for argon and krypton are also comparable to that for hydrogen in magnitude, but the cross section becomes very small at about 3 eV and this is referred to as the Ramsauer–Townsend effect. It is explained by showing that the phase shift $\delta_0$ tends to zero at this energy.

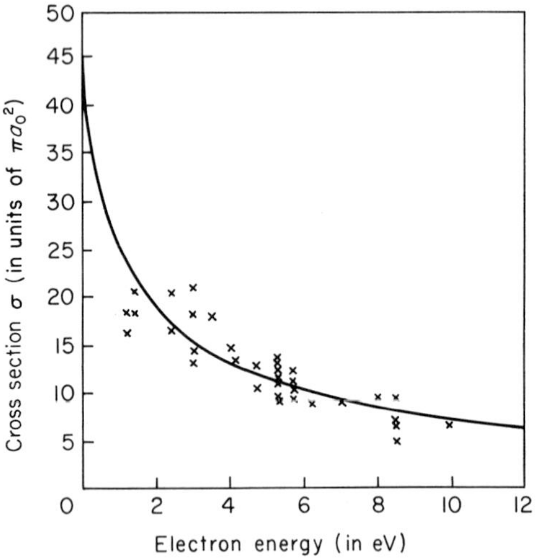


**Fig. 4.13** Total cross section for the elastic scattering of electrons by hydrogen atoms. [From B. L. Moiseiwitsch, *in* "Atomic and Molecular Processes" (D. R. Bates, ed.), p. 315, Fig. 3, Academic Press, New York, 1962.]

Inelastic cross sections are much smaller in general than elastic cross sections. The cross section for the excitation of the 2p level by electron collision is shown in Fig. 4.14[2]:

$$H(1s) + e \rightarrow H^*(2p) + e$$

This is an optically allowed transition and the cross section is comparable to the "size" of the hydrogen atom. The electric dipole matrix element

[1] See Moiseiwitsch [*18*]. Original data taken from the experimental data of Brackmann *et al.* [*18a*] and the curve is from the calculations of Temkin and Lamkin [*18b*].

[2] See Geltman [*16*] for a discussion of these curves.

occurs in both the radiation problem and the collision problem for optically allowed transitions. The 1s–2s cross section for excitation is one-fifth of that for the 1s–2p and illustrates the feature that optically allowed transitions have greater cross sections than transitions that are not allowed.

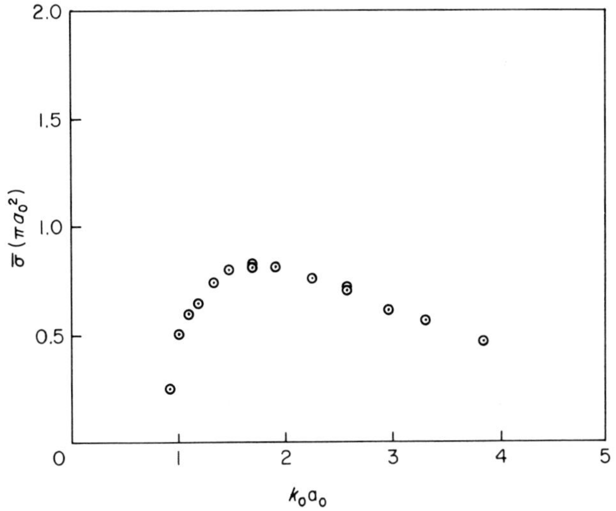


**Fig. 4.14** Total cross section for the 1s–2p excitation of hydrogen atoms by electron impact. (From S. Geltman, "Topics in Atomic Collision Theory," p. 109, Fig. 11, Academic Press, New York, 1969.)

A cross section of considerable interest is the excitation of the $^1$S and $^3$S states of He. Just above threshold, the cross section is appreciable and is attributed to the exchange of an incident electron and an atomic electron. The cross section for the $^1$S–$^3$S transition is shown in Fig. 4.15.[3] Threshold is near 19.6 eV and the cross section rises quite rapidly to values of the order of $0.04\pi a_0^2$ at 20.4 eV and falls rapidly toward zero near 21 eV.

It was indicated earlier that optically allowed cross sections can become quite large for electron excitation. This is illustrated in Fig. 4.16 for the Na 3s–3p transition. Cross sections as large as $40\pi a_0^2$ are approached.[3]

Finally, atomic hydrogen can be ionized by collision with an electron [*20*]. Ionization starts at 13.59 eV, rises to a maximum of approximately $\pi a_0^2$, and then declines slowly at higher energies.

[3] See Seaton [*19*]. Experimental data are taken from Schulz and Fox [*19a*] and the theoretical curve is taken from Massey and Moiseiwitsch [*19b*].

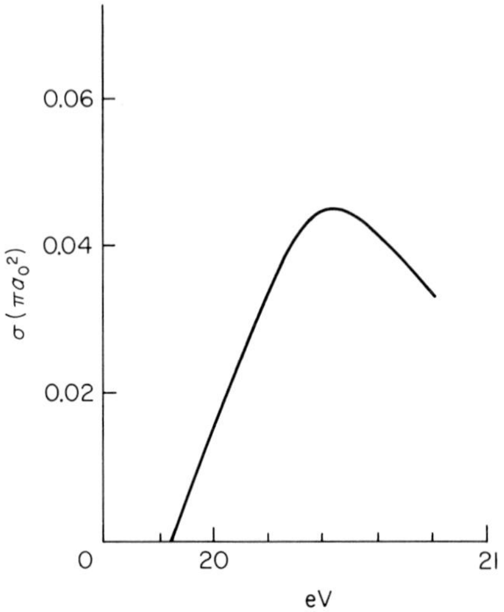


**Fig. 4.15** Cross section for He $1^1S$–$2^3S$. The large cross section just above threshold is attributed to the exchange of the incident and atomic electrons. [From M. J. Seaton, *in* "Atomic and Molecular Processes" (D. R. Bates, ed.), p. 411, Fig. 14, Academic Press, New York, 1962.]

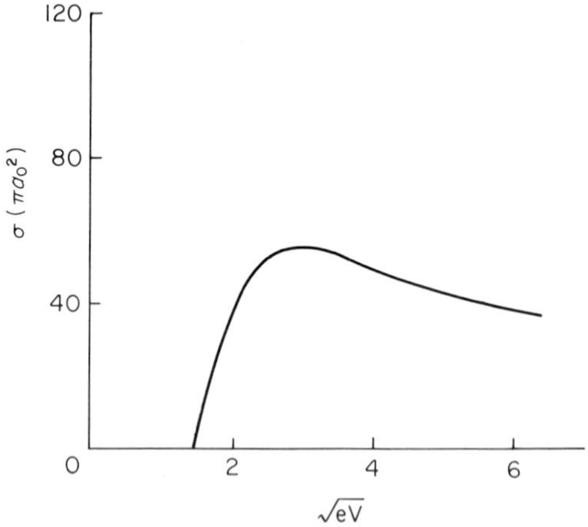


**Fig. 4.16** Cross section for Na 3s–3p by electron excitation. Solid line is experimental curve. [From M. J. Seaton, *in* "Atomic and Molecular Processes" (D. R. Bates, ed.), p. 413, Fig. 15, Academic Press, New York, 1962.]

These brief comments illustrate some of the features of electron–atom collisions. The reader is referred to the review articles for greater detail and reference to the original literature.

### *4.6.2 Atomic and Molecular Collisions*

Since both particles have internal degrees of freedom, even the collision of two hydrogen atoms is more complex than the electron–atom collision problem. These collisions do not change the nuclear properties in the range of energies of interest and again the properties of the system well before and well after the collision can be described by an equation of the form

$$\mathrm{A} + \mathrm{B} \leftrightarrow \mathrm{A}' + \mathrm{B}'$$

The equation of motion of the system is given by the Schrödinger equation

$$(H - E)\psi = 0$$

or, in detail,

$$\left\{-\hbar^2\left(\frac{1}{2M_\mathrm{A}}\nabla_\mathrm{A}^2 + \frac{1}{2M_\mathrm{B}}\nabla_\mathrm{B}^2\right) - \frac{\hbar^2}{2m}\sum_i \nabla_i^2 + \frac{e^2}{4\pi\varepsilon_0}\left[\frac{Z_\mathrm{A}Z_\mathrm{B}}{r_\mathrm{AB}} - \sum_a\left(\frac{Z_\mathrm{A}}{r_{i\mathrm{A}}} + \frac{Z_\mathrm{B}}{r_{i\mathrm{B}}}\right) + \sum_{i\neq j}\frac{1}{r_{ij}}\right] - E\right\}\psi = 0 \qquad (4.102)$$

Coordinates of the nuclei are given by $r_\mathrm{A}$ and $r_\mathrm{B}$ and of the electrons by $r_i$. The Laplacian terms give the kinetic energy of the nuclei and of the electrons, the $1/r_{i\mathrm{A}}$ and $1/r_{i\mathrm{B}}$ terms the potential energy of Coulomb interaction between the electrons and the nuclei, and the $1/r_{ij}$ terms the electron–electron Coulomb interaction energy. In any given scattering experiment, it is assumed that the system is prepared in states A and B, that is, in states of linear momentum $k_\mathrm{A}$ and $k_\mathrm{B}$ and internal states $a$ and $b$. Subsequently, a measurement is made to determine the states of the particles and are found to be $k_{\mathrm{A}'}$, $k_{\mathrm{B}'}$, $a'$, $b'$ as discussed earlier in this chapter. These statements are necessary since the equation of motion does not give any direct indication of the molecular states referred to in the reaction equation.

At relative kinetic energies of less than 100 eV, the collisions are "slow" and the Born approximation relating the initial and final states and the interaction potential cannot be used. One method of approach is to regard atoms A and B as having infinitely heavy masses and separated by a distance $R$. This approach neglects the first two terms in Eq. (4.102) and is the

Born–Oppenheimer treatment of a diatomic molecule. For each internuclear distance $R$, the allowed molecular wave function and the allowed energies are found, and this solution yields the electronic configuration for each $R$.

At very large $R$, the potential energy term can be divided into two groups, but in order to simplify the discussion, the coordinates of two hydrogen atoms as shown in Fig. 4.17 are used. The potential energy term for the two electrons is written as

$$U_d = \frac{1}{r_{1B}} + \frac{1}{r_{2A}} - \frac{1}{r_{12}} \approx \frac{1}{R^3}[\mathbf{r}_{1A}\cdot\mathbf{r}_{2B} - 3(\mathbf{r}_{1A}\cdot\hat{R})(\mathbf{r}_{2B}\cdot\hat{R})]$$

$$\approx \frac{1}{R^3}(x_1x_2 + y_1y_2 - 2z_1z_2) \tag{4.103}$$

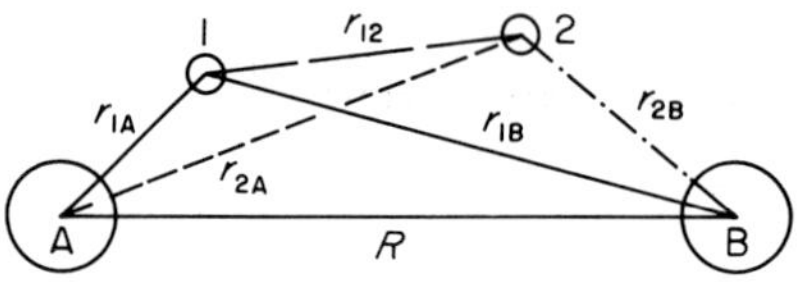


**Fig. 4.17** Coordinates for the hydrogen molecule. A and B are the positive nuclei and 1 and 2 are the electrons.

where the $z$ axis is along AB and $\mathbf{r}_{1B} = R\hat{z} - \mathbf{r}_{1A}$, $\mathbf{r}_{12} = R\hat{z} - \mathbf{r}_{1A} + \mathbf{r}_{2B}, \ldots$ . If $U_d$ is omitted, the problem separates into that of two hydrogen atoms. If both atoms are in their ground states, there is no first-order interaction since the matrix element $(1s \mid x \mid 1s)$ is zero by parity arguments. Second-order perturbation theory includes matrix elements of the type $(1s \mid x \mid 1p) \times (1p \mid x \mid 1s)$ and by approximate closure arguments [*21*], the second-order perturbation theory yields an electronic energy of [see Eq. (4.2)]

$$U_d(R) \approx -\frac{C}{R^6}$$

The $C$ coefficient includes matrix elements of the type $(1s \mid x^2 \mid 1s)$. These matrix elements also occur in determining the index of refraction of the atomic gas, and hence the name "dispersion forces" as well as the historical name of "van der Waals" forces is used. More complex atoms or molecules A and B require a sum over all terms within the parentheses in Eq. (4.103) and this sum is also dependent on the index of refraction of the species. An approximate equation is

$$U_d \approx -\frac{C}{R^6} \approx -\frac{3/2}{(4\pi\varepsilon_0)^2}\left(\frac{E_{IA}E_{IB}}{E_{IA}+E_{IB}}\right)\frac{\alpha_A\alpha_B}{R^6} \tag{4.104}$$

where $E_{IA}$ and $E_{IB}$ are the ionization energies and $\alpha_A$ and $\alpha_B$ are the polarizabilities of the atoms or molecules. The polarizability and the optical index of refraction $\mu$ of a gas are related by

$$\mu^2 - 1 = N\alpha \tag{4.105}$$

where $N$ is the number of species A or B per cubic meter.

Thermal elastic collisions with an effective potential $U_d(R) \approx -C/R^6$ which are discussed in the previous paragraph will obey classical mechanics to a large degree and the angular scattering is approximately given by Eq. (4.17) or its quantum counterpart Eq. (4.59). Only at temperatures lower than 20°K will quantum corrections become important in the transport cross sections. The arguments given here apply to all atomic and molecular collisions.

As $R$ decreases, the molecular designation for the states is used. In the collision of two particles, the two nuclei form a symmetry axis which is designated as $\hat{z}$ and the electrons can be regarded as being in an axially symmetric potential. This is already apparent in the van der Waals potential, which is proportional to $(\mathbf{r}_\alpha \cdot \mathbf{r}_\beta - 3z_\alpha z_\beta)$. Angular momentum of the electrons about the symmetry axis is a constant of motion and takes on integer values $|\Lambda| = 0, 1, 2, \ldots$. These electronic angular momentum states are designated by $\Sigma$, $\Pi$, $\Delta$, .... The simple Hamiltonian does not contain the electron spin, but the Pauli exclusion principle requires that the wavefunction be antisymmetric in the exchange of electrons. Thus the spin becomes important in selecting the allowed wavefunctions. The total spin of the electrons is characterized by $S$ and there are $2S + 1$ values for $M_S$. In this approximation, these levels are degenerate and $2S + 1$ is the multiplicity of the configuration. Thus $^3\Pi$ designates a configuration with $\Lambda = 1$ and $S = 1$. The states with $\Lambda \neq 0$ are twofold degenerate. This follows by noting that invariance under rotation by angle $\varphi$ leads to the rigid rotation wavefunctions $\exp(\pm i\Lambda_\varphi)$. $\Lambda = 0$ is a special case for more than one electron since the wave function depends on the angle $\varphi_{ij}$ between the electrons or the angle of each electron relative to $\varphi$. A plane through the $z$ axis is a symmetry plane and reflection through this plane changes the sign of the wavefunction in the same sense that $\cos\varphi_{ij}$ and $\sin\varphi_{ij}$ change sign as $y \to -y$. $\Sigma$ wave functions are classified as $\Sigma^+$ or $\Sigma^-$ as the wavefunction $\psi \to \pm\psi$ under this reflection. If the nuclei are identical, a center of symmetry occurs and the Hamiltonian is invariant under the inversion of all electron coordinates. If the wave function changes sign under inversion, it is designated as $u$ and if it does not change sign it is designated as $g$.

With these simple ideas and a knowledge of the atomic states at large $R$, curves of the potential as a function of $R$ can be labeled.

Labeling of the levels according to the states of well-separated atoms or by the symmetry of an axial system completes the easier aspects. A numerical solution is in general needed to find the electronic energy of the $^{2S+1}\Sigma_g^+$, ... states of the atoms A + B as a function of $R$. This is outside the scope of this text, but typical curves are shown in Fig. 4.18a for some

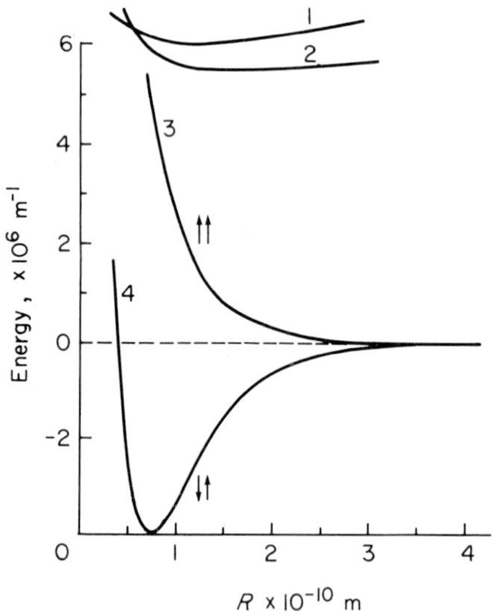


**Fig. 4.18a** Electronic potential energy curves for hydrogen molecule. (1) c $^3\Pi_u$; (2) B $^1\Sigma_u^+$, (3) b $^3\Sigma_g^+$, (4) $^1\Sigma_g^+$.

of the states of hydrogen atoms[4] and in Fig. 4.18b for oxygen atoms [*26*]. For many of the simpler systems, the electronic energy can be used for the potential curves for the interaction between the two nuclei and the vibration of these two nuclei in the potential well can be treated as a simple quantum mechanical oscillator problem with vibrational levels $(v + \frac{1}{2})h\nu_0$. From the experimental viewpoint the frequency of vibration $\nu_0$, the deviation from $\nu_0$ for higher values of $v$, and the dissociation energy can be used to infer the shape of the potential curves for bound states [*23*].

As noted earlier, all atoms attract each other according to a $1/R^6$ law at large $R$. As $R$ decreases, the Pauli exclusion principle becomes important for the electronic wavefunction and the electron–electron interaction $1/r_{ij}$ may become dominant. This occurs in hydrogen and for H($^1$S) + H($^1$S),

[4] Data taken in part from Lichten [*22*], Herzberg [*23*], Coolidge and James [*24*], and Kolos and Roothaan [*25*].

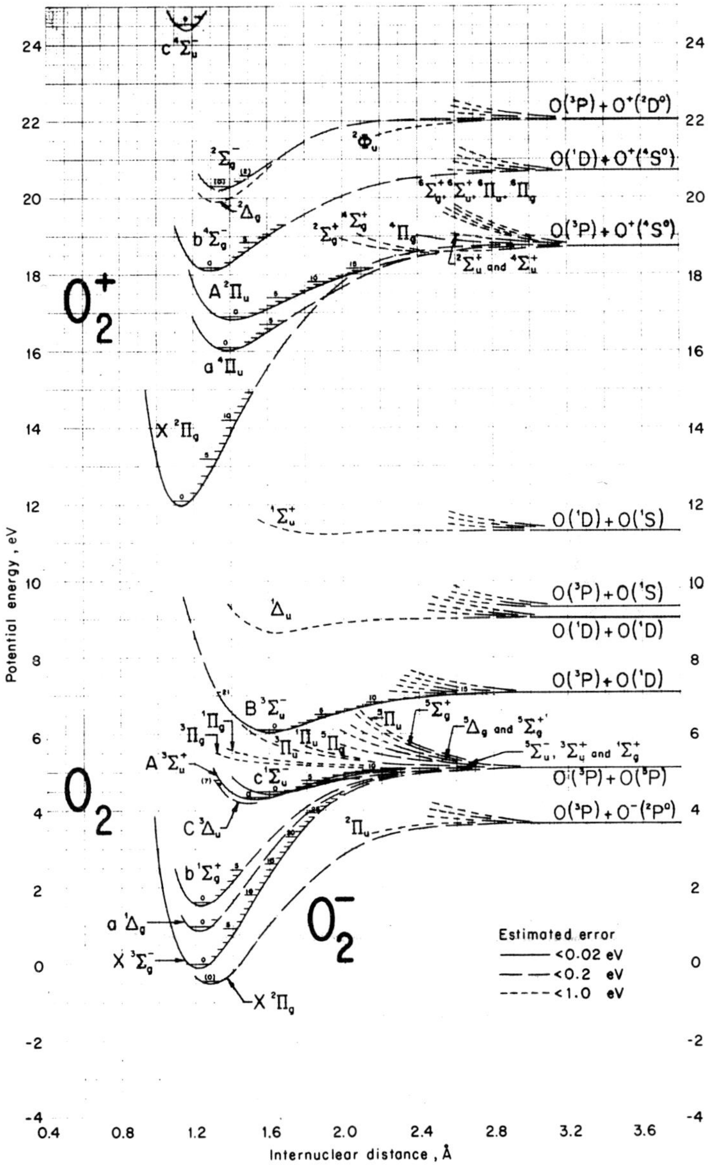


**Fig. 4.18b** Electronic potential energy curves for $O_2^-$, $O_2$, and $O_2^+$. [From F. R. Gilmore, *J. Quant. Spectrosc. Radiat. Transfer* **5**, 369 (1965).]

the wave function with parallel spins ${}^3\Sigma_u{}^+$ is strongly repulsive and the ${}^1\Sigma_g{}^+$ state with antiparallel spins is strongly attractive. This is a very strong attractive potential for hydrogen and this antiparallel spin bond forms the basis of the chemical bond. At short range, the $1/r_{ij}$ term becomes dominant and the exchange force becomes repulsive for all atomic or molecular pairs.

Interaction of molecular species, such as $H_2 + H_2$, $O_2 + O_2$, or $N_2+N_2$, in their ground states is more nearly like the interaction of two rare gas atoms. The interaction is attractive as $1/R^6$ and at small $R$, becomes strongly repulsive. A model which permits calculations is the Lennard–Jones potential

$$U(R) = 4\varepsilon\left[\left(\frac{R_0}{R}\right)^{12} - \left(\frac{R_0}{R}\right)^{6}\right] \tag{4.106}$$

where $\varepsilon$ is the depth of the attractive potential and $R_0$ is the radius at which $U = 0$. Molecular collisions which excite either rotational or vibrational levels [*27*] form an interesting subject. Some cross sections for rotational and vibrational excitation are shown in Figs. 4.19a and 4.19b for the collision of two hydrogen molecules.

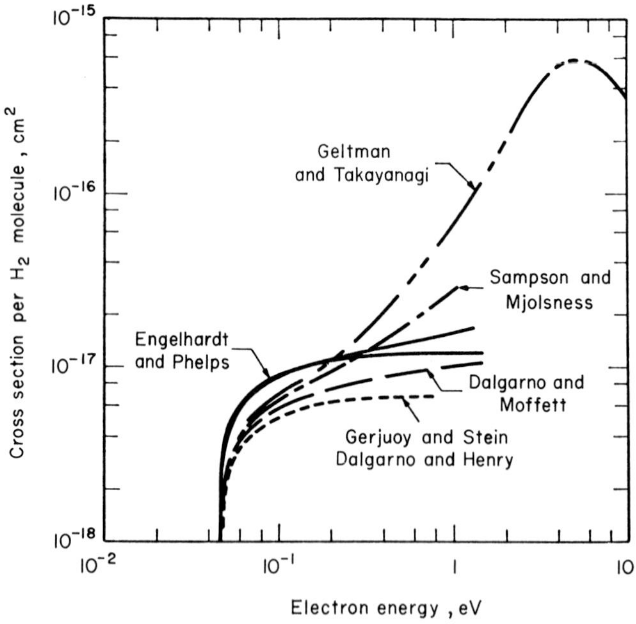


**Fig. 4.19a** Cross section for the rotational excitation of $H_2$ from the $J = 0$ to the $J = 2$ level. [From A. V. Phelps, *Rev. Mod. Phys.* **40**, 401 (1968), Fig. 1.]

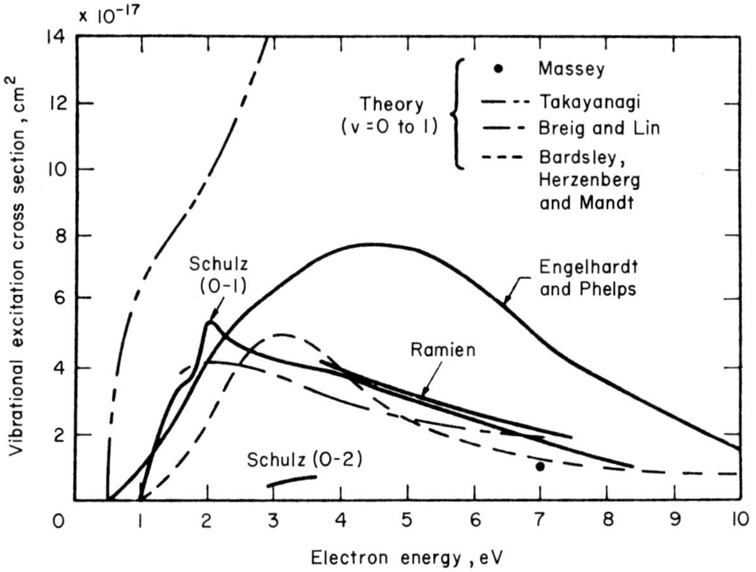


**Fig. 4.19b** Vibrational excitation cross section for $H_2$. [From A. V. Phelps, *Rev. Mod. Phys.* **40**, 403 (1968), Fig. 3.]

## 4.7 Atomic Cross Sections from Atomic Beam Experiments

Historically, the first rough measurements of atomic scattering were of the rate of deposition of silver from an atomic beam that passed through air at various pressures [*28*]. Higher angular resolution was obtained by measuring the scattering from crossed molecular beams [*29*]. Scattering measurements at small angles confirmed the quantum theory predictions that the small-angle scattering is large, but finite.

Extensive measurements were made [*30, 31*] on the collision of alkali atoms with monatomic and diatomic gas molecules.

Unfortunately, many experiments use a thermal source and the beam received at the detector must be averaged over the molecular speeds in the source. Ramsey [*32*] gives a partial discussion of the averaging and references to much of the literature.

In Table 4.1, the observed cross sections for alkali-metal–rare-gas atomic collisions and the derived van der Waals constants are given. The cross-sections are determined from the experimental value of $\lambda$ and for a hard-

**Table 4.1**

| Alkali-metal atom | Rare-gas atom | $\sigma$ ($\times 10^{-20}$ m$^2$) | $C$ ($\times 10^{-79}$ J m$^6$) | $C'$ ($\times 10^{-79}$ J m$^6$) |
|---|---|---|---|---|
| Li | He | 106 | 14.4 | 17 |
| Li | Ne | 120 | 18.7 | 32 |
| Li | Ar | 303 | 188 | 125 |
| Na | He | 130 | 17.8 | (26) |
| Na | Ne | 213 | 40.4 | (51) |
| Na | Ar | 401 | 192 | (200) |
| K | He | 165 | 31.7 | 35 |
| K | Ne | 250 | 50.5 | 68 |
| K | Ar | 580 | 356 | 260 |
| Rb | He | 152 | 25.4 | 40 |
| Rb | Ne | 268 | 49.0 | (77) |
| Rb | Ar | 572 | 249 | (290) |
| Cs | He | 162 | 29.4 | 44 |
| Cs | Ne | 287 | 56.7 | 87 |
| Cs | Ar | 572 | 235 | 325 |
| $H_2$ | $H_2$ | — | — | 11 |
| $H_2$ | He | — | — | 3.4 |
| He | He | — | — | 1.2 |

sphere model. With a short-range repulsion and a long-range attractive potential of

$$U = -\frac{C}{r^s} \tag{4.107}$$

the approximate cross section is given by [3, 33]

$$\sigma = 2\pi\left(\frac{s-3}{s-2}\right)\left(\frac{f4\pi C}{hV}\right)^{2/(s-1)} \tag{4.108}$$

where

$$f(2) = \frac{\pi}{2}, \qquad f(s) = \frac{\pi}{2}[(s-3)(s-5)\cdots 1][(s-2)(s-4)\cdots 2]^{-1} \tag{4.109a}$$

and

$$f(3) = 1, \qquad f(s) = [(s-3)(s-5)\cdots 2][(s-2)(s-4)\cdots 3]^{-1} \tag{4.109b}$$

The hard-sphere cross section is compared with this cross-section with $s = 6$ for the determination of the coefficient $C$. Then, $C$ is compared with the value $C'$ which is obtained from experimental data on atomic polarizability and which is given in detail by Eq. (4.104) (see Table 4.1).

### *4.7.1 Persistence of Velocities after a Collision*

After a collision with another molecule, the velocity of a given molecule will have an average component in the direction of its original motion. This phenomenon is known as persistence of velocities [*34*] and is shown in Fig. 4.20. Denoting the velocity of a particle labeled 1 as $\mathbf{v}_1$ before the

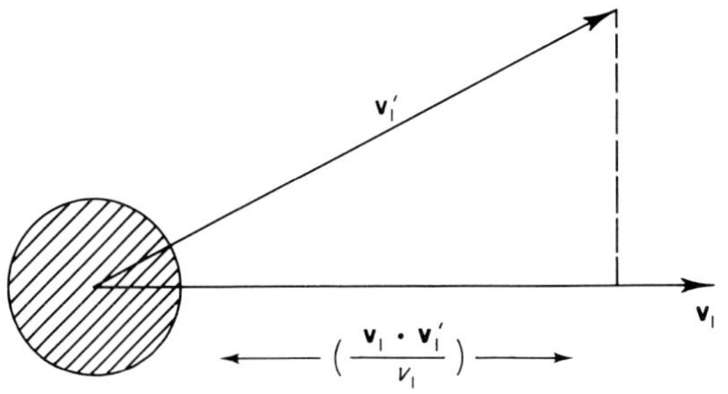


**Fig. 4.20** Persistence of velocities after a collision.

collision and $\mathbf{v}_1'$ after the collision, the component of $\mathbf{v}_1'$ along $\mathbf{v}_1$ is given by $(\mathbf{v}_1' \cdot \hat{v}_1)$. The average value of the scalar coefficient is

$$\langle \mathbf{v}_1' \cdot \hat{v}_1 \rangle = \frac{\int [(\mathbf{v}_1' \cdot \hat{v}_1)] n_2 P(v_2^2)\, d\mathbf{v}_2\, V\sigma(V, \Omega)\, d\Omega}{\int n_2 P(v_2^2)\, d\mathbf{v}_2\, V\sigma(V, \Omega)\, d\Omega} \tag{4.110}$$

and for the collision of hard spheres reduces to

$$\langle \mathbf{v}_1' \cdot \hat{v}_1 \rangle = \left[\left(\frac{m_1}{m_1 + m_2}\right) + \frac{1}{2}\left(\frac{m_2}{m_1 + m_2}\right) G(x)\right] v_1 \tag{4.111}$$

where $x = v_1/(2kT_2/m_2)^{1/2}$. Some numerical values of $G(x)$ are given in Table 4.2.

If the incident molecule is a heavy molecule, $m_1 \gg m_2$, the heavy molecule continues its path nearly undisturbed, while the light molecule moves off in a direction which is unrelated to its previous motion.

Table 4.2

| $x$ | $G(x)$ | $x$ | $G(x)$ |
|---|---|---|---|
| 0 | $-2/3$ | 2 | $-0.195$ |
| 0.5 | $-0.593$ | 3 | $-0.10$ |
| 1 | $-0.428$ | $\infty$ | 0 |

**Exercise 4.18** A beam of Rb atoms with a speed characteristic of the mean flow speed at 400°K, that is, $v_1 \approx (2kT/m)^{1/2}$, moves through He gas at a pressure of $10^{-5}$ Torr and a temperature of 300°K. What is the reduction in the velocity of the Rb atoms in the $\hat{v}_1$ direction per collision? Express as a fractional or per cent change per collision. Find the average change in linear momentum $\Delta p_1$ of the Rb atom in the $\mathbf{v}_1$ direction and $\Delta p/p_1$. Estimate the change $\Delta \mathbf{p}$ normal to $\mathbf{v}_1$. Should this change be comparable to the change along $\mathbf{v}_1$?

## 4.8 Interaction of Molecules with Solid Surfaces

When a molecule strikes a surface, it may suffer an elastic collision and rebound immediately or it may undergo an inelastic collision and exchange energy with the wall. In the latter case, the molecule may be permanently adsorbed by the wall or it may remain at the wall for only a short period of time and then return the gas. Elastic and inelastic collisions are now considered separately.

The simplest type of elastic collision is specular reflection and would occur in the classical approximation for the collision of hard spheres with a smooth, hard wall. By using the quantum approach and the de Broglie wavelength of the particles $\lambda = h/p = h/mv$, the condition for specular reflection becomes the same as in optics. The height $d$ of the surface irregularities when projected on the direction of the incident beam must be less than a wavelength. If $\beta$ is the glancing angle of the incident beam relative to the surface, as shown in Fig. 4.21, then the condition for specular reflection is

$$d \sin \beta < \lambda \tag{4.112}$$

A second condition is that the time spent by the molecule on the surface be sufficiently short that properties upon leaving are determined by the incident properties and not the last stages of surface history. The irregularities of mechanically polished surfaces are of the order of $10^{-5}$ cm, while the de Broglie wavelength for atomic hydrogen at room temperature is of the order of $10^{-8}$ cm. The heavier atoms have even shorter wavelengths. For hydrogen, the condition for reflection is $\beta < 10^{-3}$ rad, or the glancing angle must be less than a few minutes of arc. A 5% reflecting power has been observed for a hydrogen beam at an angle of $10^{-3}$ rad and at room temperature.

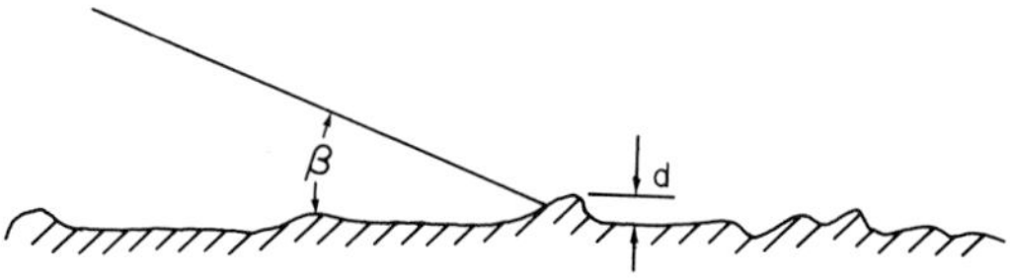


**Fig. 4.21** Reflection by a rough surface. Specular reflection occurs for $d \sin \beta < \lambda$, where $d$ is the average height of the irregularities.

Cleaved surfaces of a crystal are the smoothest obtainable surfaces, but even these surfaces are limited by the amplitude of atomic oscillations, which are of the order of $10^{-8}$ cm. Reflections for glancing angles as large as 20–30° have been observed [*35*] in the reflection of a He beam from a cleaved LiF crystal. Experiment also confirms the reduction of the amplitude of atomic oscillations with temperature reduction, and increased reflecting power is observed at the lower temperatures.

Since the surface appears to the incident beam of molecules as a two-dimensional lattice of scattering centers, diffraction phenomena by the surface are expected. The diffraction condition for a one-dimensional lattice is shown in detail in Fig. 4.22(a). For an incident beam making angles $\alpha_0$, $\beta_0$, $\gamma_0$ with the $x, y, z$ axes of the two-dimensional surface lattice as shown in Fig. 4.22(b), the diffracted beam will have maxima at angles

$$\cos \alpha - \cos \alpha_0 = \frac{n_1 \lambda}{a}, \qquad \cos \beta - \cos \beta_0 = \frac{n_2 \lambda}{a}$$

A typical result for a molecular diffraction experiment [*32, 35*] is shown in Fig. 4.22(c). The crystal is arranged so that $\beta_0$ is zero. The central peak is the specularly reflected (0, 0) beam and the side peaks are the (0, 1) beams. The breadth of the peaks is an indication of the velocity of the particles in the incident beam and have a Maxwellian distribution.

In Section 1.7, it was shown that for a gas in equilibrium with a wall, the cosine law for molecules leaving the surface followed immediately from a detailed balance argument. In Exercise 1.12, it was shown that

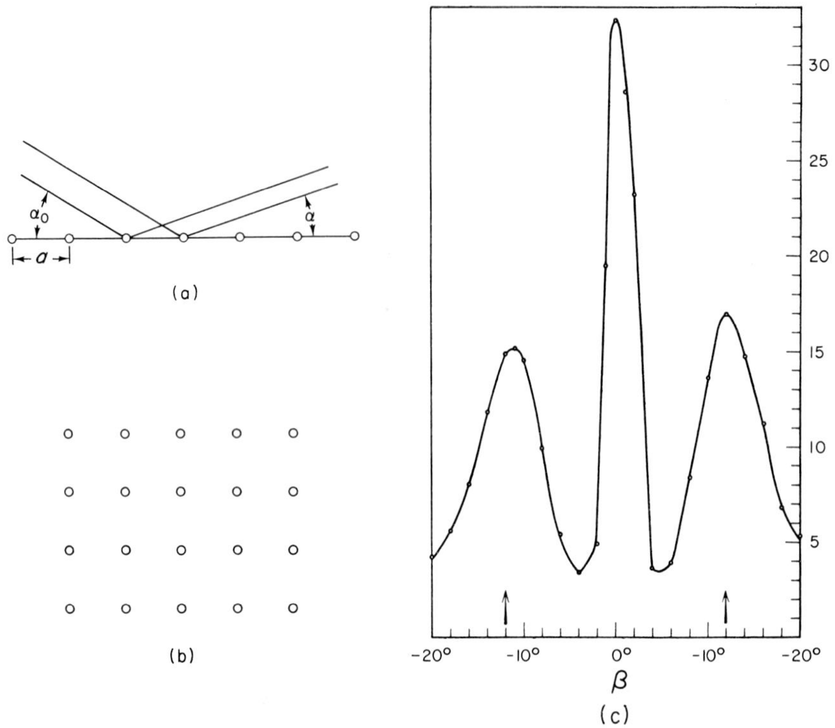


**Fig. 4.22** (a) Diffraction by a row of particles; $\cos \alpha - \cos \alpha_0 = n\lambda/a$. (b) Character of the scattering elements of a smooth, solid surface. (c) Diffraction of He at a glancing angle of 18.5° on a cleavage face of LiF. Incident beam intensity is 320 on the same scale. $\beta$ is the angle shown by the curve. [Part (c) from I. Estermann, *Z. Phys.* **61**, 95 (1930), Fig. 10.]

deviations from the cosine law violated the second law of thermodynamics. Knudsen has shown experimentally that the angular distribution of molecules absorbed for an appreciable length of time and then reevaporated from the surface also follow the cosine law and the cosine law is more general than indicated in Chapter I. Wood, in one of the earliest studies of the angular distribution of restituted molecules, allowed a Hg beam to strike a glass target at the center of a bulb and then measured the thickness of the deposit on the bulb. A similar experiment by Knudsen is considered in Exercise 4.19.

The exchange of energy between a gas and a wall is usually discussed in terms of Knudsen's thermal accommodation coefficient $\alpha'$,

$$\alpha' = \frac{E_i - E_r}{E_i - E_w} \qquad (4.113)$$

where $E_i$, $E_r$, and $E_w$ are, respectively, the average thermal energies of the gas molecules in the incident beam and in the beam returned from the wall, and of the gas molecules in thermal equilibrium with the wall. Massey and Burhop [9] have summarized experimental studies and theories regarding the accommodation coefficient for these free–free inelastic collisions. In the free–free transition, the time spent near the surface is short and absorption does not occur. Processes of this kind occur when gas molecules are exposed to hot walls.

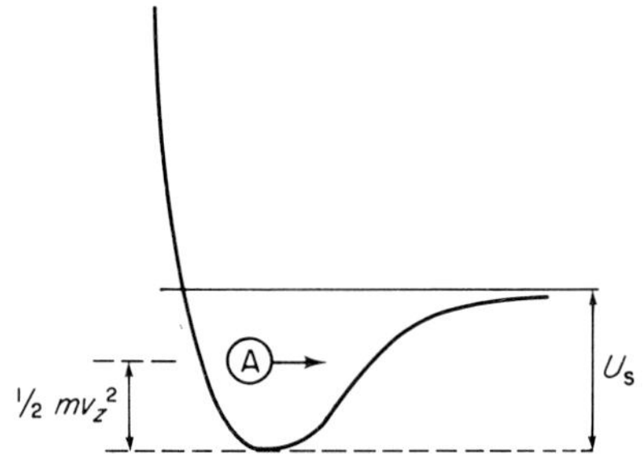


**Fig. 4.23** Potential energy curve for a gas particle near a surface. $U(r) \approx$ constant $\times\, r^{-4}$ + constant $\times\, r^{-12}$.

For processes in which the molecules condense or are absorbed by the wall and then some time later are returned to the beam by evaporation, the attractive force between the crystal surface and the incident molecule must be considered. For this purpose, a van der Waals attractive force is assumed and the crystal surface is assumed to have a potential of the form shown in Fig. 4.23. A very crude approximation to the probability of an atom leaving the potential well is given by the product of the probability of the atom A approaching the barrier with kinetic energy $\frac{1}{2}mv_z^2 > U_s$ and the number of approaches per second $\nu$. Here $\nu$ is of the order of the vibration frequency. With this model,

$$\left\{\begin{matrix}\text{probability of leaving} \\ \text{the surface per second}\end{matrix}\right\} \approx \nu \exp\left(-\frac{U_s}{kT}\right) \approx \left\{\begin{matrix}\text{probability of condensing} \\ \text{on the surface per second}\end{matrix}\right\} \tag{4.114a}$$

Equilibrium and detailed balance require the probability of an incident atom sticking to the surface be proportional to the probability of the atom leaving the surface.[5] The time spent on the surface is approximately given by the reciprocal of Eq. (4.114a):

$$\tau \approx \frac{1}{\nu} \exp\left(\frac{U_s}{kT}\right) \tag{4.114b}$$

[5] A recent discussion of this equation has been given by Allen and Fefier [36].

The probability of condensation is more correctly expressed as

$$\{\text{probability of condensation}\} \approx \frac{h\nu}{kT} \exp\left(-\frac{U_s}{kT}\right) \tag{4.114c}$$

$U_s$ and $\nu$ can be estimated from the van der Waals interaction. A summary of sticking coefficients is given by Wexler [*37*].

**Exercise 4.19** Knudsen measured the angular distribution of the restituted beams by measuring the angular distribution inside the sphere shown in Fig. 4.24. Show that the cosine law of restitution requires an even deposit inside the sphere. (Refer to Chapter I.)

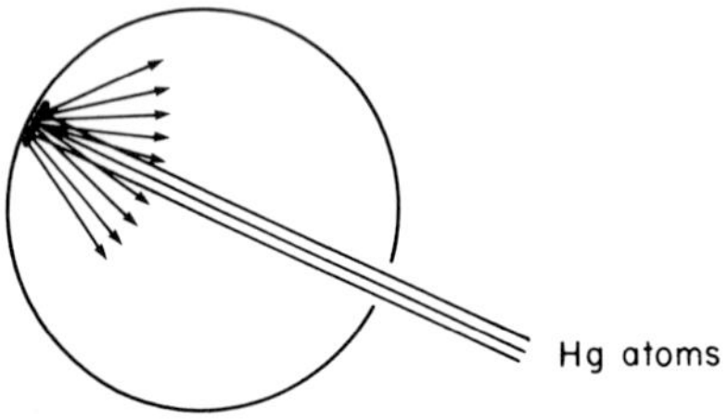


**Fig. 4.24** An experiment of Knudsen's to show that Hg atoms are returned from a surface according to the cosine law.

As an illustration of gas–surface phenomena, the surface interaction between a metal surface and a diatomic molecule is considered [*38*]. An elementary model of the energetics of the process is shown in Fig. 4.25. The curves have the following origins.

Curve A. The long range interaction between the molecule $X_2$ and the metal surface is via a van der Waals or dispersion interaction. The $1/r^6$ attraction is replaced by a $1/r^4$ attraction for the interaction between a molecule and a surface. Repulsive forces dominate at short range. Since the attractive interaction is weak, the minimum in the potential energy occurs at a distance which is large on the atomic scale. This process is usually referred to as physisorption.

Curve B. If the molecule is dissociated, then the atomic constituents can approach the wall. The attractive interaction is much stronger and is attributed to an exchange force in which the "binding orbital" is formed by the interaction of an atomic electron with the electrons of the metal. The binding is much stronger and the atom–metal distance is smaller than in *A*. This type of strong binding is referred to as chemisorption.

Finally, it can be noted that curves A and B cross at point *P*. If an incoming molecule has sufficient energy to cross the potential barrier, then chemisorption rather than physisorption can occur.

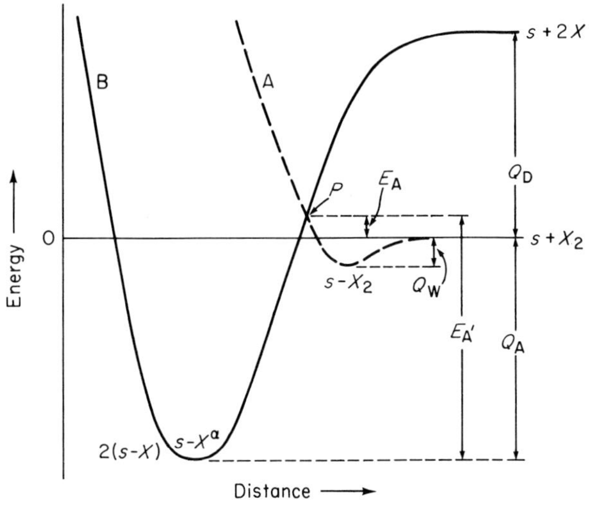


**Fig. 4.25** Potential energy curves near a surface, illustrating physisorption and chemisorption. [From H. Wise and B. J. Wood, *Advan. Atom. Mol. Phys.* **3**, 292 (1967), Fig. 1.]

The statistical thermodynamic aspects of surface adsorption are continued in Section 9.9.

An example of these processes occurs in an electrical discharge with molecular hydrogen. The discharge contains energetic electrons of the order of 10 eV which excite the unstable $^3\Sigma_u$ state,

$$H_2(^1\Sigma_g) + e \rightarrow H_2(^3\Sigma_u) + e \rightarrow H + H + e$$

with a cross section of approximately $10^{-20}$ m$^2$. This process is used for the preparation of atomic hydrogen. Certain materials enhance the recombination of atomic hydrogen to molecular hydrogen at the walls or surfaces of the enclosure. Denoting the wall recombination rate as $\gamma$, where $\gamma$ is defined as the probability that a hydrogen atom striking the surface forms molecular hydrogen and is returned to the gas as molecular hydrogen. The recombination coefficient $\gamma$ for hydrogen recombination on metals [*39*] is shown in Fig. 4.26.

Surfaces that inhibit recombination are of interest for the hydrogen maser and other experiments which require atomic rather than molecular species. Recombination coefficients on Pyrex are of the order of $4\times10^{-3}$ for hydrogen and $10^{-4}$ for oxygen and nitrogen. Various wall coatings that inhibit recombinations have been tried, and Teflon is used to inhibit hydrogen recombination at the walls in the hydrogen maser [*40*]. The hydrogen maser uses the 1420-MHz transition of hydrogen, and the energy levels for the $F = 1$ to $F = 0$ hyperfine transition are as follows:

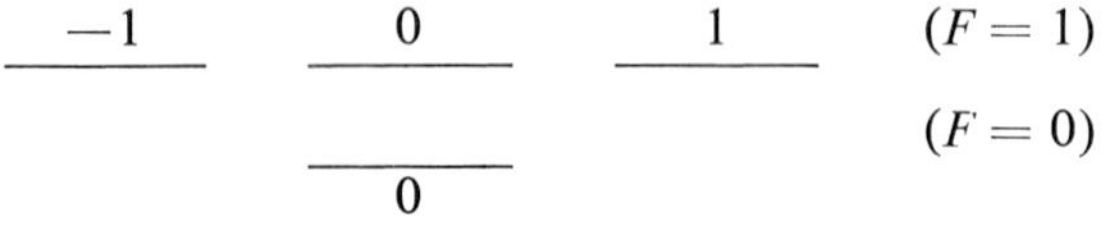


Since the hydrogen beam enters a quartz bulb and after the first wall collision is returned according to a cosine law, the atoms are confined in the enclosure. The beam has passed through an inhomogeneous magnetic field and only the (1, 1) and (1, 0) states enter the hole into the electro-

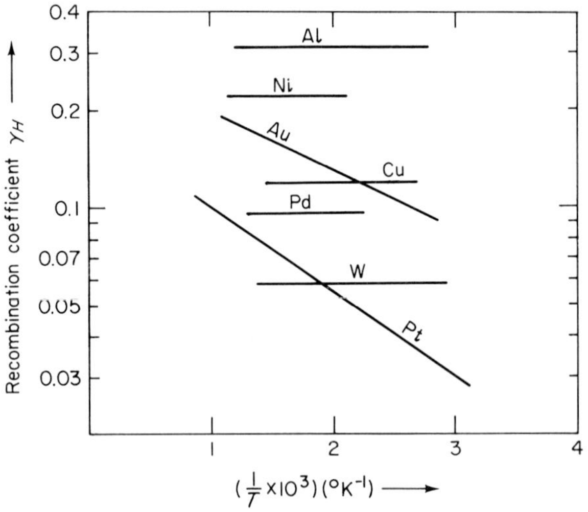


**Fig. 4.26** Recombination rate for the recombination of atomic hydrogen to molecular hydrogen on various surfaces. [From H. Wise and B. J. Wood, *Advan. Atom. Mol. Phys.* **3**, 324 (1967), Fig. 10.]

magnetic cavity. The other two states are deflected out of the beam. This system has only upper-state atoms and can produce gain. Only the (1,0) → (0, 0) transition is used in the maser. It is necessary to keep the (1, 0) state for a time interval of the order of 1–10 sec. Some $10^4$–$10^5$ wall collisions occur during this interval. Teflon coating not only inhibits the formation of molecular hydrogen by wall recombination, but also inhibits the "spin-flip" process in which a wall collision causes the transition (1,0) → (1, 1) or (1, 0) → (1, −1).

## References

1. F. London, *Z. Phys. Chem. B* **11**, 222 (1930); and W. Heitler and F. London, *Z. Phys.* **44**, 455 (1927).
1a. A. Messiah, "Quantum Mechanics," Vol. I and II. North-Holland Publ., Amsterdam, 1962.
2. *Handb. Math. Functions* (Nat. Bur. of Stand. Appl. Math. Ser.) **55**, 365 (1964).
3. H. S. W. Massey and C. B. O. Mohr, *Proc. Roy. Soc. Ser. A* **141**, 434 (1933).
4. J. O. Hirschfelder, C. R. Curtiss, and R. B. Bird, "The Molecular Theory of Gases and Liquids." Wiley, New York, 1954.
5. R. F. Post, *Rev. Mod. Phys.* **28**, 338 (1956).
6. R. H. Fowler and E. A. Guggenheim, "Statistical Thermodynamics." Cambridge Univ. Press, London and New York, 1949.
7. R. D. Present, "Kinetic Theory of Gases." McGraw-Hill, New York, 1958.
8. A. Javan, W. R. Bennett, Jr., and D. R. Herriott, *Phys. Rev. Lett.* **6**, 106 (1961).
9. H. S. W. Massey and E. H. S. Burhop, "Electronic and Ionic Impact Phenomena." Oxford Univ. Press (Clarendon), London and New York, 1952.
10. C. E. Moore, Atomic energy levels, Vol. 1. *Nat. Bur. Stand. (U.S.) Circ.* **No. 467,** 1949.
11. T. Holstein, *Phys. Rev.* **83**, 1159 (1951).
12. C. K. N. Patel, *Phys. Rev. Lett.* **13**, 617 (1964).
13. R. L. Taylor and S. Bitterman, *Rev. Mod. Phys.* **41**, 26 (1969).
14. C. B. Moore, R. E. Wood, B. Hu, and J. T. Yardley, *J. Chem. Phys.* **46**, 4222 (1967).
15. D. R. Bates, ed., "Atomic and Molecular Processes." Academic Press, New York, 1962.
16. S. Geltman, "Topics in Atomic Collision Theory." Academic Press, New York, 1969.
17. B. L. Moiseiwitsch and S. J. Smith, *Rev. Mod. Phys.* **40**, 238 (1969).
18. B. L. Moiseiwitsch, *in* "Atomic and Molecular Processes" (D. R. Bates, ed.). Academic Press, New York, 1962.
18a. R. T. Brackmann, W. L. Fite, and R. H. Neynaber, *Phys. Rev.* **112**, 1157 (1958).
18b. A. Temkin and J. C. Lamkin, *Phys. Rev.* **121**, 788 (1961).
19. M. J. Seaton, *in* "Atomic and Molecular Processes" (D. R. Bates, ed.). Academic Press, New York, 1962.
19a. G. J. Schulz and R. E. Fox, *Phys. Rev.* **106**, 1179 (1957).
19b. H. S. W. Massey and B. L. Moiseiwitsch, *Proc. Roy. Soc. Ser. A* **227**, 38 (1954).
20. M. R. H. Rudge, *Rev. Mod. Phys.* **40**, 564 (1968).
21. L. I. Schiff, "Quantum Mechanics," 3rd ed., p. 260. McGraw-Hill, New York, 1968.
22. W. Lichten, *Phys. Rev.* **120**, 848 (1960).
23. G. Herzberg, "Spectra of Diatomic Molecules." Van Nostrand-Reinhold, Princeton, New Jersey, 1950.
24. A. S. Coolidge and H. M. James, *J. Chem. Phys.* **6**, 730 (1938).
25. W. Kolos and C. C. J. Roothaan, *Rev. Mod. Phys.* **32**, 205 (1960).
26. F. R. Gilmore. Potential energy curves for $N_2$, NO, $O_2$ and corresponding ions. *J. Quant. Spectrosc. Radiat. Transfer* **5**, 369 (1965).
27. A. V. Phelps, *Rev. Mod. Phys.* **40**, 399 (1968).

28. M. Born, *Phys. Z.* **21**, 578 (1920); F. Bielz, *Z. Phys.* **32**, 81 (1925).
29. R. G. J. Fraser and L. F. Broadway, *Proc. Roy. Soc. Ser. A* **141**, 626 (1933).
30. S. Rosin and I. I. Rabi, *Phys. Rev.* **48**, 373 (1935).
31. I. Estermann, S. Stoner, and O. Stern, *Phys. Rev.* **71**, 250 (1947).
32. N. R. Ramsey, "Molecular Beams," p. 28. Oxford Univ. Press, London and New York, 1956.
33. H. W. S. Massey and C. B. O. Mohr, *Proc. Roy. Soc. Ser. A* **141**, 188 (1934).
34. S. Chapman and T. G. Cowling, "The Mathematical Theory of Non-Uniform Gases," p. 98. Cambridge Univ. Press, London and New York, 1958.
35. I. Estermann and O. Stern, *Z. Phys.* **61**, 95 (1930).
36. R. T. Allen and P. Fefier, *J. Chem. Phys.* **40**, 2810 (1964).
37. S. Wexler, *Rev. Mod. Phys.* **30**, 402 (1958); see also B. B. Dayton, *Trans. Nat. Vac. Symp.* **6**, 101 (1959).
38. H. J. Wise and B. J. Wood, *Advan. At. Mol. Phys.* **3**, 291–353 (1967).
39. B. J. Wood and H. Wise, *J. Phys. Chem.* **65**, 1976 (1961).
40. D. Kleppner, H. M. Goldenberg, and N. F. Ramsey, *Phys. Rev.* **126**, 603 (1962).

# Transport Phenomena V

## 5.1 Introduction

The elementary description of transport phenomena given in Chapter I in terms of a mass point gas requires modifications when the effect of intermolecular interactions is included. Gas properties and the velocity distribution function no longer depend only on the properties of the walls, and due to interchange of energy and momentum during collisions, an anisotropy in the velocity distribution function is expected. It is the purpose of the more general theory to determine the form of this distribution function, and it is the purpose of the elementary theory to avoid its determination. Both aspects are developed in what follows.

## 5.2 Elementary Theory of Transport

In developing the properties of transport in the free molecular flow region, it was assumed that a colliding atom was returned to the gas with a property characteristic of the wall. In the elementary theory of transport, this is modified, and the molecule is assumed to take up the properties of the gas in the vicinity of its last collision. Assume two parallel plates as shown in Fig. 5.1. At some intermediate position, the average velocity of the gas is $W_y(z)$ and the velocity at an adjacent position is $W_y(z)+\Delta z\,\partial W_y/\partial z$. Higher-order terms in the Taylor series expansion are neglected and the

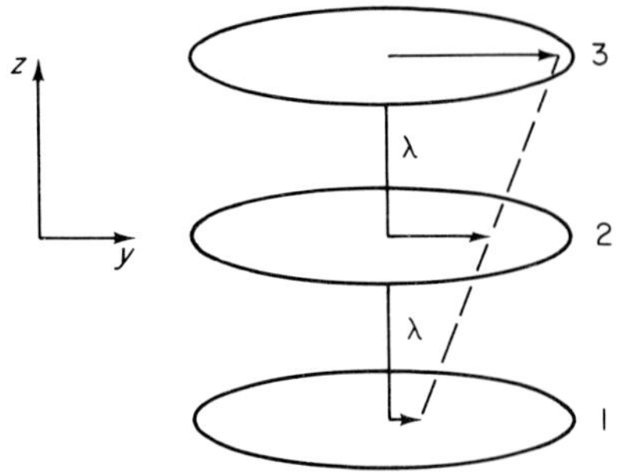


**Fig. 5.1** Transport of momentum across surface elements by the motion of particles in a gas. $W_y(z)$ is the $y$ component of the average particle velocity. (1) $W_y(z_0 - \lambda)$, (2) $W_y(z_0)$, (3) $W_y(z_0 + \lambda) \approx W(z_0) + \lambda\, \partial W_y/\partial z$.

velocity is assumed to increase linearly between surface $A$ and surface $B$. Consider some surface at position $z_0$ as shown in Fig. 5.1. *The property transported across the surface $z_0$ is characteristic of the last collision* and this occurs at distance $\lambda$, the average free path, from $z_0$. For large, parallel surfaces the average number of molecules colliding at distance $\lambda$ and then crossing $z_0$ is $\frac{1}{4}n\bar{v}$ and the momentum $mW_y$ transported across a surface with a normal in the $z$ direction is

$$P_{yz} = -\tfrac{1}{4}n\bar{v}[mW_y(z_0 + \lambda) - mW_y(z_0 - \lambda)]$$

Using the Taylor series expansion, this becomes

$$P_{yz} = -\frac{1}{2}\, nm\bar{v}\lambda \frac{\partial W_y}{\partial z} = -\eta \frac{\partial W_y}{\partial z} \quad \frac{\text{N}}{\text{m}^2} \tag{5.1}$$

for the stress across the surface at $z_0$. The coefficient of viscosity is defined as

$$\eta = \frac{1}{2}\, nm\bar{v}\lambda \quad \frac{\text{N sec}}{\text{m}^2} \tag{5.2}$$

If $\varepsilon$ is the energy transported by an atom, the heat transport can be developed in a similar manner and is given by

$$q_z = -\tfrac{1}{4}n\bar{v}[\varepsilon(z_0 + \lambda) - \varepsilon(z_0 - \lambda)]$$

$$q_z = -\left(\frac{1}{2}\, n\bar{v}\lambda \frac{\partial \varepsilon}{\partial T}\right) \frac{\partial T}{\partial z} = -K \frac{\partial T}{\partial z} \quad \frac{\text{W}}{\text{m}^2} \tag{5.3}$$

The coefficient of thermal conductivity is

$$K - \frac{1}{2}\, n\bar{v}\lambda \frac{\partial \varepsilon}{\partial T} \quad \frac{\text{W}}{\text{m }^\circ\text{K}} \tag{5.4}$$

For a monatomic gas, $\varepsilon = \frac{3}{2}kT$.

Some attempt can be made to use more appropriate averaging techniques for the elementary theory of transport, but little is gained over the very elementary development just given. For monatomic gases, the coefficient of viscosity is in reasonable accord with the theory. Heat transport is usually in very poor agreement. Table 5.1 contains some experimental data

**Table 5.1**

Thermal Conductivity and Viscosity of Gases

| Gas | $\eta(10°C)$ (N-sec/m²) ($\times 10^{-6}$) | $K$ (W/m-°K) ($\times 10^{-3}$) | $C_V$ (J/kg-mole-deg) ($\times 10^{3}$) | $KM/\eta C_V$ | $\sigma_\eta$ [$= (5\pi/32\sqrt{2})m\bar{v}/\eta$] ($\times 10^{-20}$) |
|---|---|---|---|---|---|
| He | 18.8 | 144 | 12.6 | 2.44 | 14.7 |
| Ne | 29.9 | 46.3 | 12.7 | 2.47 | 20.7 |
| Ar | 21.0 | 15.9 | 12.7 | 2.42 | 41.5 |
| $H_2$ | 8.4 | 174 | 20.1 | 2.06 | 2.32 |
| $N_2$ | 16.7 | 23.7 | 20.9 | 1.91 | 43.7 |
| $O_2$ | 19.2 | 24.0 | 20.9 | 1.92 | 40.7 |
| $CO_2$ | 13.8 | 142 | 27.8 | — | 66.0 |
| $CH_4$ | 10.3 | 30.1 | 26.8 | — | 53.5 |
| $NH_3$ | 9.2 | 21.5 | 28.6 | — | 61.8 |

for some of the more common gases. The ratio of $K/\eta$ is independent of $n\bar{v}\lambda$ and is of particular interest:

$$\frac{K}{\eta} = \frac{1}{m}\frac{\partial \varepsilon}{\partial T} = \frac{1}{M} C_V \tag{5.5}$$

$KM/\eta C_V = 1$ with the above model. The experimental value of the ratio ranges from 2.5 for monatomic gases to values of 1.9 for more complex molecules. The more precise development given later accounts for this factor of 2.5.

### 5.2.1 *Poiseuille Flow*

The free molecular flow through a cylindrical tube considered in Chapter I must be modified for intermolecular collisions. The equation for stress just given can be used for this development. Consider a long tube as shown in Fig. 5.2. The macroscopic flow velocity in the $z$ direction has a radial

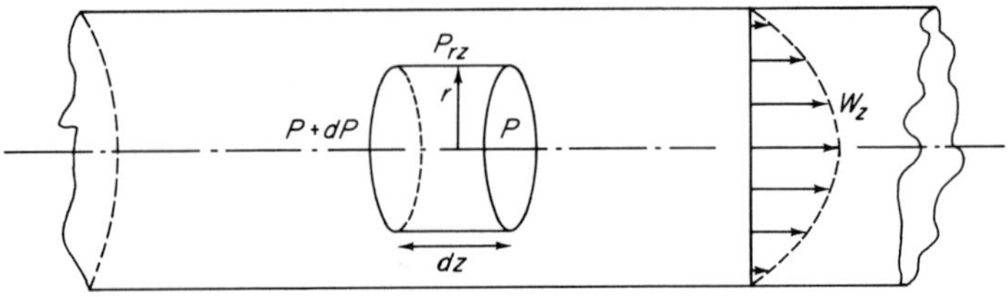


**Fig. 5.2** Velocity profile $W_z(r)$ and the stresses for Poiseuille flow in a tube.

gradient and from Eq. (5.1), the stress and the velocity gradient are related by

$$P_{rz} = -\eta \frac{\partial W_z}{\partial r} \tag{5.6}$$

The net force acting on a cylinder of length $dz$ and radius $r$ is

$$2\pi r\, dz\, P_{rz} = \pi r^2\, dP$$

and eliminating the stress between these two equations yields a relationship between the velocity gradient and the pressure gradient,

$$\frac{dW_z}{dr} = -\frac{r}{2\eta}\frac{dP}{dz} \tag{5.7a}$$

The boundary condition at the wall requires that the velocity at the wall be zero, $W_z(a) = 0$, and integration of Eq. (5.6) from $r = a$ to $r$ yields

$$W_z(r) = \frac{a^2 - r^2}{4\eta}\frac{dP}{dz} \tag{5.7b}$$

Equation (5.7b) is the well-known parabolic velocity profile across the tube diameter as shown in Fig. 5.2. The flow of molecules down the tube is

$$\begin{aligned} \dot{N}_z &= \{\text{flow in molecules per second}\} \\ &= \int_0^a n(z) W_z(r, z) 2\pi r\, dr = \frac{\pi a^4 n}{8\eta}\frac{dP}{dz} \end{aligned} \tag{5.8}$$

in terms of the pressure gradient. For steady flow at constant temperature, integration over $\dot{N}_z\, dz$ and $dP$ gives the convential expression for Poiseuille flow,

$$\{\text{flow in molecules per second}\} = \frac{\pi a^4}{16\eta kT}\frac{P_2^2 - P_1^2}{L} \tag{5.9}$$

Equation (5.8) can be compared with the earlier expression in Chapter I

[Eq. (1.59)] for free molecular flow through a long thin tube, and it can be noted that if the average path in the expression for the viscosity is limited to the size of the average step in the $z$ direction, the expressions are almost the same.

Exercise 5.1 Let $\lambda = a$ and compare the expressions for free molecular flow and Poiseuille flow through a long tube at $T =$ constant and $\eta = \frac{1}{2}nm\bar{v}\lambda$.

Exercise 5.2 In Fig. 1.14, the velocity of plate $B$ is $W_0$ and the separation of the plates is $D$. Plot the drag force per square meter as a function of the pressure, $\log P$, over a broad range of $P$.

Exercise 5.3 Two parallel plates are at temperatures $T_B$ and $T_A$ and are separated by a distance $D$. Plot the heat transport $Q/A$ as a function of pressure, $\log P$, over a broad range of $P$.

Exercise 5.4 Find numerical values for $N_2$ gas in Exercises 5.2 and 5.3. Let $W_0 = 1$ m/sec, $T_B - T_A = 10°$K, $D = 1$ cm, and $T = 300°$K.

Ans. $1.8 \times 10^{-3}$ N/m$^2$; 8–10 W/m$^2$

### 5.2.2 *Viscous Drag Force on a Sphere*

Consider a sphere placed in a medium that is moving with velocity $\mathbf{W}_0$ at large distances from the sphere, or the equivalent problem of a sphere moving through a medium with relative velocity $W_0$ (Fig. 5.3). The problem of the viscous drag force on the sphere is of considerable interest. Equation (5.1) relates the stress to the velocity gradient. This problem has axial symmetry about the $\hat{z}$ axis and a solution for the velocity which has the correct values at the surface of the sphere and at large distances is given by

$$\mathbf{W}(\mathbf{r}) = W_0\left[\hat{z}\left(1 - \frac{a}{r}\right) + \frac{1}{4}\left(\frac{a^3}{r^3} - \frac{a}{r}\right)(2\hat{r}\cos\theta + \hat{\theta}\sin\theta)\right] \qquad (5.10)$$

At large $r$, this tends to $W_0\hat{z}$, and at $r = a$, is zero. Since div $\mathbf{W} = 0$, this equation is only correct for incompressible flow. Even so, for small objects and $W_0$ not too large, the change in density $\varrho$ can be neglected. The equation is independent of time and therefore implies a steady flow and a non-turbulent flow. This velocity function also satisfies Eq. (5.57), but this

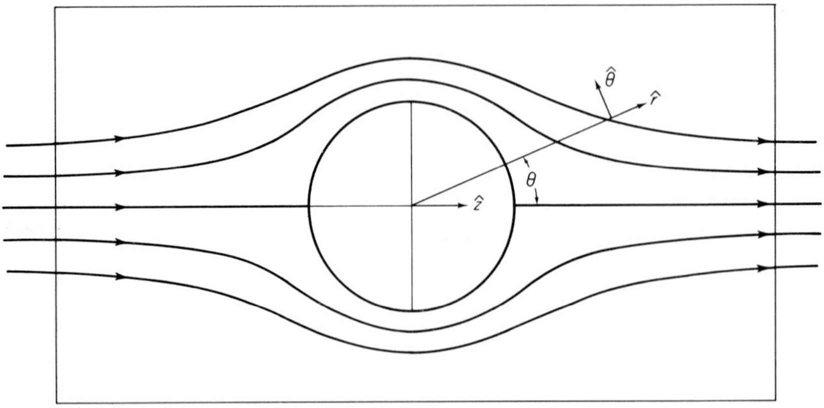


**Fig. 5.3** Velocity profile for flow around a sphere and the viscous drag force on a sphere.

detail is not pursued. The drag force in the $\hat{z}$ direction on the sphere is

$$F_z = \int [P_{zx}(\hat{x} \cdot \hat{r}) + P_{zy}(\hat{y} \cdot \hat{r}) + P_{zz}(\hat{z} \cdot \hat{r})]2\pi a^2 \sin\theta \, d\theta$$

where $\hat{r}$ is the surface normal of the sphere with an element of area $2\pi a^2 \times \sin\theta \, d\theta$. $P_{zx}, \ldots$ are determined from Eq. (5.1) and **W**. Completing the integration yields *Stoke's Law*,

$$F_z = 6\pi a \eta W_0 \tag{5.11}$$

for the viscous drag force on a sphere. If the sphere is moving with velocity $\hat{z}W_0$, then the drag force is also given by this same equation.

### 5.2.3 *Diffusion*

The elementary discussion of diffusion is based on the analysis of the problem shown in Fig. 5.4. A background gas of type B fills the tube. A particle of type A is inserted in the tube at $P$ and the probability of reaching

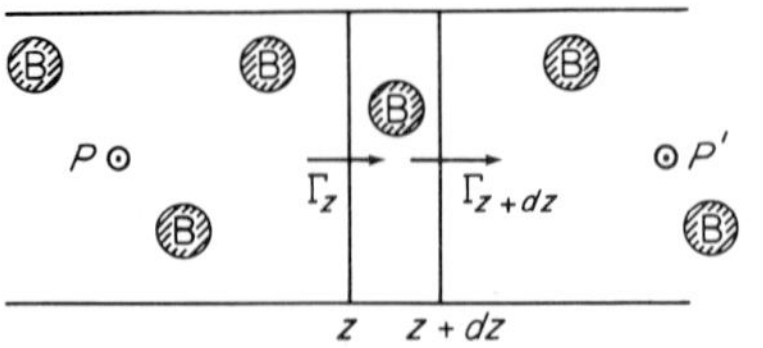


**Fig. 5.4** Flow by diffusion of particles of type A through a gas of type B molecules.

$P'$ some time later is of interest. It is assumed that there is a cross section for A–B collisions $\sigma_{AB}$ and an average path between collisions of $\lambda_{AB} = 1/n_B\sigma_{AB}$. Particle A takes an average step between collisions of $\lambda_{AB}$ and at each collision, the direction of the next step is not correlated with the previous step. All directions are equally probable after each collision for the velocity vector. Thus the problem is characteristic of the random walk treated in Chapter III. It was shown in Eq. (3.104a) that the diffusion equation follows from the probability equation and the flow is given by

$$\Gamma_z = -D \frac{\partial n_A}{\partial z} \tag{5.12a}$$

or

$$\boldsymbol{\Gamma} = -D \operatorname{grad} n_A \tag{5.12b}$$

where $n_A$ is the number of A particles per cubic meter. If only one A particle is present, then $n_A(r, t)$ becomes the probability $p_A(r, t)$. Since the system is dilute and the A atoms are independent, an increase in the number of A atoms to $N$ yields $n_A(r, t) = Np_A(r, t)$. This single-particle concept is emphasized, since it is the property of primary interest in most experiments. As the concentration of the A species increases, it is necessary to require that the pressure be uniform throughout the volume in order to minimize the viscous flow discussed in the previous section. This requires that $P = (n_A + n_B)kT =$ constant and diffusive flow of A requires a counterflow of B. In an elementary consideration of this type, the coefficient of diffusion is obtained in the same manner as that used for Eq. (3.103b),

$$D_{AB} = \frac{1}{2}\lambda_{AB}\bar{v}_A = \frac{1}{2n_B\sigma_{AB}}\bar{v}_A \tag{5.12c}$$

This is only approximate and greater precision requires the use of Eq. (5.72) or Eq. (5.73).

At a wall, species A may either be absorbed by the wall, in which case the boundary condition at the surface is

$$n_A(\mathbf{r}_s, t) = 0 \qquad \text{(absorbing surface)} \tag{5.12d}$$

or be returned to the gas, in which case the boundary condition is

$$\operatorname{grad} n_A(\mathbf{r}_s, t) - 0 \qquad \text{(restituting surface)} \tag{5.12e}$$

where $\mathbf{r}_s$ is a point on the surface.

## 5.3 Gas Dynamics

The dynamic theory of gases requires a consideration of the general theory of the transport of molecules, momentum, and energy. A transport model, which makes use of some of the ideas developed in Chapter IV and the following assumptions, is considered.

(a) Encounters occur between molecules, but such encounters are sufficiently rare that it is possible to define asymptotic values for the velocities of the particles both before and after the encounters. Each particle has a position in physical space $\mathbf{r}_j$ and an asymptotic velocity $\mathbf{v}_j$ in velocity space. The number of particles per cubic meter in an element of volume $d\mathbf{r}$ with velocity in the interval between $\mathbf{v}$ and $\mathbf{v} + d\mathbf{v}$ is defined by a distribution function

$$f(\mathbf{r}, \mathbf{v}, t)\, d\mathbf{r}\, d\mathbf{v} \tag{5.13a}$$

Between binary encounters, the particles obey the usual dynamic laws and the coordinate $\mathbf{r}$ changes to $\mathbf{r} + \mathbf{v}\,\Delta t$ and $\mathbf{v}$ changes to $\mathbf{v} + \mathbf{a}\,\Delta t$ during the time interval $\Delta t$. Here $\mathbf{a}$ is the acceleration due to external forces. If an element $\Delta x'\,\Delta v_{x'}$ is considered as shown in Fig. 5.5, the molecules that lie in element $\Delta x'\,\Delta v_{x'}$ at time $t$ will lie in the adjacent element at time $t + \Delta t$. Molecules 1, 2, 3, and 4 at the corners of $\Delta x'\,\Delta v_{x'}$ will move to position 1, 2, 3, and 4 of $\Delta x\,\Delta v_x$ during interval $\Delta t$. The length of 1–2

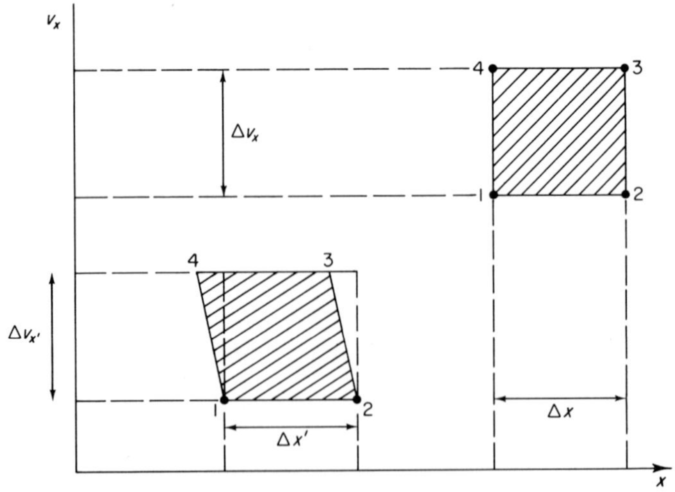


**Fig. 5.5** Flow of molecules from $\Delta x'\,\Delta v_{x'}$ to $\Delta x\,\Delta v_x$ by dynamic motion and collisions.

does not change in the $x$ direction. Similar arguments are true for the other lengths and the element $\Delta x' \, \Delta v_{x'}$ is the same size as $\Delta x \, \Delta v_x$. If the element $\Delta x \, \Delta v_x$ is now considered, in the course of time, molecules will leave this element due to dynamic motions and also by collisions. In addition, molecules will enter this same element of volume and the net gain of molecules in $\Delta x \, \Delta v_x$ in the interval $\Delta t$ is

$$\begin{aligned}\{\text{net gain in } \ \Delta x \, \Delta v_x \ \text{ in } \ \Delta t\} = {} & \{\text{number entering from } \ \Delta x' \, \Delta v_{x'}\} \\ & - \{\text{number leaving } \ \Delta x \, \Delta v_x\} \\ & + \{\text{net gain by collision}\}\end{aligned}$$

or

$$\begin{aligned}&[f(x, v_x, t + \Delta t) - f(x, v_x, t)] \, \Delta x \, \Delta v_x \\ &\quad = f(x', v_{x'}, t) \, \Delta x \, \Delta v_x - f(x, v_x, t) \, \Delta x \, \Delta v_x + \left(\frac{\partial f}{\partial t}\right)_{\text{coll}} \Delta x \, \Delta v_x \, \Delta t\end{aligned}$$

The molecules in $\Delta x' \, \Delta v_{x'}$ will enter $\Delta x \, \Delta v_x$ during the interval $\Delta t$ if, at time $t$, $x' = x - v_x \, \Delta t$ and $v_{x'} = v_x - a_x \, \Delta t$. Expanding each term in a Taylor series expansion, omitting higher-order terms, and canceling the common term $\Delta x \, \Delta v_x \, \Delta t$ permits this expression to be rewritten as a partial differential equation,

$$\frac{\partial f}{\partial t} + v_x \frac{\partial f}{\partial x} + a_x \frac{\partial f}{\partial v_x} = \left(\frac{\partial f}{\partial t}\right)_{\text{coll}} \tag{5.13b}$$

A similar argument can be made in three dimensions and the result is the sum of Eq. (5.13b) over $x, y, z$. The element of volume $\Delta \mathbf{r} \, \Delta \mathbf{v}$ is often referred to as $\eta$-space.

The second assumption now follows.

(b) The change in the distribution function with time is described by the partial differential equation

$$\frac{\partial f}{\partial t} + \sum_{x,y,z} v_x \frac{\partial f}{\partial x} + \sum_{x,y,z} a_x \frac{\partial f}{\partial v_x} = \left(\frac{\partial f}{\partial t}\right)_{\text{coll}} \tag{5.13c}$$

### 5.3.1 *Transport Properties*

The distribution function $f(\mathbf{r}, \mathbf{v}, t)$ contains the asymptotic velocity $\mathbf{v}$. Here $\mathbf{v}$ is not a function of $\mathbf{r}$ and $t$, but does depend on the inertial frame of reference. An inertial frame is introduced by considering the particles in an element of volume $\Delta\mathbf{r}$. These molecules have an average velocity $\mathbf{W}$, defined as

$$\mathbf{W}(\mathbf{r}, t) = \left(\frac{\sum n_j \mathbf{v}_j}{\sum n_j}\right)_{\text{av}} \tag{5.14a}$$

where the average indicates a sum over the particles in $\Delta\mathbf{r}$. An intrinsic velocity $\mathbf{C}$ is now introduced,

$$\mathbf{C} = \mathbf{v} - \mathbf{W}(\mathbf{r}, t) \tag{5.14b}$$

Since the element $d\mathbf{v}$ does not depend on the choice of reference for the origin of velocities $\mathbf{W}$, it follows that

$$d\mathbf{v} = d\mathbf{C}$$

In subsequent discussions, the fact that $\partial\mathbf{v}/\partial x = 0$, $\partial\mathbf{v}/\partial t = 0, \ldots$ is often used. This is not true for the intrinsic velocity $\mathbf{C}$ defined above.

In the discussion of the transport of some physical property across a surface in the gas, the observation is made as one moves along with the local velocity $\mathbf{W}$ of the gas. Thus the flow of molecules, momentum, and energy across a surface with the surface normal in the $\hat{x}$ direction and the surface moving with local velocity $\mathbf{W}(\mathbf{r}, t)$ is

$$\Gamma_x = \int d\mathbf{C}\ f C_x \qquad \frac{\text{molecules}}{\text{m}^2\ \text{sec}} \tag{5.15}$$

$$P_{xy} = \int d\mathbf{C}\ f C_x(mC_y) \qquad \frac{\text{N}}{\text{m}^2} \tag{5.16}$$

$$q_x = \int d\mathbf{C}\ f C_x\left(\frac{1}{2}\,mC^2\right) \qquad \frac{\text{W}}{\text{m}^2} \tag{5.17}$$

where

$$n = \int d\mathbf{C}\ f \tag{5.18}$$

$$\mathbf{W} = n^{-1}\int d\mathbf{C}\ f\mathbf{v} \tag{5.19}$$

$$0 = \int d\mathbf{C}\ f\mathbf{C} \tag{5.20}$$

These quantities serve as formal definitions of the physical quantities that are transported and of the physical quantities that are usually observed in experiments. Such quantities were correlated with experiments in an elementary manner in the previous section.

The change of the distribution function with time is given by Eq. (5.13c) and the change in some physical property $\phi$ of the gas in volume element $\Delta\mathbf{r}$ is the average of the equation describing $f$ or the velocities,

$$\int d\mathbf{v}_1\, \phi(\mathbf{v}_1)\left(\frac{\partial f}{\partial t} + \sum v_x \frac{\partial f}{\partial x} + \sum a_x \frac{\partial f}{\partial v_x}\right) = \int d\mathbf{v}_1\, \phi(\mathbf{v}_1)\left(\frac{\partial f}{\partial t}\right)_{\text{coll}} \tag{5.21}$$

This is the Boltzmann transport equation.

### 5.3.2 Binary Collisions

In Chapter IV, the probability of a binary encounter between two molecules was discussed and the results summarized in Eqs. (4.65) and (4.70). In the notation of this chapter, the probability of a binary encounter in volume element $\Delta\mathbf{r}$ is proportional to

$$[f(\mathbf{v}_1)\, d\mathbf{v}_1][f(\mathbf{v}_2)\, d\mathbf{v}_2][V\sigma(V, \Omega)\, d\Omega] \tag{5.22}$$

It is assumed that such a binary encounter removes the atom from element $\Delta\mathbf{r}_1\, \Delta\mathbf{v}_1$. Collisions also scatter molecules into this same element and are often referred to as "inverse collisions." A consideration of the dynamic aspects of a collision suggests that the inverse collision can be found by changing the velocities $\mathbf{v}_j$ to $-\mathbf{v}_j$ and so on in Fig. 4.1 and changing "before" to "after." This is the same as changing $t$ to $-t$. The dynamic aspects are invariant to time reversal. In classical physics, the geometric aspects are not. An inverse encounter between a ball and a cube is not invariant to time reversal. A proper quantum mechanical calculation for small molecules indicates that the cross sections for inverse encounters is the same as for the encounter, that is

$$\sigma(\mathbf{V}, \mathbf{V}') = \sigma(\mathbf{V}', \mathbf{V}) = \sigma(V, \Omega) \tag{5.23}$$

Huang [*1*] has discussed this relationship in terms of the invariance to time reversal and the invariance under rotation and reflection for quantum mechanical encounters. From Eq. (4.69) and subsequent equations, one can show for elastic collisions that

$$d\mathbf{v}_1\, d\mathbf{v}_2 = d\mathbf{G}\, d\mathbf{V} = d\mathbf{G}\, d\mathbf{V}' = d\mathbf{v}_1'\, d\mathbf{v}_2' \tag{5.24}$$

Thus the net number of particles entering $\Delta\mathbf{r}_1\,\Delta\mathbf{v}_1$ is proportional to the number scattered in from $\Delta\mathbf{v}_1'\,\Delta\mathbf{v}_2'$ minus the number scattered out of $\Delta\mathbf{v}_1\,\Delta\mathbf{v}_2$, and

$$\{\text{net gain by collisions in } \Delta\mathbf{r}_1\,\Delta\mathbf{v}_1 \text{ in } \Delta t\}$$
$$= \Delta\mathbf{r}_1\,\Delta t \int [f(\mathbf{v}_1')f(\mathbf{v}_2')\,d\mathbf{v}_2\,d\mathbf{v}_1 - f(\mathbf{v}_1)f(\mathbf{v}_2)\,d\mathbf{v}_1\,d\mathbf{v}_2][V\sigma\,d\Omega] \quad (5.25)$$

As in Chapter IV, $\mathbf{v}_2' - \mathbf{v}_1' = \mathbf{V}'$ and $\mathbf{v}_2 - \mathbf{v}_1 = \mathbf{V}$. If three-body collisions are neglected, then

$$\left(\frac{\partial f}{\partial t}\right)_{\text{coll}} d\mathbf{r}\,d\mathbf{v}_1 = d\mathbf{r}\,d\mathbf{v}_1 \int d\mathbf{v}_2\,[f(v_1')f(v_2') - f(v_1)f(v_2)][V\sigma(V,\Omega)\,d\Omega] \quad (5.26)$$

is the change in $d\mathbf{r}\,d\mathbf{v}_1$ due to binary collisions. Equation (5.21) can now be written in various forms, such as

$$\int d\mathbf{v}\phi(\mathbf{v})\left(\frac{\partial f}{\partial t} + \sum v_x \frac{\partial f}{\partial x} + \sum a_x \frac{\partial f}{\partial v_x}\right)$$
$$= \int d\mathbf{v}\,\phi(\mathbf{v})\left(\frac{\partial f}{\partial t}\right)_{\text{coll}}$$
$$= \int d\mathbf{v}_1\,d\mathbf{v}_2\,\phi(v_1)[f(v_1')f(v_2') - f(v_1)f(v_2)][V\sigma(V,\Omega)\,d\Omega]$$
$$= \frac{1}{4}\int d\mathbf{v}_1\,d\mathbf{v}_2\,[f(v_1')f(v_2') - f(v_1)f(v_2)]$$
$$\times [\phi(v_1) + \phi(v_2) - \phi(v_1') - \phi(v_2')][V\sigma(V,\Omega)\,d\Omega] \quad (5.27)$$

The $f$'s used imply $f(\mathbf{r}, \mathbf{v}_j, t)$, and $\mathbf{r}$ and $t$ are omitted to make the notation less cumbersome. The right-hand side is a definite integral and is invariant to the exchange of the indices 1 and 2 and the interchange of the primes. Proper manipulation yields the final form. This last expression is particularly convenient for a discussion of the collisional invariants $\phi = 1$, $\phi = mv_x, \ldots$, and $\phi = \frac{1}{2}mv^2$. Direct substitution yields a zero contribution for the right-hand side. Here $\phi = 1$ expresses the conservation of the number of particles in a collision, $\phi = mv_x$ expresses the conservation of linear momentum, and $\phi = \frac{1}{2}mv^2$ expresses the conservation of energy during a binary encounter.

### *5.3.3 Equilibrium Properties and the Maxwell–Boltzmann Distribution Law*

The equilibrium solution of Boltzmann's equation (5.27) is a solution for which $f$ is independent of $\mathbf{r}$ and $t$. In this section, the external force, and therefore the acceleration $\mathbf{a}$, is assumed to be zero. Equilibrium requires $\partial f/\partial t = 0$ and $\partial f/\partial x = 0$, or the left-hand side of Eq. (5.27) must be zero. The right-hand side is already zero for the collisional invariants $\phi = 1$, $m\mathbf{v}$, or $\frac{1}{2}mv^2$ and therefore the left-hand side is zero for any $\phi$ of the form

$$\begin{aligned}\phi_\alpha &= \alpha_0 + \boldsymbol{\alpha}_1 \cdot (m\mathbf{v}) + \alpha_2(\tfrac{1}{2}mv^2) \\ &= \alpha_0 - \tfrac{1}{2}\beta m(\mathbf{v} - \mathbf{W}_0)^2 = \alpha_0 - \tfrac{1}{2}\beta mC^2 \end{aligned} \tag{5.28}$$

$\alpha_0$, $\boldsymbol{\alpha}_1$, and $\alpha_2$, or $\alpha_0$, $\mathbf{W}_0$, and $\beta$, are five constants and $\mathbf{C}$ is the intrinsic velocity. $\mathbf{W}_0$ is merely the choice of an inertial frame and coincides with the velocity of the container. It is convenient to select $\mathbf{W}_0$ as zero in equilibrium problems.

Boltzmann decided to examine the average properties of $\ln f$, or

$$H = \int d\mathbf{v}\, f \ln f$$

for a distribution function that is independent of $\mathbf{r}$, and then to examine the properties of $\partial H/\partial t$. With $\phi = \ln f$, the collision term in Eq. (5.27) is of the form

$$\int d\mathbf{v}_1\, d\mathbf{v}_2 \left\{ [f(v_1')f(v_2') - f(v_1)f(v_2)] \ln\left[\frac{f(v_1)f(v_2)}{f(v_1')f(v_2')}\right]\right\}[V\sigma(V, \Omega)\, d\Omega] \tag{5.29}$$

Since the integrand can never be positive for any set of $f$'s, the integral is either negative or zero and $H$ can never increase. This is known as the Boltzmann $H$-theorem. In the earlier discussion in Chapter III of an ensemble of systems with a single constraint, the average value of $\ln p(m\mu)$ was examined, and the quantity $\sigma$ which was defined by Eq. (3.83), or $\sigma = -\sum_{m\mu} p(m\mu) \ln p(m\mu)$, has properties which are similar to those of $H$. This question will be discussed briefly in Chapter VI, but the validity of the $H$-theorem and the criticism of the $H$-theorem are not discussed in this text and other appropriate references are given [*2*]. Since the integrand can never be positive, the quantity inside the curly brackets must vanish and this requires that

$$\ln f(\mathbf{v}_1') + \ln f(\mathbf{v}_2') = \ln f(\mathbf{v}_1) + \ln f(\mathbf{v}_2)$$

If $\ln f$ is a linear combination of the collisional invariants, this condition is met, and

$$\ln f = \phi_\alpha = \alpha_0 - \tfrac{1}{2}\beta m C^2 \tag{5.30}$$

A calculation of the number density $n$ and the pressure $P$ with this distribution function yields

$$n = \int d\mathbf{C}\, f \tag{5.31}$$

$$P = P_{zz} = \int d\mathbf{C}\, f C_z(mC_z) = n\beta^{-1} \tag{5.32}$$

and suggests that agreement with the elementary gas law requires that

$$\beta^{-1} = kT \tag{5.33}$$

$\alpha_0$ is the normalization constant. This development yields a distribution law for speeds of the form

$$f_0 = n\left(\frac{m}{2\pi kT}\right)^{3/2} \exp\left(-\frac{mC^2}{2kT}\right) \tag{5.34}$$

and is referred to as the Maxwell–Boltzmann Distribution Law in honor of their original work.

### *5.3.4 External Field of Force*

If the acceleration of the particles is due to an external force which may be derived from a potential energy function $U(\mathbf{r})$,

$$ma_x = -\frac{\partial U}{\partial x} \tag{5.35}$$

a suitable distribution function can be derived. The collision term depends only on the asymptotic velocities $\mathbf{v}$ and does not depend on the potential energy. The term depending on the potential through the $a_x$ can be canceled by introducing a dependence of $n$ on $\mathbf{r}$ into $\ln f$. Thus

$$\ln f = \alpha_0 - \frac{mC^2}{2kT} - \frac{U(r)}{kT} \tag{5.36}$$

is the form of the distribution function in volume element $d\mathbf{r}$. As an example, consider the earth's gravitational field near the earth,

$$U(z) = mgz \tag{5.37}$$

where $z$ is positive in the vertical direction. Using $\mathbf{W} = 0$ and the form of $f$ given earlier, it is apparent that

$$v_z \frac{\partial f}{\partial z} + a_z \frac{\partial f}{\partial v_z} = 0 \tag{5.38}$$

and the given form of the distribution meets the necessary form of the distribution function. The probability of finding a molecule at position $dz$ with velocity in interval $dC$ is proportional to

$$dz\, d\mathbf{C} \exp\left(-\frac{\frac{1}{2}mC^2 + mgz}{kT}\right) \tag{5.39}$$

If this is applied to pressure or density, it is referred to as the barometer formula. Integration of Eq. (5.18) over velocities yields

$$n(z) = n(z_0) \exp\left(-\frac{mg(z - z_0)}{kT}\right) \tag{5.40}$$

and Eq. (5.16) for pressure yields

$$P(z) = n(z)kT \tag{5.41}$$

Exercise 5.5 Show that a spherically symmetric distribution function $f_0 = nA \exp(-\frac{1}{2}mC^2/kT)$ yields $\Gamma_x = 0$, $P_{xy} = 0$, and $q_x = 0$ in Eqs. (5.15)–(5.17). Show that $P_{xx} = nkT$.

Exercise 5.6 Show that the collision term in Eq. (5.27) is zero for $\phi = 1$, $mv_x$, $\frac{1}{2}mv^2$.

Exercise 5.7 Find the normalization constant $\alpha_0$ in Eq. (5.30).

Exercise 5.8 Show that the integrand of Eq. (5.29) is never positive and hence $\ln f$ is of the form of Eq. (5.30) if the integral is zero. Why is the expected value of the integral zero under equilibrium conditions?

Exercise 5.9 Verify Eq. (5.38) for $U = mgz$.

**Exercise 5.10** Use the barometer formula to estimate the pressure as a function of distance above the surface of the earth and compare with experimental data.

**Exercise 5.11** Using the gravitational potential $U(r) = -GmM/r$, find the probability of finding a mass $m$ at a distance of $r$ from the earth's center. This distribution assumes thermal equilibrium by gas collisions. At sufficiently large distances from the earth, the pressure is so low that the mean free path radially out may be regarded as infinite. If during the last collision the particle takes up a kinetic energy characteristic of this last collision, say 400°K, what is the probability of exceeding the escape velocity from the earth's gravitational field? Do you expect to find H, $H_2$, $^4He$, . . . , in the earth's atmosphere? The mass of the earth is $5.98 \times 10^{24}$ kg, the radius of the earth is $6.38 \times 10^6$ m, and the gravitational constant is $6.67 \times 10^{-11}$ N m²/kg, and $m'$ is the molecular weight number.

$$\text{Ans.} \quad \{\text{probability}\} \approx \exp[-(GmM/R_E)/kT] \approx \exp(-19.2m')$$

## 5.4 Nonequilibrium Properties of the Distribution Function

The previous section considered the form of the distribution function for a gas for which the exchange of energy between the molecules was due to binary collisions. As expected, the velocities of the molecules were randomly distributed in direction. The speeds of the particles had a Gaussian distribution and since the velocity orientation was random, the velocities in the $x$, $y$, $z$ directions were also Gaussian. The spherical symmetry of the distribution function gives a zero contribution for the flows $P_{xy}$, $q_z$, . . . and gives the ideal gas pressure for $P_{zz}$. If the flows are small and the deviation from equilibrium is small, one might expect the distribution function to deviate by terms linearly proportional to $P_{xy}$, $q_z$, . . . . The Gaussian form of the equilibrium distribution suggests that in forming an expansion for $f$, Hermite polynomials should yield a good approximation. Grad [3] made this suggestion, and examined a distribution function of the form

$$f = f_0\left[1 + \frac{m}{2PkT} \sum \sum (P_{xy} - \delta_{xy}P)C_xC_y - \frac{m}{PkT} \sum q_xC_x\left(1 - \frac{mC^2}{5kT}\right)\right] \tag{5.42}$$

where

$$f_0 = n\left(\frac{m}{2\pi kT}\right)^{3/2} \exp\left(-\frac{mC^2}{2kT}\right)$$

The stress term is a double sum over subscripts $(x, y, z)$ and $\delta_{xy}$ is zero unless $x = y, \ldots .$ $(P_{xx} - \delta_{xx}P) = 0$ and the zero-order term in pressure is excluded from this first-order term. The multiplying coefficient is chosen to yield an identity for the flows $P_{xy}, q_z, \ldots$ when $f$ is used in the flow equations (5.16) and (5.17). The three-dimensional Hermite polynomials are $1, C_x, C_xC_y, C_x[1 - (mC^2/5kT)], \ldots .$ Since the form suggested by Grad has intuitive appeal, it is used in the following. No attempt is made here to justify this distribution function and reference to the paper by Grad is recommended for the detailed discussion. Earlier work with other expansions was done by, for example, Chapman and Enskog, Maxwell, Hilbert, and Burnett. A general reference to this earlier work and to a systematic treatment of nonuniform gases is the treatise by Chapman and Cowling [*4*].

### *5.4.1 Viscosity and Heat Conduction*

Since the $v$'s are asymptotic velocities, the collision term may be evaluated in any inertial system. Choosing the intrinsic velocity, the binary collision integral given in Eq. (5.27) can be written as

$$\int d\mathbf{C}_1\, d\mathbf{C}_2\, \phi(\mathbf{C}_1)[f(\mathbf{C}_1')f(\mathbf{C}_2') - f(\mathbf{C}_1)f(\mathbf{C}_2)][V\sigma(V, \Omega)\, d\Omega] \tag{5.43}$$

and the $f$'s are given by Eq. (5.42). The $P_{xy}, q_z, \ldots$ are regarded as first-order terms and all products of first-order terms, being of second order, are omitted in the integrand. All terms of odd parity, that is, $C_{1x}C_{2x}C'_{1x}C'_{2y} \ldots$ in the integrand may be omitted since they yield a zero contribution. If the transport of a quantity like $C_yC_z$ is considered, the integral can be transformed to the center-of-mass system $\mathbf{G}$ and $\mathbf{V}$ and reduced to the following form. For examining stress across a surface, the flow of momentum $(mC_y)C_z$ is examined. $\phi$ is selected as

$$\phi(\mathbf{C}_1) = C_{1y}C_{1z}$$

and the integral (5.43) becomes

$$= P_{yz}\left(\frac{m}{8PkT}\right) \int d\mathbf{G}\, f_0(\mathbf{G}) \int d\mathbf{V}\, f_0(\mathbf{V})$$
$$\times [V_yV_z(V_y'V_z' - V_yV_z)][V\sigma(V, \Omega)\, d\Omega] \tag{5.44a}$$

$$= \left(-6\,\frac{n}{m}\,P_{yz}\right)B \tag{5.44b}$$

In a similar manner, for heat flow $\phi$ is selected as

$$\phi(\mathbf{C}_1) = C_{1z}\left(1 - \frac{mC_1^{\ 2}}{5kT}\right)$$

and the integral (5.43) becomes

$$[\tfrac{8}{5}Pq_z]B \tag{5.45}$$

In both cases, the important integral $B$ is

$$B = \frac{4}{15}\left(\frac{kT}{\pi m}\right)^{1/2} \int_0^\infty du\, u^7 e^{-u^2} [(1 - \cos^2\theta)\sigma(u, \theta)\, d\Omega] \tag{5.46}$$

The expression (5.44a) for stress can be reduced by the following steps. The integral over **G** can be taken immediately, but the integral over the relative velocity requires greater effort. **V** and **V**′ contact a sphere of radius $V$ and the angle between then is always $\theta$. Since $(x, y, z)$ are space-fixed directions, these are chosen as reference. The probability that **V** lies in solid angle $(\sin\beta)\, d\alpha\, d\beta$ and **V**′ lies in solid angle $d\Omega = (\sin\theta)\, d\gamma\, d\theta$ relative to **V** depends on the Eulerian angles $(\alpha, \beta, \gamma)$ in a random manner when $\sigma(\mathbf{V}, \Omega) = \sigma(V, \theta)$. These angles are shown in Fig. 5.6. Expressing **V** and **V**′ in terms of these Eulerian angles permits the integral to be reduced to

$$\int_0^{2\pi} d\alpha \int_0^{\pi} (\sin\beta)\, d\beta \int_0^{2\pi} d\gamma\, [V_y V_z(V_y{}' V_z{}' - V_y V_z)] = \frac{4}{5}\pi^2 V^4(\cos^2\theta - 1)$$

Dimensionless units $u^2 = \frac{1}{4}mV^2/kT$ are introduced to obtain the final form for $B$ given in Eq. (5.46).

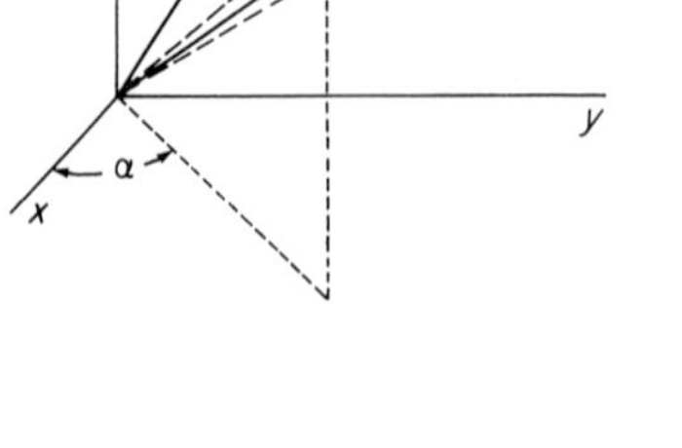

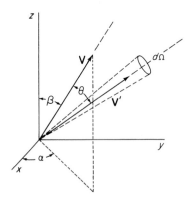

**Fig. 5.6** Reduction of integral (5.44a) to (5.46) by using the Eulerian angles $(\alpha, \beta, \gamma)$ for describing the orientation of **V** and **V**′.

The transport equation (5.27) can be compared in an approximate manner with the earlier calculations for viscosity and heat conduction. Expression (5.44b) or (5.45) is used for the collision term. The dynamic aspects of $f$, which are contained in the left-hand side of Eq. (5.27), can be further reduced.

Exercise 5.12 Show for $\phi = C_z[1 - (mC^2/5kT)]$ that the integral (5.43) can be reduced to

$$q_z\left(-\frac{m}{PkT}\right)\left(\frac{m}{5kT}\right)^2 \int d\mathbf{G}\, f_0(\mathbf{G})G_z^2 \int d\mathbf{V} \times [V_y V_z(V_y' V_z' - V_y V_z)][V\sigma(V, \Omega)\, d\Omega] \tag{5.47}$$

Note that the coefficient for $q_z$ is the same as that for $P_{yz}$ for the relative velocity term, but differs from that for $P_{yz}$ with regard to the motion of the center of mass, which is important for $q_z$.

### 5.4.1.1 Viscosity

Let $\phi = C_x C_y$. Assume that all gradients $\partial W_y/\partial z$ are small and that the flow is steady, so that $\partial f/\partial t$ is also small. Then,

$$\int d\mathbf{v}\, C_y C_z\left(\frac{\partial f}{\partial t} + \sum v_x \frac{\partial f}{\partial x}\right) \approx -\left(\frac{6nB}{m}\right)P_{yz}$$

The left-hand side reduces to

$$\frac{\partial W_y}{\partial z}\int d\mathbf{C}\, f_0 C_z^2 = \left(\frac{P}{m}\right)\frac{\partial W_y}{\partial z}$$

and with these approximations, the stress and velocity gradient are related by

$$P_{yz} = -\left(\frac{kT}{6B}\right)\frac{\partial W_y}{\partial z} = -\eta\,\frac{\partial W_y}{\partial z} \tag{5.48}$$

The transformation of the integral is considered in more detail in the next section.

### 5.4.1.2 Heat Conduction

With $\phi = C_z[1 - (mC^2/5kT)]$, the left-hand side of Eq. (5.27) can be reduced to a term proportional to $\partial T/\partial z$ and other smaller terms. Neglecting the small terms,

$$q_z = -\left(\frac{5k^2T}{8Bm}\right)\frac{\partial T}{\partial z} = -K\frac{\partial T}{\partial z} \tag{5.49}$$

In summary, the simple flow is approximated with a coefficient of viscosity

$$\eta = \frac{kT}{6B} \tag{5.50}$$

and a coefficient of thermal conductivity

$$K = \frac{5k^2T}{8mB} = \left(\frac{15k}{4m}\right)\eta \tag{5.51}$$

Since both $\eta$ and $K$ depend on $B$, the ratio $Km/\eta C_V$ becomes a constant and is in good agreement with the data for monatomic gases given in Table 5.1.

The theory developed is for monatomic gases. In general, for heat transport, it is necessary to include inelastic collisions and to consider the exchange of energy between the internal degrees of freedom for the molecules [*5*].[1]

Exercise 5.13 Show for $\sigma(V, \theta)$ = constant that $B$ reduces to

$$B = \frac{8}{15}\left(\frac{kT}{\pi m}\right)^{1/2}\sigma_0 \tag{5.52}$$

$\sigma_0$ is referred to as the hard-sphere cross section. The $\sigma_\eta$ given in Table 5.1 were evaluated by assuming that the viscosity is for a hard-sphere gas.

## 5.5 General Transport Equations

The simple expressions for transport given in the previous section do not take into account the interplay between momentum and heat transport. This can be shown in a set of equations which are correct to first order by using the expression for $f$ given by Eq. (5.42). Since the notation will soon become quite tedious, a more general notation $x_j$ is used to replace $x$ and

[1] A recent review article concerning transport coefficients and including recent references is presented by Ernest *et al.* [*6*].

$C_j$ to replace $C_x$, and so forth. The index $j$ runs over $x, y, z$. Repeated indices are summed over the repeated index; that is, $C_iC_i$ is summed over $x$, $y$, and $z$. With these ideas in mind, the simplification of Eq. (5.21) is of interest:

$$\int d\mathbf{v}\,\phi(\mathbf{v})\left(\frac{\partial f}{\partial t} + v_j\,\frac{\partial f}{\partial x_j}\right) = \{\text{collision term}\} \tag{5.53}$$

A first-order collision term proportional to $P_{ij}$ and $q_j$ was developed in the previous section. In the same sense, $\partial f/\partial t$ and $\partial f/\partial x_j$ are first-order terms and this is apparent by expanding the just given. Again $d\mathbf{v}$ may be changed to $d\mathbf{C}$ and $\partial v_j/\partial t = 0$ or $\partial v_j/\partial x_i = 0, \ldots,$ since the $\mathbf{v}$'s are asymptotic velocities and do not depend on the inertial frame. In this first-order approximation, $f$ as given in Eq. (5.42) depends on the intrinsic velocity $\mathbf{C}$,

$$\mathbf{C} = \mathbf{v} - \mathbf{W}$$

The interesting integrals for $\phi(\mathbf{C})$ are

$$\begin{aligned}\int d\mathbf{v}\,\phi(\mathbf{C})\,\frac{\partial f}{\partial t} &= \frac{\partial}{\partial t}\left[\int d\mathbf{v}\,\phi(\mathbf{v}-\mathbf{W})f\right] - \int d\mathbf{v}\,f\,\frac{\partial}{\partial t}\,[\phi(\mathbf{v}-\mathbf{W})] \\ &= \frac{\partial}{\partial t}\left[\int d\mathbf{C}\,\phi(\mathbf{C})f\right] + \int d\mathbf{C}\,f\,\frac{\partial \phi}{\partial C_j}\,\frac{\partial W_j}{\partial t}\end{aligned} \tag{5.54}$$

and

$$\begin{aligned}\int d\mathbf{v}\,\phi(\mathbf{C})\mathbf{v}_j\,\frac{\partial f}{\partial x_j} &= \frac{\partial}{\partial x_j}\left[\int d\mathbf{C}\,f\phi(\mathbf{C})(C_j + W_j)\right] \\ &\quad + \int d\mathbf{C}\,f(C_j + W_j)\,\frac{\partial \phi}{\partial C_i}\,\frac{\partial W_i}{\partial x_j}\end{aligned} \tag{5.55}$$

Here,

$$\frac{\partial \phi}{\partial t} = -\,\frac{\partial \phi}{\partial C_i}\,\frac{\partial W_i}{\partial t} \quad \text{and} \quad \frac{\partial \phi}{\partial x_j} = -\,\frac{\partial \phi}{\partial C_i}\,\frac{\partial W_i}{\partial x_j}$$

are used.

The general flow equations are now given for the collisional invariants $\phi = 1$, $mv_j$, and $\frac{1}{2}mC^2$ and for the stress $\phi = C_iC_j$, for heat transfer $\phi = C_j[1 - (mC^2/5kT)]$, and for the form of $f$ given in Eq. (5.42). With $\varrho = nm$ and $P = nkT$,

$$\tilde{P}_{ij} = (P_{ij} - \delta_{ij}P)$$

The general flow equations are given in the following section.

### 5.5.1 General Flow Equations

*Equation of Continuity,* $\phi = 1$:

$$\frac{\partial \varrho}{\partial t} + \frac{\partial(\varrho W_i)}{\partial x_i} = 0 \tag{5.56}$$

*Equation of Conservation of Linear Momentum,* $\phi = mC_i$:

$$\varrho \frac{\partial W_i}{\partial t} + \varrho W_j \frac{\partial W_i}{\partial x_j} + \frac{\partial P_{ij}}{\partial x_j} = 0 \tag{5.57}$$

*Equation of Conservation of Thermal Energy,* $\phi = \frac{1}{2}mC^2$:

$$\frac{\partial P}{\partial t} + \frac{\partial(PW_i)}{\partial x_i} + \frac{2P_{ij}}{3}\frac{\partial W_i}{\partial x_j} + \frac{2}{3}\frac{\partial q_i}{\partial x_i} = 0 \tag{5.58}$$

*Equation for Stress,* $\phi = C_iC_j$:

$$\begin{aligned}\frac{\partial \tilde{P}_{ij}}{\partial t} &+ \frac{\partial(W_r\tilde{P}_{ij})}{\partial x_r} + \frac{2}{5}\left(\frac{\partial q_i}{\partial x_j} + \frac{\partial q_j}{\partial x_i} - \frac{2\delta_{ij}}{3}\frac{\partial q_r}{\partial x_r}\right)\\ &+ \left(\tilde{P}_{ir}\frac{\partial W_j}{\partial x_r} + \tilde{P}_{jr}\frac{\partial W_i}{\partial x_r} - \frac{2\delta_{ij}\tilde{P}_{rs}}{3}\frac{\partial W_r}{\partial x_s}\right)\\ &+ P\left(\frac{\partial W_i}{\partial x_j} + \frac{\partial W_j}{\partial x_i} - \frac{2\delta_{ij}}{3}\frac{\partial W_r}{\partial x_r}\right) = -6nB\tilde{P}_{ij}\end{aligned} \tag{5.59}$$

*Equation of Heat Transport,* $\phi = C_i[1 - (mC^2/5kT)]$:

$$\begin{aligned}\frac{\partial q_i}{\partial t} &+ \frac{\partial(W_rq_i)}{\partial x_r} + \left(\frac{7q_r}{5}\frac{\partial W_i}{\partial x_r} + \frac{2q_r}{5}\frac{\partial W_r}{\partial x_i} + \frac{2q_i}{5}\frac{\partial W_r}{\partial x_r}\right)\\ &+ \frac{kT}{m}\frac{\partial \tilde{P}_{ir}}{\partial x_r} + \frac{7\tilde{P}_{ir}}{2m}\frac{\partial(kT)}{\partial x_r} - \frac{\tilde{P}_{ir}}{\varrho}\frac{\partial P_{rs}}{\partial x_s} + \frac{5kP}{2m}\frac{\partial T}{\partial x_i} = -4nBq_z\end{aligned} \tag{5.60}$$

### 5.5.2 Navier–Stokes Equations

For small stress and heat flow gradients, the dominant terms in Eq. (5.59) relate the stress to the velocity gradients, that is,

$$\tilde{P}_{ij} = (P_{ij} - \delta_{ij}P) = -\eta\left(\frac{\partial W_i}{\partial x_j} + \frac{\partial W_j}{\partial x_i} - \frac{2\delta_{ij}}{3}\frac{\partial W_r}{\partial x_r}\right) \tag{5.61}$$

when terms of the type $\partial q_i/\partial x_j$, $\partial W_i/\partial x_j$, ... are regarded as small and are omitted. Equation (5.57) for the momentum flow can now be rewritten as

$$\varrho\frac{\partial W_i}{\partial t} + \varrho W_i \frac{\partial W_j}{\partial x_i} = -\frac{\partial P}{\partial x_i} + \eta\left(\frac{\partial^2 W_i}{\partial x_j\,\partial x_j} - \frac{1}{3}\frac{\partial^2 W_j}{\partial x_i\,\partial x_j}\right) \tag{5.62}$$

This is the Navier–Stokes equation, which is frequently used for gases and liquids. **W** is the velocity field.

## 5.6 Equilibrium Mixture of Two Gases

If a gas is composed of a mixture of molecules of masses $m_A$ and $m_B$, the probability of finding a molecule of type A in $d\mathbf{r}\,d\mathbf{v}$ is proportional to $f_A(\mathbf{r}, \mathbf{v}, t)$ and that of finding one of type B is proportional to $f_B(\mathbf{r}, \mathbf{v}, t)$. These distributions are connected through the A–B collision term, and the equation for the transport of a physical property $\phi$ depends on

$$\begin{aligned}\int dv_1\,\phi_A(\mathbf{v}_1)\left(\frac{\partial f_A}{\partial t} + \sum v_{1x}\frac{\partial f_A}{\partial x}\right)\\ &= \int d\mathbf{v}_1\,d\mathbf{v}_2\,\phi_A(\mathbf{v}_1)[f_A(v_1')f_A(v_2') - f_A(v_1)f_A(v_2)](V_{AA}\sigma_{AA}\,d\Omega)\\ &\quad + \int d\mathbf{v}_1\,d\mathbf{v}_2\,\phi_A(\mathbf{v}_1)[f_A(v_1')f_B(v_2') - f_A(v_1)f_B(v_2)](V_{AB}\sigma_{AB}\,d\Omega)\end{aligned} \tag{5.63}$$

A similar equation with A and B interchanged is appropriate for $f_B$. Although the development is not given here, by using the invariance of the integrals on the right to the interchange of primes and the interchange of indices 1 and 2, the equation can be placed in a form that contains the expressions

$$[\phi_A(v_1) + \phi_A(v_2) - \phi_A(v_1') - \phi_A(v_2')]$$

and

$$[\phi_A(v_1) + \phi_B(v_2) - \phi_A(v_1') - \phi_B(v_2')]$$

and so on for B. For the collisional invariants $\phi = 1$, $\phi_A = m_A\mathbf{v}$, $\phi_B = m_B\mathbf{v}$, $\phi_A = \frac{1}{2}m_A v^2$, and $\phi_B = \frac{1}{2}m_B v^2$, these expressions are zero. By examining the properties of $\phi_A = \ln f_A$ and $\phi_B = \ln f_B$, one can show that necessary and sufficient conditions for equilibrium are

$$\begin{aligned}\phi_A(\mathbf{v}) &= \ln f_A = \alpha_A - \tfrac{1}{2}m_A(\mathbf{v} - \mathbf{W})^2 = \alpha_A - \tfrac{1}{2}m_A C^2\\ \phi_B(\mathbf{v}) &= \ln f_B = \alpha_B - \tfrac{1}{2}m_B(\mathbf{v} - \mathbf{W})^2 = \alpha_B - \tfrac{1}{2}m_B C^2\end{aligned} \tag{5.64}$$

Again, $\mathbf{C}$ is the intrinsic velocity and for equilibrium problems, the velocity of the container $\mathbf{W}$ may be selected as zero. $\alpha_{\mathrm{A}}$ and $\alpha_{\mathrm{B}}$ ensure normalization.

The flow expressions given by Eqs. (5.15)–(5.20) remain valid for each species. It is necessary to sum over the indices A and B. Thus

$$n_{\mathrm{A}} = \int d\mathbf{C}\, f_{\mathrm{A}}$$

$$P_{\mathrm{A}} = (P_{\mathrm{A}})_{zz} = \int d\mathbf{C}\, f_{\mathrm{A}} C_z (mC_z) = n_{\mathrm{A}} kT$$

$$P = P_{\mathrm{A}} + P_{\mathrm{B}}$$

and so on. The pressure is additive as for the mass point gas, and, as discussed in Chapter I, the average kinetic energy per molecule is given by

$$\left\{\begin{matrix}\text{average translational kinetic} \\ \text{energy per molecule}\end{matrix}\right\} = \langle \tfrac{1}{2} m_{\mathrm{A}} C^2 \rangle = \langle \tfrac{1}{2} m_{\mathrm{B}} C^2 \rangle = \tfrac{3}{2} kT$$

and is independent of the mass of the molecule.

Exercise 5.14 Show by direct substitution of the $f_{\mathrm{A}}$ and $f_{\mathrm{B}}$ [given by Eq. (5.64)] into Eq. (5.63) that this form is sufficient for equilibrium.

Exercise 5.15 Show for a binary mixture that $P_{xx} = P = (n_{\mathrm{A}} + n_{\mathrm{B}})kT$ and that

$$\{\text{average translational energy}\} = \tfrac{3}{2} kT$$

and is independent of the gas species.

## 5.7 Diffusion

The two constituents A and B of a gas mixture are assumed to be diffusing relative to one another if, across a surface moving with the average velocity $\mathbf{W}$, there is a net transport of species A or B across the surface. $\mathbf{W}$ is defined as

$$\mathbf{W} = \left( \frac{\sum n_i \mathbf{v}_i}{\sum n_i} \right)_{\mathrm{av}} = \frac{\int d\mathbf{v}\, (f_{\mathrm{A}} \mathbf{v}_{\mathrm{A}} + f_{\mathrm{B}} \mathbf{v}_{\mathrm{B}})}{n_{\mathrm{A}} + n_{\mathrm{B}}} \tag{5.65}$$

for the binary gas mixture. The flow of particles across the surface is given by

$$\Gamma_{Ax} = \int d\mathbf{C}\, f_A(\mathbf{C})C_x \tag{5.66a}$$

$$\Gamma_{Bx} = \int d\mathbf{C}\, f_B(\mathbf{C})C_x \tag{5.66b}$$

For small flows, the anisotropy in the distribution function is assumed to be proportional to the flow and the $f_A$ and $f_B$ are expanded in terms of the Hermite polynomials,

$$f_A = f_A{}^0\left(1 + \frac{m_A}{P_A} \sum \Gamma_{Ax} C_{Ax} + \cdots\right) \tag{5.67a}$$

and $f_B$ follows by changing A to B. The stress and heat conduction terms are omitted in order to simplify the development. This neglects very interesting cross terms relating heat flow, stress, and diffusion.

Equation (5.63) and a similar equation with A and B interchanged are used as the basis for the discussion of diffusion. In Eq. (5.63), $\phi_A = C_{1z}$ is selected as the physical property of interest. $\phi_B = C_{2z}$ is selected in the equation for $f_B$. On the right-hand side of Eq. (5.63), only the cross term for A–B collisions remains. The term for A–A or B–B collisions is zero, since for equal masses these terms may be expressed as a collisional invariant. The dynamic term for $f_A$ on the left is reduced in a procedure similar to that used for viscosity and heat conduction. Thus

$$\int d\mathbf{v}_1\, C_{1z}\left(\frac{\partial f_A}{\partial t} + \sum v_{1x} \frac{\partial f_A}{\partial x}\right) = \frac{\partial}{\partial z} \int d\mathbf{C}_1\, f_A C_{1z}^2 = \frac{1}{m_A} \frac{dP_A}{dz} \tag{5.67b}$$

where the transformations given by Eqs. (5.54) and (5.55) are used. The expression is further reduced by assuming the diffusion is steady, so that $\partial f/\partial t = 0$, and that $\partial W_i/\partial x_j$ terms are small. With this very considerable simplification, the diffusion equation for one-dimensional steady diffusion is

$$\begin{aligned}
\frac{1}{m_A} \frac{dP_A}{dz} &\approx \int d\mathbf{C}_1\, d\mathbf{C}_2\, C_{1z}[f_A(C_1')f_B(C_2') - f_A(C_1)f_B(C_2)](V_{AB}\sigma_{AB}\, d\Omega) \\
&\approx \int d\mathbf{C}_1\, d\mathbf{C}_2\, f_A{}^0(C_1)f_B{}^0(C_2)C_{1z} \\
&\quad \times \left[\Gamma_{Az} \frac{m_A}{P_A}(C_{1z}' - C_{1z}) + \Gamma_{Bz} \frac{m_B}{P_B}(C_{2z}' - C_{2z})\right](V_{AB}\sigma_{AB}\, d\Omega)
\end{aligned}$$

which, upon transforming to the center-of-mass coordinates, becomes

$$\frac{1}{m_A}\frac{dP_A}{dz} = \left(\frac{m_A}{m_A + m_B}\right) n_A n_B \left(\frac{\mu}{2\pi kT}\right)^{3/2} \left(\frac{\Gamma_{Az}}{P_A} - \frac{\Gamma_{Bz}}{P_B}\right)$$
$$\times \int d\mathbf{V} \left[\exp\left(-\frac{\mu V^2}{2kT}\right)\right] \mu [V_z(V_z' - V_z)](V\sigma_{AB}\, d\Omega) \quad (5.68)$$

Following the procedure used for viscosity, the integral over $V_z(V_z' - V_z)V$ can be reduced to

$$B_d = \frac{8}{3}\left(\frac{2kT}{\pi\mu}\right)^{1/2} \int_0^\infty d\xi\, e^{-\xi^2}\, \xi^5 (1 - \cos\theta)\sigma_{AB}(\xi, \theta)\, d\Omega \quad (5.69)$$

and the equation for diffusion is

$$\frac{\partial P_A}{\partial z} \approx \mu kT B_d n_A n_B \left(\frac{\Gamma_{Bz}}{P_B} - \frac{\Gamma_{Az}}{P_A}\right) \quad (5.70)$$

A similar expression for B is obtained by interchanging A and B. The approximations are equivalent to assuming that $\partial P/\partial z = 0$, or $P =$ constant, and that the net flow of particles into volume $\Delta\mathbf{r}$ is zero. Hence

$$\Gamma_{Az} + \Gamma_{Bz} = 0$$

and the diffusion equation can be written as

$$\Gamma_{Az} = -nD_{AB}\frac{1}{P}\frac{\partial P_A}{\partial z} = -D_{AB}\frac{\partial n_A}{\partial z} \quad (5.71)$$

where

$$D_{AB} = \frac{kT}{n\mu B_d} \quad (5.72)$$

Exercise 5.16 Show for hard spheres that

$$D_{AB} = \frac{3\pi}{32}\frac{1}{n\sigma_{AB}}\left(\frac{8kT}{\pi\mu}\right)^{1/2} = \frac{3\pi}{32}\left(\frac{1}{n\sigma_{AB}}\right)\bar{v} \quad (5.73)$$

## 5.8 Sound Waves in Gases

An elementary derivation of the small motion of particles in a gaseous medium in which the average particle velocity is zero is as follows. When the particles in the region between $z$ and $z + \Delta z$ are displaced to the region

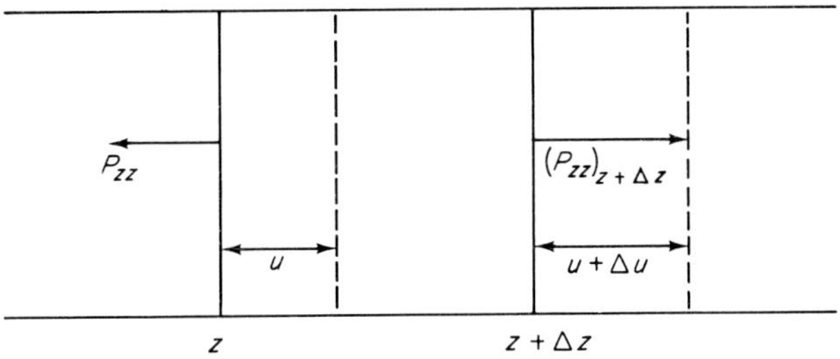


**Fig. 5.7** Displacement of the surfaces from $z$ to $z+u$ and $z+\Delta z$ to $z+\Delta z + u + \Delta u$ by the stresses $P_{zz}(z)$ and $P_{zz}(z+\Delta z)$.

between $z+u$ and $z+\Delta z+u+\Delta u$ by the stresses shown in Fig. 5.7, the equation of motion follows from Newton's Second Law $F_z = ma_z$; or for a tube of cross section $A$,

$$\varrho_0 A\, \Delta z \frac{\partial^2 u}{\partial t^2} = A[(P_{zz})_{z+\Delta z} - (P_{zz})_z]$$

Expanding the term on the right and canceling common terms yields

$$\varrho_0 \frac{\partial^2 u}{\partial t^2} = \frac{\partial P_{zz}}{\partial z} = -\frac{\partial P}{\partial z} \tag{5.74}$$

for the relationship between acceleration and stress. The equation of continuity, or the equation for the conservation of the number of particles in the volume element, is

$$\varrho_0 A\, \Delta z = \varrho(z, t) A(\Delta z + \Delta u)$$

or

$$\varrho(z, t) = \varrho_0 \left(1 + \frac{\partial u}{\partial z}\right)^{-1} \tag{5.75}$$

Further progress requires a consideration of the properties of the gas within the element of volume $A\,\Delta z$. If the particles do not exchange energy with the surroundings, then the process can be regarded as adiabatic and the pressure $P$ is a function of the density $\varrho$. In order for the gradient of the stress or pressure to have any physical significance, it is necessary that $\Delta z$ be many average paths, that is, $\Delta z \gg \lambda$. This is emphasized since the equation of state for a gas is, to a good approximation, given by the $P/\varrho^\gamma =$ constant and the collisions seem unnecessary. With $P$ a function of $\varrho$, it follows that

$$\frac{\partial P}{\partial z} = \left(\frac{dP}{d\varrho}\right)_S \frac{\partial \varrho}{\partial z} \tag{5.76}$$

Combining Eqs. (5.74)–(5.76) and regarding $|\partial u/\partial z| \ll 1$ yields the wave equation

$$\frac{1}{v^2}\frac{\partial^2 u}{\partial t^2} - \frac{\partial^2 u}{\partial z^2} = 0 \tag{5.77}$$

For a nearly ideal gas,

$$v^2 = \left(\frac{dP}{d\varrho}\right)_S = \left(\frac{\gamma kT}{m}\right)^{1/2} \tag{5.78}$$

for the velocity of sound. If $\partial/\partial z$ is taken of Eq. (5.77), then with Eq. (5.75),

$$\varrho_0 \frac{\partial u}{\partial z} = \varrho(z, t) - \varrho_0 = \Delta\varrho$$

yields a wave equation for the waves of changing density, that is, $u \to \varrho$. Using Eq. (5.76) permits $\Delta\varrho$ to be replaced by the change in pressure $\Delta P$ or wave equation for the change in pressure,

$$\frac{1}{v^2}\frac{\partial^2 P}{\partial t^2} - \frac{\partial^2 P}{\partial z^2} = 0 \tag{5.79}$$

$\partial^2/\partial z^2$ can be replaced by $\nabla^2$ for three-dimensional problems. The average energy flow is $\frac{1}{2}(\Delta P)(\dot{u})$. In order to appreciate the magnitude of the radiated power, the radiation by a spherical source is given. If the radius of a sphere undergoes a displacement

$$r = a + A \cos \omega t$$

then

$$\begin{Bmatrix}\text{power radiated by} \\ \text{a spherical source}\end{Bmatrix} = \left(\frac{\varrho 2\pi a^4 A^2}{v}\right)\omega^4 \quad \text{W} \tag{5.80}$$

where

$$\{\text{wavelength}\} = \frac{v}{\nu} \gg a$$

For wavelengths large compared to the size of the radiating system, the radiated power is proportional to the frequency to the fourth power. Spherical objects also scatter radiation with this same dependence.

It should be noted that at large amplitudes of oscillation, that is, $\Delta P = 0.1P$, the velocity of sound depends on the amplitude. Thus as the wave progresses, second harmonics are generated.

If the gas is moving, that is, the oscillations are generated in moving air or the wind is blowing, then it is necessary to use the Navier–Stokes equations (5.61) and (5.62). Treating again the one-dimensional problem, let

$$W_z = W_0 + \dot{u}$$

and ignore the viscous terms. Here $W_0$ is the speed of the moving gas. Then the equation for small disturbances is given by

$$\varrho_0 \frac{\partial^2 u}{\partial t^2} + \varrho_0 W_0 \frac{\partial^2 u}{\partial t\, \partial z} = -\frac{\partial P}{\partial z} \tag{5.81a}$$

and the equation of continuity by

$$\frac{\partial \varrho}{\partial t} + \varrho_0 \frac{\partial^2 u}{\partial z\, \partial t} + W_0 \frac{\partial \varrho}{\partial z} = 0 \tag{5.81b}$$

Combining these two equations yields the density equation. Again, $\varrho$ is replaced by $P$, and

$$\frac{\partial^2 P}{\partial t^2} + 2W_0 \frac{\partial^2 P}{\partial z\, \partial t} - v^2 \frac{\partial^2 P}{\partial z^2} = 0 \tag{5.82}$$

This has an approximate solution of the form

$$g[z \mp vt - W_0 t] \tag{5.83}$$

and suggests that the wave propagates with a velocity

$$\frac{\Delta z}{\Delta t} = v \pm W_0$$

where the upper sign is along $+\hat{z}$ and the lower along $-\hat{z}$. Thus a small disturbance in a gas moves with velocity $v$ for an observer moving with the gas and $v \pm W_0$ for a fixed observer. This is quite different from the propagation of light waves, which have the same velocity for both observers.

**Exercise 5.17** Find the velocity of sound in He, $N_2$, and air at 300°K and compare with the average particle speed $\bar{v}$.

**Exercise 5.18** A sphere with a radius of 2 cm oscillates at 500 Hz. Let the amplitude of oscillation be 0.02 mm. Find the radiated power.

## References

1. K. Huang, "Statistical Mechanics." Wiley, New York, 1963.
2. D. ter Haar, "Elements of Statistical Mechanics." Holt, New York, 1954.
3. H. Grad, *Commun. Pure Appl. Math.* **2**, 331 (1949); **5**, 257 (1952).
4. S. Chapman and T. G. Cowling, "The Mathematical Theory of Non-Uniform Gases." Cambridge Univ. Press, London and New York, 1958.
5. J. O. Hirschfelder, C. R. Curtiss, and R. B. Bird, "The Molecular Theory of Gases and Liquids." Wiley, New York, 1954.
6. M. H. Ernst, L. K. Haines, and J. R. Dorfman, *Rev. Mod. Phys.* **41**, 296 (1969).

# Statistical Thermodynamics VI

## 6.1 Introduction

Before discussing the nonequilibrium properties of gases in greater detail, the equilibrium properties of matter are considered. Equilibrium properties must obey the laws of thermodynamics and these laws are reviewed in this chapter in order to clearly state the notation which is used for these thermodynamic variables. Statistical laws which are consistent with the thermodynamic laws are then developed. Specific examples are discussed in Chapters VII and VIII.

## 6.2 Laws of Thermodynamics

In their least abstract form, the apparent thermodynamic coordinates for a thermodynamic system are the pressure $P$ and the volume $V$. These macroscopic quantities are readily obvious to the senses and can be measured in a convenient manner. Volume and pressure can be varied for the system and the properties of the system become apparent in the functional relationship between $P$ and $V$. Other similar thermodynamic coordinates can occur, but these are considered with specific examples. The mechanical work done by the system on changing the pressure and volume is given by

$$dW = P\,dV \tag{6.1}$$

in differential form. Heat transfer can also occur between two systems. Experimental evidence indicates that if two systems $A$ and $B$ are brought into contact via a diathermic wall, that is, a wall which permits only the flow of heat, the systems come into equilibrium and the thermodynamic coordinates $P_A$, $V_A$, $P_B$, and $V_B$ tend to constant values. This is regarded as a state of thermodynamic equilibrium between the two systems. If system $A$ is brought into thermal contact with $C$ and system $B$ is also brought into thermal contact with $C$ and the systems are allowed to come into thermal equilibrium, it is experimentally observed that no further change in $P_A$, $V_A$, $P_B$, or $V_B$ occurs when systems $A$ and $B$ are brought into thermal contact. This observation requires the existence of a single-valued function of the thermodynamic variables of the system, that is,

$$T = T_A(P_A V_A) = T_B(P_B V_B) = T_C(P_C V_C)$$

and this function is the temperature $T$. It is often suggested that this observation be referred to as the Zeroth Law of Thermodynamics.

*Zeroth Law of Thermodynamics*

> Two systems in thermal equilibrium with a third are in thermal equilibrium with each other.

It is further implied that a functional relationship exists connecting $T$, $P$, and $V$,

$$T = T(P, V) \qquad \text{or} \qquad f(P, V, T) = 0 \tag{6.2}$$

and these functions are the equations of state for the system. If Eq. (6.2) is plotted in three-dimensional space with $P$, $V$, $T$ as the coordinate axes, a surface is generated. Such a surface is shown in Fig. 9.1 for $H_2O$. All possible values of $P$, $V$, $T$ lie on this surface. Various slopes of this surface are measurable and of interest. If the functional dependence of the pressure is denoted by $P(V, T)$, then

$$dP = \left(\frac{\partial P}{\partial V}\right)_T dV + \left(\frac{\partial P}{\partial T}\right)_V dT \tag{6.3}$$

and these can be related to the measurable quantities

$$\{\text{isothermal bulk modulus}\} = -V\left(\frac{\partial P}{\partial V}\right)_T \tag{6.4a}$$

$$\left\{\begin{matrix}\text{isobaric coefficient of}\\ \text{volume expansion}\end{matrix}\right\} = \frac{1}{V}\left(\frac{\partial V}{\partial T}\right)_P \tag{6.4b}$$

$$\left\{\begin{matrix}\text{coefficient of pressure increase}\\ \text{in a constant-volume process}\end{matrix}\right\} = \frac{1}{P}\left(\frac{\partial P}{\partial T}\right)_V \tag{6.4c}$$

For constant pressure, or $dP = 0$, the relationship between these three quantities is given by

$$\left(\frac{\partial P}{\partial T}\right)_V = -\left(\frac{\partial P}{\partial V}\right)_T\left(\frac{\partial V}{\partial T}\right)_P \tag{6.4d}$$

### 6.2.1 First Law of Thermodynamics

Overall conservation of energy is expected for the thermodynamic system. Even so, it is found experimentally that if the system is placed inside an adiabatic wall, that is, a wall which does not permit transfer of heat, the state of the system is independent of the manner in which work is done on the system. Starting with initial values of $P_i$, $V_i$, $T_i$, the final values $P_f$, $V_f$, $T_f$ are exactly the same for adiabatic work done by applying pressure and changing the volume, or by doing adiabatic work by stirring, or by other means. This implies that the effect of adiabatic work is independent of the path, and for adiabatic work,

$$(-W_{i\to f})_{\text{adiabatic}} = U_f - U_i \tag{6.5a}$$

or

$$(-dW)_{\text{adiabatic}} = dU \tag{6.5b}$$

$U$ is a property of the system alone and is denoted as the internal energy. It is a property of the system in the same sense as $P$, $V$, and $T$, and $U$ is a thermodynamic coordinate. $dU$ is an exact differential and is path-independent. Equations (6.5a) and (6.5b) are statements of the First Law of Thermodynamics, which can be stated as follows:

*First Law of Thermodynamics*

> If a thermally insulated system is caused to change from an initial state to a final state by adiabatic means only, the work done is the same for all adiabatic paths connecting the two states.

Heat $Q$ can also be added to the system and conservation of energy requires that in changing a system from an initial state to a final state,

$$Q = U_f - U_i + W \tag{6.6a}$$

where $Q$ is the heat added "to" the system and $W$ is the work done "by" the system. Both $Q$ and $W$ depend on the path for the process. A differential form for this energy conservation is

$$\delta Q = dU + \delta W \tag{6.6b}$$

or

$$\delta Q = \delta U + P\, dV \tag{6.6c}$$

for a simple system. $\delta Q$ and $\delta W$ are used to note that the path must be specified. The First Law of Thermodynamics can be used for both reversible and irreversible changes in the system.

Addition of heat at constant volume is used to define the heat capacity of the system $C_V$, and

$$C_V = \left(\frac{\delta Q}{\delta T}\right)_V = \left(\frac{\partial U}{\partial T}\right)_V \tag{6.7}$$

### 6.2.2 Second Law of Thermodynamics

Attempts to construct engines which do work $W$ from a supply of heat $Q_1$ indicate that the efficiency of such an engine is always less than one,

$$\{\text{efficiency}\} = \frac{W}{Q_1} = 1 - \frac{Q_2}{Q_1} < 1 \tag{6.8a}$$

In the discussion of an engine, it is implied that the working substance of the engine is brought back to its initial state after a complete cycle and the change in internal energy of the working substance is zero in a complete cycle, $U = 0$, and

$$W = Q_1 - Q_2 \tag{6.8b}$$

In thermodynamic considerations, heat $Q_1$ is supplied by an abstract hot reservoir and heat $Q_2$ is rejected to a cold reservoir. From a more direct point of view, $Q_1$ is the heat supplied by the fuel which one buys and $W$ is the work which one sells, and one might wish $W = Q_1$ for maximum profit. Experimental evidence indicates that this is not possible, and these experiments suggest the following two statements of the Second Law of Thermodynamics.

*Kelvin–Planck Statement of the Second Law*

> It is impossible to construct an engine that, operating in a cycle, will produce no effect other than the extraction of heat from a reservoir and the performance of an equivalent amount of work.

*Clausius Statement of the Second Law*

> It is impossible to construct a device that, operating in a cycle, will produce no effect other than the transfer of heat from a cooler to a hotter body.

Symbolic diagrams of these forbidden engines and refrigerators are shown in Fig. 6.1. It should be noted that if Fig. 6.1b were not forbidden, one could buy a window air conditioner which would cool a room without an electrical or other connection for a source of power. It would operate forever at no expense. Even casual everyday experience suggests that this free cooling is not possible. Figure 6.1a indicates that engines that reject no heat and use all the fuel for work are not possible, and again our daily experience with automobiles indicates that this seems reasonable.

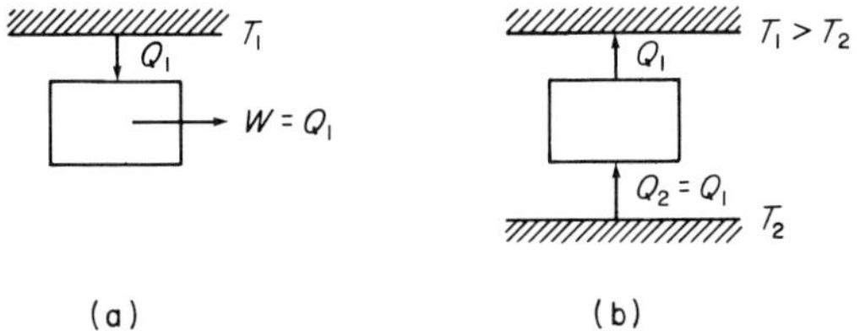


**Fig. 6.1** (a) Forbidden heat engine. (b) Forbidden refrigerator.

Carnot was one of the first engineers to consider this problem of engine efficiency and his investigation led him to suggest what is known as the Carnot cycle for a gas. This cycle is more general and an ideal engine has the following properties:

(a) The working substance is changed from temperature $T_2$ to $T_1$ by adiabatic work being done on it.

(b) Heat $Q_1$ is added isothermally at temperature $T_1$ to the working substance.

(c) The working substance is changed from temperature $T_1$ to $T_2$ by doing adiabatic work.

(d) Heat $Q_2$ is rejected isothermally by the working substance at temperature $T_2$, until the working substance returns to its initial state.

Each of these steps must proceed in a reversible manner for the ideal engine. The reversed cycle is a refrigerator; and a symbolic engine and refrigerator are shown in Fig. 6.2. Here, $T_1$ and $T_2$ denote reservoirs which supply the heat in a reversible manner. $W$ is the work which is supplied by the engine or the work supplied to the refrigerator. Since the working substance is carried through a complete cycle, $U = 0$, and $W = Q_1 - Q_2$. This is an ideal cycle for an ideal engine and Carnot concluded that no real engine could be more efficient than this ideal engine. A statement of Carnot's theorem is as follows:

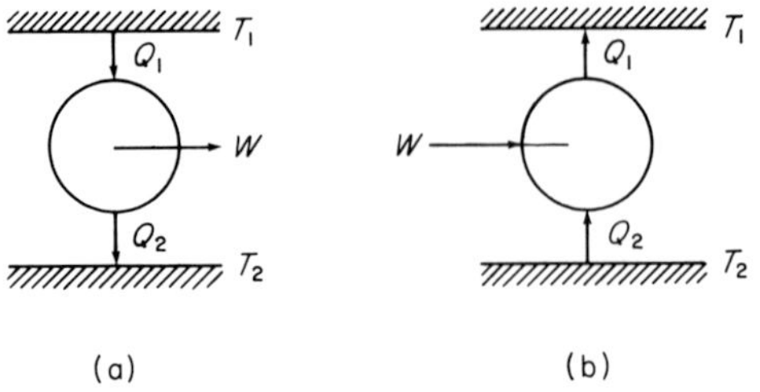


**Fig. 6.2** (a) Carnot engine. (b) Carnot refrigerator, or reversed Carnot engine.

*Carnot's Theorem*

> No engine operating between two given reservoirs can be more efficient than a Carnot engine operating between the same two reservoirs.

An engine with an efficiency greater than Carnot's engine violates the Second Law of Thermodynamics. This can be shown by using this engine to drive the Carnot engine as a refrigerator. This is always possible since the Carnot engine is reversible. If these two systems are connected together and placed between reservoirs at temperatures $T_1$ and $T_2$ and the output work $W$ is all used to drive the Carnot refrigerator, the combined system has no input work, extracts heat from the cold reservoir, and rejects an equivalent amount of heat to the hot reservoir. This violates Clausius's statement of the Second Law.

The efficiency of the Carnot engine $\eta_{\mathrm{R}}$ is independent of the working substance and is a function only of the heat extracted from the hot reservoir $Q_1$ at temperature $T_1'$ and the heat rejected to the cold reservoir $Q_2$ at temperature $T_2'$. Thus

$$\eta_{\mathrm{R}} = 1 - \frac{Q_2}{Q_1} = \eta_{\mathrm{R}}(T_1', T_2')$$

or

$$\frac{Q_1}{Q_2} = f(T_1', T_2') = \frac{\psi(T_1')}{\psi(T_2')}$$

and the ratio of the heats extracted and rejected is a function of the reservoir temperatures. By suitable arguments using a third reservoir, one can show that this function is the ratio of two functions. Thus the ratio of the heats extracted and rejected by a Carnot engine can be used to define the Kelvin temperature:

$$\frac{Q_1}{Q_2} = \frac{T_1}{T_2} \tag{6.9}$$

One must define the scale and this is done by defining the triple point of water as

$$\{\text{temperature of the triple point of water}\} = 273.16°\text{K} \quad (6.10)$$

### 6.2.3 *Entropy*

The Second Law of Thermodynamics implies the existence of another internal property of the system or state function for the system. Consider the reversible cycle shown in Fig. 6.3; for convenience, the cycle is given

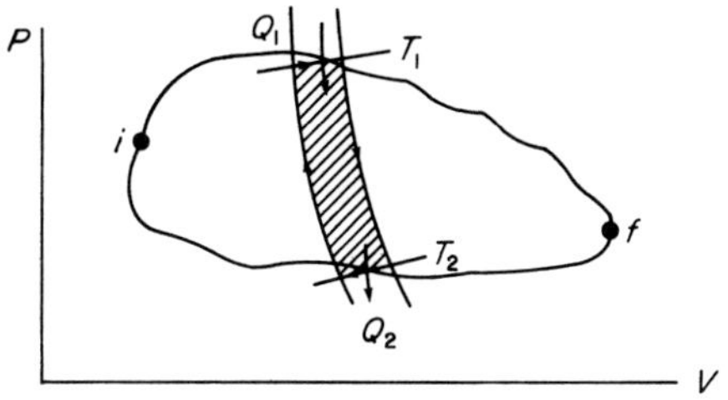


**Fig. 6.3** *i–i* is a closed path for a substance in the *P*, *V* diagram and can be generated by a set of Carnot cycles. The curve bounding the cross-hatched area is one of these Carnot cycles.

in the thermodynamic *P*, *V*, *T* coordinates. Such a cycle can be subdivided into reversible Carnot cycles in the manner shown in the figure. For each small cycle, $(Q_1/T_1) + (Q_2/T_2) = 0$, where the algebraic sign is included in $Q_1$ and $Q_2$. Around a completely reversible path from *i* to *i*, one has

$$0 = \sum \frac{Q_i}{T_i} = \oint_R \frac{dQ}{T}$$

and this is independent of the path selected from *i* to *i*. By selecting various paths which pass through *i* and *f*, it then follows that the integral from *i* to *f* is path-independent and a function of the end points,

$$\int_i^f \frac{dQ}{T} = S_f - S_i \quad (6.11a)$$

or

$$\frac{dQ_R}{T} = dS \quad (6.11b)$$

This new state function is the entropy and is denoted by *S*. Entropy is an internal function of the system. In any real process, the heat transfer to a reservoir is always less than for the reversible system,

$$\delta Q \leq T\,dS \quad (6.12)$$

where the equality implies a reversible process. This can be shown by constructing an engine with $\delta Q > \delta Q_R = T\,dS$ and then by using this engine to drive a Carnot refrigerator. The combined system violates the Second Law of Thermodynamics.

If heat is added to a system at constant volume, then

$$\delta Q = T\,dS = dU \qquad (\delta W = 0) \tag{6.13}$$

follows from the definition of entropy and the First Law. Entropy can be regarded as a function of the other thermodynamic variables, and for $S(U, V)$,

$$dS = \left(\frac{\partial S}{\partial U}\right)_V dU + \left(\frac{\partial S}{\partial V}\right)_U dV \tag{6.14a}$$

Comparison with Eq. (6.13) yields the relationship

$$\left(\frac{\partial S}{\partial U}\right)_V = \frac{1}{T} \tag{6.14b}$$

and relates the entropy, energy, and temperature.

It is desirable to show that the entropy $S(U, V)$ is a maximum for a system with a fixed volume and energy. This corresponds directly to one of the statistical systems to be considered later. Entropy is an additive quantity like volume and internal energy. The entropy of two systems at constant volume is

$$S(U) = S_1(U_1) + S_2(U_2)$$

and the internal energy

$$U = U_1 + U_2 = \text{constant}$$

Consider first the two systems in equilibrium. If the entropy is a maximum, then

$$0 = \frac{dS}{dU_1} = \frac{dS_1}{dU_1} + \frac{dS_2}{dU_2}\,\frac{dU_2}{dU_1} = \frac{dS_1}{dU_1} - \frac{dS_2}{dU_2} = \frac{1}{T_1} - \frac{1}{T_2}$$

and

$$T_1 = T_2$$

Temperature equality follows for two systems in equilibrium if the entropy is a maximum. If the two systems are not in equilibrium but form a closed system and gradually come into equilibrium, their common entropy should increase with time,

$$0 < \frac{dS}{dt} = \frac{dS_1}{dt} + \frac{dS_2}{dt} = \frac{dS_1}{dU_1}\frac{dU_1}{dt} + \frac{dS_2}{dU_2}\frac{dU_2}{dt}$$

Total energy is conserved and $(dU_1/dt) + (dU_2/dt) = 0$. Direct substitution then yields

$$0 < \frac{dS}{dt} = \left(\frac{1}{T_1} - \frac{1}{T_2}\right)\frac{dU_1}{dt}$$

Let the temperatures be such that $T_1 > T_2$; then

$$\frac{dU_1}{dt} < 0 \qquad \text{and} \qquad \frac{dU_2}{dt} > 0$$

and energy passes from the hotter body to the colder body, in accord with experience.

The equality of pressures for bodies in equilibrium can also be derived from the condition that the entropy be a maximum. Internal energy and volume are used again as the independent variables, and in differential form,

$$T\,dS = dU + P\,dV \qquad \text{(always)}$$

If two parts with volumes $V_1$ and $V_2$ form a closed system of volume

$$V = V_1 + V_2 = \text{constant}$$

then the requirement that the entropy be a maximum with respect to a variation of the volume at constant total energy $U$ yields

$$0 = \frac{\partial S}{\partial V_1} = \frac{\partial S_1}{\partial V_1} + \frac{\partial S_2}{\partial V_2}\frac{\partial V_2}{\partial V_1} = \frac{P_1}{T_1} - \frac{P_2}{T_2}$$

Pressure and entropy are related at constant internal energy by

$$\frac{P}{T} = \left(\frac{\partial S}{\partial V}\right)_U$$

Since from the previous proof the temperatures are equal, the temperatures and pressures for two parts in thermal equilibrium are related by

$$T_1 = T_2 \qquad \text{and} \qquad P_1 = P_2$$

It must be emphasized that the thermodynamic laws apply to any type of matter. If a collection of "junk" is placed inside an enclosure, it obeys the laws of thermodynamics. The form or the character of the matter is not important unless this property is measured. For most matter, the *ultimate mechanical variables* are pressure $P$ and volume $V$. The internal properties of the matter require for their description the existence of an equation of state $T(P, V)$, an internal energy function $U$, and an entropy function $S$. These properties become apparent in experimental measurements and require the laws of thermodynamics for their description as the *ultimate mechanical attributes of matter.*

## 6.3 Thermodynamic Functions

Other functions describing the internal state of a system can now be introduced. Since such functions depend only on the initial and final states of the system and not upon the path, they can be extremely useful for the discussion of real physical systems. Thus in the flow of gas through a porous plug, or a throttling process, the function $U + PV$ remains constant. This particular quantity is referred to as the enthalpy or heat function,

$$H = U + PV \tag{6.15}$$

If a given mass of gas passes through a porous plug, or a throttling process occurs, for which $Q = 0$ and $W = P_f V_f - P_i V_i$, then

$$\mathrm{H_i} = \mathrm{H_f}$$

Between neighboring equilibrium states,

$$dH = T\,dS + V\,dP \tag{6.16}$$

and the change in enthalpy can be directly related to heat capacity at constant pressure,

$$C_\mathrm{P} = \left(\frac{\delta Q}{\delta T}\right)_\mathrm{P} = \left(\frac{\partial H}{\partial T}\right)_\mathrm{P} \tag{6.17}$$

Processes at constant temperature or at constant volume are described best by the Helmholtz function $A$,

$$A = U - TS \tag{6.18}$$

Thus the isothermal work done on a system is given by the change in $A$,

$$dA = -S\,dT - P\,dV \tag{6.19}$$

with $dT = 0$. The Helmholtz function is constant for an isothermal, constant-volume process, that is, $dA = 0$. As noted later in this chapter, the Helmholtz function plays a fundamental role in statistical mechanics.

Processes at constant temperature or constant pressure are described by the Gibbs function,

$$G = H - TS = U + PV - TS = A + PV \tag{6.20}$$

For an infinitesimal reversible process,

$$dG = V\,dP - S\,dT \tag{6.21}$$

and for a reversible process at constant temperature and constant pressure, $dT = dP = 0$, the change in the Gibbs function is

$$dG = 0 \qquad \text{or} \qquad G = \text{constant} \tag{6.22a}$$

This occurs in a system with various phases in equilibrium, and the Gibbs functions for equilibrium between a gas phase and a liquid phase are equal, and along the vaporization curve,

$$G_l = G_v \tag{6.22b}$$

Along the sublimation curve,

$$G_s = G_v \tag{6.22c}$$

and at the triple point, $G_s = G_l = G_v$. The Gibbs function is of importance in chemical reactions occurring at constant $P$ and $T$ and in statistical mechanics when the number of particles are important.

### *6.3.1 Dependence of Thermodynamic Quantities on the Number of Particles*

It is inherent in the use of thermodynamic coordinates that these coordinates be regarded as macroscopic variables. Quantities like $U$, $S$, $H$, $A$, and $G$ are additive quantities and any system is the sum of its individual parts. These individual parts are macroscopic and a $U_n$, $S_n$, ... can be given for each part,

$$U = \sum_n U_n, \qquad S = \sum_n S_n,$$

and so forth. If the system is formed with $N$ particles, a change in the number of particles by $\Delta N$ implies that the change in $U$ is linear in $N$, that is, $\Delta N/N = \Delta U/U$. Internal energy is a function of the additive functions $S$ and $V$ and is of the form

$$U = U(S, V) = Nf\left(\frac{S}{N}, \frac{V}{N}\right) \tag{6.23a}$$

Since temperature is constant throughout a body, the Helmholtz function is of the form

$$A = A(V, T) = NF\left(\frac{V}{N}, T\right) \tag{6.23b}$$

and the Gibbs function is

$$G = N\mu(P, T) \tag{6.23c}$$

In a physical system in which the number of particles is allowed to change, the change in $N$ is another independent thermodynamic variable, and for an infinitesimal change in energy,

$$dU = \left(\frac{\partial U}{\partial S}\right)_{V,N} dS + \left(\frac{\partial U}{\partial V}\right)_{S,N} dV + \left(\frac{\partial U}{\partial N}\right)_{S,V} dN \tag{6.24}$$

Comparison with earlier expressions permits $T$ and $P$ to be used as coefficients for $dS$ and $dV$. The coefficient of $dN$ is denoted as the chemical potential,

$$\mu = \left(\frac{\partial U}{\partial N}\right)_{S,V} \tag{6.25}$$

Equation (6.24) can now be written as

$$dU = T\,dS - P\,dV + \mu\,dN \tag{6.26a}$$

The other differential expressions of interest are

$$dA = -S\,dT - P\,dV + \mu\,dN \tag{6.26b}$$

and

$$dG = -S\,dT + V\,dP + \mu\,dN \tag{6.26c}$$

From these equations, it follows that the chemical potential is also given by

$$\mu = \left(\frac{\partial A}{\partial N}\right)_{V,T} = \left(\frac{\partial G}{\partial N}\right)_{P,T} \tag{6.27}$$

Finally, it can be noted that $G$ is linear in $N$ and

$$G = N\mu \tag{6.28}$$

One example of the usefulness of the chemical potential is in the study of the equilibrium of a system or body in an external field. Consider the system in an external field which is constant in time and then consider the equilibrium between neighboring volumes. Again assume that the entropy $S = S_1 + S_2$ should be a maximum for equilibrium between these two parts. Earlier considerations for the exchange of energy between the two parts with the constraint $U = U_1 + U_2 =$ constant indicated that the entropy is a maximum for $T = T_1 = T_2$. If now the exchange of particles occurs between the two parts, the requirement that $S$ be a maximum subject to the constraint

$$N = N_1 + N_2 = \text{constant}$$

yields,

$$0 = \frac{\partial S}{\partial N_1} = \frac{\partial S_1}{\partial N_1} - \frac{\partial S_2}{\partial N_2} = -\frac{\mu_1}{T_1} + \frac{\mu_2}{T_2} \tag{6.29a}$$

or

$$\mu_1 = \mu_2$$

Equation (6.26a) can be used to show that

$$\left(\frac{\partial S}{\partial N}\right)_{U,T} = -\frac{\mu}{T} \tag{6.29b}$$

Hence for equilibrium in an external field,

$$\mu = \text{constant} \tag{6.30a}$$

In a gravitational field, let $\mu_0$ denote the chemical potential of a molecule in the absence of the field and $u(x, y, z)$ the gravitational energy per molecule. Then

$$\mu_0(P, T) + u(x, y, z) = \text{constant} \tag{6.30b}$$

throughout the system. For a system at the surface of the earth, $u = mgz$ and $\mu_0(P, T) + mgz =$ constant. Since

$$d\mu - -s\,dT + v\,dP \tag{6.31}$$

where $s$ is the entropy per molecule and $v$ the volume per molecule, a system in equilibrium obeys the equation

$$v\, dP = -mg\, dz \tag{6.32}$$

Further integration requires a knowledge of $v(P)$.

## 6.4 Statistical Physics

Statistical physics is the study of the laws governing the behavior and properties of macroscopic systems that are made up of a large number of atoms or molecules. Thermodynamics is used as a guide and indicates that only a few coordinates such as the previously used thermodynamic coordinates are needed to describe a system with a large number of particles. The individual motions of the atoms or molecules and the laws that govern these motions have very little direct bearing on the general laws for the macroscopic properties. Thus a system with $N$ particles requires $3N$ coordinates for its specification. Yet, for this same system, the thermodynamic coordinates $P$, $V$, $T$, $U$, and $S$ can be used to describe the macroscopic properties at equilibrium.

A general feature of an assembly of particles is the invariance under the choice of the coordinate system with respect to displacement and to rotation. In the absence of external forces, this invariance yields a constant linear momentum which depends on the total mass, and yields a constant angular momentum for the entire system. The internal properties of the system depend only on the relative positions of the particles. If the system is isolated and its properties are invariant under the choice of the origin of the time coordinate, the energy is a constant of the motion. These are the only seven independent additive integrals of motion for a mechanical system, and since they can be generated by coordinate displacements, they are valid in either classical or quantum mechanics. Linear and angular momenta and the kinetic energy of the center of mass are connected with the motion as a whole. *Only the internal energy, which depends on the relative motions of the particles, is needed for the discussion of a closed system of particles.* Thus the internal energy of a closed system acquires a unique role in statistical physics. It is the only additive constant of motion of interest. Linear and angular momenta and the energy of the center of mass are often excluded by placing the system inside a rigid box and using coordinates in which this box is at rest.

### 6.4.1 Ensembles of Statistical Mechanics

Consider a system of $N$ particles enclosed in a fixed volume $V$. Let $H_N$ be the Hamiltonian that describes this system, and then in principle the time-independent Schrödinger equation

$$H_N \psi = E\psi \tag{6.33a}$$

can be solved for this mechanical system. Let

$$\psi_{\alpha\mu} = | \psi_{\alpha\mu}) = | \alpha\mu) \tag{6.33b}$$

be one of the many ways of denoting the allowed wave functions for this system and let $E_\alpha$ denote the allowed values of the energy. Here $\mu$ is an index to denote the number of states with the same energy. The solution of such a problem is in general quite complex and only for a few ideal cases can these allowed states be found in detail. For our purposes, only the existence of a set of states of given energy is needed. Since the Schrödinger equation is a wave equation, an infinite amount of time would be required to prepare and then measure the energy of such a system. It is inherent in such a system that the accuracy to which the energy can be measured in time interval $\Delta t$ is given by the uncertainty relationship $\Delta E \, \Delta t \approx h$. Thus, even under the best conditions, the energy is unknown by this amount. In real physical systems of interest, heat exchange with the surroundings must also be considered. The introduction of the interactions with the surroundings is a problem of considerable difficulty, and ensembles are often introduced to avoid the need for a direct knowledge of these interactions.

Consider an ensemble of systems of the type just discussed and shown schematically in Fig. 6.4. Each cell has a Hamiltonian $H$, a fixed volume $V$, and a fixed number of particles $N$. These need not be the same in detail

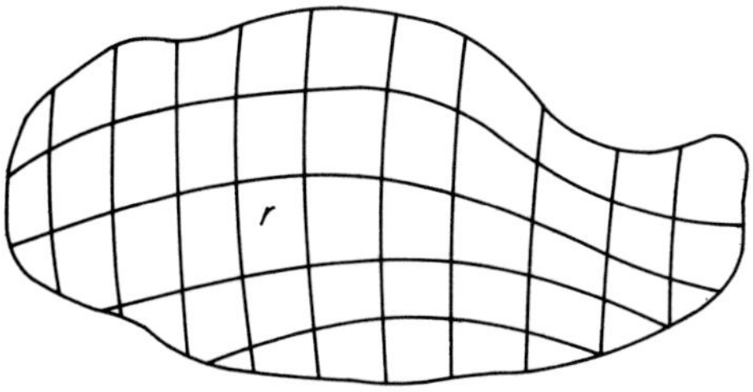


**Fig. 6.4** Ensemble of systems with Hamiltonian $H_r$, volume $V_r$, and number of particles $N_r$ in thermal contact. Energy of ensemble is $\mathscr{U}$ and is a constraint on the subsystems.

and the separate systems are denoted by the label $r$. Exchange of energy between these systems is permitted and the entire ensemble of systems is regarded to have a total energy

$$\mathscr{U} = \sum_{r=1}^{\mathscr{N}} E_r \tag{6.34}$$

This forms a large, closed system of fixed volume and fixed energy. By increasing the size, the uncertainty in this energy $\mathscr{U}$ can be made arbitrarily small. Such a system is regarded as a *microcanonical ensemble* in this text. From the statistical point of view, this is a system with a single constraint. The number of allowable configurations of this constrained ensemble and the probability of a particular value of $E_r$ are of interest. This is similar to the problem of conditional probability in Chapter III. Even though these systems can exchange energy, they are physically distinct systems for a short period of time and during this time, the energy is characteristic of the energy states $E_\alpha$.

Each system in the microcanonical ensemble is regarded in the same sense as a die with an infinite number of faces. Each face is characterized by the indices $\alpha\mu$ which indicate one of the previous states $\psi_{\alpha\mu}$. Let $g(\alpha)$ denote the number of states with the same energy $E_\alpha$. In the absence of energy exchange, each system would have its own characteristic energy $E_r, E_{r'}, \ldots$. In the presence of this single constraint, there is a conditional probability for this energy. Label these energy levels as points on a line separated by a distance $\varepsilon$ and select $\varepsilon$ sufficiently small so that all energy levels are represented by $m\varepsilon$. Thus the energy

$$E = m\varepsilon, \qquad m = 0, 1, 2, \ldots, \infty$$

$$\begin{aligned} g(m) &= \{\text{number of states with } \; E_\alpha\} && \text{if } \; E_\alpha = m\varepsilon \\ &= 0 && \text{if } \; E_\alpha \neq m\varepsilon \end{aligned}$$

and not all points on the line correspond to allowed energies. These points have weight zero. As in the example of the die, it is assumed that some distribution law governs this system and since there is only a single constraint, it is of the form

$$y(m) = g(m)\frac{e^{-\beta m\varepsilon}}{Z(\beta)} \tag{6.35a}$$

Each state has equal weight in the absence of the constraint. Since our system is always found in one of its allowed energies,

$$\sum_{m=0} y(m) = 1 \tag{6.35b}$$

and therefore for the $r$th system,

$$Z(\beta) = \sum_m g(m)e^{-\beta m\varepsilon} = \sum_\alpha g(\alpha)e^{-\beta E_\alpha} \tag{6.35c}$$

If the total energy of the ensemble is given by

$$\mathscr{U} = \sum_{r=1}^{\mathscr{N}} E_r = \sum_{r=1}^{\mathscr{N}} m_r\varepsilon_r \tag{6.36}$$

where $E_r$ is the energy of the subsystem with the label $r$, then the distribution law for $\mathscr{N}$ mutually independent random variables $E_r$ yields for the probability of total energy $\mathscr{U}$.

$$\begin{aligned} P(\mathscr{U}) &= \sum_{\text{all}} \left[\prod_{r=1} y(m_r\varepsilon_r)\ \delta\left(\mathscr{U} - \sum_{r=1} m_r\varepsilon_r\right)\right] \\ &= \frac{G(\mathscr{U}, \mathscr{N})e^{-\beta\mathscr{U}}}{\prod Z_r(\beta)} \end{aligned} \tag{6.37}$$

This is similar to Eq. (3.63) and entails a similar counting over all allowed configurations. As in Section 3.7, the characteristic function

$$\varphi_r(a) = Z_r(\beta - ia)/Z_r(\beta)$$

can be introduced and then the characteristic function $\Phi(a) = \Pi\, \varphi_r(a)$ is expanded as in Eqs. (3.64a)–(3.67). Equation (3.68) gives the number of arrangements subject to the constraint

$$G(\mathscr{U}, \mathscr{N}) = \frac{\prod_r Z_r(\beta) \exp \beta\mathscr{U}}{(2\pi D_{\mathscr{N}})^{1/2}} \exp\left[\frac{-(\mathscr{U} - \overline{\mathscr{U}})^2}{2D_{\mathscr{N}}}\right] \tag{6.38}$$

The average energy $\overline{\mathscr{U}}$ is

$$\overline{\mathscr{U}} = -\sum_{r=1}^{\mathscr{N}} \frac{d(\ln Z_r)}{d\beta} \tag{6.39}$$

and the parameter $\beta$ is determined by the requirement that

$$\mathscr{U} = \overline{\mathscr{U}} = -\sum_r \frac{d(\ln Z_r)}{d\beta} \tag{6.40}$$

This equation has a single root $\beta$ and differs from the development for the die. The number of energy levels $E_\alpha$ is infinite and $\beta \geq 0$, while the die problem permitted both positive and negative values of $\beta$. With this value of $\beta$, the number of permitted configurations is

$$G(\bar{\mathscr{U}}, \mathscr{N}) = \left[\prod_r Z_r(\beta)\right](\exp \beta\bar{\mathscr{U}})(2\pi D_{\mathscr{N}})^{-1/2} \tag{6.41}$$

$D_{\mathscr{N}}$ is the dispersion and is given by Eq. (3.66b), or

$$D_{\mathscr{N}} = \sum_r \frac{d^2(\ln Z_r)}{d\beta^2} > 0 \tag{6.42}$$

and increases linearly with $\mathscr{N}$.

Since $G$ is a very large number, usually $\ln G$ is of greater interest, and

$$\ln G(\bar{\mathscr{U}}, \mathscr{N}) = \sum_r^{\mathscr{N}} \ln Z_r(\beta) + \beta\bar{\mathscr{U}} + O(\ln \mathscr{N}) \tag{6.43}$$

Since both the sum and $\bar{\mathscr{U}}$ are linear in $\mathscr{N}$, the term $\ln D_{\mathscr{N}}$, which is of order $\ln \mathscr{N}$, can be omitted. For this microcanonical ensemble with a single constraint $\mathscr{U}$, the logarithm of the number of permitted configurations is, to a very high degree of approximation, an additive function and is denoted as the entropy of the microcanonical ensemble,

$$\mathscr{S}(\bar{\mathscr{U}}, \mathscr{N}) = \ln G(\bar{\mathscr{U}}, \mathscr{N}) = \sum_r \ln Z_r(\beta) + \beta\bar{\mathscr{U}} \tag{6.44}$$

$\mathscr{S}$ has a maximum value and this follows from Eq. (6.40) and the discussion of Eq. (3.69). $\beta$ is a property of every system of the ensemble and can be compared with the temperature,

$$\frac{\partial \mathscr{S}}{\partial \bar{\mathscr{U}}} = \beta = \frac{1}{kT} \tag{6.45}$$

### 6.4.2 Canonical Ensemble

The canonical ensemble is one of the systems in the previous discussion, or a set of subsystems. For convenience, let it be one of the systems with $N$ particles, volume $V_N$, Hamiltonian $H_N$, and energy levels $E_\alpha$. The probability that this system has energy $E_\alpha$ is of primary interest. *It is assumed that each allowed configuration is equally probable* and therefore the probability that the selected system has energy $E_\alpha$ is

$$p(E_\alpha) = \frac{g(\alpha) G(\bar{\mathscr{U}} - E_\alpha; \mathscr{N} - 1)}{G(\bar{\mathscr{U}}, \mathscr{N})} \tag{6.46}$$

where $g(\alpha)$ is the number of configurations of the systems with energy $E_\alpha$, $G(\bar{\mathscr{U}}, \mathscr{N})$ is the number of configurations of the microcanonical ensemble with the constraint $\bar{\mathscr{U}}$ on the energy for $\mathscr{N}$ systems, and $G(\bar{\mathscr{U}} - E_\alpha, \mathscr{N} - 1)$ is the number of configurations with the constraint $\bar{\mathscr{U}} - E_\alpha$ excluding the system under consideration. These $G$'s follow from Eq. (6.41) and the probability is in accordance with the development of Eq. (3.74), in which each allowed configuration was given equal weight. It follows that

$$p(E_\alpha) = \frac{g(\alpha)e^{-\beta E_\alpha}}{Z(\beta)} = \frac{g(\alpha)e^{-\beta E_\alpha}}{Q_N} \tag{6.47a}$$

is the conditional probability for energy $E_\alpha$. The probability for a particular state $\alpha\mu$ in this particular system is

$$p(\alpha\mu) = \frac{p(E_\alpha)}{g(\alpha)} = \frac{e^{-\beta E_\alpha}}{Q_N} \tag{6.47b}$$

$Z(\beta)$ is the generating function for this distribution and is henceforth denoted by $Q_N = Z(\beta)$. With this notation, the generating function for the canonical ensemble is

$$Q_N = \sum_\alpha g(\alpha)e^{-\beta E_\alpha} = \sum_{\alpha\mu} e^{-\beta E_{\alpha\mu}} \tag{6.48}$$

where the sum over $\alpha$ is over energy levels and the sum over $\alpha\mu$ is the sum over states.

The average value of the energy of the system is given by

$$U_N = \frac{\sum_\alpha E_\alpha g(\alpha)e^{-\beta E_\alpha}}{\sum_\alpha g(\alpha)e^{-\beta E_\alpha}} = -\frac{\partial(\ln Q_N)}{\partial\beta} \tag{6.49}$$

A second average of interest is the average of $\ln p(\alpha\mu)$:

$$\sigma_N = -\sum_{\alpha\mu} p(\alpha\mu) \ln p(\alpha\mu) = \sum_{\alpha\mu} \frac{e^{-\beta E_{\alpha\mu}}}{Q_N} (\ln Q_N + \beta E_{\alpha\mu}) \tag{6.50a}$$

$$\sigma_N = \ln Q_N + \beta U_N \tag{6.50b}$$

It can be noted that if $\sigma_N$ is computed for each system, then $\mathscr{S}$ is given by this sum, that is,

$$\mathscr{S}(\bar{\mathscr{U}}, \mathscr{N}) \approx \sum_r (\sigma_N)_r = \sum_r (\ln Q_N + \beta U_N)_r \tag{6.51}$$

As noted in Eqs. (3.82)–(3.84), this technique can be used to estimate the number of allowed configurations $G = \exp \mathscr{S}$ for a given value of $\beta$. Changing $\beta$ to $1/kT$ and taking the units so that $S_N = k\sigma_N$, it follows that

$$-kT \ln Q_N = U_N - TS_N \tag{6.52}$$

and the generating function of the distribution is the Helmholtz function

$$A_N = -kT \ln Q_N \tag{6.53}$$

Entropy is given by

$$S_N = -\frac{\partial A_N}{\partial T} \tag{6.54}$$

This entire development is at constant volume. If a change in volume as well as temperature occurs, then

$$dA_N = +\left(\frac{\partial A_N}{\partial T}\right)_{\text{V}} dT + \left(\frac{\partial A_N}{\partial V}\right)_{\text{T}} dV = -S_N\, dT - P\, dV \tag{6.55}$$

since

$$\left(\frac{\partial A_N}{\partial V}\right)_{\text{T}} = -kT \frac{1}{Q_N} \sum \left[-\beta \left(\frac{\partial E_\alpha}{\partial V}\right)_{\text{T}} g(\alpha) e^{-\beta E_\alpha}\right] = \left(\frac{\partial U_N}{\partial V}\right)_S = -P$$

The subscript $N$ denoting an $N$-particle system is usually suppressed.

### *6.4.3 Grand Canonical Ensemble*

Exchange of particles as well as energy can occur for most real physical systems and this statistical system is referred to as the grand canonical ensemble. This statistical system can be generated by placing small holes in the walls in Fig. 6.4. Particles can be transferred throughout the entire microcanonical ensemble, but for short periods of time, a given volume $V$ contains $N$ particles and is described by the Hamiltonian function $H_N$. During this short period, its average properties are described by the generating function $Q_N$ of the canonical ensemble. Another measurement would yield a different $N$ and the generating function for the distribution of particles in this cell is

$$\mathscr{Q} = \sum_{N=0} z^N Q_N = \mathscr{Q}(zVT) \tag{6.56}$$

Fixing the number of particles is a constraint on the microcanonical ensemble. The parameter $z$ is related to this constraint and its thermodynamic counterpart will become apparent in the subsequent development. At constant volume $V$ and temperature $T$, the average number of particles in this volume is

$$\bar{N} = \frac{\sum N z^N Q_N}{\sum z^N Q_N} = z\left[\frac{\partial(\ln \mathscr{Q})}{\partial z}\right]_{\mathrm{V,T}} \tag{6.57}$$

and the average energy at constant $z$ and $V$ is

$$\bar{U} = \frac{\sum z^N U_N Q_N}{\sum z^N Q_N} = -\left[\frac{\partial(\ln \mathscr{Q})}{\partial \beta}\right]_{z,\mathrm{V}} \tag{6.58}$$

Since $Q_N = e^{-\beta A_N}$ and $\mathscr{Q}$ can be written as

$$\mathscr{Q} = \sum_{N=0} z^N e^{-\beta A_N} \tag{6.59}$$

the pressure $P = -\partial A_N/\partial V$ is related to the average pressure in the grand canonical generating function by

$$P = \frac{1}{\beta}\left(\frac{\partial(\ln \mathscr{Q})}{\partial V}\right)_{z,\mathrm{T}} \tag{6.60}$$

Although this is the average pressure, the symbol for average is usually omitted. This function can be integrated at constant $z$ and $T$ to yield the equation of state

$$\frac{PV}{kT} = \ln \mathscr{Q} \tag{6.61}$$

This development and the generating function suggest that $z^N Q_N$ is proportional to the probability that a particular cell contains $N$ particles. Let $p(N)$ be the probability that the $r$th cell has $N$ particles,

$$p(N) = \frac{z^N Q_N}{\mathscr{Q}} \tag{6.62a}$$

In a slight change of notation, let the indices $a_N$ denote the allowed states for $N$ particles in volume $V$ and then the generating function $Q_N = \sum e^{-\beta E(a_N)}$ is summed over the states with this index. Let the probability of a state with index $a_N$ be given by

$$p(a_N) = \frac{z^N e^{-\beta E(a_N)}}{\mathscr{Q}} \tag{6.62b}$$

so that

$$p(N) = \sum_{a_N} p(a_N)$$

Again it should be emphasized that $a_N$ refers to a state of the $N$-particle system in volume $V$ and not an energy level. The average value of $-\ln p(a_N)$ yields the entropy,

$$\begin{aligned}
\sigma &= -\sum_N \left[\sum_{a_N} p(a_N) \ln p(a_N)\right] \\
&= -\sum_N \left\{\sum_{a_N} p(a_N)[N \ln z - \beta E(a_N) - \ln \mathscr{Q}]\right\} \\
&= -\sum_N [p(N)(N \ln z - \beta U_N - \ln \mathscr{Q})] \\
&= -\bar{N} \ln z + \beta \bar{U} + \ln \mathscr{Q} \qquad (6.63a)
\end{aligned}$$

For this particular cell, the entropy $S = k\sigma$ and

$$\frac{\bar{S}}{k} = -\bar{N} \ln z + \beta(\bar{U} + PV) \qquad (6.63b)$$

or

$$\bar{N}kT \ln z = \bar{U} + PV - T\bar{S} = \bar{G} = \bar{N}\mu \qquad (6.63c)$$

where $z$ is usually called the "fugacity" and $\mu$ the "chemical potential," and then

$$\mu = kT \ln z \qquad (6.63d)$$

The development of this section implies that $\mu$ and $z$ refer to a particular chemical species and only one such species is present. If more than one species is present, then a $z$ or $\mu$ is needed for each type. This will be discussed in greater detail in the examples of the next chapter. Phases are discussed in Chapter IX.

Fixing the total energy $\mathscr{U}$ of the microcanonical ensemble introduced a constraint. It was shown how to introduce a single parameter $\beta$ and the canonical ensemble to estimate the number of allowed arrangements for the microcanonical ensemble. In principle, this could be repeated for the constraint introduced by fixing the total number of particles. A single parameter $z$ and the grand canonical ensemble can then be used to estimate the number of configurations which are allowed by both the energy and number constraints.

### 6.4.5 Density Matrix for the Canonical Ensemble

The density matrix of quantum mechanical system is often written as the operator

$$\varrho = \sum_{a} | a) P_a (a | \tag{6.64}$$

where $P_a$ is the probability of the state $| a)$, or $| \alpha\mu)$ in our previous example. Average values of operators $F$ are given by

$$\text{tr } \varrho F = \sum_{aa'} (a | \varrho | a')(a' | F | a) = \sum P_a (a | F | a) \tag{6.65}$$

The notation $(a' | a) = \delta_{aa'}$ is used. For the canonical ensemble, $P_a = Q^{-1} \times e^{-\beta E_a}$ and the density matrix for an $N$-particle canonical ensemble can be written as

$$\varrho_N = \frac{e^{-\beta H_N}}{Q_N} = e^{\beta(A_N - H_N)} \tag{6.66}$$

With this notation, the normalization condition yields

$$\text{tr } \varrho_N = 1 \tag{6.67}$$

and the generating function for the canonical ensemble is given by the equivalent forms

$$Q_N = \sum_a e^{-\beta E_a} = \sum_a (a | e^{-\alpha H_N} | a) = \text{tr } e^{-\beta H_N} \tag{6.68}$$

The other thermodynamic variables are given by

$$U_N = \text{tr } \varrho_N H_N \tag{6.69a}$$

$$A_N = -kT \ln(\text{tr } e^{-\beta H_N}) \tag{6.69b}$$

$$S_N = -\left(\frac{\partial A_N}{\partial T}\right)_V \tag{6.69c}$$

$$\vdots$$

#### 6.4.5.1 The "Slatersum"

A sum introduced into quantum statistics by Slater [1] and often referred to as the "Slatersum" is often used. The $N$-particle generating function can be written in the following alternate ways given by Eq. (6.68). A complete set of functions that do not yield the allowed values of energy can also be used. If $\varphi_b$ denotes a set of functions which spans the space of the $N$-

particle system and $\psi_a$ denotes the set of functions for which $H_N\psi_a = E_a\psi_a$, then $\psi_a$ can be expanded in terms of the other complete set $\varphi_b$,

$$\psi_a = \sum_b | \varphi_b)(\varphi_b | \psi_a)$$

Direct substitution into Eq. (6.68) yields

$$Q_N = \sum_b (\varphi_b | e^{-\beta H_N} | \varphi_b) = \mathrm{tr}\, e^{-\beta H_N} \tag{6.70}$$

This a manifestation of the invariance of the trace and any complete set of functions which spans the space of the $N$-particle system can be used for the evaluation of the generating function $Q_N$.

## References

1. J. Slater, *Phys. Rev.* **38**, 237 (1931).

## General References

*Macroscopic Thermodynamics*

Callen, H. B., "Thermodynamics." Wiley, New York, 1960.
Fermi, E., "Thermodynamics." Dover, New York, 1957.
Guggenheim, E. A., "Thermodynamics," 4th ed. Wiley (Interscience), New York, 1960.
Kirkwood, J. G., and Oppenheim, I., "Chemical Thermodynamics." McGraw-Hill, New York, 1961.
Pippard, A. B., "The Elements of Classical Thermodynamics." Cambridge Univ. Press, London and New York, 1957.
Zemansky, M. W., "Heat and Thermodynamics," 4th ed. McGraw-Hill, New York, 1957.

*Statistical Mechanics*

Davidson, N., "Statistical Mechanics." McGraw-Hill, New York, 1962.
Fowler, R. H., "Statistical Mechanics," 2nd ed. Cambridge Univ. Press, London and New York, 1955.
Fowler, R. H., and Guggenheim, E. A., "Statistical Thermodynamics," revised ed. Cambridge Univ. Press, London and New York, 1949.
Hill, T. L., "An Introduction to Statistical Thermodynamics." Addison-Wesley, Reading, Massachusetts, 1960.
Huang, K., "Statistical Mechanics." Wiley, New York, 1963.
Landau, L. D., and Lifshitz, E. M., "Statistical Physics." Addison-Wesley, Reading, Massachusetts, 1959.

Reif, F., "Fundamentals of Statistical and Thermal Physics." McGraw-Hill, New York, 1965.

Rushbrooke, G. S., "Introduction to Statistical Mechanics." Oxford Univ. Press, London and New York, 1949.

ter Haar, D., Foundations of statistical mechanics. *Rev. Mod. Phys.* **27**, 289 (1955).

ter Haar, D., "Elements of Statistical Mechanics." Holt, New York, 1954.

Tolman, R. C., "The Principles of Statistical Mechanics." Oxford Univ. Press, London and New York, 1958.

Tribus, M., "Thermostatics and Thermodynamics." Van Nostrand-Reinhold, Princeton, New Jersey, 1961.

Wilks, J., "The Third Law of Thermodynamics." Oxford Univ. Press, London and New York, 1961.

# Molecular Systems at Low Densities VII

## 7.1 Nearly Ideal Gases

The statistical properties of nearly ideal gases, or molecular gases at low densities, are considered in this chapter. If a single molecule or atom is placed in volume $V$ and the exchange of energy with the wall is small, the allowed states for this atom follow from a solution of the Schrödinger equation. Wall interactions are neglected in this solution. The translational coordinates of the center of mass can always be separated from the rotational motion and the internal coordinates of the system. Energy exchange does not occur between the translational and internal coordinates and even for this single molecule these states are statistically independent. The one-particle partition or generating function becomes the product of generating functions. As discussed in Chapter III, statistical independence requires that the generating function be a product of generating functions. This very basic concept is used in every development in this chapter. Thus the canonical generating function for the one-molecule system is

$$Q_1 = \{Q_{\mathrm{t}}\}\{Q_{\mathrm{int}}\} \tag{7.1}$$

For monatomic atoms at low temperatures, $Q_{\mathrm{int}} = 1$ and only $Q_t$ remains. $Q_{\mathrm{int}}$ for molecules and excited atoms will be discussed as special examples.

If two identical molecules are placed in the volume $V$ and the volume $V$ is very large, the interaction between the two molecules can be made negligibly small. One might expect the two molecules to be statistically independent and the generating function $Q_2$ for two molecules to be of the form $Q_2 = Q_1^2$. This is the combination rule used for dice in Chapter III. One of the remarkable aspects of quantum mechanics is that this procedure overcounts the number of allowed states. The wave function for the two-particle system must be either symmetric or antisymmetric under the exchange of two identical particles. For two particles with coordinates $\mathbf{r}_1$ and $\mathbf{r}_2$ and in states $\alpha$ and $\beta$, the wave function must be of the form

$$\tfrac{1}{2}[\psi_\alpha(1)\psi_\beta(2) \pm \psi_\alpha(2)\psi_\beta(1)] \qquad \text{(only one state permitted)}$$

The two states $\psi_\alpha(1)\psi_\beta(2)$ and $\psi_\alpha(2)\psi_\beta(1)$ are not permitted. This constraint on the allowed states is called Fermi–Dirac statistics for the negative sign and Bose–Einstein statistics for the plus sign. Thus, for this nearly ideal system,

$$Q_2 = \frac{Q_1^2}{2!}$$

This argument can be extended to an $N$-particle system with coordinates $\mathbf{r}_1, \ldots, \mathbf{r}_N$ and states $\alpha, \ldots, \nu$. There are $N!$ ways of distributing $N$ particles among these states, but only one of these $N!$ states is allowed. Thus

$$Q_N = \frac{Q_1^N}{N!} \tag{7.2}$$

is the generating function for $N$ noninteracting atoms in volume $V$.

For molecular gases at low densities, the generating function for $N$ particles in volume $V$ is approximately

$$Q_N = \frac{Q_1^N}{N!} \approx \left(\frac{eQ_1}{N}\right)^N \tag{7.3a}$$

$$\approx \left[\left(\frac{e}{N}\right)Q_t\right]^N [Q_{\text{int}}]^N \tag{7.3b}$$

Stirling's expansion $\ln N! \approx N \ln N - N$, or $N! \approx e^{-N}N^N$, is used to form Eq. (7.3b). The Helmholtz function is, from Eq. (6.53),

$$A(VTN) = -kT \ln Q_N \approx -NkT \ln \frac{eQ_1}{N} \tag{7.4a}$$

The statistical independence of the internal and translational motions allows $A$ to be written as a sum of independent parts, or

$$A = A_t + A_{\text{int}} = -NkT\left(\ln \frac{eQ_t}{N} + \ln Q_{\text{int}}\right) \tag{7.4b}$$

The chemical potential is used in this chapter and can be obtained from

$$\mu = \left(\frac{\partial A}{\partial N}\right)_{\mathrm{V,T}} \qquad \text{or} \qquad A_{N+1} - A_N$$

$$= -kT \ln \frac{Q_1}{N} \tag{7.5a}$$

or from the grand partition or generating function. From Eq. (6.56),

$$\mathcal{Q} = \sum_{N=0} z^N Q_N \approx \sum \frac{(zQ_1)^N}{N!} \approx e^{zQ_1} \tag{7.6}$$

and from Eq. (6.61),

$$\ln \mathcal{Q} = \frac{PV}{kT} \approx zQ_1 \tag{7.7}$$

Equation (6.57) relates $\mathcal{Q}$ to the average number of particles

$$\bar{N} = z\left(\frac{\partial(\ln \mathcal{Q})}{\partial z}\right)_{\mathrm{V,T}} = zQ_1 \tag{7.8}$$

and is of course the Ideal Gas Law discussed in Chapter I. Since $z = e^{\mu/kT}$, the chemical potential is again

$$\mu = -kT \ln \frac{Q_1}{N} \tag{7.5b}$$

Only the one-molecule partition function is needed for the discussion of low-density molecular gases, and this is shown in greater detail in Chapter VIII.

### 7.1.1 *Translational Generating Function*

Schrödinger's equation for spinless particles

$$-\frac{\hbar^2}{2M} \nabla^2 \psi + V\psi = E\psi \tag{7.9}$$

is readily solved for one molecule of mass $M$ in a box with rigid walls, and the wave function is of the form

$$\psi_k = \frac{1}{L^{3/2}} \exp i\mathbf{k} \cdot \mathbf{r} \tag{7.10a}$$

$$E_{\mathbf{k}} = \frac{\hbar^2 k^2}{2M} \tag{7.10b}$$

Traveling-wave solutions are used rather than standing waves, and with the periodic boundary conditions $\psi(x + L) = \psi(x), \ldots,$ for $y$ and $z$, the allowed values of $k_x$ are

$$k_x = \frac{2\pi n_x}{L}, \qquad n_x = 0, \pm 1, \pm 2, \ldots \tag{7.10c}$$

and $k_y$ and $k_z$ follow by using $n_y$ and $n_z$. These states are now quantized and the translational partition function

$$Q_t = \sum_{\mathbf{k}} \exp(-\beta E_{\mathbf{k}}) \tag{7.11}$$

is found by summing over all integral values of the indices $k_x, k_y, k_z$. It is convenient to change the sum to an integral and, using Eq. (2.15) for the number of states in the range $\Delta\mathbf{k}$,

$$\sum_{\Delta\mathbf{k}} \to L^3 \frac{dk_x\, dk_y\, dk_z}{(2\pi)^3} \qquad \text{or} \qquad L^3 \frac{k^2\, d\Omega\, dk}{(2\pi)^3} \tag{7.12}$$

$L^3 = V$ is used in most subsequent equations. Comparison with classical expressions can be made by the substitutions

$$p_x = \hbar k_x \qquad \text{and} \qquad E_{\mathbf{k}} = \frac{p^2}{2M} \tag{7.13}$$

Transforming to continuous variables,

$$\begin{aligned} Q_t &= V \int_{-\infty}^{\infty} \left\{ \exp\left[ - \frac{\beta\hbar^2(k_x{}^2 + k_y{}^2 + k_z{}^2)}{2M} \right] \right\} \frac{dk_x\, dk_y\, dk_z}{(2\pi)^3} \\ &= V \left( \frac{2\pi MkT}{h^2} \right)^{3/2} = \frac{V}{\lambda^3} \end{aligned} \tag{7.14}$$

and $\beta^{-1} = kT$, where $k$ is the Boltzmann constant in the last equation. $\lambda$ is a thermal de Broglie wavelength defined by Eq. (8.16) and $\lambda = 1.74 \times 10^{-9}(M'T)^{-1/2}$, where $M'$ is the molecular weight number.

Using Eq. (7.14) for $Q_{\cdot}$ and Eq. (7.5) for the chemical potential, the chemical potential for a molecular gas becomes

$$\mu = -kT \ln\left\{ \left[ \frac{V}{N} \left( \frac{2\pi MkT}{h^2} \right)^{3/2} \right] Q_{\text{int}} \right\} \tag{7.15}$$

## 7.2 Photon Gas

A photon gas is an ideal gas in that photons do not interact; a photon gas is the description of the electromagnetic radiation inside an enclosure. In the same manner as discussed in Section 2.2, a cavity of volume $V$ with perfectly conducting walls (or almost perfectly conducting walls) is analyzed in terms of its normal modes. These normal modes are independent. Let an index $a$ denote the normal mode. The energy associated with this normal mode is

$$E_{n_a} = h\nu_a(n_a + \tfrac{1}{2}), \qquad n_a = 0, 1, \ldots \tag{7.16}$$

where $\nu_a$ is the natural frequency of oscillation of the mode. The zero-point energy is usually omitted in calculations involving an infinite number of modes. If the walls are almost perfectly conducting and a small piece of matter is placed in the cavity, energy exchange can occur. Each mode can exchange energy with the walls or with the small piece of matter. Since the modes do not exchange energy, or the photons of different modes do not interact, the modes can be regarded as statistically independent and the generating function is a product of generating functions,

$$Q = \prod_a Q_a \tag{7.17}$$

Again the index $a$ labels all modes of the system. Since the total number of photons is not a constraint, the chemical potential for photons is zero and the canonical ensemble can be used as a generating function. Each mode is in thermal equilibrium with the walls and the small piece of matter, or, in the frequently used terminology, "speck of dust" at temperature $T$. The single-mode generating function is

$$Q_a = \sum_{n_a=0} e^{-\beta n_a h\nu_a} = \frac{1}{1 - e^{-h\nu_a/kT}} \tag{7.18}$$

where the simplification results by using

$$\sum_{n=0}^{\infty} x^n = \frac{1}{1-x} \tag{7.19}$$

All of the thermodynamic quantities can be defined for this mode and can be derived from the free energy,

$$A_a = -kT \ln Q_a = +kT \ln(1 - e^{-h\nu_a/kT}) \tag{7.20}$$

and for all modes, as

$$A = \sum_a A_a = -kT \sum_a \ln Q_a \tag{7.21}$$

The average number of thermal quanta associated with a mode is

$$\langle n_a \rangle = Q_a^{-1} \sum n_a e^{-h\nu_a n_a/kT}$$

or

$$\bar{n}_a = \langle n_a \rangle = \frac{1}{e^{h\nu_a/kT} - 1} \tag{7.22}$$

and this justifies the form of Eq. (2.13). This is called the *Planck distribution*.

Plane-wave expansions are more convenient and for a large volume $V$, the indices $a$ can be changed to $k$. Again, $k$ represents four indices and $\hat{k}$ denotes the direction of the plane wave, $h\nu_k$ the energy, $p$ the polarization, and $\hbar\mathbf{k}$ the momentum of the photon. The arguments used in developing Eqs. (2.15)–(2.17) can be used to show that the radiation is isotropic and the average number of thermal photons with energy $h\nu$ is

$$\bar{n}_\nu = \frac{1}{e^{h\nu/kT} - 1} \tag{7.23a}$$

This is Eq. (2.20) and, with the number of states in $d\Omega\, d\nu$, or Eq. (2.14), makes possible the description of most of the properties of thermal radiation discussed in Chapter II.

The use of the number of photons in state $a$ obscures the statistics used in the Planck distribution. Photons have angular momentum in units of $\pm\hbar$ and are intrinsic spin-1 particles which obey Bose–Einstein statistics. The wave functions must be symmetric in the exchange of photon coordinates, and the use of photon numbers $n_a$, $n_k$, $n_\nu$, ... with values between 0 and $\infty$ ensures that only this symmetric state is kept. The remaining configurations are not allowed.

**Exercise 7.1** Find the entropy of state $a$ from Eq. (7.20) and show that it can be written as

$$\frac{S_a}{k} = \ln(1 + \bar{n}_a) + \left(\frac{h\nu_a}{kT}\right)\bar{n}_a \tag{7.23b}$$

**Exercise 7.2** Find the total free energy $A/V$, the energy $U/V$, the entropy $S/V$, and the heat capacity $C_V$ for thermal radiation. Let the sum over states be transformed to an integral by Eq. (2.14). Use the definite integrals introduced in Chapter II.

Exercise 7.3 Use the free energy $A$ to find the pressure exerted by thermal radiation and compare with the results given in Chapter II.

## 7.3 Bose and Fermi Distributions

If a particle of mass $M$ is placed in volume $V$, the Schrödinger equation for its motion is given by Eq. (7.9) and this equation can be solved for the state $\psi_{\alpha\mu}$ and energy $\varepsilon_\alpha$. If a second particle is placed in the enclosure, the interaction between the particles in a large enclosure is small. Ideal Bose and Fermi gases are assumed to have zero intermolecular interaction. Even so, these molecules are not statistically independent and the two-particle generating function is not $Q_2 = Q_1{}^2$. This ideal problem can be reformulated so that the occupation numbers for the one-particle states $\psi_{\alpha\mu}$ are specified. Such a reformulation is similar to the normal-mode expansion for the electromagnetic field, and for particles is referred to as second quantization. The state of an $N$-particle system is given by the number of particles $n_a$ in the state $\psi_a$ with energy $\varepsilon_a$; $\alpha\mu$ has been replaced by $a$ for convenience in notation. Then,

$$N = \sum_a n_a \tag{7.24a}$$

$$E = \sum n_a \varepsilon_a \tag{7.24b}$$

The grand canonical generating or partition function can be written as

$$\mathscr{Q} = \sum_N z^N Q_N = \sum_N \delta\left(\sum n_a - N\right) \prod_{n_a} e^{\beta(\mu - \varepsilon_a)n_a}$$

but this formulation misses the point. There is no term in the Hamiltonian that causes a direct interaction between particles, and only particles in the same state interact or interfere. In principle, it is possible to bring each state independently into equilibrium with a thermal and particle reservoir. These states are now statistically independent and the generating function for the grand canonical ensemble is

$$\mathscr{Q} = \prod_a \mathscr{Q}_a \tag{7.25}$$

where $\mathscr{Q}_a$ is the generating function for the state $a$. Our vessel of volume $V$ has small holes in it and can exchange particles with the surroundings.

The average properties of the state $a$ are generated by

$$\mathcal{Q}_a = \sum_{n_a} z^{n_a} e^{-\beta n_a \varepsilon_a} = \sum_{n_a} e^{+\beta(\mu-\varepsilon_a)n_a} \tag{7.26}$$

where the sum is over the allowed number of particles in this particular state. The index $a$ refers to the exact allowed states and only in very special examples can it be replaced by the plane-wave index $k$.

Let $F$ denote the total angular momentum of a particle or the particle spin; then we have the situation shown in Table 7.1, where particle refers

**Table 7.1**

| Particle spin $F$ | Symmetry of wavefunction under exchange of identical particles | Statistics | Occupation number |
|---|---|---|---|
| Half-integer | Antisymmetric | Fermi–Dirac | $n_a = 0, 1$ |
| Integer | Symmetric | Bose–Einstein | $n_a = 0, 1, \ldots, \infty$ |

to a nucleus, atom, or molecule. The particles must be identical or indistinguishable and only the translational aspects need be considered.

Fermi–Dirac statistics limit the number of particles in a state to zero or one, and the generating function for this state is

$$\mathcal{Q}_a = 1 + ze^{-\beta\varepsilon_a} = 1 + e^{\beta(\mu-\varepsilon_a)} \tag{7.27}$$

The average number of particles in this state is

$$\langle n_a \rangle = z\left[\frac{\partial(\ln \mathcal{Q}_a)}{\partial z}\right]_{\mathrm{V,T}} \tag{7.28}$$

or

$$\langle n_a \rangle = \frac{1}{e^{(\varepsilon_a-\mu)/kT} + 1} \tag{7.29}$$

and this is the Fermi distribution.

Bose–Einstein statistics permit any positive value of the occupation number $n_a$ and, from Eq. (7.26),

$$\mathcal{Q}_a = \frac{1}{1 - ze^{-\beta\varepsilon_a}} \tag{7.30}$$

This is evaluated by using the sum given by Eq. (7.19). The average number of particles in the state is given by Eq. (7.28),

$$\langle n_a \rangle = \frac{1}{e^{(\varepsilon_a - \mu)/kT} - 1} \tag{7.31}$$

and this is the Bose distribution.

In either example, the thermal properties of all states follow from Eq. (7.25):

$$\frac{PV}{kT} = \ln \mathcal{Q} = \sum_a \ln \mathcal{Q}_a \tag{7.32a}$$

and

$$N = z\left(\frac{\partial(\ln \mathcal{Q})}{\partial z}\right)_{V,T} = \sum_a \langle n_a \rangle \tag{7.32b}$$

$\mu$ is determined from a knowledge of $N$. In this chapter, the assumption of low densities implies

$$\langle n_a \rangle \ll 1$$

and therefore

$$\langle n_a \rangle \approx e^{(\mu - \varepsilon_a)/kT} = z e^{-\varepsilon_a/kT} \tag{7.33}$$

for *either statistics*. Equation (7.32) can be solved for $z$ or $\mu$, and

$$z = \left(\frac{1}{N}\sum_a e^{-\varepsilon_a/kT}\right)^{-1} = \frac{N}{Q_1} \tag{7.34}$$

or

$$\mu = -kT \ln \frac{Q_1}{N} \tag{7.35}$$

and is the same as Eq. (7.5b). The Ideal Gas Law $PV = NkT$ also follows from a combination of these equations. For the purposes of this chapter, Eq. (7.33) implies that the probable number of molecules in the translational state with momentum $\hbar\mathbf{k}$ and energy $\hbar^2 k^2/2M$ must be very much less than one.

$Q_t$ is given by Eq. (7.14) and with $Q_{\text{int}} = 1$ and spin zero,

$$z^{-1} = e^{-\mu/kT} \approx \frac{V}{N}\left(\frac{2\pi MkT}{h^2}\right)^{3/2} = \frac{1}{n\lambda^3} \tag{7.36}$$

for molecular gases at low densities.

**Exercise 7.4** What is the value of $z$ for helium and argon gases at a pressure of 100, 1, $10^{-6}$ Torr and a temperature of 300°K?

**Exercise 7.5** Find the average number of particles in a state with energy $kT$ in helium gas at a temperature of 300°K and a pressure of 100 Torr.

### *7.3.1 The Gibbs Paradox*

Equations (7.4) and (7.14) can be used to obtain the free energy of the molecular gas at low densities:

$$A = -NkT\left\{\ln\left[\left(\frac{eV}{N}\right)\left(\frac{2\pi MkT}{h^2}\right)^{3/2}\right] + \ln Q_{\text{int}}\right\} \tag{7.37}$$

Suppose gas is placed in a container with two separate volumes $V$ as shown in Fig. 7.1. The volumes are separated by a membrane and the gas in each

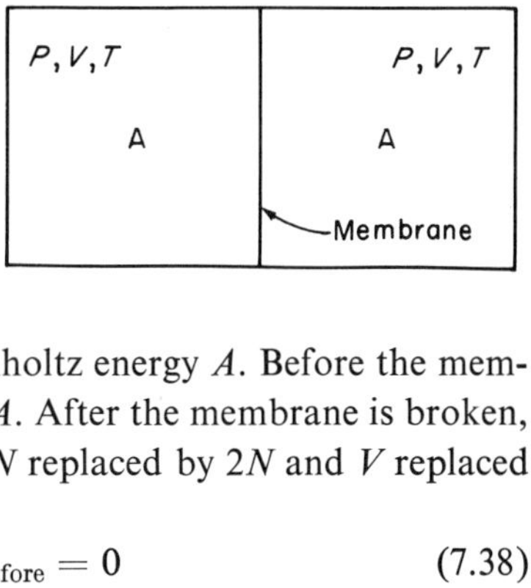


**Fig. 7.1** The Gibbs paradox.

volume is described by $P$, $V$, $T$ and a Helmholtz energy $A$. Before the membrane is broken, the Helmholtz energy is $2A$. After the membrane is broken, the value of $A$ is given by Eq. (7.37) with $N$ replaced by $2N$ and $V$ replaced by $2V$, and

$$\Delta A = A_{\text{after}} - A_{\text{before}} = 0 \tag{7.38}$$

If the $N!$ is removed from Eq. (7.3a), the term $e/N$ is replaced by 1 in Eq. (7.37) and the Helmholtz free energy of mixing is $2NkT \ln 2$. Since the Helmholtz free energy and the entropy are additive thermodynamic variables, the insertion and removal of the membrane should not change the entropy or the free energy. The use of the Fermi or Bose distribution removes the paradox, which would occur if the allowable state were classical or the counting of configurations were used as with the dice in Chapter III.

**Exercise 7.6** Find the entropy of mixing from the Helmholtz free energy, $A_N = -kT \ln Q_N$, with and without the $N!$ term in Eq. (7.3a).

## 7.4 Chemical Reactions

Chemical reactions are usually written in the form

$$\nu_A A + \nu_B B \rightleftharpoons \nu_C C + \nu_D D \tag{7.39}$$

and at equilibrium, the equation for chemical equilibrium, or the Gibbs–Duhem relation, is

$$\sum_A \nu_A \mu_A = 0 \tag{7.40}$$

where $\nu_A = \pm\{\text{integer}\}$ and $\mu_A$ is the chemical potential for constituent A. This expression is developed in Chapter IX. For molecular gases at low densities, this can be written in terms of the single-molecule generating function by the use of Eq. (7.5) for the chemical potential. Thus Eq. (7.40) becomes

$$\sum_A \ln\left(\frac{Q_A}{N_A}\right)^{\nu_A} = 0 \tag{7.41a}$$

or

$$\prod_A \left(\frac{Q_A}{N_A}\right)^{\nu_A} = 1 \tag{7.41b}$$

Introducing the translational generating function and $n_A = N/V$ yields

$$\prod_A \left[\frac{1}{n_A}\left(\frac{2\pi m_A kT}{h^2}\right)^{3/2} (Q_{\text{int}})_A\right]^{\nu_A} = 1 \tag{7.42}$$

All canonical generating functions $Q_A$ must have the same reference energy, or the same zero of energy. Some examples are considered in the following sections.

## 7.5 Thermal Ionization and Excitation of Atoms

This problem was first considered by Saha (see Saha and Srivastava [*1*]) in 1921 and has been subsequently used in problems with regard to astrophysics, the electrical conductivity of flames, the formation of the electrical arc, the formation of the ionosphere, and the determination of the electron affinity of the halogens. Some of these topics are considered in detail in this section.

As an initial and rather simple example, consider gaseous sodium in thermal equilibrium with its surroundings. One possible process is the excitation of the sodium atom to one of its excited states,

$$\mathrm{Na} \leftrightarrow \mathrm{Na}^*$$

Equation (7.42) yields for the relative number density

$$\frac{n_{\mathrm{Na}^*}}{n_{\mathrm{Na}}} = \frac{Q_{\mathrm{Na}^*}}{Q_{\mathrm{Na}}} = \frac{g^*}{g} \exp\left(-\frac{\varepsilon^*}{kT}\right) \tag{7.43}$$

where $\varepsilon^*$ is the excitation energy, $g^*$ the number of distinct states with energy $\varepsilon^*$, and $g$ the number of distinct states of the ground state. For sodium, the ground state and the levels that give rise to the yellow sodium $D_1$ and $D_2$ spectral lines are

$$\begin{array}{llll} \mathrm{Na} & {}^2S_{1/2} & Q_{\mathrm{Na}} = 2(2I+1) & \\ \mathrm{Na}^* & {}^2P_{1/2} & Q_{\mathrm{Na}^*} = 2(2I+1)\exp\left(-\dfrac{\varepsilon^*}{kT}\right) & D_1 = 0.5896 \quad \mu\mathrm{m} \\ \mathrm{Na}^* & {}^2P_{3/2} & Q_{\mathrm{Na}^*} = 4(2I+1)\exp\left(-\dfrac{\varepsilon^*}{kT}\right) & D_2 = 0.5890 \quad \mu\mathrm{m} \end{array}$$

Since the nuclear spin degeneracy $(2I + 1)$ is common to both the excited state and the ground state, it is often omitted. The total number of states or wave functions with each level is $(2J + 1)(2I + 1)$, where $J$ is the total electronic angular momentum of the atom. The factor $g^*/g$ is twice as large for the $^2P_{3/2}$ state as for the $^2P_{1/2}$ state and the number of atoms in the $^2P_{3/2}$ energy level is twice as large as in the $^2P_{1/2}$. This suggests that the intensity from a thermal source is twice as large for the $D_2$ spectral line as for the $D_1$ line. This is in accord with thermal emission data and indicates the correctness of the counting procedure.

A second process in the thermal gas is the ionization of neutral sodium,

$$\mathrm{Na} \leftrightarrow \mathrm{Na}^+ + \mathrm{e}$$

Direct application of Eq. (7.42) yields

$$\frac{n_{\mathrm{Na}^+} n_{\mathrm{e}}}{n_{\mathrm{Na}}} = \left(\frac{2\pi kT}{h^2}\right)^{3/2} \left(\frac{m_{\mathrm{Na}^+} m_{\mathrm{e}}}{m_{\mathrm{Na}}}\right)^{3/2} \frac{Q_{\mathrm{Na}^+} Q_{\mathrm{e}}}{Q_{\mathrm{Na}}} \tag{7.44a}$$

The internal generating functions include the excited states of Na or

$Na^+$, and

$$Q_{Na} = \sum_a \exp\left(-\frac{\varepsilon_a}{kT}\right) = (2I+1)\left[2 + \sum_a{}' \exp\left(-\frac{\varepsilon_a}{kT}\right)\right]$$
$$\approx (2I+1)2$$
$$Q_{Na^+} = (2I+1)\left[\exp\left(-\frac{\varepsilon_I}{kT}\right)\right]\left[1 + \sum_{a'}{}' \exp\left(-\frac{\varepsilon_{a'}^+}{kT}\right)\right]$$
$$= (2I+1)\exp\left(-\frac{\varepsilon_I}{kT}\right)$$
$$Q_e = 2$$

Since electrons have intrinsic spin $\frac{1}{2}$, $Q_e = 2$. $\varepsilon_I$ is the ionization energy and the ionized state of $Na^+$ is $^1S_0$. Excited states of $Na^+$ have much higher energies and can be neglected. Again, the higher excited states of Na contribute a negligible amount and can be omitted. Nuclear spin occurs in both $Q_{Na+}$ and $Q_{Na}$ and can be omitted. With these approximations, the relative number of particles per cubic meter becomes

$$\frac{n_{Na^+} n_e}{n_{Na}} = \left(\frac{2\pi m_e kT}{h^2}\right)^{3/2} \frac{g_e g_+}{g} \exp\left(-\frac{\varepsilon_I}{kT}\right) \tag{7.44b}$$

Again, $g$ refers to the electronic angular momentum $(2J+1)$ and $g_e g^+/g = 2 \times \frac{1}{2} = 1$ for sodium. It should be noted that electrons can be produced by other processes in the gas and $n_e$ is the number of electrons per cubic meter due to all processes. Any of the variables can be changed to pressure by introducing the Ideal Gas Law $P = nkT$. For electrons, $P_e = n_e kT$ and then Eq. (7.44b) can be written as

$$\ln\left(\frac{n_{Na^+}}{n_{Na}} P_e\right) = -\frac{\varepsilon_I}{kT} + \frac{5}{2}\ln T + \ln\left[\left(\frac{2\pi m_e}{h^2}\right)^{3/2} k^{5/2}\right] \tag{7.44c}$$

An increase in electron pressure can decrease the concentration of $n_{Na+}$ in the gas. The electron mass and temperature play very important roles in this expression and in some temperature regions, the number of ionized sodium atoms can exceed the number of excited atoms even though $\varepsilon^* < \varepsilon_I$.

Since local thermal equilibrium may exist in the photosphere and low chromosphere of the sun, this work can now be applied to estimate the ratio of the ionized atoms to the neutral atoms in this region of the sun's atmosphere [2]. The electron contributions of the six elements Si, Mg, Fe, C, Na, and Al are considered; the ground state multiplets of $X_0$, the ionization potential $\varepsilon_I$ for the formation of $X^+$,

$$X_0 \rightleftharpoons X^+ + e$$

and the multiplets of $X^+$ are given in Table 7.2. Equation (7.44) can be rewritten for this more general example as

$$\ln \frac{n_+}{n_0} = -\frac{\varepsilon_I}{kT} - \ln P_e + \frac{5}{2} \ln T + \ln i_0 + \ln\left(\frac{Q_e Q_+}{Q_0}\right) \tag{7.45}$$

where $i_0 = (2\pi m_e/h^2)^{3/2} k^{5/2}$ and $Q_{X^+} = Q_+ \exp(-\varepsilon_I/kT)$. Here $n_+$ is the number of ionized atoms per cubic meter, $n_0$ the number of neutral atoms, and $P_e$ the electron gas pressure. $Q_e = 2$ is the generating function for the electron. For Mg, the ground state of the neutral atom is $^1S$ and the ground state of the ionized atom $Mg^+$ is $^2S$ and $Q_e Q_+/Q_0 = 2 \cdot (2/1) = 4$. The ground state of Si is $^3P$ with the $J = 0, 1, 2$ energy levels at 0, 77, and 223 cm$^{-1}$, respectively. Thus the generating function

$$Q_{Si} = 1 + 3 \exp\left(-\frac{111}{T}\right) + 5 \exp\left(-\frac{321}{T}\right) + \cdots \approx 9 \tag{7.46}$$

is approximately nine at temperatures of the order of 5000°K. Ionized Si and the other atoms are treated in a similar manner for the determination of $Q_e Q_+/Q_0$. For this particular calculation the electron pressure is assumed to be $P_e = 0.4$ N/m$^2$ and the local temperature is 5200°K. The ratio of $n_+/n_0$ for the various atoms is shown in Table 7.2. It may be noted that the number of neutral Na and Cs atoms is small and this suggests that the Fraunhofer absorption spectra of these lines will be weak in this region of the sun's atmosphere.

**Table 7.2**

Ratio of Ionized Atoms to Neutral Atoms in the Sun's Photosphere as Calculated from Eq. (7.45) with $T = 5200$°K and Electron Pressure of $P_e = 0.4$ N/m$^2 \approx 3 \times 10^{-3}$ Torr

| Atom | $\varepsilon_I$ (eV) | $X_0$ | $X^+$ | $\frac{Q_e Q_+}{Q_0}$ | $\frac{n_+}{n_0}$ |
|---|---|---|---|---|---|
| Si | 8.12 | $^3P$ | $^2P$ | $2 \cdot (6/9)$ | 3.1 |
| Fe | 7.83 | $^5D$ | $^6D$ | $2 \cdot (30/25)$ | 11 |
| Mg | 7.61 | $^1S$ | $^2S$ | $2 \cdot (2/1)$ | 29 |
| C | 11.2 | $^3P$ | $^2P$ | $2 \cdot (6/9)$ | 0.0033 |
| Al | 5.96 | $^2P$ | $^1S$ | $2 \cdot (1/6)$ | 100 |
| Na | 5.12 | $^2S$ | $^1S$ | $2 \cdot (1/2)$ | 1800 |
| Cs | 3.87 | $^2S$ | $^1S$ | $2 \cdot (1/2)$ | 30,000 |
| $H^-$ | 0.75 | $^1S$ | $^2S$ | $2 \cdot (2/1)$ | $10^8$ |

It is shown in Section 7.5.2 that the electron affinity of hydrogen favors the production of $H^-$

$$H^- \rightleftharpoons H + e$$

and this reaction uses up the electrons released by the ionization of Si, Mg, and so forth. Equation (7.45) can be used to determine the ratio $n_H/n_{H^-}$, and this value is included in Table 7.2. Although the fraction of $H^-$ is small, the absorption by $H^-$ is responsible for most of the opacity in this region of the sun's atmosphere and the formation of $H^-$ is the primary source of the visible radiation.

Detailed models of the sun's atmosphere are necessary for the determination of the temperature and electron pressure in each region of the photosphere and chromosphere and this subject is of considerable interest [*2*].

Statistical weights play an important role in these calculations. $Q_e Q_+/Q_0 = 4$ for Mg and $Q_e Q_+/Q_0 = 1/3$ for Al and the effect of the statistical weight is comparable to an increase in the ionization potential of Al by 1 eV at 5200°K.

Exercise 7.7 Assume that $n_{Na^+} = n_e$ in sodium gas. Compare $n_{Na^+}$ with the number of atoms $n_{Na*}$ in the $^2P_{3/2}$ state at 1000, 3000, 4000, 5000, and 6000°K. Let $n_{Na} = 10^{20}$ per cubic meter.

### *7.5.1 Quenching of Spectral Lines*

If Ca is heated in a vacuum furnace, the degree of excitation of any particular level is

$$\frac{n_{Ca*}}{n_{Ca}} = \frac{g^*}{g} \exp\left(-\frac{\varepsilon^*}{kT}\right) \tag{7.47}$$

and the emission lines at $h\nu = \varepsilon^*$ are directly proportional to $n_{Ca*}$ and the spontaneous emission lifetime. The ground state of Ca is a $4s^2\,{}^1S_0$ state and the first emission state is the 4s4p $^3P$ state at 0.6573 μm, or 1.88 eV. This line appears first as the temperature is raised. The next spectral line to appear is the $^1S_0$—$^1P$ at 0.4227 μm. At 2000°K, the 4s $^2S_{1/2}$—4p $^2P$ lines of $Ca^+$ appear and are very weak, but become strong as the temperature is further increased. The spectral lines become strong as the temperature is further increased. The spectral lines of $Ca^+$ are readily quenched [*1*, *3*] by the addition of Cs. The reactions of interest are

$$Ca \leftrightarrow Ca^+ + e \qquad (\varepsilon_I = 6.1 \quad eV)$$
$$Cs \leftrightarrow Cs^+ + e \qquad (\varepsilon_I = 3.8 \quad eV)$$

Equation (7.44b) can be used for Cs and Ca, or for each species,

$$\frac{n_e n_+}{n} = \left(\frac{2\pi m_e kT}{h^2}\right)^{3/2} \left[\exp\left(-\frac{\varepsilon_I}{kT}\right)\right] \frac{g_e g_+}{g} \tag{7.48}$$

At 2000°K, the ratio $n_e n_+/n$ is $5.5 \times 10^{11}$ for Ca and $4.8 \times 10^{16}$ for Cs. Since $n_e$ is common to both species, the ratio

$$\frac{n_{Ca^+}}{n_{Ca}} = \frac{n_{Cs^+}}{n_{Cs}} \left[4 \exp\left(-\frac{\varepsilon_{Ca}^I - \varepsilon_{Cs}^I}{kT}\right)\right] \tag{7.49}$$

follows and the fraction of ionized Ca is considerably less than the fraction of ionized Cs. At 2000°K, the term in parentheses is of the order of $10^{-7}$. An alternate approach is to observe that $n_e n_{Ca+}$ is constant and as Cs is added and as the electron concentration is increased, the amount of $Ca^+$ must decrease.

Exercise 7.8 Show that the values just given are correct at 2000°K for $n_e n_+/n$ for Ca and Cs separately. Find $(n_{Ca+}/n_{Cs+})/(n_{Ca}/n_{Cs})$ at 2000, 2500, and 3000°K.

### 7.5.2 *Electron Affinity of Hydrogen and Halogens*

The electron affinity of hydrogen [*4*] is particularly interesting and the energy level scheme for the reaction

$$H + e \leftrightarrow H^- + h\nu$$

is shown in Fig. 7.2. Due to the astrophysical importance of this reaction, the absorption coefficients of the bound–free and the free–free transitions have been examined in considerable detail [5]. These calculations are sufficiently accurate that there can be no doubt that absorption by $H^-$ is responsible for most of the opacity, or deviation from blackbody radiation, in the visible region of the solar spectrum. By the arguments of Chapter II, it follows that the primary source of visible radiation from the sun must be due to the formation of $H^-$ ions by the reaction given here. This same process is important in the emission from a majority of the stars in the visible spectral region. It would seem that $H^-$ has a single stable level below the ionization continuum at $\varepsilon_I = 0.75$ eV or 1.6502 μm.

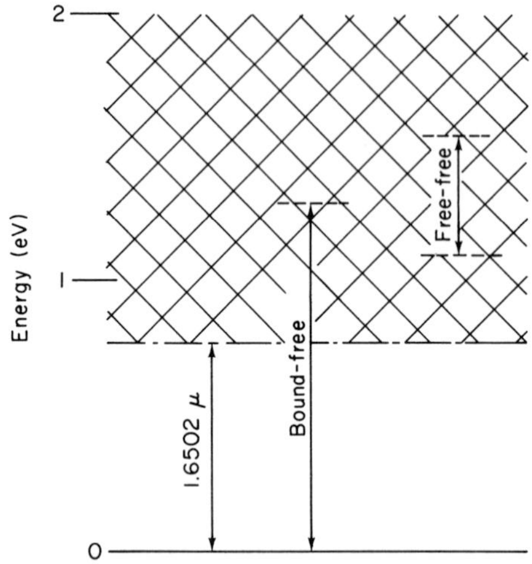


**Fig. 7.2** Energy level diagram of $H^-$ ion. Bound–free and free–free transitions are shown.

The affinity of the halogens [*6*] with a $p^5$ configuration to form the negative ion with the closed shell $p^6$ configuration is expressed by the reaction

$$\mathrm{Cl} + \mathrm{e} \leftrightarrow \mathrm{Cl}^- + \varepsilon_\mathrm{A}$$

where $\varepsilon_\mathrm{A}$ is the electron affinity or binding energy [*4*]. One method for determining this quantity experimentally is to heat NaCl in a vacuum chamber. The following reactions occur:

$$\begin{aligned} \mathrm{NaCl} &\leftrightarrow \mathrm{Na} + \mathrm{Cl} - \varepsilon_\mathrm{D} \\ \mathrm{NaCl} &\leftrightarrow \mathrm{Na}^+ + \mathrm{Cl}^- - \varepsilon_Q \\ \mathrm{Na} &\leftrightarrow \mathrm{Na}^+ + \mathrm{e} - \varepsilon_\mathrm{I} \\ \mathrm{Cl} + \mathrm{e} &\leftrightarrow \mathrm{Cl}^- + \varepsilon_\mathrm{A} \end{aligned}$$

Conservation of energy requires

$$\varepsilon_\mathrm{A} = \varepsilon_\mathrm{D} + \varepsilon_\mathrm{I} - \varepsilon_Q$$

The dissociation energy $\varepsilon_\mathrm{D}$ is obtained from chemical or spectroscopic data and the ionization energy of Na is known. Only the heat of formation of NaCl from $\mathrm{Na}^+$ and $\mathrm{Cl}^-$ is needed to determine the electron affinity. Using Eq. (7.42) and writing $n = P/kT$, yields the expression

$$\ln \frac{P_{\mathrm{Na}^+} P_{\mathrm{Cl}^-}}{P_{\mathrm{NaCl}}} = -\frac{\varepsilon_Q}{kT} + \frac{5}{2} \ln T + \left[ \ln \frac{(2\pi)^{3/2} k^{5/2}}{h^3} \right] \frac{m_{\mathrm{Na}^+} m_{\mathrm{Cl}^-}}{m_{\mathrm{NaCl}}} + \ln \left( \frac{Q_{\mathrm{NaCl}}}{Q_{\mathrm{Na}^+} Q_{\mathrm{Cl}^-}} \right)_{\mathrm{int}} \qquad (7.50)$$

As shown in a subsequent section

$$Q_{\mathrm{NaCl}} = Q_r Q_v = \frac{8\pi^2 IkT}{h^2}(1 - e^{-h\nu/kT})^{-1}$$

$Q_{\mathrm{Na^+}} = 1$ and $Q_{\mathrm{Cl^-}} = 1$ since these are closed-shell configurations, and $\varepsilon_Q$ is included earlier in the expression. The $Na^+$ and $Cl^-$ concentrations are measured by measuring the positive and negative effusion currents. Measuring these and the pressure over a range of temperatures yields $\varepsilon_Q$. Such experiments yield values of $\varepsilon_A$ as follows:

| | F | Cl | Br | I |
|---|---|---|---|---|
| $\varepsilon_A$ (eV) | 3.44 | 3.60 | 3.35 | 3.05 |

## 7.6 Thermodynamic Properties of Heteronuclear Diatomic Molecules

Heteronuclear diatomic molecules are described by three translational, two rotational, and one vibrational degrees of freedom. A quantum mechanical solution yields a set of energy levels

$$E_{k,J,v} = E_k + BJ(J+1) + h\nu_0(v + \tfrac{1}{2}), \quad J = 0, 1, 2, \ldots; \; v = 0, 1, 2, \ldots \tag{7.51}$$

The quantized translational kinetic energy has been discussed previously. The rotational kinetic energy in $BJ(J+1)$, where $B$ is the rotational constant,

$$B = \frac{\hbar^2}{2I} \tag{7.52a}$$

and

$$I = \frac{m_a m_b}{m_a + m_b} r_e^2 \tag{7.52b}$$

$I$ is the moment of inertia of the molecule, $m_a$ and $m_b$ the atomic masses, and $r_e$ the equilibrium internuclear distance; $J$ is the rotational quantum number and there are $2J + 1$ states for each value of $J$; $\nu_0$ is the vibrational frequency and $v$ is the vibrational quantum number; and $E_k$ is the previously discussed translational energy. Typical internal energy levels are shown in Fig. 7.3.

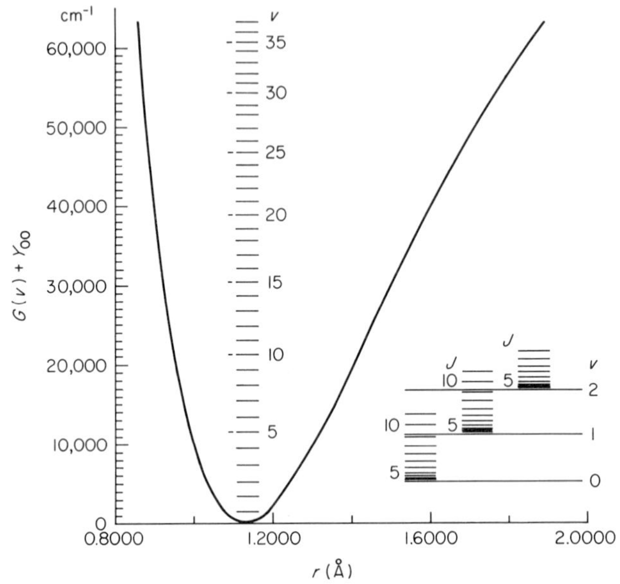


**Fig. 7.3** Potential energy curve for the ground electronic state X $^1\Sigma^+$ of $^{12}C^{16}O$ molecule. Accurate determination of the energy levels to very high $v$ values by observation and measurement of the CO laser emissions with precision spectrographs enabled the use of the Rydberg–Klein–Rees method to evaluate the turning points $r_{min}$ and $r_{max}$ to a high degree of accuracy. The vibrational and rotational energy levels with their appropriate spacing are superimposed on this potential energy curve. [This curve was kindly supplied by K. N. Rao and the data for this figure was taken from A. W. Mantz, J. Watson, K. N. Rao, D. L. Albritton, A. L. Schmeltekopf, and R. N. Zare, *J. Molec. Spectrosc.* **39**, 180 (1971).]

A free molecule does not exchange energy between the rotational and vibrational motions. If the nucleus has spin, this motion is usually separable. Since these systems are noninteracting and statistically independent, the generating function is a product of the generating functions,

$$Q_1 = Q_t Q_v Q_r Q_n \tag{7.53}$$

where $Q_v$ is the vibrational, $Q_r$ the rotational, and $Q_n$ the nuclear generating or partition functions. The translational partition function is now written with the total mass of the molecule as

$$Q_t = V\left(\frac{2\pi(m_A + m_B)kT}{h^2}\right)^{3/2} \tag{7.54}$$

The vibrational generating function is

$$Q_v = \sum_v \exp\left(-\frac{h\nu_0(v + \frac{1}{2})}{kT}\right) \tag{7.55}$$

Usually, the zero of energy is taken at the first energy level $v = 0$, and from Eq. (7.19), the generating function is

$$Q_v = \frac{1}{1 - e^{-h\nu_0/kT}} \tag{7.56}$$

No simple reduction is possible for the rotational generating function and

$$Q_r = \sum_{J=0} (2J + 1)e^{-J(J+1)B/kT} \tag{7.57}$$

At high temperature,

$$Q_r \approx \frac{kT}{B}, \qquad kT \gg B \tag{7.58}$$

The nuclear generating function is just the total number of wave functions associated with each value of $J$, and

$$Q_n = (2I_a + 1)(2I_b + 1) \tag{7.59}$$

If $N$ such molecules are placed in volume $V$, the thermodynamic properties follow from

$$A = -NkT \ln \frac{eQ_1}{N} \tag{7.60a}$$

$$= -NkT\left(\ln \frac{eQ_t}{N} + \ln Q_v + \ln Q_r + \ln Q_n\right) \tag{7.60b}$$

$$= A_t + A_v + A_r + A_n \tag{7.60c}$$

The additive thermodynamic variable $A$ is the sum of the independent contributions. This is true for the remaining additive thermodynamic variables, $U$, $S$, $C_V$, and so forth. Consider first the pressure. Only $A_t$ depends on volume, and from Eq. (7.14),

$$P = -\left(\frac{\partial A}{\partial V}\right)_{T,N} = \frac{N}{V} kT = nkT \tag{7.61}$$

and the Ideal Gas Law follows. The entropy is given as

$$S_t = -\left(\frac{\partial A_t}{\partial T}\right)_{V,N} = Nk \ln \frac{eQ_t}{N} + \frac{3}{2} Nk \tag{7.62}$$

The heat capacity at constant volume is

$$(C_V)_t = \left(T \frac{\partial S_t}{\partial T}\right)_{V,N} = \frac{3}{2} Nk \tag{7.63}$$

Internal energy is given by $A = U - TS$, or

$$U_t = \tfrac{3}{2}NkT \tag{7.64}$$

and is of course in agreement with the work in Chapter I.

### 7.6.1 Rotational Aspects

Rotational motion cannot be simplified and either numerical evaluation of

$$\begin{aligned} A_r &= -NkT \ln Q_r \\ &= -NkT \ln \sum_{J=0} (2J+1)e^{-J(J+1)B/kT} \end{aligned} \tag{7.65}$$

is necessary or the discussion is limited to the high- and low-temperature regions. A plot of the rotational heat capacity per molecule is shown in Fig. 7.4. The asymptotic value at high temperatures follows from

$$\begin{aligned} A_r &\approx -NkT \ln \frac{kT}{B} \\ S_r &\approx +Nk \ln \frac{kT}{B} + Nk \qquad T > \frac{B}{k} \\ (C_V)_r &\approx Nk \end{aligned} \tag{7.66}$$

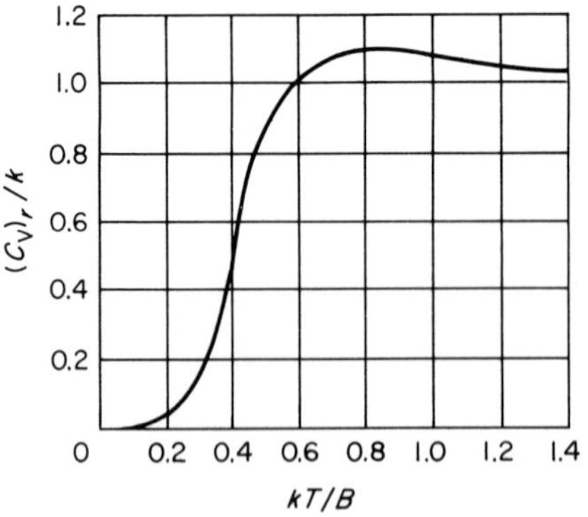


**Fig. 7.4** Rotational heat capacity per diatomic molecule with rotational constant $B$.

At low temperatures, only the first two terms are kept and

$$\left.\begin{aligned} A_r &= -NkT\ln(1 + 3e^{-2B/kT}) \\ &\approx -3NkTe^{-2B/kT} \\ S_r &\approx 3Nke^{-2B/kT}\left(1 + \frac{2B}{kT}\right) \end{aligned}\right\} \qquad T \ll \frac{B}{k} \tag{7.67a}$$

Some values of the rotational constant $B/k$ are given in Table 7.3.

**Table 7.3**

Rotational and Vibrational Constants for Some Diatomic Molecules

| Molecule | $B/k$ | $h\nu_0/k$ |
|---|---|---|
| $H_2$ | 85.5 | 6320 |
| $D_2$ | 42.5 | 4480 |
| HD | 63.5 | 5500 |
| $O_2$ | 2.085 | 2260 |
| $N_2$ | 2.39 | 3380 |
| NO | 2.965 | 2740 |
| CO | 2.765 | 3120 |
| $Cl_2$ | 0.3465 | 810 |
| $Br_2$ | 0.1165 | 470 |
| $I_2$ | 0.054 | 310 |
| $Li_2$ | 0.98 | 500 |
| $Na_2$ | 0.2235 | 230 |
| $K_2$ | 0.081 | 140 |
| HCl | 15.25 | 4300 |
| NaCl | — | 546 |

**Exercise 7.9** Find the rotational heat capacity at low temperatures.

$$\text{Ans.} \quad \frac{(C_V)_r}{Nk} = 3\left(\frac{2B}{kT}\right)^2 e^{-2B/kT} \tag{7.67b}$$

### *7.6.2 Vibrational Aspects*

Equation (7.55) can be used to obtain the vibrational contribution to the free energy, and

$$A_V = +NkT \ln(1 - e^{-h\nu_0/kT}) \tag{7.68a}$$

$$S_V = -Nk \ln(1 - e^{-h\nu_0/kT}) + \frac{Nh\nu_0}{T(e^{h\nu_0/kT} - 1)} \tag{7.68b}$$

$$(C_V)_V = Nk\left(\frac{h\nu_0}{kT}\right)^2 \frac{e^{h\nu_0/kT}}{(e^{h\nu_0/kT} - 1)^2} \tag{7.68c}$$

A plot of the heat capacity is given in Fig. 7.5.

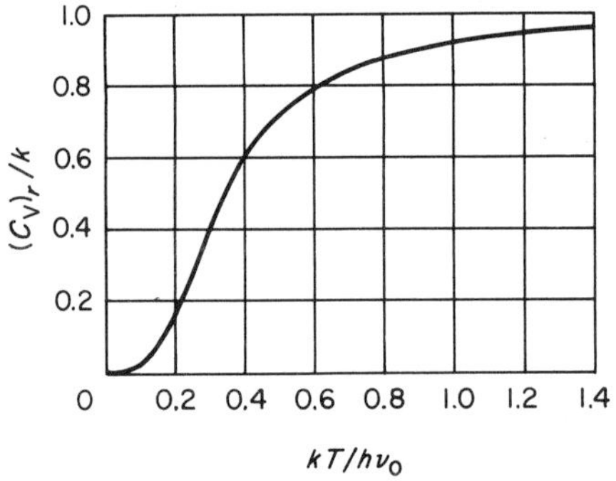


**Fig. 7.5** Vibrational heat capacity per diatomic molecule with natural frequency of vibration of $\nu_0$.

At high temperatures, or $kT \gg h\nu_0$, the heat capacity tends toward the constant value of $Nk$, and at low temperatures, the heat capacity tends toward zero:

$$\frac{(C_V)_v}{N} \to \begin{cases} k, & kT > h\nu_0 \\ 0, & kT \ll h\nu_0 \end{cases} \tag{7.69}$$

Some values of $h\nu_0/k$ were given in Table 7.3.

**Exercise 7.10** Plot the contributions of the rotational and vibrational modes of HCl to the entropy and heat capacity over a broad temperature range.

## 7.7 Homonuclear Diatomic Molecules

Homonuclear diatomic molecules illustrate the importance of the quantum method of counting states. The overall wave function must be symmetric or antisymmetric under the exchange of identical nuclei. The molecules of primary interest are $H_2$, $D_2$, and $T_2$ and are now considered in detail. The proton or hydrogen nucleus has a spin of $I = \frac{1}{2}$ and is a Fermi–Dirac particle. Under exchange of the two protons in the hydrogen molecule, the overall wavefunction must be antisymmetric. Hydrogen, deuterium, and tritium have $^1\Sigma_g^+$ molecular electronic states and the overall wavefunctions without the nuclear wavefunction are symmetric for even values of $J$ and antisymmetric for odd values of $J$ under inversion and reflection. These labels are shown in Table 7.4 for $H_2$ and $D_2$. For identical nuclei,

**Table 7.4**

| | $H_2$ | | | $D_2$ | | |
|---|---|---|---|---|---|---|
| $J$ | $\psi_s$ | $\psi_n$ | $\psi_T$ | $\psi_s$ | $\psi_n$ | $\psi_T$ |
| 4 | *s* | *a* | *a* | *s* | *s* | *s* |
| 3 | *a* | *s* | *a* | *a* | *a* | *s* |
| 2 | *s* | *a* | *a* | *s* | *s* | *s* |
| 1 | *a* | *s* | *a* | *a* | *a* | *s* |
| 0 | *s* | *a* | *a* | *s* | *s* | *s* |

there are $\frac{1}{2}(2I + 1)(2I)$ antisymmetric nuclear wavefunctions and $\frac{1}{2}(2I + 1) \times (2I + 2)$ symmetric ones. This is readily apparent for $H_2$. Let $\alpha$ denote the spin value of $m_I = +\frac{1}{2}$ and $\beta$ denote $m_I = -\frac{1}{2}$. The four possible combinations for the nuclear wavefunctions are

$$\left.\begin{array}{c}\alpha(1)\beta(2)\\ \alpha(2)\beta(1)\\ 2^{-1/2}[\alpha(1)\beta(2) + \alpha(2)\beta(1)]\end{array}\right\} \quad \text{symmetric}$$

$$2^{-1/2}[\alpha(1)\beta(2) - \alpha(2)\beta(1)] \qquad \text{antisymmetric}$$

These must be combined with the spatial wavefunction $\psi_s$ so that the total wavefunction $\psi_T = \psi_s\psi_n$ is antisymmetric (a) for $H_2$, that is, "*a*."

Deuterium has a spin $I = 1$ and is a Bose–Einstein particle and the total wavefunction of the molecule must be symmetric under the exchange of identical nuclei; $\psi_T = \psi_s\psi_n$ is symmetric (*s*).

In both $H_2$ and $D_2$, the coupling to the nuclear spin states is very weak and there is only a very small tendency for the nuclear wavefunction to change its symmetry. As a result, the molecules with even values of $J$ are almost distinct from those with odd $J$. The conversion may take days or longer unless a catalyst is used. For this reason, the even and odd values of $J$ are given different names. The states with the larger number of nuclear levels are referred to as *ortho* and the others as *para*. For $H_2$ and $D_2$, the number of nuclear wavefunctions and the permitted $J$ values are given in Table 7.5.

**Table 7.5**

| | $H_2$ and $T_2$, $I = \frac{1}{2}$ | $D_2$, $I = 1$ |
|---|---|---|
| *para*: $\frac{1}{2}(2I + 1)(2I)$ | 1 (even $J$) | 3 (odd $J$) |
| *ortho*: $\frac{1}{2}(2I + 1)(2I + 2)$ | 3 (odd $J$) | 6 (even $J$) |

Since, under most experimental conditions, the transformation between the *ortho* and *para* molecules is so slow that they can be regarded as distinct species, the generating function is the product of the generating functions, or the Helmholtz free energy is the sum of the *ortho* and *para* Helmholtz free energies, and

$$(H_2)_{para} \nleftrightarrow (H_2)_{ortho}$$

$$Q_N \approx \frac{Q_p^{N_p}}{N_p!} \frac{Q_o^{N_o}}{N_o!} \tag{7.70a}$$

or

$$A = A_{para} + A_{ortho} \tag{7.70b}$$

Only the rotational and nuclear generating functions are different and combining $Q_r Q_n$ yields

$$\textit{para}: \qquad Q_{r,p} = \tfrac{1}{2}(2I + 1)(2I) \sum_{\text{even}} (2J + 1)e^{-J(J+1)B/kT} \tag{7.71a}$$

$$\textit{ortho}: \qquad Q_{r,o} = \tfrac{1}{2}(2I + 1)(2I + 2) \sum_{\text{odd}} (2J + 1)e^{-J(J+1)B/kT} \tag{7.71b}$$

for the *ortho* and *para* molecules. The free energies are

$$A_p = -N_p kT\left\{\ln\left[\left(\frac{eV}{N_p}\right)\left(\frac{2\pi MkT}{h^2}\right)^{3/2}\right] - \ln(1 - e^{-h\nu_0/kT}) + \ln Q_{r,p}\right\} \tag{7.72}$$

and $A_o$ follows by using $N_o$ and $Q_{r,o}$. "Even" and "odd" must be interchanged for $D_2$.

If sufficient time is allowed for equilibrium to occur between the species, then the chemical potentials must be equal and

$$(H_2)_{ortho} \leftrightarrow (H_2)_{para}$$

and

$$\mu_o = \mu_p \tag{7.73}$$

Equation (7.15) can be used for the number of each species of molecule and

$$\frac{n_p}{n_o} = \frac{Q_{\mathrm{r},p}}{Q_{\mathrm{r},o}} \tag{7.74}$$

where $Q_{\mathrm{r},p}$ and $Q_{\mathrm{r},o}$ are given by Eq. (7.71). At high temperatures, the sums over even and odd $J$ are equal and the relative concentration is determined by the nuclear spin states,

$$\left(\frac{n_p}{n_o}\right)_{\mathrm{H}_2} = \frac{1}{3}, \qquad \left(\frac{n_p}{n_o}\right)_{\mathrm{D}_2} = \frac{6}{3} \qquad (kT \gg B) \tag{7.75a}$$

At very low temperatures, only the first terms in the sum are needed and

$$\left(\frac{n_o}{n_p}\right)_{\mathrm{H}_2} = 9e^{-2B/kT}, \qquad \left(\frac{n_p}{n_o}\right)_{\mathrm{D}_2} = \frac{9}{6}\, e^{-2B/kT} \tag{7.75b}$$

Normally, during a measurement of the heat capacity, the ratio of *ortho* to *para* remains fixed and the thermodynamic properties follow from

$$A = A_p + A_o \tag{7.76a}$$

where $A_p$ and $A_o$ are given by Eq. (7.72). The other additive thermodynamic properties follow in the usual manner,

$$S = S_p + S_o \tag{7.76b}$$

$$C_\mathrm{V} = (C_\mathrm{V})_p + (C_\mathrm{V})_o \tag{7.76c}$$

The heat capacities of various mixtures [7] are shown in Fig. 7.6. It should be noted that the Helmholtz free energy and the entropy contain an entropy-of-mixing term. If two systems not at equilibrium are brought into equilibrium via a catalyst or by waiting for a sufficient period of time, heat will be generated during the conversion process.

Experimental evidence indicates that pure *para*-hydrogen is converted at about 0.2%/day into the equilibrium concentration at 300°K. Pure *para*-hydrogen can be produced by passing liquid hydrogen at 20°K over a paramagnetic material. *Ortho*-hydrogen can be produced by passing hydrogen gas at 20°K and 50 Torr over $\gamma$-alumina. The *ortho*-hydrogen

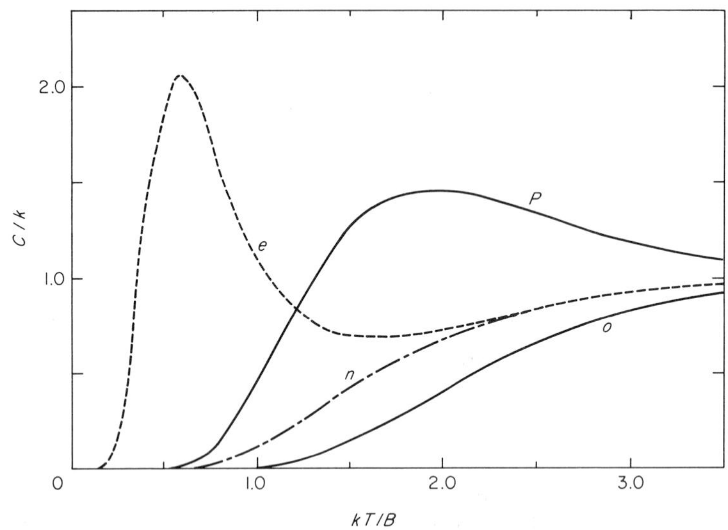


**Fig. 7.6** Rotational heat capacity of molecular hydrogen gas. The solid curves are the rotational heat capacities of pure *para*-hydrogen and pure *ortho*-hydrogen and can be used to determine the heat capacities of nonequilibrium mixtures. The dashed curve (– – –) is the heat capacity of an equilibrium mixture and the nuclear statistical weight introduces a considerable deviation from the heat capacity curve for a heteronuclear molecule such as that shown in Fig. 7.4. The dot-dashed curve –·–·– is for normal hydrogen, that is a 25% *para* and a 75% *ortho* mixture and is in good agreement with experimental heat capacity. This agreement confirms that the system must be treated as a nonequilibrium mixture [7].

is in the $J = 1$ state and this is absorbed strongly by the $\gamma$-alumina surface. The *para* or $J = 0$ state is not absorbed and separation of *ortho*- and *para*-hydrogen is possible [*8*].

Exercise 7.11 What is the difference in entropy for $N$ molecules of $H_2$ at 100°K with $n_p/n_o = \frac{1}{3}$ and with $n_p/n_o$ at the equilibrium value?

## 7.8 Equipartition Theorem

The classical, or nonquantum, treatment uses the classical Hamiltonian for a diatomic molecule. Such a Hamiltonian can be written as

$$H = \frac{P_X^2 + P_Y^2 + P_Z^2}{2M} + \frac{p_\varphi^2}{2I \sin^2 \theta} + \frac{p_\theta^2}{2I} + \frac{p^2}{2\mu} + \frac{1}{2} Kq^2 \qquad (7.77)$$

The equipartition theorem states that each squared momentum $p_i$ and each squared coordinate $q_i$ in the Hamiltonian has an average energy of $\frac{1}{2}kT$. For the canonical ensemble, the probability of a given state is proportional to $e^{-H/kT}$ and the generating function is

$$Q_1 = \frac{1}{h^6} \int e^{-H/kT} [(dX\, dY\, dZ\, dP_X\, dP_Y\, dP_Y)(d\varphi\, d\theta\, dp_\varphi\, dp_\theta)(dp\, dq)]$$

$$= Q_t Q_r Q_v \tag{7.78}$$

where the translational integral is integrated over volume $V$ and the translational momenta $P_X$, $P_Y$, $P_Z$ are integrated between $\pm\infty$. The rotational angles $\varphi$ and $\theta$ are integrated from 0 to $2\pi$ and 0 to $\pi$, respectively, and the conjugate angular momenta $p_\varphi$ and $p_\theta$ are integrated between $\pm\infty$. The vibrational coordinate $q$ and the conjugate momentum $p$ are integrated between $\pm\infty$. The semiquantal treatment divides the equation by $h^6$ to obtain the appropriate size for the cells in phase space. The result of this calculation is

$$Q_1 = V\left(\frac{2\pi MkT}{h^2}\right)^{3/2}\left(\frac{8\pi^2 IkT}{h^2}\right)\left(\frac{kT}{h\nu_0}\right) \tag{7.79}$$

where $\nu_0 = (2\pi)^{-1}(K/\mu)^{1/2}$ is the natural frequency of oscillation. The average energy can be computed from $A = U - TS = -NkT \ln(e/N)Q_1$, and this average energy is

$$U_t = \tfrac{3}{2}kT \tag{7.80a}$$

$$U_r = kT \tag{7.80b}$$

$$U_v = kT \tag{7.80c}$$

These correspond to the high-temperature values in the earlier discussion in this chapter. The rule of equipartition is that for each squared $p_i$ or $q_i$ in the Hamiltonian, a contribution of $\frac{1}{2}kT$ occurs in the average energy. The three translational terms contribute $\frac{3}{2}kT$, the two rotational terms $kT$, and the two vibrational terms $kT$. The heat capacity at constant volume of $O_2$ or $N_2$ is $\frac{7}{2}k$, in disagreement with experiment. Since the vibrational modes are not excited at 300°K, the quantum treatment, in which the vibrational contribution to the heat capacity at this temperature is zero, is needed.

**Exercise 7.12** Show that direct integration of Eq. (7.78) yields Eq. (7.79). Be sure to integrate over momentum first.

Exercise 7.13 Show that $\partial(\ln Q_1)/\partial\beta$ yields the average value of $\langle H\rangle$.

Exercise 7.14 Split $Q_1$ into three separate integrals $Q_t$, $Q_r$, and $Q_v$ and show that $-\partial(\ln Q)/\partial\beta = \langle H_t\rangle + \langle H_r\rangle + \langle H_v\rangle$. Evaluate and show that $\langle H_t\rangle = \frac{3}{2}kT$, $\langle H_r\rangle = kT$, and $\langle H_v\rangle = kT$.

### 7.8.1 *Number of Atoms or Molecules in a Given State*

The generating function $Q_1$ describes each state of an atom or molecule that is accessible. In the example of *ortho-* and *para-hydrogen*, some states were inaccessible and required special treatment. The probability of a thermally accessible state is given by

$$p_a = \frac{e^{-\beta\varepsilon_a}}{Q_1} \tag{7.81}$$

Even so, this statement requires some care in its use. For translational motion, the development leading to Eq. (7.33) is most convenient, and

$$\langle n_k\rangle = \frac{Ne^{-\beta\varepsilon_k}}{Q_t} \tag{7.82}$$

where $k$ is one of the triple indices describing the state. Usually, the number in range $dk_x\, dk_y\, dk_z = k^2\, d\Omega\, dk = d\mathbf{k}$ is more convenient, and

$$\langle n_k\rangle \frac{d\mathbf{k}}{(2\pi)^3} = \frac{N}{V}\left(\frac{h^2}{2\pi mkT}\right)^{3/2}\left[\exp\left(-\frac{\beta\hbar^2k^2}{2m}\right)\right]\frac{d\mathbf{k}}{(2\pi)^3} \tag{7.83}$$

This expression can be compared with the results of Chapters I and IV with the transformation

$$\frac{\hbar^2k^2}{2m} \to \frac{p^2}{2m} = \frac{1}{2}mv^2$$

Direct substitution yields Eq. (1.47), that is,

$$P(v^2)\, d\mathbf{v} = \left(\frac{m}{2\pi kT}\right)^{3/2}\left[\exp\left(-\frac{mv^2}{2kT}\right)\right] d\mathbf{v} \tag{7.84}$$

for the probability of velocity $\mathbf{v}$. This is the justification for the Boltzmann distribution used for the velocities in the earlier chapters.

Other quantities are less troublesome and the ratio of the probabilities follows from Eq. (7.81), and

$$\frac{p_{a'}}{p_a} = \frac{n_{a'}}{n_a} = \exp\left(-\frac{\varepsilon_{a'} - \varepsilon_a}{kT}\right) \tag{7.85}$$

where $a'$ and $a$ denote *states*. For atoms, the ratio is usually in terms of the quantum numbers used to label states. Thus in thermal equilibrium, the label $a'$ or $a$ for an atom may be

$$\begin{aligned} a &\rightarrow \alpha LSIM_L M_S M_I \\ &\rightarrow \alpha LS'IJM_J M_I \\ &\rightarrow \alpha LSIJFM_F \end{aligned}$$

There are $2L+1$ values of $M_L$, $2S+1$ values of $M_S$, $2I+1$ values of $M_I$, $2J+1$ values of $M_J$, and $2F+1$ values of $M_F$. In a very strong magnetic field, each of these $(2L+1)(2S+1)(2I+1)$ states will have distinct energies. In zero magnetic field, the states with the same $M_J$ and $M_I$ can be degenerate and often in the literature, these levels are grouped together. Thus

$$\begin{aligned} \frac{n(J')}{n(J)} &= \frac{(2J'+1)(2I+1)}{(2J+1)(2I+1)} \exp\left(-\frac{\varepsilon_{J'} - \varepsilon_J}{kT}\right) \\ &= \frac{g_{J'}}{g_J} \exp\left(-\frac{\varepsilon_{J'} - \varepsilon_J}{kT}\right) \end{aligned} \tag{7.86}$$

$g_J = 2J+1$ is the number of states with a given value of $J$ and $n(J)$ is the number of atoms in states with label $|\,\alpha LSIJ)$. Since the nuclear spin $I$ does not change as the other indices $\alpha$, $L$, and $S$, change each set of atomic levels is multiplied by $2I+1$ and this term cancels. Atoms with hyperfine splitting of the energy levels require the indices $M_F$ and $F$, and the nuclear spin does not cancel. Thus the 1s $^2S_{1/2}$ state of atomic hydrogen has a hyperfine splitting of approximately 1420 MHz and these states are labeled $|\,1s\,^2S_{1/2}FM_F)$ where $F=1$ for the upper hyperfine level and $F=0$ for the lower one. Thus

$$\frac{n(1M_F)}{n(00)} = e^{-\Delta/kT} \tag{7.87}$$

where $\Delta = h(1420 \times 10^6)$. Yet the number of atoms in the upper three levels is almost three times larger at normal temperatures.

This counting of states is particularly important in lasers. A gas will amplify radiation if the number of atoms in an upper state exceeds the number in a lower state. For convenience, let $I = 0$. Then, laser action occurs when $n(\alpha' J' M') > n(\alpha J M)$, where the primed indices are the upper state. The number of atoms in an energy level is not the quantity of importance. Thus hydrogen gas amplifies when

$$n(1M_F) > n(00)$$

and leads to devices like the hydrogen maser. A thermal hydrogen gas at 300°K has almost three times as many atoms in the upper $F = 1$ level as in the lower $F = 0$ energy level, but absorbs the 1420-MHz radiation since the number in each $| 1M_F)$ state is less than in the $| 00)$ state.

Equation (7.85) can also be used for molecules. The labels for the diatomic molecules were given in the previous sections as $vJM$. A set of rotational states is rather interesting in that, for a given value of $v$,

$$n(J) = \frac{(2J+1)}{Q_r} e^{-BJ(J+1)/kT} \tag{7.88}$$

Since the number of states increases as $2J + 1$, it is possible for the higher-$J$ energy levels to have more molecules in them than the lower ones. For the reasons given earlier concerning states, a thermal system is still absorbing. As a thermal emitter, the line intensity will follow $n(J)$ and have a maximum value at the maximum of $n(J)$. This is shown in Fig. 7.7.

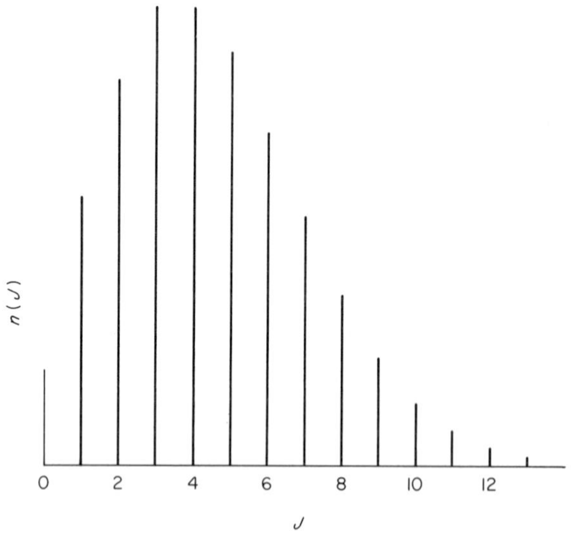


**Fig. 7.7** The probability $n(J)$ of a molecule being in rotational state $J$ is given by Eq. (7.88). The data which is shown in this plot is for $kT = 32B$.

**Exercise 7.15** Show for a thermal molecular gas that the maximum of $n(J)$ occurs near $(kT/2B)^{1/2} - \frac{1}{2}$.

## 7.9 Gases at Very High Temperatures

Pair production in a thermal gas at extremely high temperatures can be treated in the same sense as a chemical reaction. At equilibrium, the reaction for pair production and annihilation is

$$e^+ + e^- \rightleftarrows \gamma\text{'s} \tag{7.89}$$

where $e^+$ and $e^-$ denote the positron and electron, respectively, and $\gamma$ denotes one or more thermal photons. Since the chemical potential of the photon gas is zero, the equation for the chemical potentials is

$$\mu_+ + \mu_- = 0 \tag{7.90a}$$

or

$$\frac{Q_+}{N_+}\frac{Q_-}{N_-} = 1 \tag{7.90b}$$

In the nonrelativistic range of energies, $kT < 0.1mc^2$, Eq. (7.42) yields

$$n_+n_- = 4\left(\frac{2\pi mkT}{h^2}\right)^3 \exp\left(-\frac{2mc^2}{kT}\right) \tag{7.91}$$

The factor of four is due to the two spin states of the positron and of the electron; $2mc^2$ is the equivalent of the heat of reaction and the chemical potential now includes the rest energies of the particles. In the relativistic region, the chemical potential for relativistic particles is needed and this is given by Eqs. (7.27) and (7.32):

$$\ln \mathcal{Q} = \sum_k \ln(1 + ze^{-\beta\varepsilon_k}) \tag{7.92}$$

where

$$\varepsilon_k^2 = c^2(\hbar^2k^2 + m^2c^2) = c^2(p^2 + m^2c^2)$$

is the relativistic energy of the particles. The sum can be changed to an integral in the usual manner and with a sum over the two spin states, becomes

$$\ln \mathcal{Q} = 2V\int \frac{4\pi k^2\, dk}{(2\pi)^3} \ln(1 + ze^{-\beta\varepsilon_k}) \tag{7.93}$$

The fugacity $z$ is given by Eq. (6.57), and

$$N = 2V \int \frac{4\pi k^2\, dk}{(2\pi)^3} \left( \frac{1}{z^{-1} e^{\beta \varepsilon_k} + 1} \right) \tag{7.94}$$

Consider now a gas in which the initial number of electrons per unit volume is $n_i$. Then

$$n_- = n_i + n_+ \tag{7.95}$$

and in the nonrelativistic region, Eq. (7.91) yields the number of positrons per cubic meter as

$$n_+ = -\frac{1}{2} n_i + \left[ \frac{n_i^2}{4} + 4\left( \frac{2\pi mkT}{h^2} \right)^3 \exp\left( -\frac{2mc^2}{kT} \right) \right]^{1/2} \tag{7.96}$$

At a temperature of $kT \approx 0.1mc^2$, the term $(2\pi mkT/h^2)^3 \approx 10^{69}$ becomes dominant and pair production is the primary source of positrons and electrons, or

$$n_+ \approx n_-, \qquad kT > 0.1mc^2 \tag{7.97}$$

For convenience, it may be noted that

$$mc^2 = 0.51 \times 10^6 \text{ eV} \qquad \text{and} \qquad \frac{mc^2}{k} = 6 \times 10^9 \text{ °K}$$

and that pair production is the primary source of electrons and positrons at temperatures above $10^8$ °K.

Pair production is dominant in the extremely high-temperature region of $kT > mc^2$. Since $n_+ \approx n_-$, the chemical potentials become almost equal and therefore almost zero,

$$\mu_+ \approx \mu_- \approx 0 \qquad \text{or} \qquad z_\pm \approx 1 \tag{7.98}$$

This is expected since there is very little constraint on the number of particles. Replacing $z$ by 1 in Eqs. (7.73) and (7.94) and by assuming that in the extreme relativistic region, $\varepsilon = cp = \hbar kc$, the integrals can be performed in the same manner as for blackbody radiation. These calculations yield

$$n_+ = n_- = 0.183\left( \frac{2\pi kT}{hc} \right)^3 \tag{7.99}$$

$$\begin{aligned} \frac{U^+}{V} = \frac{U^-}{V} &= \left[ \frac{(56/120)\pi^5 k^4}{h^3 c^3} \right] T^4 \\ &= \frac{7}{8} \text{ (thermal radiation)} \end{aligned} \tag{7.100}$$

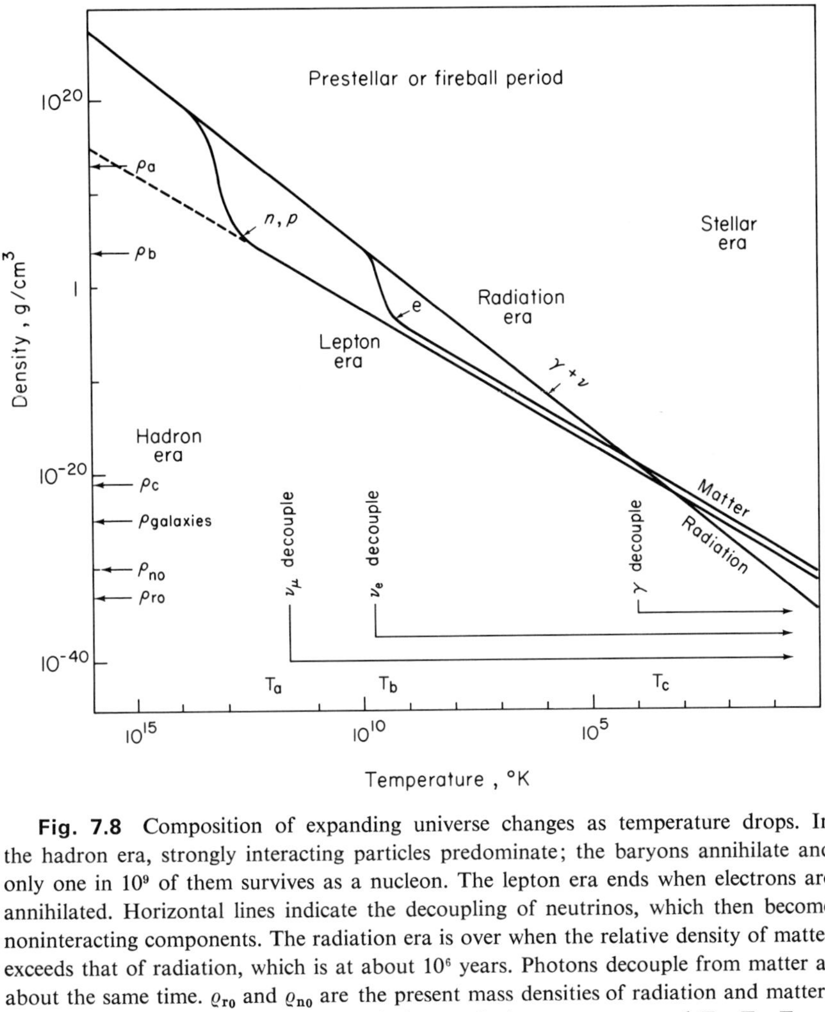


**Fig. 7.8** Composition of expanding universe changes as temperature drops. In the hadron era, strongly interacting particles predominate; the baryons annihilate and only one in $10^9$ of them survives as a nucleon. The lepton era ends when electrons are annihilated. Horizontal lines indicate the decoupling of neutrinos, which then become noninteracting components. The radiation era is over when the relative density of matter exceeds that of radiation, which is at about $10^6$ years. Photons decouple from matter at about the same time. $\varrho_{ro}$ and $\varrho_{no}$ are the present mass densities of radiation and matter; $\varrho_{galaxies}$ is the mean density of moderately large galaxies; $\varrho_a$, $\varrho_b$, $\varrho_c$, and $T_a$, $T_b$, $T_c$ are the densities and temperatures at the end of the hadron, lepton, and radiation eras, respectively. We are in the stellar era. [From E. R. Harrison, *Phys. Today* **21**, 35 (1968), Fig. 3.]

This type of calculation has become of interest in the discussion of the "primeval fireball" and the present thermal 3°K background radiation.

Some of the highly speculative discussions concerning "the early universe" are summarized in Fig. 7.8.[1] In the very earliest stages of the "prime-

[1] Harrison [9] gives a very interesting discussion and a list of 66 references concerning the various "eras."

val fireball" model, or prior to $10^{-44}$ sec, densities as large as $10^{94}$ g/cm$^3$ and energies per particle of $10^{29}$ eV are postulated. As the temperature decreases with expansion, the hadrom era begins and hadrons, leptons, and photons are the constituents. As the expansion continues and the temperature drops toward $10^{12}$ °K, the hadron era ends when the pions annihilate and only a relatively few strongly interacting particles remain. The pion annihilation is similar to the electron–positron annihilation, but the rest mass is larger and corresponds to a temperature characteristic of the rest mass of the pions, or 100 MeV. The era lasts for the order of $10^{-4}$ sec. The lepton era begins at a density of less than $10^{14}$ g/cm$^3$ and a temperature of less than $10^{12}$ °K, and lasts for a period of the order of 10 sec. The particle population consists of positive and negative muons, muon neutrinos and antineutrinos, positive and negative electrons, and electron neutrinos and antineutrinos. The lepton era ends with the annihilation of electron–positron pairs at a temperature of the order of 1 MeV, or $10^{10}$ °K, and a density of less than $10^4$ g/cm$^3$. With the decline of the electrons, the electron neutrino is lost in the equilibrium. The radiation era is entered and annihilation processes cease with further expansion as the temperature falls from $10^{10}$ to 3000°K. This era lasts about $10^6$ yr, during which time helium is produced. Below 3000°K, electrons and nuclei combine and radiation is decoupled from matter. The density decreases to $10^{-21}$ g/cm$^3$. The stellar era begins and aggregations of matter are formed into celestial structures during the ensuing $10^{10}$ yr. The temperature of the thermal radiation decreases to the present value of approximately 3°K [*10*]. The suggested average density of the present universe is $10^{-30}$ g/cm$^3$, or $10^{-6}$ nucleons/cm$^3$, or one nucleon per cubic meter. The density of the blackbody radiation is of the order of $10^{-33}$ g/cm$^3$.

An elementary particle reaction of interest at high temperatures and high densities is the decay of the neutron

$$\mathrm{n} \leftrightarrows \mathrm{p} + \mathrm{e}^- + \bar{\nu}_\mathrm{e} \tag{7.101}$$

into a proton, negative electron, and an electron antineutrino. The chemical potentials are related by

$$\mu(\mathrm{n}) = \mu(\mathrm{p}) + \mu(\mathrm{e}^-) + \mu(\bar{\nu}_\mathrm{e}) \tag{7.102}$$

and this equation is discussed in greater detail in Chapter VIII where we discuss very dense matter. Equilibrium requires special consideration since the neutrino absorption cross section is extremely small. Particle–antiparticle processes require

$$\mu(e^+) + \mu(e^-) = 0 \tag{7.103}$$

$$\mu(p^+) + \mu(p^-) = 0 \tag{7.104}$$

and so forth. The chemical potential for neutral mesons and photons is zero,

$$\mu(\gamma) = \mu(\pi^\circ) = \mu(K^\circ) = 0 \tag{7.105}$$

Since it is usually assumed that the number of neutrinos and number of antineutrinos are equal, the chemical potential of the neutrinos is assumed to be zero [*10*]:

$$\mu(\nu_e) = \mu(\bar{\nu}_e) = \mu(\nu_\mu) = \mu(\bar{\nu}_\mu) = 0 \tag{7.106}$$

Experimental proof of this statement would occur if there should exist a $\beta$-decay or a $\mu$-decay process between the same initial and final products A and B *with* or *without* neutrino emission, respectively. Then, $\mu(A) = \mu(B) + \mu(e)$ and $\mu(A) = \mu(B) + \mu(e) + \mu(\nu_e)$ and the chemical potential of the neutrino could be shown to be zero [*10*].

Exercise 7.16 Use Eq. (7.91) and the information that $n_+ \approx n_-$ to find the number of pairs at a temperature of 51 keV or $6 \times 10^8$ °K.

Exercise 7.17 Find the number of photons and pairs per cubic meter at temperatures of 1 MeV and 100 MeV. Using $E = mc^2$, find the equivalent mass per cubic meter which is due to the pairs and the thermal radiation. Note that the energy density of radiation and pairs is comparable above 1 MeV, but the pair energy falls rapidly below 0.5 MeV, or $5 \times 10^9$ °K.

Ans. {density} $\approx 10^8$ and $10^{16}$ kg/m$^3$

Exercise 7.18 (a) Why does the reaction

$$\pi^\circ \leftrightarrows \gamma + \gamma \quad \text{or} \quad \gamma\text{'s} \tag{7.107}$$

imply $\mu(\pi^\circ) = 0$?

(b) Why does the reaction

$$K^\circ \leftrightarrows \pi^\circ\text{'s} \tag{7.108}$$

imply that $\mu(K^\circ) = 0$?

(c) Why does

$$\pi^\pm \leftrightarrows e^\pm + \nu \tag{7.109}$$

imply $\mu(\pi^\pm) \neq 0$?

## 7.10 Paramagnetism

The interaction of a molecule having permanent magnetic moment $\boldsymbol{\mu}$ with an external magnetic field $\mathbf{B}$ is given by

$$H_1 = -\boldsymbol{\mu} \cdot \mathbf{B} \tag{7.110}$$

For a large class of problems, the magnetic moment is proportional to the angular momentum $\mathbf{J}$,

$$\boldsymbol{\mu} = \left(\frac{\mu_J}{J}\right)\mathbf{J} = \hbar\gamma_J\mathbf{J} \tag{7.111}$$

and this is assumed to be the appropriate form in this discussion. Other examples will be discussed later. The energy levels are given by

$$E_1(Jm) = \hbar\gamma_J m B_0 \tag{7.112}$$

where $m = J,\ J-1, \ldots, -J$ and $\mathbf{B} = B_0\hat{z}$. The generating function describing this molecule in thermal equilibrium with its surroundings is

$$Q_{\text{int}} = \sum_{m=-J}^{J} \exp\left(-\frac{\hbar\gamma_J m B_0}{kT}\right) \tag{7.113}$$

It is never possible to regard electric or magnetic dipoles as noninteracting molecules. The dipole fields fall off as $r^{-3}$ and since the number of molecules in an enclosure increases as $r^3$, the magnetic interaction must be included. Fortunately, this can be accomplished to a good approximation by replacing $B_0$ in Eq. (7.113) by an effective or local magnetic field. The interaction Hamiltonian of an ensemble of permanent magnetic dipoles is given by

$$H_N = -\sum_{\alpha=1}^{N} \boldsymbol{\mu}_\alpha \cdot (\mathbf{B}_e + \mathbf{B}_1 + \mathbf{B}_2 + \mathbf{B}_3) \tag{7.114a}$$

where

$$\mathbf{B}_1 + \mathbf{B}_2 + \mathbf{B}_3 = \frac{\mu_0}{4\pi}\sum_\beta \left[\frac{(3\boldsymbol{\mu}_\beta \cdot \mathbf{r}_{\alpha\beta})\mathbf{r}_{\alpha\beta}}{r_{\alpha\beta}^5} - \frac{\boldsymbol{\mu}_\beta}{r_{\alpha\beta}^3}\right] \tag{7.114b}$$

is the interaction with the other dipoles. This field is treated classically and divided into three parts. At the $\alpha$th site, a fictitious sphere of radius $r_0$ is placed around the site and the sum is replaced by an integral over the volume outside the sphere plus a sum over the material inside the sphere. In transforming the sum to an integral, let $\mathbf{M}$ denote the magnetization per unit volume and replace the quantity inside the brackets in Eq.(7.114b) by grad$(\mathbf{M} \cdot \mathbf{r})/r^3$,

$$\sum_{\beta} \rightarrow \int_{r_0}^{V} \text{grad}\left(\frac{\mathbf{M} \cdot \mathbf{r}}{r^3}\right) d\mathbf{r} + \sum_{\beta} \{\text{small sphere}\}$$

The volume integral can be transformed into a surface integral over the outer surface of the sample and the inner sphere of radius $r_0$,

$$\int_{r_0}^{V} \text{grad}\left(\frac{\mathbf{M} \cdot \mathbf{r}}{r^3}\right) d\mathbf{r} = \int_{\text{outer}} \left(\frac{\mathbf{M} \cdot \mathbf{r}}{r^3}\right) \hat{n}\, dA + \int_{\text{inner}} \left(\frac{\mathbf{M} \cdot \mathbf{r}}{r^3}\right) \hat{n}\, dA$$

If the sample is ellipsoidal in shape and the magnetization **M** is along one of the principal axes of the ellipsoid, then these surface integrals can be evaluated, and

$$\begin{aligned} \mathbf{B}_1 &= -\mu_0 N_1 \mathbf{M} \qquad \text{outer surface} \\ \mathbf{B}_2 &= +\mu_0 N_2 \mathbf{M} \qquad \text{inner surface} \end{aligned} \tag{7.115}$$

and the signs are selected in accordance with the direction of the surface normals. The constant $N$ is the demagnetization factor and its value for some simple shapes is given in Table 7.6. A sum of the demagnetization factors for the three principal axes must meet the condition that $N_x + N_y + N_z = 1$ and this is quite apparent for the symmetric shapes given in Table 7.6. If cgs units are used, the sum is $4\pi$. The demagnetization factor for an ellipsoid of revolution is

$$N_{\parallel} = \frac{1 - e^2}{e^2}\left[\frac{1}{2e} \ln\left(\frac{1+e}{1-e}\right) - 1\right] \tag{7.116}$$

for **M** parallel to the axis of revolution, and for **M** perpendicular to the axis of revolution, it can be obtained from $2N_{\perp} + N_{\parallel} = 1$. If $c$ is the length

**Table 7.6**

| Shape | Axis | $N$ |
|---|---|---|
| Sphere | Any | $\frac{1}{3}$ |
| Thin slab | Normal | 1 |
| Thin slab | In-plane | 0 |
| Long circular cylinder | Longitudinal | 0 |
| Long circular cylinder | Transverse | $\frac{1}{2}$ |

of the axis of revolution and $a$ is the transverse length, then $e^2 = 1 - (a/c)^2$. The magnetic field inside a real cavity whose shape is described by $N_2$ for a sample whose external shape is described by $N_1$ is

$$\mathbf{B}_\mathrm{i} = \mathbf{B}_\mathrm{e} - \mu_0 N_1 \mathbf{M} + \mu_0 N_2 \mathbf{M} \tag{7.117}$$

The magnetic field (which is used in Maxwell's equations) inside bulk material is the field inside a needle-shaped cavity, $N_2 = 0$, and is given by

$$\mathbf{B} = \mathbf{B}_\mathrm{e} - \mu_0 N_1 \mathbf{M} \tag{7.118}$$

The contribution to the field by the magnetic dipoles inside the cavity is a more complex problem. If all the dipoles $\boldsymbol{\mu}_\beta$ are parallel, then $\mathbf{B}_3 = 0$ for simple cubic, body-centered cubic, and face-centered cubic lattices, and for isotropic distributions in gases or random distributions in solids. Deviations from $\mathbf{B}_3 = 0$ are considered in a later section and the field at the molecular site is given by the Lorentz local field

$$\mathbf{B}_\mathrm{loc} = \mathbf{B}_\mathrm{e} + \mu_0(\tfrac{1}{3} - N_1)\mathbf{M} \tag{7.119}$$

for the remaining discussion. Within this approximation, all molecules inside an ellipsoidal sample experience the same magnetic field $\mathbf{B}_\mathrm{loc}$ and the problem of $N$ molecules in an ellipsoidal shape reduces to a product of one-molecule generating functions,

$$Q_N \approx \frac{Q_\mathrm{t}^N Q_\mathrm{int}^N}{N!} \tag{7.120a}$$

for gases. For solids,

$$Q_N = Q_\mathrm{int}^N \tag{7.120b}$$

and for both gases and solids,

$$Q_\mathrm{int} = Q_\mathrm{1M} \tag{7.121}$$

where $B_0$ is replaced by $B_\mathrm{loc}$ in Eq. (7.113).

The magnetic work done on a system by changing the external magnetic field is given by

$$dW = -\mathscr{M}\, dB_\mathrm{e} \tag{7.122}$$

and this term should be added to Eq. (6.19) for the magnetic contribution to the Helmholtz function. This definition implies that $B_\mathrm{e}$ is the magnetic field in the absence of the material and the magnetization $\mathscr{M}$ is parallel to

$\mathbf{B}_e$. This limits the discussion to ellipsoids with $\mathbf{B}_e$ and $\mathscr{M}$ along an axis of the ellipsoid. If $N$ molecules are placed in volume $V$, the thermodynamic relationships follow from

$$A = -NkT \ln Q_{1M} \tag{7.123a}$$

$$\mathscr{M} = VM = -\left(\frac{\partial A}{\partial B_e}\right)_T \tag{7.123b}$$

$$S = -\left(\frac{\partial A}{\partial T}\right)_{B_e} \tag{7.123c}$$

The discussion is now limited to spherical samples and it is assumed that the Lorentz local field and the external field are equal,

$$B_{loc} \approx B_e = B$$

The generating function $Q_{int}$ given by Eq. (7.113) can be summed directly by noting that

$$\sum_{m=-J}^{J} x^m = \frac{x^{(2J+1)/2} - x^{-(2J+1)/2}}{x^{1/2} - x^{-1/2}}$$

or

$$Q_{1M} = \frac{\sinh[\frac{1}{2}(2J+1)\hbar\gamma_J B/kT]}{\sinh(\frac{1}{2}\hbar\gamma_J B/kT)} \tag{7.124}$$

This can be used with Eqs. (7.123) to find $A$, $M$, and $S$. A plot [*11*] of $S$ against $T$ is shown in Fig. 7.9 for a paramagnetic salt with $J = \frac{5}{2}$.

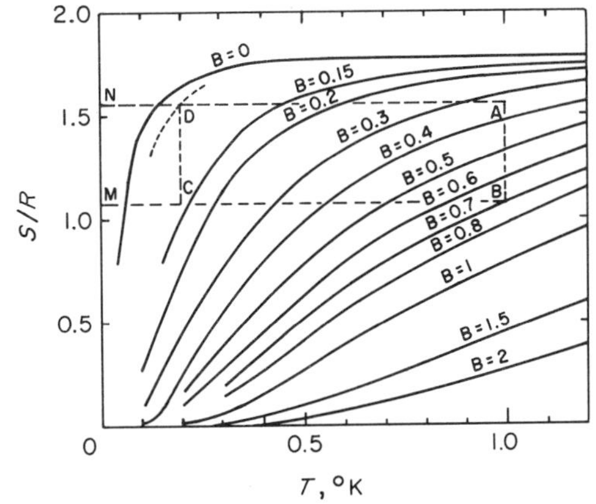


**Fig. 7.9** Entropy–temperature diagram for iron ammonium alum. [From C. V. Heer, C. B. Barnes, and J. G. Daunt, *Rev. Sci. Instrum.* **25**, 1088 (1954), Fig. 2.]

For convenience, let

$$x = \frac{\hbar\gamma B}{kT}$$

and then for small $x$,

$$Q_{1M} = \sum_{m=-J}^{J} [1 - mx + \tfrac{1}{2}(mx)^2 + \cdots]$$
$$\approx (2J + 1)[1 + \tfrac{1}{6}J(J + 1)x^2 + \cdots] \tag{7.125a}$$

where $\sum m^2 = \frac{1}{3}(2J + 1)J(J + 1)$ is used. The free energy is approximately

$$A \approx -NkT[\ln(2J + 1) + \tfrac{1}{6}J(J + 1)x^2] \tag{7.125b}$$

and the entropy is

$$S \approx Nk[\ln(2J + 1) - \tfrac{1}{6}J(J + 1)x^2] \tag{7.125c}$$

At $x = 0$, the entropy is $Nk \ln(2J + 1)$ and is in accord with the concept that the entropy is the positive logarithm of the number of orientations of the $N$ spin systems, or $(2J + 1)^N$. As the temperature is decreased, the second term provides the correction and leads to a heat capacity of $C_B = T(\partial S/\partial T)_B$, or

$$\frac{C_B}{Nk} = \frac{J(J + 1)x^2}{3} \propto \left(\frac{B}{T}\right)^2 \tag{7.126}$$

Heat capacity increases as $(B/T)^2$ for small $x$ and approaches $k$ per molecule as the saturation region of $x = 1$ is reached. The magnetization is given by

$$M = \frac{\mathscr{M}}{V} = \frac{(N/V)(\hbar\gamma_J)^2 J(J + 1)B}{3kT} \tag{7.127}$$

and the magnetic susceptibility by

$$\chi = \frac{M}{H} = \frac{(N/V)(\hbar\gamma_J)^2 J(J + 1)\mu_0}{3kT} \propto \frac{1}{T} \tag{7.128}$$

For small values of $x$, the magnetic susceptibility is proportional to $1/T$ and this is referred to as Curie's Law. At low temperature, this $1/T$ dependence is used as the principle for a low-temperature thermometer. Atomic spin systems are used in the temperature range of 0.01–1°K and nuclear spin systems at lower temperatures.

As $x$ approaches unity, the system saturates and the approximation given by Eq. (7.125) is no longer valid. Equation (7.124) must be used. The saturation is apparent in the entropy curves in Fig. 7.9.

Atomic gases are paramagnetic whenever a nonzero value of $J$ occurs for the atom. Thus with the $|\alpha LSJM)$ designation, it is apparent that paramagnetic states can occur for both the ground states and excited states of the atoms. A few gaseous molecules have magnetic moments of this same order of magnitude, including $O_2$, NO, $NO_2$, $Cl_2O$, and $F_2O$. These molecules have unpaired electrons in the ground state and the magnetic moments are large compared to molecules in the $^1\Sigma$ state. Nuclear paramagnetism occurs in all gases in which the nucleus has a magnetic moment $I \neq 0$. Nuclear magnetic moments are in general $10^{-3}$ of atomic moments.

### *7.10.1 Paramagnetic Ions in Crystals*

The paramagnetic ions in crystals can be divided into the iron group and the rare earth or lanthanide group. The iron group ions have the configuration $3d^n$. Hund's rule for the multiplets of the $3d^n$ free ions states that the multiplet of lowest energy has the maximum value of $2S + 1$ permitted by the Pauli exclusion principle and of these, the state with maximum $L$ has the lowest energy. Configurations $d^1$ through $d^5$ have total angular momenta $J = |L - S|$ and $d^6$ through $d^{10}$ have $J = |L + S|$. Configurations and the ground-state multiplet of the free ions are given in Tables 7.7 and 7.8. When the free ion of the iron group is placed in a

**Table 7.7**

Effective Magneton Numbers for Iron Group Ions[a]

| Ion | Configuration | Basic level | $p$(calc) $= g[J(J+1)]^{1/2}$ | $p$(calc) $= 2[S(S+1)]^{1/2}$ | $p$(exp) |
|---|---|---|---|---|---|
| $Ti^{3+}$, $V^{4+}$ | $3d^1$ | $^2D_{3/2}$ | 1.55 | 1.73 | 1.8 |
| $V^{3+}$ | $3d^2$ | $^3F_2$ | 1.63 | 2.83 | 2.8 |
| $Cr^{3+}$, $V^{2+}$ | $3d^3$ | $^4F_{3/2}$ | 0.77 | 3.87 | 3.8 |
| $Mn^{3+}$, $Cr^{2+}$ | $3d^4$ | $^5D_0$ | 0 | 4.90 | 4.9 |
| $Fe^{3+}$, $Mn^{2+}$ | $3d^5$ | $^6S_{5/2}$ | 5.92 | 5.92 | 5.9 |
| $Fe^{2+}$ | $3d^6$ | $^5D_4$ | 6.70 | 4.90 | 5.4 |
| $Co^{2+}$ | $3d^7$ | $^4F_{9/2}$ | 6.63 | 3.87 | 4.8 |
| $Ni^{2+}$ | $3d^8$ | $^3F_4$ | 5.59 | 2.83 | 3.2 |
| $Cu^{2+}$ | $3d^9$ | $^2D_{5/2}$ | 3.55 | 1.73 | 1.9 |

[a] From Charles Kittel, "Introduction to Solid State Physics," 3rd ed., Chap. 14, Table 2; Copyright 1966, Wiley, New York. Used by permission of John Wiley & Sons, Inc.

crystal, the crystalline electric field quenches the orbital angular momentum $L$. Consider as an example $Cr^{3+}$ in $Al_2O_3$ or ruby. As a free ion, $Cr^{3+}$ has a $^4F_{3/2}$ ground state. The cubic electric field of $Al_2O_3$ interacts strongly with the d configuration and only four degenerate states form the lowest energy level. In general, the cubic electric fields leave the ground state with a $2S+1$ degeneracy and $S$ can be used as a quantum number. An axial electric field also occurs in ruby. Spin–orbit interaction and the axial crystalline field cause further splitting of these four energy levels. The axial crystalline field has a potential of the form $U = 3z^2 - r^2$ and without knowing further details, the Wigner–Eckart theorem requires that between states of equal $S$, this operator have the same matrix elements as the operator $3S_z^2 - S_{op}^2$. Thus the effective Hamiltonian in an axial crystal field is of the form

$$(3z^2 - r^2) \rightarrow 3S_z^2 - S(S+1) \tag{7.129}$$

where $\hat{z}$ is the symmetry axis of the crystal. The same argument can be used to show that the magnetization operator $\boldsymbol{\mu}$ has axial symmetry and between states of equal $S$,

$$\frac{\boldsymbol{\mu}}{\mu_B} = g_\perp(\hat{x}S_x + \hat{y}S_y) + g_\parallel \hat{z}S_z \tag{7.130}$$

Thus the effective Hamiltonian of the $Cr^{3+}$ ion in $Al_2O_3$ is

$$H = D[S_z^2 - \tfrac{1}{3}S(S+1)] + \mu_B[g_\perp(S_xB_x + S_yB_y) + g_\parallel S_zB_z] \tag{7.131}$$

where $2D = 11{,}593$ MHz and $g_\perp \approx g_\parallel = 1.986$. Here $\mu_B$ is the Bohr magneton. Effective Hamiltonians can be constructed for other iron group ions in a similar manner. This particular transition is used in the microwave ruby maser, and the energy level separation for various orientations of the magnetic field have been tabulated [*12*]. In zero magnetic field, the states split into two energy levels $\pm\frac{3}{2}$ and $\pm\frac{1}{2}$ separated by 11,593 MHz. The low-temperature thermodynamic properties follow from the one-ion generating function

$$Q_{1M} = 2(1 + e^{-2D/kT}) \tag{7.132}$$

The Wigner–Eckart theorem [*13*]

$$(\alpha jm \mid T_M{}^K \mid \alpha jm') = \text{constant} \times (jKmM \mid jm') \tag{7.133}$$

provides a very useful theorem for constructing effective Hamiltonians. A cubic potential is given by

$$V_4 = A\{(x^4 + y^4 + z^4) - \tfrac{3}{5}r^4\} = R(r)[Y_0{}^4 + (\tfrac{5}{14})^{1/2}(Y_4{}^4 + Y_{-4}^4]$$
$$\rightarrow [T_0{}^4 + (\tfrac{5}{14})^{1/2}(T_4{}^4 + T_{-4}^4)]$$

and can be represented by spherical harmonics $Y_M^L$ or tensor operators $T_M^K$. If $j = S = \frac{3}{2}$, the Clebsch–Gordon coefficient is zero since $\frac{3}{2} + \frac{3}{2} < 4$ and the triangular rule is not obeyed. If an axial potential is included, $V_{\text{axial}} = 3z^2 - r^2 = f(r)Y_0^2 \rightarrow T_0^2$, the Clebsch–Gordon coefficients are proportional to the matrix elements of $3S_z^2 - S_{op}^2$. If the crystalline field is trigonal to tetragonal, the dominant correction is of the above form. The $Cr^{3+}$ ion in chromium potassium alum is in a cubic plus trigonal field and is described by $2D/k = 0.25°$K. Iron ammonium alum, $Fe(NH_4)(SO_4)_2 \cdot 12H_2O$ with a molecular weight of 482.2 and a density of 1.70, has a ground-state described by $S = \frac{5}{2}$. Since the lowest multiplet is a $^6S$, the orbital angular momentum is zero. Even so, the cubic potential interacts with the $S = \frac{5}{2}$ states. Clebsch–Gordon coefficients are nonzero between all values $m + m' = \pm 4$ since the triangular rule $\frac{5}{2} + \frac{5}{2} = \{\text{an integer}\} \geq 4$ is possible. The matrix elements connect the states $\pm\frac{5}{2}$ with $\pm\frac{3}{2}$ and these four states have an energy greater than the $\pm\frac{1}{2}$ ground states. A trigonal field is present and this field further splits the four energy states. Such problems require a complete diagonalization of the energy matrix to find the position of the energy levels. The cubic plus trigonal field can be written as $H = A(x^4 + y^4 + z^4) + B(3z^3 - r^2)$ and the Clebsch–Gordon coefficients can be obtained from tables to solve for the energies. Assuming the trigonal field to be zero, the heat capacity is approximately given by assuming that the four energy levels $\pm\frac{3}{2}$ and $\pm\frac{5}{2}$ are degenerate and are $\delta/k = 0.20°$K above the $\pm\frac{1}{2}$ ground state. The zero-field entropy shown in Fig. 7.9 is determined with this value of $\delta$.

Figure 7.9 also shows the cycle followed in an adiabatic demagnetization experiment. This method was first proposed by Debye and Giauque in 1926. An extensive summary of almost all experimental investigations prior to 1956 is given in a review article by de Klerk [*14*]. The basic Carnot cycle is the path *ABCDA*. Isothermal magnetization occurs from *A* to *B* and the heat of magnetization is transferred to the surroundings. Usually, the surroundings or heat reservoir is a $^4$He bath at approximately 1°K, a $^3$He bath at 0.5°K, or another paramagnetic salt. Thermal contact is via an exchange gas of helium or a thermal valve. A superconducting metal has a much lower thermal conductivity in the superconducting state than in the normal state and can be used as a heat valve or switch [*15*]. Lead is frequently used and has a switching ratio of 100 to 1 at 1°K and 1000 to 1 at 0.4°K. The paramagnetic salt is isolated by removing the exchange gas, or, for a lead thermal switch, by removing the magnetic field at the lead. The magnetic field on the salt is lowered and since the salt is thermally isolated, the process is adiabatic demagnetization and is indicated by the constant-entropy path *BC*. The path *CD* is the isothermal heating of the salt and this heat is supplied by the experimental sample that is being cooled. *DA* is the adiabatic magnetization of the salt to complete the cycle. A magnetic refrigerator [*11*] which used these thermal valves and the cycle shown in Fig. 7.9 was developed for continuous cooling between 0.2 and 1°K.

If the energy of a particular state is denoted by the index *m* for a particular ion, then the probability of the ion being in this state is given by

$$p(E_m) = \frac{e^{-\beta E_m}}{\sum_m e^{-\beta E_m}} \tag{7.134}$$

and the entropy of the system is given by

$$S = -Nk \sum p(E_m) \ln p(E_m) \tag{7.135}$$

Demagnetization at constant entropy illustrates with great clarity some aspects of a process at constant entropy. Since $B$ changes at constant $S$, $dS = 0$ for the process. This is possible if

$$p(E_m) = \text{constant}$$

or if the probability of a particular state being occupied remains constant during a constant-entropy process and this can occur only if

$$\left(\frac{E_m}{T}\right)_\text{i} = \left(\frac{E_m}{T}\right)_\text{f} \tag{7.136}$$

If the splitting due to the crystal field or, eventually, magnetic interactions can be ignored, this implies

$$\frac{B_\text{i}}{T_\text{i}} = \frac{B_\text{f}}{T_\text{f}} \tag{7.137}$$

and provides a simple relationship between magnetic field and temperature. If $B_\text{f} = 0$, this does not imply an infinitely low temperature. The magnetic interaction must be included and $B_\text{f} \to B_\text{eff}$ rather than zero and $T_\text{f} \to T_\text{eff}$ and limits the lowest obtainable temperatures with reasonable magnetic fields. Nondiluted paramagnetic salts are useful at temperatures above 0.005°K and nuclear spin systems [*16*] above $10^{-6}$ °K. Whenever spin–spin interactions become important, the simple model used for the local field cannot be used. The energy levels of the $N$-body problem are necessary and $Q_N$ is no longer given by $Q_1{}^N$.

**Table 7.8**

Effective Magneton Numbers $p$ for Trivalent Lanthanide Group Ions (Near Room Temperature)[a]

| Ion | Configuration | Basic level | $p(\text{calc}) = g[J(J+1)]^{1/2}$ | $p(\text{exp})$, approximate |
|---|---|---|---|---|
| $Ce^{3+}$ | $4f^1 5s^2 p^6$ | $^2F_{5/2}$ | 2.54 | 2.4 |
| $Pr^{3+}$ | $4f^2 5s^2 p^6$ | $^3H_4$ | 3.58 | 3.5 |
| $Nd^{3+}$ | $4f^3 5s^2 p^6$ | $^4I_{9/2}$ | 3.62 | 3.5 |
| $Pm^{3+}$ | $4f^4 5s^2 p^6$ | $^5I_4$ | 2.68 | — |
| $Sm^{3+}$ | $4f^5 5s^2 p^6$ | $^6H_{5/2}$ | 0.84 | 1.5 |
| $Eu^{3+}$ | $4f^6 5s^2 p^6$ | $^7F_0$ | 0 | 3.4 |
| $Gd^{3+}$ | $4f^7 5s^2 p^6$ | $^8S_{7/2}$ | 7.94 | 8.0 |
| $Tb^{3+}$ | $4f^8 5s^2 p^6$ | $^7F_6$ | 9.72 | 9.5 |
| $Dy^{3+}$ | $4f^9 5s^2 p^6$ | $^6H_{15/2}$ | 10.63 | 10.6 |
| $Ho^{3+}$ | $4f^{10} 5s^2 p^6$ | $^5I_8$ | 10.60 | 10.4 |
| $Er^{3+}$ | $4f^{11} 5s^2 p^6$ | $^4I_{15/2}$ | 9.59 | 9.5 |
| $Tm^{3+}$ | $4f^{12} 5s^2 p^6$ | $^3H_6$ | 7.57 | 7.3 |
| $Yb^{3+}$ | $4f^{13} 5s^2 p^6$ | $^2F_{7/2}$ | 4.54 | 4.5 |

[a] From Charles Kittel, "Introduction to Solid State Physics," 3rd ed., Chap. 14, Table 1; Copyright, 1966, Wiley, New York. Used by permission of John Wiley & Sons, Inc.

The rare earth or lanthanide group ions are given in Table 7.8. Electrons in the unfilled $4f^n 5s^2 p^6$ configuration determine the multiplets of the rare earth ions. Hund's rule applies in the selection of the ground-state multiplet. Since these ions only exist in very special gaseous environments, they are usually of interest as ions in crystalline lattices. Since the $f^n$ shell is well isolated by the $5s^2 p^6$ shells, spin–orbit coupling provides a stronger influence than the crystalline field and $J$ remains a good quantum number in the $|LSJM\rangle$ labeling of the states. The crystalline field acts as a perturbation and the Wigner–Eckart theorem given by Eq. (7.133) can now be used with $j = J$. These ions in host lattices of yttrium aluminum garnet, $Y_3Al_5O_{12}$, are referred in the literature as "YAG." Both YAG and rare earth ions in glass have been of considerable interest for solid-state lasers in the optical spectral region.[2]

Cerium magnesium nitrate, $Ce_2Mg_3(NO_3)_{12} \cdot 24H_2O$, with a molecular weight of 714.8 and a density of 2.0, has been of particular interest in low-temperature research. The basic multiplet of $Ce^{3+}$ is a $^2F_{5/2}$ multiplet and

[2] A comprehensive review of all aspects of lasers and masers is given in the work edited by Kleen and Muller [*17*].

the strong anisotropy of the crystalline potential splits the $J = \frac{5}{2}$ state into $\pm\frac{5}{2}, \pm\frac{3}{2}, \pm\frac{1}{2}$, or three energy levels. This anisotropy of the potential energy is quite apparent in the crystals; they grow in a flat, hexagonal shape. The $\pm\frac{5}{2}$ and $\pm\frac{3}{2}$ energy levels are so far displaced from the $\pm\frac{1}{2}$ level that only the lowest level is of interest in low-temperature experiments. A crystalline field cannot remove the $\pm\frac{1}{2}$ degeneracy and these are usually referred to as Kramer's doublets. Although splitting is not possible, the crystalline field can interact via the spin–orbit interaction and $\boldsymbol{\mu}$ has axial symmetry. Between states of $S = \frac{1}{2}$, the magnetic moment has the form of Eq. (7.130). Thus in a magnetic field, the $Ce^{3+}$ ions respond as

$$H_1 = \mu_B[g_\perp(S_x B_x + S_y B_y) + g_\parallel S_z B_z] \qquad (7.138)$$

where $\hat{z}$ is parallel to the axis of symmetry,

$$g_\parallel = 0.25, \qquad g_\perp = 1.84, \qquad \text{and} \qquad S = \tfrac{1}{2}.$$

The response of this crystal in a magnetic field is highly anisotropic. It has been suggested and shown possible to cool an anisotropic salt by rotation [*18*]. If the external magnetic field is along the $\hat{z}$ symmetry axis of the crystal, the energy level separation is given by

$$E_m = g_\parallel \mu_B m B$$

and if the field is perpendicular to the axis of symmetry, it is given by

$$E_m = g_\perp \mu_B m B$$

During a constant-entropy process, the probability $p(E_m)$ remains constant. Thus as the field is rotated from $\hat{z}$ to $\hat{x}$, or the crystal is rotated through 90° about an axis perpendicular to the axis of symmetry,

$$\left(\frac{E_m}{T}\right)_z = \left(\frac{E_m}{T}\right)_x \qquad (7.139a)$$

or

$$\frac{g_\parallel}{T_\parallel} = \frac{g_\perp}{T_\perp} \qquad (7.139b)$$

and the salt heats or cools depending on the sense of the rotation.

Since $Ce^{3+}$ is an effective spin-$\frac{1}{2}$ system at low temperatures, the lower level has a twofold degeneracy which is not affected by a crystalline field. The spacing between $Ce^{3+}$ ions is large in the salt and magnetic interactions are not important until temperatures of the order of 2 mdeg are reached. This salt obeys Curie's Law down to these very low temperatures and is widely used as a low-temperature thermometer. Anomalies have appeared in the thermometry at 2 mdeg and it is at these temperatures that magnetic interactions are important. It has been suggested that the salt becomes antiferromagnetic at this temperature [*19*].

Exercise 7.19 Plot the entropy and heat capacity of $N$ $Cr^{3+}$ ions in $Al_2O_3$ from 0.05 to 1°K. What are the high-temperature and low-temperature entropies? Plot $S/Nk$ and $C/Nk$ against $T$.

Exercise 7.20 Chromium potassium alum, $CrK(SO_4)_2 \cdot 12H_2O$, with a molecular weight of 499.4 and a density of 1.83, is a much-used salt for adiabatic demagnetization. The energy level splitting of the $Cr^{3+}$ ion is $2D/k = 0.25$°K. Find the magnetic susceptibility and the heat capacity above 0.25°K.

Exercise 7.21 What is the approximate single-ion generating function for $Fe^{3+}$ in iron alum?

Ans. $Q_{1M} = 2 + 4e^{-\delta/T}$

## References


1. M. N. Saha and B. N. Srivastava, "A Treatise on Heat." Hafner, New York, 1958.
2. C. de Jager, ed., "The Structure of the Quiet Photosphere and the Low Chromosphere," Springer-Verlag, Berlin and New York, 1968; J. N. Xanthalkis, ed., "Solar Physics," Wiley (Interscience), New York, 1967.
3. B. N. Srivastava, Proc. Roy. Soc. *Ser.* A. **176**, 343 (1940); B. N. Srivastava and A. S. Bhatnagar, *Proc. Nat. Acad. Sci. India* **15** (1946).
4. G. Herzberg, *Trans. Roy. Soc. Can. Sect.* IV **5**, 3 (1967).
5. S. Chandrasekhar and M. K. Krogdahl, *Astrophys. J.* **98**, 205 (1943); L. R. Henrich, *Ibid.* **99**, 59, 318 (1944); S. Chandrasekhar, *Ibid.* **100**, 176 (1944); **102**, 223, 395 (1945); **104**, 430 (1946); **128**, 114 (1958); W. Lochte-Holtgreven and W. Nossen, *Z. Phys.* **133**, 124 (1952); L. M. Branscomb and S. J. Smith, *Phys. Rev.* **98**, 1028 (1955).
6. M. N. Saha and A. N. Tandon, *Proc. Nat. Inst. Sci. India* **3**, 287 (1937); see also Saha and Srivastava [*1*].
7. A. Farkas, "Orthohydrogen and Parahydrogen, and Heavy Hydrogen" Cambridge Univ. Press, 1935; H. W. Wooley, R. B. Scott, and F. G. Brickwedde, *Nat. Bur. Std. J. Res.* **41**, 379 (1948) (RP 1932).
8. N. Wakao, J. M. Smith, and P. W. Selwood, *J. Catal.* **1**, 62 (1962); D. White and G. Grenier, *J. Chem. Phys.* **40**, 3015 (1964).

9. E. R. Harrison, *Phys. Today* **21**, 31 (June 1968).
10. B. K. Harrison, K. S. Thorne, M. Wakano, and J. A. Wheeler, "Gravitational Theory and Gravitational Collapse." Univ. of Chicago Press, Chicago, Illinois, 1965; L. Oster, *Phys. Rev. Lett.* **23**, 987 (1969).
11. C. V. Heer, C. B. Barnes, and J. G. Daunt, *Rev. Sci. Instrum.* **25**, 1088 (1954).
12. J. Weber, *Rev. Mod. Phys.* **108**, 537 (1959); see J. E. Geusic and H. E. D. Scovil, *Rep. Progr. Phys.* **27**, 241 (1964) for other maser materials.
13. A. Messiah, "Quantum Mechanics," Vol. I and II. North-Holland Publ., Amsterdam, 1962.
14. D. de Klerk, Low temperature physics II. "Handbuch der Physik" (S. Flügge, ed.), Vol. XV, pp. 38–209. Springer-Verlag, Berlin and New York, 1956.
15. C. V. Heer and J. G. Daunt, *Phys. Rev.* **76**, 854 (1949).
16. N. Kurti, F. N. H. Robinson, F. E. Simon, and D. A. Spohr, *Nature* (*London*) **178**, 950 (1956); E. V. Osgood and J. M. Goodkind, *Phys. Rev. Lett.* **18**, 894 (1967).
17. W. Kleen and R. Muller, eds., "Laser." Springer-Verlag, Berlin and New York, 1969.
18. C. D. Jeffries, *Cryogenics* **8**, 41 (1963); A. Abragam, *Ibid.* **3**, 42 (1963); F. N. Robinson, *Phys. Rev. Lett.* **4**, 180 (1963); K. H. Langley and C. D. Jeffries, *Ibid.* **13**, 808 (1964).
19. D. J. Abeshouse, G. O. Zimmerman, D. R. Kelland, and E. Maxwell, *Phys. Rev. Lett.* **23**, 308 (1969).

# Nonideal and Real Gases

# VIII

## 8.1 Nonideal Gases

Molecular gases at low densities were considered in Chapter VII and only the one-particle generating function $Q_1$ was necessary for the discussion. The next higher approximation will be considered in this chapter and the two-particle generating function $Q_2$ is needed for this discussion. This development will be given in the next section and then, in subsequent sections, comparison with experimental work will be discussed. Also in subsequent sections, the degenerate ideal Bose and Fermi gases will be discussed at greater length. Electrons in metals are considered as a degenerate Fermi gas. The discussion is extended to the relativistic Fermi gas and then very dense matter and neutron stars are considered as examples of degenerate Fermi gases. The last section of the chapter discusses the degenerate Bose–Einstein gas and compares some of its properties with liquid $^4$He.

## 8.2 Cluster Expansion

In the previous chapter, gases at low densities were discussed and it was observed that in the grand canonical generating function

$$\mathcal{Q} = \sum_N z^N Q_N$$

only $Q_1$ was needed for gases at low densities. The statistical variable of primary use is $\ln \mathcal{Q}$ and a power series expansion in $z$ seems desirable. As $z$ tends toward zero, the generating function $\mathcal{Q}$ tends toward one:

$$\mathcal{Q} \xrightarrow[z \to 0]{} Q_0 \equiv 1 \tag{8.1}$$

Then

$$\ln \mathcal{Q} = \ln\left(1 + \sum_N{}' z^N Q_N\right) \approx \sum_{n=1}^{\infty} a_n z^n \tag{8.2}$$

is a suitable power series expansion and in mathematical statistics this is referred to as the *cumulant expansion.* In statistical physics, it is more commonly known as the *cluster expansion.* The coefficients $a_n$ are given by MacLaurin's expansion of $\ln \mathcal{Q}$, or

$$a_n = \frac{1}{n!}\left(\frac{\partial^n(\ln \mathcal{Q})}{\partial z^n}\right)_{z=0} \tag{8.3}$$

and in detail,

$$a_1 = Q_1 \tag{8.4a}$$

$$a_2 = Q_2 - \tfrac{1}{2}Q_1^2 \tag{8.4b}$$

$$a_3 = Q_3 - Q_2Q_1 + \tfrac{1}{3}Q_1^3 \tag{8.4c}$$

$$a_4 = Q_4 - Q_3Q_1 - \tfrac{1}{2}Q_2^2 + Q_2Q_1^2 - \tfrac{1}{4}Q_1^4 \tag{8.4d}$$

$$\vdots$$

The idea of a cluster follows by noting that if a system is composed of independent clusters for short periods of time, then $\mathcal{Q}$ must be a product of generating functions, that is,

$$\mathcal{Q} \approx \prod_{n=1}^{\infty} \mathcal{Q}_n \tag{8.5a}$$

and

$$\ln \mathcal{Q} = \sum_{n=1}^{\infty} \ln \mathcal{Q}_n = \sum_{n=1} a_n z^n \tag{8.5b}$$

If the single molecules are of species A and form a cluster of two molecules of species $A_2$, at equilibrium,

$$A + A \rightleftarrows A_2 \tag{8.6a}$$

$$\mu_A + \mu_A = \mu_{A_2} \tag{8.6b}$$

where $\mu_A$ is the chemical potential of species A and $\mu_{A_2}$ the chemical potential of the two-particle species $A_2$. Equation (8.6b) can be written in terms of the fugacity $z$ as

$$z_{A_2} = z_A{}^2 \tag{8.6c}$$

and the fugacity of the two-particle cluster is just the square of the fugacity of the one-particle cluster. Thus the coefficient $a_1$ of $z$ is interpreted as the one-particle cluster, the coefficient $a_2$ of $z^2$ as the two-particle cluster, and the coefficient $a_n$ of $z^n$ as the $n$-particle cluster.

Since in the development given in this text, $Q_N$ refers to the solution of the $N$-particle Schrödinger equation for volume $V$, the cluster expansion given by Eqs. (8.2) and (8.3) is quite general. This very simple development yields the same result as the much more extensive procedure of Kahn and Uhlenbeck [*1*] in their quantum statistical mechanical development and as the classical statistical mechanical development of Ursell and Mayer.[1]

## 8.3 Equation of State and Second Virial Coefficient

Except for special examples, only $Q_2$ is discussed in this chapter. Higher approximations require a knowledge of the solution of the three-body problem. With these restrictions, the approximate equation of state is obtained from Eqs. (8.2), (8.4a), and (8.4b):

$$\frac{PV}{kT} = \ln \mathcal{Q} \approx za_1 + z^2a_2 \tag{8.7}$$

The fugacity $z$ is determined from the average number of particles in volume $V$,

$$N = z\left[\frac{\partial(\ln \mathcal{Q})}{\partial z}\right]_{\mathrm{V,T}} \approx za_1 + 2z^2a_2 \tag{8.8a}$$

This equation can be inverted by an iterative expansion and

$$z \approx \frac{N}{a_1}\left[1 - 2\left(\frac{N}{a_1}\right)\left(\frac{a_2}{a_1}\right)\right] \tag{8.8b}$$

becomes a polynomial in $N$. Direct substitution in Eq. (8.7) yields an equation of state,

$$\frac{PV}{kT} \approx N\left(1 - \frac{N}{a_1}\,\frac{a_2}{a_1}\right) \tag{8.9}$$

[1] For an extended discussion and references to original literature, see Mayer and Mayer [*2*].

It is more common to expand the pressure in a power series in the number density $(N/V)$ and this expansion is of the form

$$P = \frac{NkT}{V}\left[1 + \left(\frac{N}{V}\right)B(T) + \left(\frac{N}{V}\right)^2 C(T) + \cdots\right] \tag{8.10a}$$

where $B(T)$ is the second virial coefficient, $C(T)$ the third virial coefficient, and so on. The second virial coefficient $B(T)$ can be correlated with the cumulant coefficients by noting from Eq. (8.9) that

$$B(T) = -V\left(\frac{a_2}{a_1^2}\right) = -V\frac{Q_2 - \frac{1}{2}Q_1^2}{Q_1^2} \tag{8.10b}$$

With this notation,

$$\frac{PV}{kT} = \ln \mathcal{Q} \approx N\left[1 + \frac{N}{V}B(T)\right] \tag{8.11a}$$

and

$$z \approx \frac{N}{Q_1}\left[1 + \frac{2N}{V}B(T)\right] \tag{8.11b}$$

The internal energy is given by

$$\begin{aligned} U &= -\left[\frac{\partial(\ln \mathcal{Q})}{\partial\beta}\right]_{z,V} \\ &\approx -N\frac{\partial(\ln Q_1)}{\partial\beta} - N^2\frac{\partial(a_2/a_1^2)}{\partial\beta} \\ &\approx -N\frac{\partial(\ln Q_1)}{\partial\beta} + \frac{N^2}{V}\frac{\partial B}{\partial\beta} \end{aligned} \tag{8.11c}$$

The internal energy is directly related to the coefficient $B(T)$. The other thermodynamic quantities can be determined from $U$. From the definition of the canonical ensemble, $Q_1$ and $Q_2$ are functions of $V$ and $T$.

### 8.3.1 *Experimental Determination of the Second Virial Coefficient*

There appear to be two important methods for determining the second virial coefficient, or the degree of nonideality of a molecular gas. The first method is the direct determination of the equation of state as implied by Eq. (8.10a),

$$\frac{PV}{kT} = N\left\{1 + \frac{N}{V}B(T) + \cdots\right\}$$

It is not possible to fit all experimental data with a simple $B(T)$ function, but it is possible to give the experimental data in terms of reduced coordinates that fit a large variety of gases [*3*]. Curves for a number of frequently used gases are shown in Fig. 8.1. A Lennard–Jones 6–12 interaction poten-

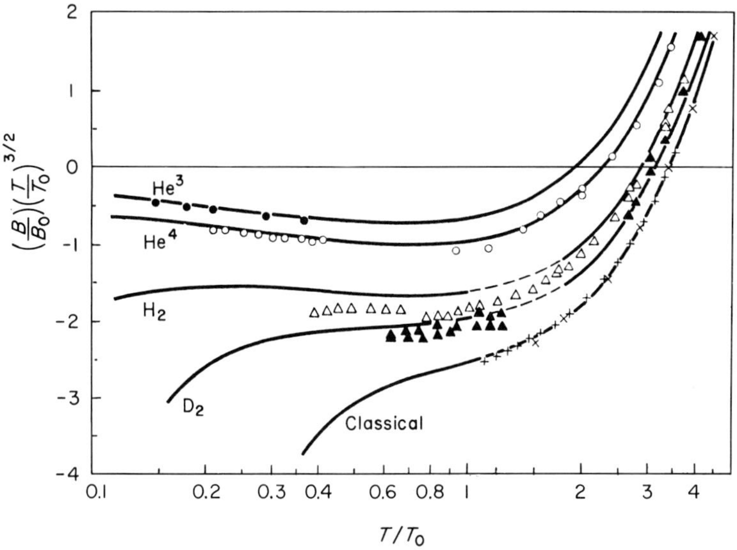


**Fig. 8.1** The reduced second virial coefficient for the Lennard–Jones 6–12 potential [*3*]. [From E. A. Mason and T. H. Spurling, The virial equation of state, *in* "International Encyclopedia of Physical Chemistry and Chemical Physics" (J. S. Rowlinson, ed.), p. 46, Fig. 2.1. Pergamon, Oxford, 1969.]

tial is used and the constants $T_0$ and $B_0$ are given later in Table 8.2 and discussed in greater detail in Section 8.6. The value of the second virial coefficient is of the order of $10^{-28}$ m$^3$ and the second virial correction $nB$ is of the order of $10^{-3}$ for a number density $n = 10^{25}/\text{m}^3$, or a density characteristic of a gas under standard temperature and pressure conditions. Further details regarding the equations of state of real gases are given in the "American Institute of Physics Handbook" [*3a*]. Corrections for deviations from the Ideal Gas Law are necessary when gas thermometers are used. Except at very low temperatures, the correction for He is quite small, and He is the most ideal gas for use at normal temperatures.

The second method for measuring the second virial coefficient involves the expansion of the gas at constant enthalpy. This is usually regarded as the expansion of a gas through a nozzle or a porous plug. The pressure on one side of the nozzle is $P_i$ and as a volume of gas $V_i$ passes through the nozzle, or the volume changes from $V_i$ to 0, the work done on the gas is $P_i V_i$. The work done by the gas leaving the other side of the nozzle and increasing its volume from 0 to $V_f$ is $P_f V_f$. Heat transfer to the gas as it passes through the nozzle is small and in the ideal nozzle $Q = 0$. By the First Law of Thermodynamics,

$$0 = U_f - U_i + (P_f V_f - P_i V_i)$$

or

$$H_i = H_f$$

The change in temperature of the gas at constant enthalpy can be obtained from Eq. (6.16) and the second $T\,dS$ relationship,

$$T\,dS = C_P\,dT - T\left(\frac{\partial V}{\partial T}\right)_P dP$$

Combining these two relationships yields

$$dH = C_P\,dT + \left[V - T\left(\frac{\partial V}{\partial T}\right)_P\right] dP$$

For a process at constant enthalpy, $dH = 0$ and solving for $dT/dP$ at constant enthalpy yields the Joule–Thomson coefficient

$$\{\text{Joule–Thomson coefficient}\} = \left(\frac{\partial T}{\partial P}\right)_H = \frac{1}{C_P}\left[T\left(\frac{\partial V}{\partial T}\right)_P - V\right] \tag{8.12a}$$

Equation (8.10a) can be expanded in terms of the volume to yield, to this degree of approximation,

$$V \approx \frac{NkT}{P}\left(1 + \frac{P}{kT}B\right) \quad \text{and} \quad T\left(\frac{\partial V}{\partial T}\right)_P = \frac{NkT}{P} + NT\frac{\partial B}{\partial T}$$

Finally, the Joule–Thomson coefficient is

$$\left(\frac{\partial T}{\partial P}\right)_H = \frac{1}{C_P/N}\left(-B + T\frac{\partial B}{\partial T}\right) \tag{8.12b}$$

in terms of the second virial coefficient. In this text, $P$ is in newtons per square meter and $C_P/N$ is the heat capacity per molecule at constant pressure in joules per degree Kelvin.

An approximate curve which describes the Joule–Thomson coefficient for a number of common gases [6] is shown in Fig. 8.2. Again the Lennard–

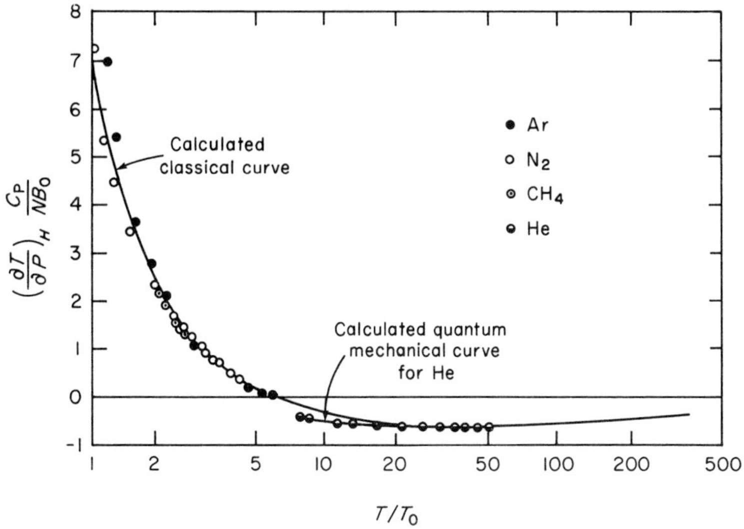


**Fig. 8.2** The reduced Joule–Thomson coefficient for the Lennard-Jones 6–12 potential [*6*]. (From J. O. Hirschfelder, C. F. Curtiss, and R. B. Bird, "Molecular Theory of Gases and Liquids," Copyright 1954, Wiley, New York. Used by permission of John Wiley & Sons, Inc.)

Jones 6–12 interaction potential is used and the two parameters are given in Table 8.2 (Section 8.6). Reduced units are used for the Joule–Thomson coefficient and an order-of-magnitude estimate can be made by observing that for $N_2$, the coefficient is given by

$$\frac{B_0}{C_P} \approx \frac{B_0}{\frac{7}{2}k} \approx 21.9 \times 10^{-7} \quad \frac{°\mathrm{K}}{(\mathrm{N/m^2})} \quad \text{or} \quad 0.22 \quad \frac{°\mathrm{K}}{\mathrm{atm}}$$

Thus $N_2$ has a coefficient of $(\partial T/\partial P)_H \approx +0.22°\mathrm{K/atm}$ at 300°K and cools upon expanding through a nozzle. This increases to the order of 1.5°K/atm at 100°K. As the temperature increases, the Joule–Thomson coefficient passes through zero and this is referred to as the inversion temperature. Heating occurs for expansion at constant enthalpy above the inversion temperature and cooling occurs below the inversion temperature. The inversion temperature occurs at $T/T_0 = 6.47$ for the Lennard-Jones 6–12 model.

The Joule–Thomson expansion has provided a very convenient method for liquefying the common gases. Since cooling always occurs below the inversion temperature, the low-pressure gas leaving the expanding nozzle can be used to cool the incoming high-pressure gas. A properly designed

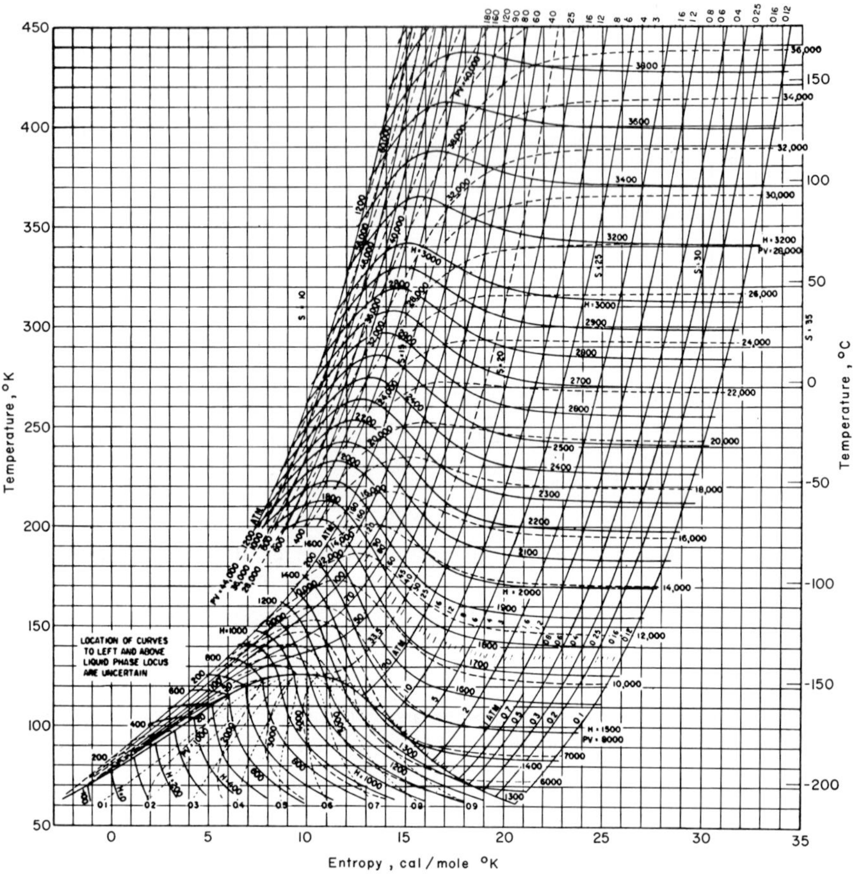


**Fig. 8.3** Temperature-entropy chart for nitrogen. (From R. B. Scott, "Cryogenic Engineering," p. 278, Fig. 9.7, Van Nostrand–Reinhold, Princeton, New Jersey, 1959.)

system decreases the temperature of the low-pressure side of the nozzle until a fraction of the substance is liquefied.[2] A counterflow heat interchanger and nozzle or expansion valve are not discussed here. Figure 8.3 illustrates some of the data available for common gases. [5] The change in the state functions' entropy and enthalpy can be obtained from this chart, on which processes that occur at constant enthalpy or constant entropy are easily followed. In an indirect manner, it can also be observed that the simple model of a gas in terms of the second virial coefficient becomes incorrect at high particle densities.

[2] A discussion of liquefaction and reference to original literature are given by Scott [*4*].

## 8.4 Second Virial Coefficients for Classical, Fermi–Dirac, and Bose–Einstein Gases

### *8.4.1 An Ideal Classical Gas*

An ideal classical gas is a gas of mass point particles which do not interact with each other and which obey the statistics $Q_N = Q_1{}^N/N!$. A true ideal classical or Boltzmann gas would obey the statistics $Q_N = Q_1{}^N$ and so the definition used here deviates from the counting procedure for statistically independent and distinguishable dice. With the definition

$$Q_N = \frac{Q_1{}^N}{N!} \tag{8.13a}$$

the cluster or cumulant coefficients are

$$a_1 = Q_1, \qquad a_n = 0 \quad (n \neq 1) \tag{8.13b}$$

Quantities like $Q_2 - \frac{1}{2}Q_1{}^2$ are zero with this substitution. This is of course expected since

$$\mathcal{Q} = \sum z^N \frac{Q_1{}^N}{N!} = e^{zQ_1}$$

and

$$\ln \mathcal{Q} = zQ_1$$

The usual Ideal Gas Laws follow immediately,

$$z = \frac{N}{Q_1} \tag{8.14a}$$

$$PV = NkT \tag{8.14b}$$

$$U = -N\frac{\partial(\ln Q_1)}{\partial\beta} \tag{8.14c}$$

It is somewhat surprising to note that this equation is valid for all $z$ and that it agrees with the cluster expansion, since all of the cumulant coefficients except $a_1$ are zero. Most of the calculations given in this chapter imply small $a_n$ coefficients for $n > 1$. The deviation of $a_n$ from zero is a measure of the nonideality of the gas and the term $Q_2 - \frac{1}{2}Q_1{}^2$ is examined in considerable detail.

### 8.4.2 *Second Virial Coefficient for Fermi–Dirac and Bose–Einstein Gases*

The solution to the one-particle Schrödinger equation was given in Chapter VII and it was shown that for a monatomic gas without spin,

$$Q_1 = V\left(\frac{2\pi MkT}{h^2}\right)^{3/2} = \frac{V}{\lambda^3} = a_1 \tag{8.15}$$

A variable $\lambda$ with the dimension of length is often used; this is a thermal de Broglie wavelength and is defined as

$$\lambda = \left(\frac{h^2}{2\pi MkT}\right)^{1/2} = 1.74\times 10^{-9}(M'T)^{-1/2} \quad \text{m} \tag{8.16}$$

where $M'$ is the molecular weight number. If the particles have internal spin, then Eq. (8.5) must be summed over the spin variable. The determination of $Q_2$ requires greater effort. The two-particle Schrödinger equation was given in Chapter IV by Eq. (4.22). Although the interaction potential $U(r)$ is zero, the effect of rigid walls which fix the volume $V$ must be included. A suitable solution to Eq. (4.22) is

$$\psi_{k_1k_2}(\mathbf{r}_1, \mathbf{r}_2) = \frac{1}{\sqrt{2}}\,[\psi_{k_1}(\mathbf{r}_1)\psi_{k_2}(\mathbf{r}_2) \pm \psi_{k_1}(\mathbf{r}_2)\psi_{k_2}(\mathbf{r}_1)] \tag{8.17}$$

$$\psi_{k_1}(r) = \frac{1}{\sqrt{V}}\exp i\mathbf{k}_1\cdot\mathbf{r} \tag{8.18a}$$

$$k_{x1} = \frac{2\pi n_{x1}}{L}, \qquad n_{x1} = 0, \pm 1, \pm 2 \tag{8.18b}$$

$L^3$ is replaced by $V$. $k_1$ and $k_2$ imply the triple set of integers given by $n_{x1}, n_{x2}, \ldots$ for $y$ and $z$. The plus sign refers to Bose–Einstein (BE) particles of zero spin. Fermi–Dirac (FD) particles of spin $\frac{1}{2}$ require four-index $k_1$ and $k_2$, three space indices $n_{x1}, n_{x2}, \ldots,$ and a spin index of $\pm\frac{1}{2}$. These wavefunctions are either symmetric or antisymmetric under the exchange of two identical particles. The energy of a state with indices $k_1$ and $k_2$ is

$$E_2 = E_{k_1k_2} = \frac{\hbar^2}{2M}(k_1^2 + k_2^2) \tag{8.19}$$

In forming $Q_2$, it was agreed to weight each state equally and in forming the canonical generating function

$$Q_2 = \sum_{\text{states}} e^{-E_2/kT} \tag{8.20}$$

some care is needed in specifying the states. First, only one distinct state occurs for a given $k_1 \neq k_2$ and only one-half of the double sum over $k_1$ and $k_2$ should be used. For Fermi–Dirac statistics, no state occurs with $k_1 = k_2$ and for Bose–Einstein statistics, the full sum occurs for $k_1 = k_2$. This is quite apparent from Eq. (8.17). $Q_2$ can now be written as the sum

$$\begin{aligned} Q_2 &= \frac{1}{2}\sum_{k_1 k_2} \exp\left[-\beta\left(\frac{\hbar^2}{2M}\right)(k_1^2 + k_2^2)\right] \mp \frac{1}{2}\sum_{k} \exp\left[-\beta\left(\frac{\hbar^2}{2M}\right)(2k^2)\right] \\ &= \frac{Q_1^2}{2} \mp \frac{Q_1}{2^{5/2}} \qquad \begin{pmatrix}\text{FD}\\ \text{BE}\end{pmatrix} \end{aligned} \tag{8.21}$$

The double sum is takėn over all states and the single sum takes care of the states with $k_1 = k_2 = k$. These sums are evaluated by changing the sums to integrals according to

$$\sum_{k} \to \sum_{s} V \frac{4\pi k^2\, dk}{(2\pi)^3}$$

where the sum on $s$ is over the spin states. The cumulant coefficient $a_2$ for the Fermi–Dirac and Bose–Einstein gases is given by

$$a_2 = Q_2 - \frac{Q_1^2}{2} = \mp \frac{Q_1}{2^{5/2}} \qquad \begin{pmatrix}\text{FD}\\ \text{BE}\end{pmatrix} \tag{8.22}$$

It is interesting to observe that the entire contribution to the correction $a_2$ is due to the single sum in Eq. (8.21). The much larger double sum cancels the other term in $a_2$. This is in a certain sense expected since the nature of the statistics is only important for identical states.

The laws governing the behavior of the two gases are

$$\ln \mathcal{Q} \approx \left(z \mp \frac{z^2}{2^{5/2}}\right) Q_1 \tag{8.23a}$$

$$\frac{PV}{kT} \approx N\left(1 \pm \frac{N}{2^{5/2} Q_1}\right) \tag{8.23b}$$

$$z = \frac{N}{Q_1}\left(1 \pm \frac{N}{2^{3/2} Q_1}\right) \tag{8.23c}$$

where the upper signs are for FD and the lower for BE statistics.

The second virial coefficient for the two gases is given by Eq. (8.10b), or

$$B(T) = \pm \frac{\lambda^3}{2^{5/2}} \qquad \begin{pmatrix} \text{FD} \\ \text{BE} \end{pmatrix} \tag{8.24}$$

This is a purely quantum mechanical effect and is intimately related to the commutative relations. Since it comes from the single sum rather than the double sum, careful consideration in the counting of states is necessary in subsequent sections to avoid missing this term. The quantum correction that is introduced by statistics appears as an attractive potential for BE statistics and as a repulsive potential for FD statistics.

The exact equations for the ideal BE and FD gases given by Eqs. (7.27)–(7.33) must yield these same results. Expansion of Eq. (7.32b) about $z = 0$ yields

$$N = z \sum_a e^{-\beta\varepsilon_a} \mp z^2 \sum_a e^{-2\beta\varepsilon_a} \qquad \begin{pmatrix} \text{FD} \\ \text{BE} \end{pmatrix} \tag{8.25a}$$

It is convenient to perform the sum over states and

$$\begin{aligned} \sum_a \exp(-\beta\varepsilon_a) &= \sum_s V \int_0^\infty \frac{4\pi k^2\, dk}{(2\pi)^3} \exp\left(-\frac{\beta\hbar^2 k^2}{2M}\right) = Q_1 \\ &= \frac{(2s+1)V}{\lambda^3} \end{aligned} \tag{8.25b}$$

and

$$\sum_a \exp(-2\beta\varepsilon_a) = \frac{Q_1}{2^{3/2}} \tag{8.25c}$$

In the previous notation,

$$N \approx zQ_1 \mp \frac{z^2 Q_1}{2^{3/2}} \tag{8.26}$$

and is the same as Eq. (8.23c) to the same degree of approximation. For these ideal FD and BE gases, the calculation of $Q_2$ and the expansion of the exact solution in terms of statistically independent normal modes yield the same thermodynamic functions.

**Exercise 8.1** Show that the internal energy for the FD and BE gases is given for a monatomic gas by

$$U \approx \frac{3}{2} NkT\left(1 \pm \frac{N\lambda^3}{2^{5/2} V}\right) \tag{8.27a}$$

Exercise 8.2 Does a Fermi–Dirac gas heat or cool upon expansion at constant enthalpy? What about a Bose–Einstein gas?

Exercise 8.3 Show that Eq. (7.32a) yields

$$\frac{PV}{kT} = \ln \mathcal{Q} \approx zQ_1 \mp z^2 \frac{Q_1}{2^{5/2}} \tag{8.27b}$$

and is in accord with Eq. (8.23b).

Exercise 8.4 Show that $Q_1$ is the number of accessible states in volume $V$ for a particle of mass $M$. How many states are accessible if the energy constraint is removed?

Exercise 8.5 Determine $Q_1$ for Ne and He at a temperature of 300°K.

## 8.5 Second Virial Coefficient for Molecular Gases with Interaction

This section considers the calculation of the second virial coefficient for molecular gases by the "phase-space method" and by a general quantum mechanical analysis of the type used for scattering in Chapter IV. The simpler phase-space method is considered first.

### 8.5.1 Phase-Space Method

Many important features of the problem of determining the virial coefficients can be seen by the use of the phase-space method. Consider the one-particle Hamiltonian for a classical particle in a box,

$$H_1 = \frac{p^2}{2M} = E_1 \tag{8.28a}$$

$Q_1$ follows from

$$Q_1 = \sum_{\text{states}} e^{-\beta E_1} = \frac{1}{h^3} \int d\mathbf{r}\, d\mathbf{p}\, e^{-\beta H_1} = \frac{V}{\lambda^3} \tag{8.28b}$$

The average energy in this approach is taken as the classical energy $E_1$

and the states are counted in a somewhat quantal fashion, that is,

$$\sum_{\text{states}} \to \frac{d\mathbf{r}\, d\mathbf{p}}{h^3} \quad \text{or} \quad \frac{d\mathbf{r}\, d\mathbf{k}}{(2\pi)^3}$$

Spin must be included as a separate sum. *A volume in phase space is denoted by* $d\mathbf{r}\, d\mathbf{p}$ *and one state is associated with each phase-space volume* $h^3$. This counting procedure agrees with the number of plane-wave states in volume $V$ with $\mathbf{p} = \hbar\mathbf{k}$. The element of space volume $d\mathbf{r}$ is integrated over the volume $V$ and the momentum over all values of momentum $\mathbf{p}$.

A two-particle generating function can be obtained in a similar manner. Let the two-particle Hamiltonian be written in the form used in Chapter IV for the collision of two classical particles,

$$H_2 = \frac{p_1{}^2}{2M} + \frac{p_2{}^2}{2M} + U(|\mathbf{r}_1 - \mathbf{r}_2|) = E_2$$

$$= \frac{P^2}{4M} + \frac{p^2}{M} + U(r) \tag{8.29a}$$

where the second equation is in the center-of-mass coordinate system. Again, let the average energy of a state be given by the classical value $E_2$ and sum over the allowed states,

$$Q_2 = \sum_{\text{states}} e^{-\beta H_2} = \frac{1}{2h^6} \int d\mathbf{r}_1\, d\mathbf{r}_2\, d\mathbf{p}_1\, d\mathbf{p}_2\, e^{-\beta H_2} \tag{8.29b}$$

Again, a state occurs for each phase-space volume $h^3$. Even so, there is an overcounting of states by a factor of two in this double integral in the same manner as that discussed in Section 8.3 and the factor of $\frac{1}{2}$ is placed in front of the double integral. If $U = 0$, direct evaluation of the double integral yields $Q_2 = \frac{1}{2}Q_1{}^2$ and $a_2 = 0$ in accord with the earlier discussion. Furthermore, the integral over $p_1$ and $p_2$ can be taken and in general,

$$Q_2 = \tfrac{1}{2}\lambda^{-6} \int d\mathbf{r}_1\, d\mathbf{r}_2\, e^{-\beta U} \tag{8.30}$$

It is now convenient to use $d\mathbf{r}_1\, d\mathbf{r}_2 = d\mathbf{R}\, d\mathbf{r}$, integrate $d\mathbf{R}$ over the volume $V$, and to replace $d\mathbf{r}$ by $4\pi r^2\, dr$. With these various substitutions, the cumulant coefficient $a_2$ is

$$a_2 = Q_2 - \tfrac{1}{2}Q_1{}^2 = \tfrac{1}{2}\lambda^{-6} V \int_0^\infty 4\pi r^2\, dr\, (e^{-\beta U(r)} - 1) \tag{8.31}$$

and the second virial coefficient is, from Eq. (8.10b),

$$B(T) = 2\pi \int_0^\infty r^2\, dr\, (1 - e^{-U(r)/kT}) \tag{8.32}$$

This is probably the most convenient form for the second virial coefficient and could in principle be determined numerically for any potential $U(r)$.

The calculation of the second virial coefficient is quite direct by the phase-space method and raises a question concerning the degree of validity of the approach. This can be examined most directly by the use of the definition of the generating function as the trace of an operator as given by Eq. (6.68). The generating function for the one-particle Hamiltonian is

$$\begin{aligned} Q_1 &= \operatorname{tr} \exp -\beta H_1 = \sum_k (k \mid \exp -\beta H_1 \mid k) \\ &\to \int \frac{d\mathbf{r}\, d\mathbf{k}}{(2\pi)^3} \psi_k^*(\mathbf{r}) \psi_k(\mathbf{r}) \exp\left(-\frac{\beta \hbar^2 k^2}{2M}\right) = \frac{V}{\lambda^3} \end{aligned} \tag{8.33}$$

where $H_1 = -(\hbar^2/2M)\nabla^2$. This is of course too simple to illustrate the point. The two-particle generating function is

$$\begin{aligned} Q_2 &= \operatorname{tr} e^{-\beta H_2} = \sum_{k_1 k_2} (\psi_{k_1 k_2} | \, e^{-\beta H_2} | \, \psi_{k_1 k_2}) \\ &= \sum_{k_1 k_2} \int d\mathbf{r}_1\, d\mathbf{r}_2\, \psi^*_{k_1 k_2} e^{-\beta H_2} \psi_{k_1 k_2} \end{aligned} \tag{8.34}$$

where $\psi_{k_1 k_2}(r_1, r_2)$ is given by Eq. (8.17), and $H_2$ is given by Eq. (4.22) or by the operator form of Eq. (8.29a). This wave function is not a solution to $H_2$, but the invariance of the trace permits one to use any complete set of basis functions which span the same coordinate space. Only if $H_2 \psi$ yields an allowed value of the energy does one know how to evaluate the exponential. If $H_2$ is written as a sum of two operators,

$$H_2 = T_2 + U, \qquad \text{where} \quad T_2 = -\frac{\hbar^2}{2M} [\nabla_1{}^2 + \nabla_2{}^2]$$

and if it is assumed that these two quantities are statistically independent, then

$$e^{-\beta H_2} \to e^{-\beta U} e^{-\beta T_2}$$

or it can be written as a product of generating functions. Since

$$(\exp -\beta T_2) \psi_{k_1 k_2} = \text{constant} \times \left\{ \exp\left[-\beta \frac{\hbar^2}{2M} (k_1{}^2 + k_2{}^2)\right]\right\} \psi_{k_1 k_2}$$

Eq. (8.34) reduces to Eq. (8.30) upon summing over $k_1$ and $k_2$. *In classical statistical mechanics, the kinetic energy and the potential energy are statistically independent.* If the same statistical independence is assumed in the quantum solution, Eq. (8.34), then the phase-space method can be used. Fortunately, this applies to a very large group of problems. Its general validity is not discussed here.

### 8.5.2 Pair Distribution Function

In the cumulant expansion, $z^2$ was interpreted as the chemical potential of a pair. The total number of particles was given by Eq. (8.8a). The term $za_1$ is the number of one-particle clusters, $z^2a_2$ the number of two-particle clusters, $z^3a_3$ the number of three-particle clusters, and so on. To the degree of approximation used in developing ln $\mathcal{Q}$ through $z^2$, the number of pairs is given by

$$z^2a_2 \approx \frac{N^2a_2}{a_1^2} = \frac{N^2(Q_2 - \frac{1}{2}Q_1^2)}{Q_1^2}$$

$$\approx -\frac{N^2}{V} B(T)$$

and is just $-N^2/V$ times the second virial coefficient. Then, in Eq. (8.32), the quantity under the integral sign is interpreted as the number of pairs in volume $V$, that is,

$$n_2(r)\, dr = \frac{N^2}{V} (e^{-\beta U} - 1)2\pi r^2\, dr$$

$$= \begin{Bmatrix} \text{number of pairs between } r \text{ and } r + dr \\ \text{with } N \text{ molecules in volume } V \end{Bmatrix} \tag{8.35a}$$

The number of pairs between $r$ and $r + dr$ must be interpreted in the following manner. The probability of one molecule being at $r = 0$ while a second molecule is between $r$ and $r + dr$ and forming a pair is given by

$$\frac{n_2(r)\, dr}{N} = \frac{N}{V} (e^{-\beta U} - 1)2\pi r^2\, dr \tag{8.35b}$$

It should be noted that a second molecule can be between $r$ and $r + dr$ and not form a pair. This occurs in an ideal gas, or a gas with $U = 0$, and in any gas for states with large values of angular momentum. Later in this section, it will be shown that collisions that do not have a phase shift or have a small phase shift do not contribute to the pair distribution function. A similar interpretation could be introduced into the calculations by means of the "Slatersum." If the expansion is continued to $z^3 a_3$, then the number of three-particle clusters can be determined in the same manner.

The pair distribution function can be negative and the interpretation of this feature also requires care. The term $n_2(r)$ is the number of pairs greater than or less than the number of pairs occurring for a random distribution and can be either positive or negative.

### *8.5.3 Van der Waals Equation and the Virial Theorem of Clausius*

No discussion of the virial coefficients is complete without a discussion of the equation of state deduced by van der Waals and the virial theorem of Clausius. Van der Waals realized that a proper equation of state must take into account both the size of the molecules and the attractive force between molecules and suggested an equation of the type

$$\left(p + \frac{a}{V^2}\right)(V - b) = NkT \tag{8.36}$$

The derivation of this equation is based on two-particle encounters and its form is somewhat fortuitous. $b$ is a measure of the reduction in volume due to the molecular size and hence increases the pressure. Molecular attractive forces reduce the pressure by the number of two-particle bonds and is proportional to $(N/V)^2$.

Clausius introduced a very powerful method for the calculation of the equation of state. Let the motion of a molecule be governed by the classical equation of motion

$$F_x = M \frac{d^2x}{dt^2}$$

Multiplying this equation by $x$ yields

$$\frac{xF_x}{2} = \frac{1}{2} \frac{d(Mx\dot{x})}{dt} - \frac{M\dot{x}^2}{2}$$

If a long time average is taken, the term $\langle x\dot{x} \rangle$ is zero and this follows by noting that $\dot{x}$ is as often positive as negative for a particle inside an enclosure. Since $x$ and $\dot{x}$ are statistically independent, $\langle x\dot{x} \rangle = \langle x \rangle \langle \dot{x} \rangle = 0$, and

$$-\tfrac{1}{2}\langle xF_x \rangle = \langle \tfrac{1}{2} M\dot{x}^2 \rangle = \tfrac{1}{2}kT$$

If this expression is summed over $x$, $y$, $z$, and all $N$ molecules in volume $V$, the kinetic energy and the *virial of force* are related by

$$\sum_{n}^{N} \langle \tfrac{1}{2} M v_n^2 \rangle = -\tfrac{1}{2} \sum_{n}^{N} \langle \mathbf{r}_n \cdot \mathbf{F}_n \rangle \tag{8.37a}$$

Equipartition gives a value of $\frac{3}{2}NkT$ for the kinetic energy. The forces on the molecules may be divided into forces between molecules and external forces. The external force exerted by the walls is the pressure, $P\hat{n}\,dA$, and the integral over the surface of the container yields

$$-\int \mathbf{r} \cdot (P\hat{n}\,dA) = 3PV$$

If the forces between molecules are derivable from a potential function $U(r_1, \ldots, r_N)$, then the equation of state is

$$PV = NkT - \frac{1}{3} \sum_{n;x,y,z}^{N} \left\langle x_n \frac{\partial U}{\partial x_n} \right\rangle = NkT - \frac{1}{3} \sum_{N} \langle (\mathbf{r} \cdot \operatorname{grad} U) \rangle \tag{8.37b}$$

For zero intermolecular forces, this is the Ideal Gas Law. A method of averaging is needed for the last term and an ensemble average rather than a time average is used. For hard spheres of diameter or distance of closest approach $r_0$, the probability that particle 1 lies in $d\mathbf{r}_1$ and particle 2 in $d\mathbf{r}_2$ is proportional to $(d\mathbf{r}_1/V)(d\mathbf{r}_2/V)$ subject to the restriction that no two particles are closer than $r_0$. The volume element can be changed to $d\mathbf{r}\,d\mathbf{R}$ and integrated over $d\mathbf{R}$ to yield $4\pi r^2\,dr/V$ as the probability that the second particle lies between $r$ and $r + dr$. A force only exists upon impact for a hard sphere and an evaluation of this force would require an examination of the change in momentum upon collision. This is avoided by examining the more general problem and regarding $e^{-\beta U}$ as the probability of two particles having relative potential energy $U$. The virial of force for two molecules in volume $V$ is

$$-\frac{1}{3} \int 4\pi r^2\,dr\, e^{-\beta U} \left( r \frac{\partial U}{\partial r} \right) = \frac{1}{\beta} \int_0^\infty (1 - e^{-\beta U}) 4\pi r^2\,dr$$

per pair, or one-half this value per molecule. If $N$ particles are placed in the volume $V$, the number is increased by $N^2$ and the second virial coefficient is the same as that given by Eq. (8.32).

Both the classical virial and quantum mechanical virial have been extensively developed and reference to much earlier work as well as extensive discussions is given by Hirschfelder *et al.* [*6*].

Exercise 8.6 Find $B(T)$ for a van der Waals gas.

### 8.5.4 Quantum Mechanical Evaluation of the Second Virial Coefficient

According to the rules given earlier, the two-particle generating function follows from

$$Q_2 = \sum_{\text{states}} e^{-\beta E_{2a}} \tag{8.38a}$$

where $E_{2a}$ are the energies permitted by the exact solution to the two-particle Schrödinger equation (4.22). It was noted in Section 4.3 that the center of mass always has states $K$. For a box of volume $V$, these are just the previously used three indices. According to Eq. (4.25), the angular momentum can be given in terms of the spherical harmonic indices $l$ and $m$. Only the relative motion equation (4.26a) remains to be solved with the supplemental condition (4.26b). The allowed values of the energy are

$$E_2 = \frac{\hbar^2 K^2}{4M} + E_{nlm} \tag{8.38b}$$

where the index $n$ is used to describe the solution to the radial equation (4.26a). Summing over the index $K$ yields

$$Q_2 = 2^{3/2} Q_1 \sum_{n,l} (2l+1) e^{-\beta E_{nl}} \tag{8.38c}$$

and only even or odd values of $l$ are permitted by the restriction that the wave functions must be symmetric or antisymmetric under the exchange of identical particles. The term $2^{3/2}Q_1$ follows from the integration of

$$(V 4\pi K^2\, dK) \exp(-\beta\hbar^2 K^2/4M)$$

and the $2^{3/2}$ occurs since the total mass $2M$ occurs rather than the mass $M$.

In the limit that $U(r) \to 0$, $Q_2$ must be the same as given by Eq. (8.21), or

$$Q_2{}^0 - \frac{1}{2} Q_1{}^2 = \mp \frac{Q_1}{2^{5/2}}$$

In order to explore this in greater detail, Eq. (8.17) is written as

$$\begin{aligned}\psi_{k_1k_2}(\mathbf{r}_1, \mathbf{r}_2) &= \psi_{Kk}(\mathbf{Rr})\\ &= \frac{1}{\sqrt{2V}} (\exp i\mathbf{K}\cdot\mathbf{R})[(\exp + i\mathbf{k}\cdot\mathbf{r}) \pm (\exp - i\mathbf{k}\cdot\mathbf{r})]\end{aligned} \tag{8.39}$$

where

$$\mathbf{K} = \mathbf{k}_1 + \mathbf{k}_2, \qquad \mathbf{R} = \frac{1}{2}(\mathbf{r}_1 + \mathbf{r}_2), \qquad E_{nlm} \to \frac{\hbar^2 k^2}{M}$$

$$\mathbf{k} = \frac{1}{2}(\mathbf{k}_1 - \mathbf{k}_2), \qquad \mathbf{r} = (\mathbf{r}_1 - \mathbf{r}_2)$$

Only one state is formed for $a$ given $k_1$ and $k_2$ and a factor of one-half is needed with the double sum. For this same reason, $k$ is limited to the half-space, or one-half the sum is needed for the full space. Even so, the state with $k_1 = k_2$ or $k = 0$ requires the same care as in the earlier section. There is no $k = 0$ state for FD statistics and one state for BE statistics.

The second virial coefficient which is defined by Eq. (8.10b) can now be written for spin-zero particles as

$$B(T) = -\frac{2^{3/2} V}{Q_1} \sum_{n,l}{}' (2l+1)[(\exp -\beta E_{nl}) - (\exp -\beta E_{nl}^{\circ})] \pm \frac{\lambda^3}{2^{5/2}} \quad (8.40)$$

The restrictions on $l$ in this sum are given at the end of this section. If the volume $V$ is assumed to have a radius $R$, then the requirement that the wave function be zero at the wall is

$$0 = u_l(R) = \sin(kR - \tfrac{1}{2} l\pi + \delta_l) \quad (8.41a)$$

and the discrete values of $k$ are

$$kR - \tfrac{1}{2} l\pi + \delta_l = n\pi, \qquad n = 0, 1, 2, \ldots, \infty \quad (8.41b)$$

Discrete values of the energy are given by

$$E_{nl} = \frac{\hbar^2 k^2}{M} = \frac{\hbar^2}{MR^2}\left(n\pi + \frac{1}{2} l\pi - \delta_l\right)^2 \quad (8.41c)$$

The sum over $n$ is usually changed to an integral. The interval $\Delta k$ between $n$ and $n+1$ follows from

$$\frac{R + (d\delta_l/dk)}{\pi} \Delta k = \Delta n \quad (8.41d)$$

with $\Delta n = 1$; and the sum over $n$ is changed into an integral by using

$$\sum_{\Delta n} \to \frac{R + (d\delta_l/dk)}{\pi} dk \quad (8.41e)$$

The density of states between the first and second sums changes from

$d\delta_l/dk$ to $d\delta_l^0/dk$. Since the phase shift for free particles is $\delta_l^0 = 0$, the second virial coefficient for spin-zero particles reduces to

$$B(T) = -(2^{3/2}\lambda^3)\Big[\sum_l{}' (2l+1)\int_0^\infty \Big[\exp\Big(-\frac{\beta\hbar^2k^2}{M}\Big)\Big]\Big(\frac{d\delta_l}{dk}\Big)\frac{dk}{\pi}$$
$$+ \sum_b (2l+1)\exp -\beta E_b\Big] \pm \frac{\lambda^3}{2^{5/2}} \tag{8.42}$$

and this equation relates the second virial coefficient to the phase shifts of Chapter IV for the scattering coefficient. Bound states must be given special consideration and these terms must be included as a separate sum in Eq. (8.42). Since $E_b < 0$ for bound states, the exponential is positive, and as the temperature is lowered, the bound-state terms dominate the integral over the $k^2 > 0$ or collision terms. Such terms correspond to diatomic molecules and their creation and destruction must occur at the walls or during three-body collisions.

Generalizations of the phase-shift method are of considerable interest and attempts are being made to use these methods for nuclear species [*7*]. The phase-shift method can be related to the $S$-matrix for scattering, and an $S$-matrix formulation for virial expansions and for statistical mechanics is being explored [*8*].

It is interesting to show that the phase shift $\delta_l$ determined by the WKB method yields a second virial coefficient which is in accord with other approximate methods. Equation (4.59) gives the first approximation to the phase shift by the WKB method. Taking the derivative with respect to $k$ and inserting the expression for $d\delta_l/dk$ into Eq. (8.42) yields

$$B(T) = -(2^{5/2}\lambda^3)\int d\eta\, d\xi\, e^{-\alpha\eta}\Big[\Big[\int_{a_m}^R dr\Big(\eta - \frac{\xi}{r^2} - V\Big)^{-1/2}$$
$$- \int_{a_0}^R dr\Big(\eta - \frac{\xi}{r^2}\Big)^{-1/2}\Big]$$

where

$$\eta = k^2, \qquad \xi = \Big(l + \frac{1}{2}\Big)^2, \qquad \alpha = \frac{\beta\hbar^2}{M}, \qquad V = \frac{MU}{\hbar^2}$$

and the limit $R \to \infty$ is taken at the end of the calculation. If the region of integration is limited in the $\xi, \eta, r$ space to the volume bounded by the surfaces $\eta - (\xi/r^2) - V = 0$, $\eta = 0$, $\xi = \frac{1}{2}, \ldots$, the triple integrals can be approximated by

$$\int d\eta\, d\xi\, dr \to \int_0^\infty dr \int_V^\infty d\eta \int_{1/2}^{r^2(\eta - V)} d\xi$$

Upon integration,

$$B(T) \approx 2\pi \int_0^\infty r^2\, dr\, [1 - \exp(-\beta U)] \exp\left(-\frac{\beta \hbar^2}{Mr^2}\right) \tag{8.43}$$

and is almost the same as Eq. (8.32). The term $\exp(-\beta\hbar^2/Mr^2)$ or $\exp(-\lambda^2/2\pi r^2)$ excludes the region near $r = 0$ and is introduced by replacing $l(l+1)$ by $\xi = (l + \frac{1}{2})^2$. This substitution was discussed earlier and improves the WKB method for small values of $l$. The region excluded is of the order of the de Broglie wavelength and suggests that hard-sphere repulsions with a range less than $\frac{1}{2}\lambda$ are not important. This term is unimportant for all atoms except helium below 4°K and can be replaced by one. Higher-order corrections to the virial coefficient require a phase-shift expression which contains higher-order terms. Thus Eq. (4.59) for the phase shift yields the classical expression for the virial coefficient. Kahn [9] used the phase-shift method for discussing the bound states and the quantum corrections to the virial coefficients.

Helium-3 is a spin-$\frac{1}{2}$ particle and $^4$He is a spin-0 particle. The second virial coefficient for $^4$He is given by Eq. (8.42) with even $l$ and the lower sign. Both even and odd values of $l$ are permitted for higher values of spin and this is similar to the counting of states for *ortho–para-hydrogen. Ortho* and *para* unbound states are formed as two $^3$He molecules collide. The fraction of each type of state is given for spin $I$ in Table 8.1. In Eq. (8.42), the sum over even $l$ is multiplied by the term in the even column, and similarly for odd $l$. For $^3$He, the even-$l$ sum is multiplied by one-fourth and the odd $l$ sum by three-fourths. For large $l$, this has negligible effect on the result.

**Table 8.1**

| Spin $I$ | Even $l$ | Odd $l$ |
|---|---|---|
| Half-integer | $\frac{I}{2I+1}$ | $\frac{I+1}{2I+1}$ |
| Even-integer | $\frac{I+1}{2I+1}$ | $\frac{I}{2I+1}$ |

## 8.6 Intermolecular Potentials

Intermolecular potentials require detailed quantum mechanical calculations and these calculations are usually complicated and not highly accurate. Even so, considerable information follows from simple potentials which have the general features of the more detailed calculations. Some of these potentials were discussed in Chapter IV. A general feature of the potential between two neutral atoms or molecules is a strong repulsive potential when the wavefunctions of the molecules overlap. This short-range potential has an exponential shape of the type given by Eq. (4.3). An attractive potential occurs between molecules proportional to $r^{-6}$ and was discussed as a London dispersion potential in Eq. (4.104). Potentials which express the general features described above are given by the following models and are illustrated in Fig. 8.4.

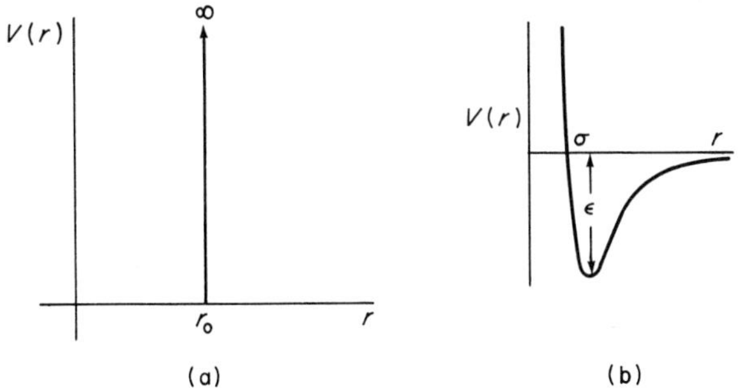


**Fig. 8.4** (a) Hard-sphere repulsive potential. (b) Lennard–Jones 6–12 potential.

### *8.6.1 Second Virial Coefficient for Hard Spheres*

A potential function, which has some features of the hard-core repulsion and which is used for exploratory calculations because of its simplicity, is shown in Fig. 8.4(a). Its mathematical form is

$$U(r) = \begin{cases} \infty, & r \leq r_0 \\ 0, & r > r_0 \end{cases} \tag{8.44a}$$

Equation (8.32) can be integrated directly for the second virial coefficient. Equation (8.24) gives the quantal correction, and the second virial coefficient is

$$B(T) = \frac{2\pi r_0{}^3}{3} \pm \frac{\lambda^3}{2^{5/2}} \tag{8.44b}$$

The magnitudes of these two terms can be compared by noting that for $^{20}Ne$ at 300°K, $\lambda = 2.2 \times 10^{-11}$ m. Here $r_0$ is of the order of $6 \times 10^{-10}$ m, and the first term is approximately 500 times larger than the second. Only for $H_2$ and He at low temperatures does this second term become important. This virial coefficient for hard spheres is almost independent of temperature and is in disagreement with experiment.

One can estimate the second virial coefficient by the phase-shift method using Eq. (8.42). The phase shift is given by Eq. (4.47b), and with this phase shift, Eq. (8.42) yields

$$B(T) = -2^{3/2}\lambda^3 \int_0^\infty \frac{dk}{\pi} \int_0^{kr_0} l\,dl\left[(-kr_0)\left(1 - \frac{1}{2}\left(\frac{l}{kr_0}\right)^2 - \frac{1}{24}\left(\frac{l}{kr_0}\right)^4\right]\right.$$
$$\times\left[\exp\left(-\frac{\beta\hbar^2k^2}{M}\right)\right] \pm \frac{\lambda^3}{2^{5/2}}$$
$$= \frac{2\pi r_0{}^3}{3} \pm \frac{\lambda^3}{2^{5/2}}$$

The phase-space method, the "Slatersum," and the phase-shift method for evaluation of the second virial coefficient are in excellent agreement for hard spheres.

In the van der Waals equation (8.36), the repulsive term is apparent as $(V - b)$, where $b$ is regarded as a measure of the reduction in volume. From the above calculation, it is apparent that

$$b = NB(T) = N\tfrac{2}{3}\pi r_0{}^3$$

is the excluded volume. It takes a certain amount of geometric construction to obtain this typical $\frac{2}{3}\pi r_0{}^3$ as the excluded volume per hard sphere of radius $r_0/2$.

### *8.6.2 The Lennard–Jones Potential*

The Lennard–Jones potential, which is shown in Fig. 8.4(b) and which has the form

$$U(r) = 4\varepsilon\left[\left(\frac{r_0}{r}\right)^{12} - \left(\frac{r_0}{r}\right)^6\right] \tag{8.45a}$$

provides a realistic model of the interaction potential for simple molecules. Here, $\varepsilon$ is the depth of the attractive potential well and $r_0$ is the radius at which $U = 0$. The minimum in $U$ occurs at $r = 2^{1/6}r_0$. The $r^{-6}$ term is the attractive dispersion or London term. The $r^{-12}$ is chosen for mathematical convenience and provides a reasonably strong repulsion. Equation (8.32) can be integrated by parts to yield

$$B = -\frac{2\pi\beta}{3}\int_0^\infty r^3\,dr\,\frac{dU}{dr}\,e^{-\beta U} \tag{8.45b}$$

For the Lennard–Jones potential, the term in $r^{-12}$ is left in the exponential and the $r^{-6}$ term in the exponential is expanded as a power series expansion. The infinite series can be integrated. Even so, a large number of terms may be needed in the expansion and these terms have been tabulated by Hirschfelder *et al.* [*6*] and reduced data are shown in Fig. 8.1 and Table 8.2 for common gases.

**Table 8.2**

Constants for Lennard–Jones 6–12 Potential Derived from Second Virial Coefficients[a]

| Gas | $r_0$ ($10^{-10}$ m) | $\varepsilon/k = T_0$ | $B_0 = \frac{2}{3}\pi r_0^3$ ($10^{-30}$ m$^3$) |
|---|---|---|---|
| He | 2.56 | 10.22 | 35.2 |
| Ne | 2.75 | 35.6 | 43.6 |
| Ar | 3.41 | 120 | 83 |
| Xe | 4.10 | 221 | 147 |
| $N_2$ | 3.70 | 95.1 | 106 |
| $O_2$ | 3.46 | 118 | 85.7 |
| $CH_4$ | 3.82 | 148 | 118 |
| $CO_2$ | 4.48 | 189 | 188 |

[a] After Hirschfelder *et al.* [*6*, pp. 1110, 1111].

At low $kT/\varepsilon$, the attractive force between molecules is dominant and $B(T)$ is negative. At higher temperatures, the energy of collision is so large that the attractive part of the potential is almost ignored during the collision and the hard-core repulsion is the dominant feature. Then, $B(T)$ takes on a positive value. As $T$ continues to increase, the colliding molecules penetrate further than $r_0$ and $B(T)$ has a maximum near $kT/\varepsilon = 25$. A slight decrease occurs with further increase of $T$. At the lower temperature, deviations occur for the heavier gases. Serious deviations occur for $H_2$ below 75°K and He below 40°K. Numerical evaluation of the phase shifts for small $l$ are needed. It can be noted in Fig. 8.1 that good agreement is obtained between experiment and the quantum calculations. Quantum calculations [*10*] require a good solution to the radial equation (4.26).

These effects are made more apparent by writing the radial wave equation in dimensionless form [*11*].

$$\frac{d^2(r^*u_l)}{d[(r^*)^2]} + \left\{(k^*)^2 - \frac{l(l+1)}{(r^*)^2} + \frac{16\pi^2}{(\Lambda^*)^2}\left[\frac{1}{(r^*)^{12}} - \frac{1}{(r^*)^6}\right]\right\}(r^*u_l) = 0 \tag{8.46}$$

where

$$r^* = \frac{r}{r_0}, \qquad k^* = kr_0, \qquad \Lambda^* = \frac{h}{r_0(m\varepsilon)^{1/2}}$$

Some values of $\Lambda^*$ are given in Table 8.3. As $\Lambda^*$ decreases, the importance of the potential energy is enhanced in the radial equation and the calculations by the WKB and the phase-space methods become quite good. As $\Lambda^*$ increases, the potential energy term is diminished in importance and small values of $l$ become important. $\Lambda^*$ was used in a quantum mechanical law of corresponding states to predict the liquefaction temperature and vapor pressure of $^3$He before the experimental measurements were made [*12*].[3]

**Table 8.3**

Values for the Quantum Parameter $\Lambda^*$

| Gas | $\Lambda^*$ | Gas | $\Lambda^*$ |
|---|---|---|---|
| $^3$He | 3.08 | $H_2$ | 1.73 |
| $^4$He | 2.67 | $CH_4$ | 0.24 |
| Ne | 0.59 | $N_2$ | 0.23 |
| Ar | 0.19 | | |

**Exercise 8.7** Compare $\pi r_0^2$ with the viscosity cross section for $N_2$ and Ar.

**Exercise 8.8** Find the second virial coefficient for the square-well potential,

$$U(r) = \begin{cases} \infty, & r < r_0 \\ -\varepsilon, & r_0 < r < \alpha r_0 \\ 0, & r > \alpha r_0 \end{cases} \tag{8.47a}$$

$$\text{Ans.} \quad B(T) = \frac{2\pi r_0^3}{3}[1 - (\alpha^3 - 1)(e^{\varepsilon/kT} - 1)] \pm \frac{\lambda^3}{2^{5/2}} \tag{8.47b}$$

[3] Experimental values were determined by Abraham *et al.* [*12a*].

This model has the hard-core repulsion of the rigid sphere and an attractive core of strength $\varepsilon$ out to a separation $\alpha r_0$. Again, this simple model has some of the desired features and often is in agreement with experimental data since there are three adjustable parameters.

Exercise 8.9 Compare the inversion temperature of $NH_3$ with the experimental values. Use $r_0 = 2.9\times 10^{-10}$ m, $\alpha = 1.27$, $\varepsilon/k = 692$.

## 8.7 Degenerate Fermi–Dirac Gas—Electrons in Metals and in Very Dense Matter

It was emphasized in Section 7.3 that a Fermi gas can have no more than one particle in a given state. Even though there is no interaction potential energy between the particles, the costraint of no more than one particle per state, or the "Pauli exclusion principle," introduces a second virial coefficient for these gases. This aspect was treated in Section 8.3 and this treatment was suitable for gases for which the probability of a state being occupied was small. The constraint has the effect of a repulsion between the particles. As the density increases, higher-order terms would be needed in the virial expansion. But at high density and low temperature, a much simpler method can be used. The generating function for a particular state is given by Eq. (7.27), and for a set of states in thermal equilibrium,

$$\ln \mathscr{Q} = \sum_a \ln \mathscr{Q}_a = \sum_a \ln(1 + ze^{-\beta\varepsilon_a}) \tag{8.48a}$$

and other thermodynamic properties follow from $\ln \mathscr{Q}$. In a system in which there is no interaction between particles, plane-wave states can be used for the space part. If there is no spin interaction, these states have a $2s + 1$ spin degeneracy for half-integer values of s. A sum over states can be replaced by

$$\sum_{\Delta a} \to (2s + 1)V4\pi k^2 \frac{dk}{(2\pi)^3}$$

and $\ln \mathscr{Q}$ becomes

$$\ln \mathscr{Q} = (2s + 1)V \int_0^\infty \frac{4\pi k^2\, dk}{(2\pi)^3} \ln(1 + ze^{-\beta\varepsilon_k}) \tag{8.48b}$$

In general, the number of states less than $k$ is given by

$$N(k) = (2s + 1)V \frac{4\pi}{3} \left(\frac{k}{2\pi}\right)^3 \tag{8.49a}$$

and is limited by $Vk^3$. Most of the interesting features of a Fermi gas are due to this limitation. $z$ must take on a value such that no more than one particle is in a given state.

For a nonrelativistic gas, the energy $\varepsilon_k$ and wave number $k$ are related by

$$\varepsilon_k = \frac{\hbar^2 k^2}{2M}$$

and the number of states with energy less than $\varepsilon$ is

$$N(\varepsilon) = (2s + 1)\left[V \frac{4\pi}{3}\left(\frac{2M}{h^2}\right)^{3/2}\right]\varepsilon^{3/2} \tag{8.49b}$$

Thermodynamic aspects follow from ln $\mathcal{Q}$ or,

$$\ln \mathcal{Q} = KV \int_0^\infty d\varepsilon \sqrt{\varepsilon} \ln(1 + ze^{-\beta\varepsilon})$$

where

$$K = (2s + 1)2^{5/2}\pi \frac{M^{3/2}}{h^3} \tag{8.50}$$

is used. As $T \to 0$ or $\beta \to \infty$, each particle is found in its state of lowest energy and the number of particles equals the number of states $N(\varepsilon) = N$. This limit defines the degenerate Fermi gas and the chemical potential is given by the value of $\varepsilon$ that includes $N$ states or from Eq. (8.49b):

$$\mu_0 = \varepsilon_F = \frac{h^2}{2M}\left[\frac{3N}{4\pi V(2s + 1)}\right]^{2/3} = \left(\frac{3N}{2KV}\right)^{2/3} \tag{8.51}$$

The importance of $\mu_0$ or the Fermi energy is quite apparent in the average occupation number for a state, and from Eq. (7.29),

$$\langle n_a \rangle = \left\{\begin{matrix} 1, & \varepsilon_a < \mu_0, \\ 0, & \varepsilon_a > \mu_0, \end{matrix}\right\} \qquad T = 0 \tag{8.52}$$

Before pursuing this problem further, a general theorem which is due to Sommerfeld is introduced. If $\varphi(x)$ is a smooth function with $(d\varphi/dx)_{x=0} = 0$, then the definite integral

$$\int_0^\infty dx \frac{d^2\varphi(x)}{dx^2} \ln(1 + e^{\alpha - x}) = \int_0^\infty \frac{dx}{e^{-\alpha + x} + 1} \frac{d\varphi}{dx}$$

$$= \varphi(\alpha) + 2 \sum_{n=1}^{\infty} c_{2n} \left(\frac{d^{2n}\varphi}{dx^{2n}}\right)_{x=\alpha} \tag{8.53a}$$

and is evaluated by two integrations by parts and then expanded about $x = \alpha$. The value of $c_n$ is

$$c_n = \sum_{k=1} (-)^{k+1} k^{-n} \tag{8.53b}$$

and special values of $c_n$ are

$$c_2 = \frac{\pi^2}{12}, \qquad c_4 = \frac{7\pi^4}{720}$$

The almost $\delta$-functionlike character of $(e^{x-\alpha} + 1)(e^{\alpha-x} + 1)$ at $x = \alpha$ is used to simplify this expansion in many developments.

This theorem permits the generating function for an almost degenerate Fermi gas to be written as

$$\ln \mathcal{Q} \approx KV\beta^{-3/2}\left[\frac{4}{15} (\ln z)^{5/2} + \frac{\pi^2}{6} (\ln z)^{1/2}\right] \tag{8.54}$$

Exercise 8.10 Show that the chemical potential is

$$\mu = \frac{1}{\beta} \ln z \approx \mu_0\left[1 - \frac{\pi^2}{12}\left(\frac{kT}{\mu_0}\right)^2\right] \tag{8.55a}$$

Exercise 8.11 Show that the average energy per particle is

$$\frac{U}{N} \approx \frac{3}{5} \mu_0\left[1 + \frac{5\pi^2}{12}\left(\frac{kT}{\mu_0}\right)^2\right] \tag{8.55b}$$

Exercise 8.12 Show that the average heat capacity at constant volume per particle is

$$\frac{C_V}{N} \approx \frac{\pi^2}{2} \frac{k^2T}{\mu_0} = \frac{\pi^2}{2} k \frac{T}{T_F} \tag{8.55c}$$

and the entropy per particle is

$$\frac{S}{N} \approx \frac{\pi^2}{2} \frac{k^2T}{\mu_0} = \frac{\pi^2}{2} k \frac{T}{T_F} \tag{8.55d}$$

Exercise 8.13 Show that the equation of state

$$PV \approx \frac{2N}{5} \mu_0\left[1 + \frac{5\pi^2}{12}\left(\frac{kT}{\mu_0}\right)^2\right] = \frac{2U}{3} \tag{8.55e}$$

**Exercise 8.14** Use the condition expressed by Eq. (8.52) to derive in an elementary manner for $T \to 0$ the expressions for $\mu_0$ and $U$. Then find the equation of state $PV$.

### *8.7.1 Elementary Theory of Electrons in Metals*

In an elementary theory of electrons in metals, the electrons are regarded as a noninteracting Fermi gas. Local charge equality permits, in an approximate sense, the charge of the electrons to be ignored and the valence electrons are treated as an uncharged Fermi gas in volume $L^3$. A typical electron density in metals yields a Fermi temperature or chemical potential of $10^4$–$10^5$ °K and at conventional temperatures, the gas is highly degenerate. Equation (8.55c) suggests a linear temperature dependence for the heat capacity and this is in fair agreement with the experimental data given in Table 8.4.

Since the simplicity of the theory depends on using the same set of basic plane-wave modes for the $N$ particles as for the one-particle problem and then assuming that these states are statistically independent, the extension of the theory to include interactions and yet maintain this one-particle simplicity has been given considerable theoretical attention.[4] Probably the simplest assumption that can be made is that if the plane-wave method has any real validity, then the interaction of the plane wave with the remaining assembly of scatterers can be treated by introducing a pseudo-index of refraction.

For electrons, this is referred to as the effective-mass model and the energy and $k$ values are related by

$$\varepsilon_k = \frac{\hbar^2 k^2}{2M^*} \tag{8.56a}$$

This implies a group velocity for the wave

$$v = \frac{\partial(\varepsilon/\hbar)}{\partial k} = \frac{\hbar k}{M^*}$$

Its most direct effect on the thermodynamic properties is through the change in the density of states. Equation (8.49b) is multiplied by $(M^*/M)^{3/2}$. The group of equations describing the thermodynamic properties have

$$\mu_0 \to \mu_0 \frac{M}{M^*} \quad \text{or} \quad T_F \to T_F \frac{M}{M^*} \tag{8.56b}$$

[4] See, for example, Pines [*13*] and Hugenholtz [*13a*].

**Table 8.4**

Experimental Electronic Heat Capacities of Metals[a]

| Metal | $\gamma^b$ | Metal | $\gamma^b$ |
|---|---|---|---|
| Li | 1.65 | Ti | 3.34 |
| Na | 1.38 | Zr | 2.80 |
| K | 2.1 | Hf | 2.15 |
| Rb | 2.6 | | |
| Cs | 4.0 | V | 9.82 |
| | | Nb | 7.81 |
| Be | 0.171 | Ta | 6.02 |
| Mg | 1.23 | | |
| Ca | 2.9 | Cr | 1.42 |
| Sr | 3.6 | Mo | 1.84 |
| Ba | 2.7 | W | 1.00 |
| Cu | 0.695 | $\alpha$-Mn | 12.7 |
| Ag | 0.647 | $\gamma$-Mn | 9.2 |
| Au | 0.70 | Re | 2.26 |
| Zn | 0.653 | Fe | 4.78 |
| Cd | 0.687 | Ru | 3.00 |
| Hg | 1.82 | Os | 2.3 |
| Al | 1.36 | Co | 4.5 |
| Ga | 0.598 | Rh | 4.7 |
| In | 1.63 | Ir | 3.2 |
| Tl | 1.47 | | |
| | | N | 7.05 |
| Sn | 1.77 | Pd | 9.40 |
| Pb | 2.99 | Pt | 6.55 |
| As | 0.192 | Th | 4.4 |
| Sb | 0.110 | $\alpha$-U | 9.8 |
| Bi | 0.008 | Np | 14.2 |
| | | Pu | 15.9 |
| Sc | 10.7 | | |
| Y | 8.2 | | |
| La(dhcp) | 9.4 | | |
| La(fcc) | 11.5 | | |

[a] This heat capacity data was kindly furnished by N. E. Phillips. A more detailed discussion of the data is given by N. E. Phillips, *Crit. Rev. Solid State Sci.* **2**, 467 (1971); Chemical Rubber Company Press, Cleveland, Ohio.

[b] The values of $\gamma$ are in milli-Joules per gram-mole-degree Kelvin$^{-2}$.

Some values for the effective mass for the alkali metals are as follows:

| | Li | Na | K | Rb | Cs |
|---|---|---|---|---|---|
| $\frac{M^*}{M}$ | 2.20 | 1.24 | 1.23 | 1.38 | 1.75 |

In crystalline material, the energy $\varepsilon_k$ must exhibit the crystal symmetry. The symmetry aspects will be most important for the larger values of $k$ or the de Broglie wavelengths which are characteristic of the dimensions of a unit cell. As a first approximation, an effective-mass tensor is introduced and

$$\varepsilon_k \approx \alpha_{xy} k_x k_y + \cdots \tag{8.57}$$

where $\alpha_{xy}$ exhibits the crystal symmetry and is a function of $k$. A more detailed band theory of solids is necessary to develop the energy dependence in detail [*14*]. The transformation of the sum in Eq. (8.48a) to an integral requires greater effort and is discussed in Exercise 8.15.

Equation (8.48b) can be written in the density-of-states model as

$$\begin{aligned} \ln \mathcal{Q} &= \int_0^\infty d\varepsilon\, g(\varepsilon) \ln(1 + z e^{-\beta\varepsilon}) \\ &\cong \beta\varphi(\mu) + \frac{\pi^2}{6\beta}\, g(\mu) \end{aligned} \tag{8.58a}$$

where $g(\varepsilon) = d^2\varphi(\varepsilon)/d\varepsilon^2$ in Eq. (8.53a). At $T = 0$, the constraint on the number of particles yields

$$N = \varphi'(\mu_0) = \int_0^{\mu_0} g(\varepsilon)\, d\varepsilon$$

where the prime denotes the derivative with respect to $\mu$. The thermodynamic properties follow most conveniently by expanding $\ln \mathcal{Q}$ about $\mu_0$,

$$\begin{aligned} \ln \mathcal{Q} \approx{}& \beta\Big[\varphi(\mu_0) + N(\mu - \mu_0) + \frac{1}{2}\, g(\mu_0)(\mu - \mu_0)^2\Big] \\ &+ \frac{\pi^2}{6\beta}\, [g(\mu_0) + g'(\mu_0)(\mu - \mu_0)] \end{aligned} \tag{8.58b}$$

Thus

$$\mu = \mu_0 - \frac{\pi^2}{6\beta^2}\, \frac{g'}{g}$$

is the chemical potential and the energy is

$$U = \varphi(\mu_0) - N\mu_0 - \frac{\pi^2}{6\beta^2}\, g(\mu_0)$$

The measurable parameter is the heat capacity

$$\frac{C_V}{N} = \frac{\pi^2}{3} k^2 g(\mu_0) T \tag{8.59}$$

and depends on the density of states at the Fermi surface. In the band model of the alkali metals, the Fermi surface is well away from the band edge and $g(\mu_0)$ is expected to have the significance of measuring this density of states by the heat capacity method. In divalent materials, the Fermi surface may intersect the band edge and some areas on the surface of constant energy are forbidden. This will give greater weight to the density of states of the remaining regions and some average value of $g(\mu_0)$ [*14*]. Heat capacity yields an average value for $g(\mu_0)$ and $g(\mu_0)$ can be replaced by $M^*/M$ for a heat capacity linear in temperature. Other experimental methods must be used to examine the density of states at the Fermi surface [*15*].

Exercise 8.15 Show that the number of states between the surface $\varepsilon =$ constant and $\varepsilon + d\varepsilon =$ constant in $k_x, k_y, k_z$ space is given by

$$g(\varepsilon) = (2s+1)\frac{V}{(2\pi)^3}\int_{\varepsilon=\text{constant}} \frac{dS}{|\operatorname{grad} \varepsilon|} \tag{8.60a}$$

where

$$\hat{n} = \frac{(\sum \hat{x}\,\partial E/\partial k_x)}{|\sum \hat{x}\,\partial E/\partial k_x|} = \frac{\operatorname{grad} \varepsilon}{|\operatorname{grad} \varepsilon|}$$

is the normal to surface and the element of surface area is

$$k^2\, d\Omega = (\hat{n} \cdot \hat{k})\, dS$$

Start the development with the definition of the number of states with energy less than $\varepsilon$ as

$$N(\varepsilon) = (2s+1)V\int_{\varepsilon=\text{constant}} \frac{k^2\, d\Omega\, dk}{(2\pi)^3} \tag{8.60b}$$

and the definition of the density of states

$$g(\varepsilon) = \frac{dN(\varepsilon)}{d\varepsilon} \tag{8.60c}$$

Show the important features on a sketch in $k$-space.

Exercise 8.16 Find $g(\varepsilon)$ for $\varepsilon = \hbar^2 k^2/2m$ and for $\varepsilon = \hbar kc$.

### *8.7.2 Relativistic Degenerate Fermi Gas*

A relativistic Fermi gas of electrons at high temperatures was considered in Section 7.10. The relativistic Fermi gas at low temperatures and high pressures is now considered. Equation (8.48a) for $\ln \mathcal{Q}$ remains appropriate, with

$$\varepsilon_k^2 = c^2(\hbar^2 k^2 + m^2 c^2) \tag{8.61}$$

In the extreme relativistic region, in which the rest energy is a small fraction of the total energy, the energy is approximately $cp$, or

$$\varepsilon_k \approx \hbar kc \tag{8.62a}$$

and the number of states with energy less than $k$ or $\varepsilon$ follows from Eq. (8.49a) or Eq. (8.49b). The chemical potential at $T = 0$ is

$$\mu_0 = \left[\frac{3N}{4\pi V(2s+1)}\right]^{1/3} hc \tag{8.62b}$$

The generating function given by Eq. (8.48b) can, with Eq. (8.62a), be written as

$$\ln \mathcal{Q} = (2s+1)\frac{4\pi}{h^3c^3}\frac{V}{\beta^3}\left[\frac{1}{12}(\ln z)^4 + \frac{\pi^2}{6}(\ln z)^2\right] \tag{8.63a}$$

As $\beta \to \infty$, the requirement for $N$ yields the value of $\mu_0$ given in Eq. (8.62b). The various thermodynamic quantities of interest are

$$\mu \approx \mu_0\left[1 - \frac{\pi^2}{3}\left(\frac{kT}{\mu_0}\right)^2\right] \tag{8.63b}$$

$$U \approx \frac{3N\mu_0}{4}\left[1 + \frac{2\pi^2}{3}\left(\frac{kT}{\mu_0}\right)^2\right] \tag{8.63c}$$

$$PV \approx \frac{U}{3} \tag{8.63d}$$

Note that $\mu_0$ is a function of the number density $(N/V)^{1/3}$ and this must be included in a discussion of the energy, entropy, and heat capacity per particle. The equation of state indicates that the pressure is one-third of the energy and this is characteristic of a relativistic gas. It can be shown [*16*] by a general use of Eq. (8.48b) that this is true for all temperatures and is the maximum pressure a macroscopic body can have for a given $U$.

Exercise 8.17 Show that the heat capacity and entropy are given by

$$C \approx \frac{N\pi^2 k^2 T}{\mu_0} = S \tag{8.63e}$$

### 8.7.3 Very Dense Matter

A Fermi gas has the unusual property of becoming more "ideal" as the number density increases [*16*]. Since the number of accessible states is proportional to $Vk^3$, the energy must increase as the number density increases, and finally the kinetic energy is much more important than the binding energy. This occurs first for the valence electrons in metals and the number density of the electrons is such that the Pauli exclusion principle requires a kinetic energy greater than the binding energy to the singly ionized atomic sites. The next region of interest is the Thomas–Fermi model for atomic electrons. Equation (8.51) can be used for the number of states with energy less than $\mu_0 = \varepsilon_F$ and this must equal the number of electrons. The binding energy of the atomic electrons is approximately $Ze^2/4\pi\varepsilon_0 a$, where $a$ is an average electron–nucleus distance, and as

$$\mu_0 \gg \frac{Ze^2}{4\pi\varepsilon_0 a}$$

the Fermi gas of electrons becomes more ideal. With $n_e$ as the number density of electrons and $a \sim (Z/n_e)^{1/3}$, this condition and the condition for $\mu_0$ indicate that a degenerate Fermi gas of atomic electrons becomes nearly ideal for

$$n_e \gg \left(\frac{me^2}{\pi\varepsilon_0 h^2}\right)^3 Z^2 \tag{8.64}$$

For average nuclei, this corresponds to a density of

$$\varrho \gg 20Z^2 \quad \frac{\text{g}}{\text{cm}^3}$$

and from Eq. (8.55e), a pressure of the order of

$$P \gg 10^9 Z^{10/3} \quad \text{atm}$$

is needed to produce this condition. This very large pressure is expected since the atomic volume is being reduced by a factor of $10^2$–$10^4$.

As the pressure is increased further, the average energy per electron $\mu_0$ becomes of the order of $mc^2$ and the electron gas becomes a degenerate relativistic Fermi gas. From Eq. (8.62b), this occurs for an electron density of

$$n_e > \frac{8\pi}{3}\left(\frac{mc}{h}\right)^3 \approx \frac{8}{\lambda_e{}^3} \approx \frac{10^{36}}{\text{m}^3} \tag{8.65}$$

or when the electron–electron distance is of the order of the Compton wavelength $\lambda_e$. This corresponds to an electron pressure of the order of

$$P_e > \tfrac{1}{4}n_e mc^2 \approx 10^{17} \quad \text{atm}$$

Although the pressure is due to the electrons, the density is due to the nuclei. With an average nuclear mass of $A/Z$ per electron, the density of the substance is of the order of

$$\varrho \approx 10^6 \quad \frac{\text{g}}{\text{cm}^3}$$

Densities of this order of magnitude occur in the white dwarf stars. For example [*17*], Sirius B has an average density of $6.8 \times 10^4$ g/cm$^3$ and Van Maanen No. 2 has an average density of $6.8 \times 10^6$ g/cm$^3$. Further increase in pressure requires placing an electron in a volume smaller than its Compton wavelength. This is also the region in which electron capture by the nucleus becomes important and this must be included as the pressure increases.

The reaction equation for electron capture is

$$\text{A}_Z + \text{e}^- \rightleftarrows \text{A}_{Z-1} + \nu$$

where $\text{A}_Z$ denotes the nucleus of mass A and charge $Z$, $\text{e}^-$ is the electron, and $\nu$ is the neutrino. The cross section for neutrino absorption is very small and the evaporation of neutrinos from the substance provides continuous cooling. Thermal equilibrium can only occur for zero temperature and further calculations are made with this assumption. The chemical potential of the neutrino is taken as zero. The chemical potential of a nucleus is almost entirely due to the binding energy of the nucleons $-\varepsilon(A, Z)$ and the equation for the chemical potentials for chemical equilibrium is

$$-\varepsilon(\text{A}, Z) + \mu_e = -\varepsilon(\text{A}, Z - 1)$$

For a particular nucleus, the difference in binding energy is known and from Eq. (8.62b) and the value of $\mu_e$ given here,

$$n_e \approx \frac{8\pi}{3}\left(\frac{\mu_e}{hc}\right)^3 \tag{8.66}$$

For typical nuclei, $\mu_e$ is of the order of a few million electron volts for electron capture and $n_e \approx 10^{37}/\text{m}^3$. This implies that the number density of electrons will remain constant at this value so long as these particular nuclei are dominant [*18*]. Since the pressure and number density of electrons are related by Eq. (8.63d), or $P \approx \frac{1}{4}n_e\mu_e$, the pressure remains constant. During further compression, both electron number and pressure remain constant as the volume decreases and the total number of electrons is decreased by electron capture.

Further increase in pressure causes further electron capture and an increase in the ratio $A/Z$ until the nuclei become unstable and disintegrate. As this condition is reached, the capture of an electron is accompanied by the production of one or more neutrons and the number of neutrons becomes comparable to the number of electrons. This occurs at a value of $\mu_e \approx 20$ MeV, or an electron concentration of $10^{41}/\text{m}^3$, a pressure of the order of $10^{24}$ atm, and a density of the order of $3 \times 10^{11}$ g/cm$^3$. From this region, the gas can be considered as a degenerate Fermi gas of neutrons. Neutrons are in chemical equilibrium with the producing species and the chemical potential $\mu_n$ can in principle be obtained from a knowledge of these reactions. The neutron number density is given by Eq. (8.51), or

$$n_n \approx \frac{8\pi}{3}\left(\frac{2m_n\mu_n}{h^2}\right)^{3/2} \tag{8.67}$$

and a pressure of $P_n \approx \frac{2}{5}n_n\mu_n$. It is often noted that in this region, the pressure increases as the number density, or the density to the 5/3 power,

$$P \propto n_n^{5/3} \quad \text{or} \quad \varrho_n^{5/3}$$

Further decrease in volume to the region in which the neutron number density approaches the volume per particle of $\lambda_n^{-3}$, where $\lambda_n$ is the Compton wavelength, causes the neutron gas to become an extreme relativistic neutron gas and this occurs for a number density of

$$n_n > \frac{8\pi}{3}\left(\frac{m_n c}{h}\right)^3 \approx \frac{8}{\lambda_n{}^3} \approx \frac{10^{45}}{\text{m}^3} \tag{8.68}$$

This distance is comparable to the internucleon distance in nuclei and further decrease in volume is equivalent to compressing the elementary particles to a density greater than that of the nucleus or $\varrho > 10^{15}$ g/cm$^3$.

**Exercise 8.18** (a) Show that the chemical potential of a positive electron gas is given by

$$\mu(e^+) = -\mu_0$$

in the nonrelativistic region, where $\mu_0$ is given by Eq. (8.55a).

(b) Show that the chemical potential of a nucleus is approximately the chemical potential of the free particles minus the binding energy. Since the translational chemical potential is small, the primary contribution is the binding energy $-\varepsilon(A, Z)$.

**Exercise 8.19** Consider the reaction

$$n \rightarrow p + e^- + \bar{\nu}_e$$

and assume that the chemical potential of the electron antineutrino is zero. Assume extreme relativistic degeneracy so that the calculation of the chemical potentials is carried out at zero temperature. Since $n_p = n_{e^-}$ for charge equality, show that $n_p/n_n = \frac{1}{8}$.

### *8.7.4 Neutron Stars*

Matter at high density has been discussed in the previous section, and during the past 40 years there has been considerable speculation about the existence of stars at these high densities. As noted earlier, the white dwarf stars are characteristic of the densities which occur when the electrons are relativistic and form a degenerate Fermi gas. Neutron stars are suggested at the densities at which the neutrons are relativistic and form a degenerate Fermi gas, that is, by Eq. (8.51), at densities of the order of $6 \times 10^{15}$ g/cm$^3$.

The Schwarzschild metric, which is used to discuss many problems in general relativity, possesses a singularity for certain critical masses. In special relativity, the transverse Doppler effect term $[1 - (v^2/c^2)]^{-1}$ plays a dominant role in the Lorentz transformation at high velocities. In general relativity, this is in a certain sense replaced by $[1 - (2\varphi/c^2)]$, where $\varphi$ is the gravitational potential. This analogy can be made by observing that the motion of a test mass $m$ about mass $M$ is given by $mv^2/R = GmM/R^2$ for a circular orbit and the velocity of light is approached as $v^2 = GM/R \rightarrow c^2$.

For a body of mass $M$, the gravitational potential $\varphi = GM/R$ and a singularity occurs when $1 - (2GM/Rc^2) = 0$. For a star with the mass of the sun, the Schwarzschild radius occurs at a radius of 26 km. This corresponds to a density of $10^{16}$ g/cm$^3$ and is comparable to the density of a relativistic degenerate neutron gas. The physical meaning of this Schwarzschild radius is still under discussion and many questions are raised concerning light signals from such sources [*18*]. Even so, the density and radius are characteristic of the neutron star.

Astrophysicists and astronomers suggest that every year or so one of the $10^{11}$ stars in our galaxy finishes burning its nuclear fuel and dies.[5] Depending upon its initial mass, it may expire peacefully as a white dwarf and then slowly cool to a black dwarf. However, for stars with masses greater than $1.2M_\odot$, where $M_\odot$ is the mass of our sun, the electron degeneracy pressure which stabilizes the white dwarf is no longer sufficient to balance the gravitational compression. Such stars collapse violently and are the suggested origin of supernova explosions, which occur at the rate of one event every 100 years. In the supernova explosion in the Crab nebula in 1054, a large fraction of the initial stellar mass imploded inward and the remainder is a rapidly expanding cloud which is a strong source of x rays and visible radiation. One of the possible fates of the imploding material is to form a neutron star and if the mass of the contracting material is less than $2M_\odot$, theoretical calculations with various equations of state suggest that this can occur. A typical radius is of the order of 10 km.

The discovery of pulsars suggests that these are neutron stars. Pulsars are characterized by periodic short bursts of polarized meter-wavelength radio noise. The period between pulses is extremely regular and a change in period per period of less than one part in $10^{15}$ seems to occur. Periods between pulses range from a few seconds to 0.033 sec for the pulsar in the Crab nebula. This short-period pulse from the Crab nebula occurs for both radio and optical radiation. Such short periods, with their remarkable regularity, imply a small source of large density and only the neutron star seems capable of providing this regularity and range of periods. The discovery of the pulsars has given further impetus to the theoretical studies of neutron stars. Enormous magnetic fields of the order of $10^{10}$ G or $10^6$ Wb/m$^2$ are suggested for these stars and models which consider the mantle as a nuclear superfluid are considered. Since these studies combine statistical mechanics, solid-state physics, general relativity, and nuclear physics, the original references are suggested [*20*].

[5] See Ruderman [*19*] for references to original discussions.

## 8.8 Degenerate Bose–Einstein Gas

The ideal Bose–Einstein gas has a very interesting feature as the density increases and the temperature decreases. Again, the nature of the statistics is only important for particles in the same state and each state can be treated as an independent thermodynamic system. The average number of particles in a state is given by [from Eq. (7.31)]

$$\bar{N}_a = \frac{1}{z^{-1}e^{\beta\varepsilon_a} - 1}$$

The lowest state is the zero of energy $\varepsilon_0 = 0$, and as $z \to 1$, the average number of particles in the lowest state $N_0 \to \infty$. All other values remain finite and the lowest state becomes the dominant term. This phenomenon of the lowest state being preferred by all particles is known as *Bose–Einstein condensation* and is now examined in greater detail.

For a gas at low density, the value of $z$ is much less than unity and the smallest value of $z$ is zero. From the above discussion, it is apparent that the fugacity is limited to values in the range $0 < z \leq 1$ for a Bose–Einstein gas for which all states are in equilibrium with a thermal and particle reservoir and therefore with each other. An equation of state follows from Eq. (7.32a),

$$\begin{aligned} \beta PV = \ln \mathcal{Q} &= -\sum_a \ln(1 - ze^{-\beta\varepsilon_a}) \\ &= -\ln(1-z) - KV\int_0^\infty d\varepsilon\sqrt{\varepsilon}\,\ln(1 - ze^{-\beta\varepsilon}) \end{aligned} \tag{8.69}$$

where the procedure for writing the sum as an integral over energy is that used in developing Eqs. (8.48a,b) and (8.50). Since the term with $\varepsilon = 0$ is omitted by the integral, this term is included as a separate term. The value of $K$ is given by Eq. (8.50) with s = 0 for $^4$He and for even values of s for other Bose–Einstein particles. It is now convenient to limit the discussion to s = 0 and to write $\ln \mathcal{Q}$ so that the integral is in dimensionless form,

$$\ln \mathcal{Q} = -\ln(1-z) + \frac{V}{\lambda^3}\, g_{5/2}(z) \tag{8.70}$$

where

$$g_{5/2}(z) = -\frac{2}{\sqrt{\pi}}\int_0^\infty dx\sqrt{x}\,\ln(1 - ze^{-x}) = \sum_{n=1}^{\infty}\frac{z^n}{n^{5/2}} \tag{8.71a}$$

and $g_{5/2}(1) = 1.342$. Another useful quantity is

$$g_{3/2}(z) = z\frac{\partial g_{5/2}}{\partial z} = \sum_{n=1} \frac{z^n}{n^{3/2}} \tag{8.71b}$$

and these quantities have been tabulated [*21*]. The fugacity $z$ is related to the particle number by

$$N = z\left[\frac{\partial(\ln \mathscr{Q})}{\partial z}\right]_{V,\beta} = \frac{z}{1-z} + \frac{V}{\lambda^3} g_{3/2}(z) \tag{8.72}$$

As $z$ approaches unity, $g_{3/2}$ becomes the Riemann zeta function $\zeta(\frac{3}{2})$ and

$$g_{3/2}(1) = 2.612$$

is the maximum value of this function. For $n = N/V > 2.612/\lambda^3$, the fugacity $z$ must adjust so that the remaining particles go into the lowest state. The critical value at which an appreciable filling of the lowest state begins or at which the onset of Bose–Einstein condensation occurs is

$$\lambda_c^3 n = 2.612 \tag{8.73}$$

Again, $\lambda^2 = h^2/2\pi mkT$ and $n$ is the number density $n = N/V$. Regarding $n$ as the average number density of particles and $n_0$ as the average number density in the lowest state, Eq. (8.72) can be written as

$$n = n_0 + 2.612\lambda^{-3} \tag{8.74a}$$

for $z = 1$. Rewriting $2.612\lambda^{-3}$ as $n(T/T_c)^{3/2}$ yields an equation for $n_0$

$$n_0 = n\left[1 - \left(\frac{T}{T_c}\right)^{3/2}\right] \tag{8.74b}$$

where $T_c$ depends on $n$ in the manner indicated by Eq. (8.73). A plot of $n_0/n$ as a function of $T/T_c$ is shown in Fig. 8.5.

Other thermodynamic functions follow from $\ln \mathscr{Q}$ and the internal energy is given by

$$U = \frac{3V}{2\lambda^3} g_{5/2}(z)kT \tag{8.75a}$$

At $z = 1$, the internal energy falls rapidly since there is a negligible amount of energy associated with the lowest state. The Gibbs function is given as

$$G = NkT \ln z \tag{8.75b}$$

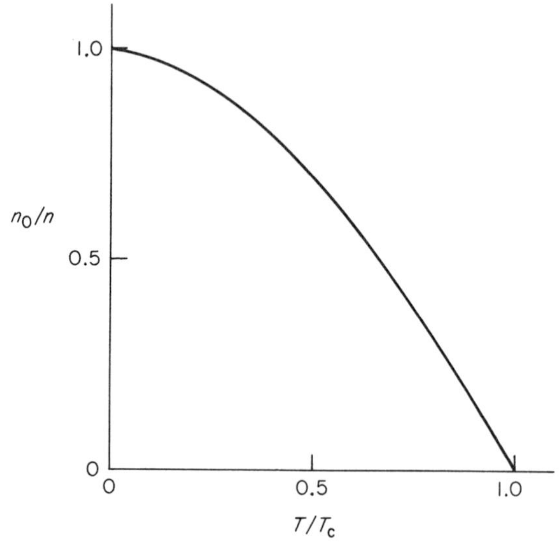


**Fig. 8.5** Plot of $n_0/n$, the fraction of the Bose–Einstein atoms in the lowest translational state, as a function of the reduced temperature $T/T_c$ in Eq. (8.74b). (From F. London, "Superfluids," Vol. II, Fig. 21; Copyright 1954, Wiley, New York. Used by permission of John Wiley & Sons, Inc.)

and since $z$ is assumed known in terms of the average number $N$ by Eq. (8.72), the entropy is most conveniently obtained from $G = U + PV - TS$, or

$$\frac{S}{Nk} = \frac{5V}{2N\lambda^3} g_{5/2}(z) - \ln z \tag{8.75c}$$

For $T < T_c$, the entropy is determined by $(\lambda_c/\lambda)^3$ or $(T/T_c)^{3/2}$. Heat capacity calculations require greater care since the volume and number are held constant during the differentiation of $U$ or $S$ with respect to $T$. For $T < T_c$, the heat capacity is given by

$$\frac{C_V}{Nk} = \frac{15V}{4N\lambda^3} g_{5/2}(1) = \left(\frac{T}{T_c}\right)^{3/2} \tag{8.76a}$$

For $T > T_c$, the relationship $g_{3/2}(z) = N\lambda^3/V$ permits the relationship

$$dg_{3/2} = \frac{dg_{3/2}}{dz} dz = g_{1/2}\, d(\ln z) = \frac{\lambda^3}{V} dN - \frac{N\lambda^3}{V^2} dV + \frac{3N\lambda^2}{V} d\lambda$$

to be formed. Since

$$\left(\frac{\partial g_{5/2}}{\partial T}\right)_{N,V} = \frac{dg_{5/2}}{dz}\left(\frac{\partial z}{\partial T}\right)_{N,V} = g_{3/2}(z)\left(\frac{\partial(\ln z)}{\partial T}\right)_{N,V}$$

the derivative relationship needed for the heat capacity follows from

$$\left[\frac{\partial(\ln z)}{\partial T}\right]_{N,V} = -\frac{3g_{3/2}}{2g_{1/2}}\frac{1}{T}$$

Direct differentiation of $U$ at constant $N$ and $V$ yields with these substitutions

$$\frac{C_V}{k} = \frac{15V}{4\lambda^3} g_{5/2}(z) - \frac{9Ng_{3/2}(z)}{4g_{1/2}(z)} \tag{8.76b}$$

Since $g_{1/2}(1) = \infty$, this agrees with the earlier expression for $z = 1$. A plot of the heat capacity as a function of temperature is shown in Fig. 8.6.

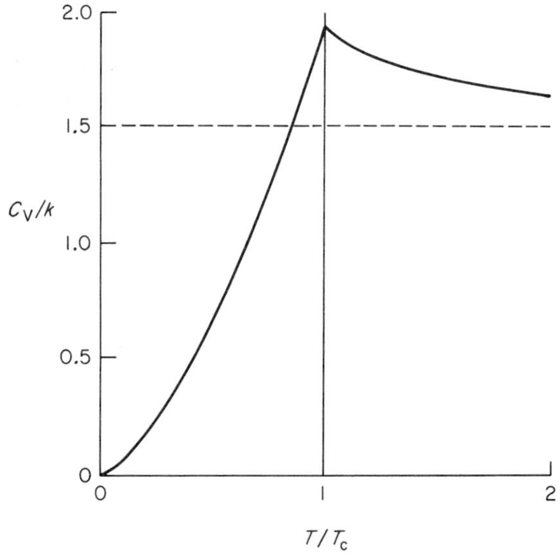


**Fig. 8.6** Heat capacity of a Bose–Einstein gas as a function of the reduced temperature. (From F. London, "Superfluids," Vol. II, Fig. 20; Copyright 1954, Wiley, New York. Used by permission of John Wiley & Sons, Inc.)

The question now arises as to the interpretation of the Bose–Einstein condensation phenomenon. The author prefers the interpretation as a "condensation in momentum space" since the wavefunction of the ground state in the simple "naive" gas model is $\sin(\pi x/L)\sin(\pi y/L)\sin(\pi z/L)$, where $L$ is a typical length in the cube. Particles in this state prefer to be away from the walls. The limit $L \to \infty$ can be considered and this decreases the surface effects, but it does not limit the extent of the wavefunction to distances of less than $L$. Huang [*22*] prefers to consider the problem as a real separation into a condensed phase with zero volume per particle and a gas phase with a finite volume per particle. Since the initial problem assumed that a finite volume $V$ was accessible to all noninteracting particles and the states were determined on this basis, physical separation is not obvious and other references are suggested for further consideration of this point [*21*].

If $N$ particles are placed in volume $V$, then, as $V$ is decreased at constant temperature, the pressure becomes almost constant as the volume is decreased below the critical volume per particle, which is given by Eq. (8.73) with $T$ fixed. Particles move into the lowest state after the critical volume per particle is reached, and, since particles in the lowest state do not exert a thermal pressure, the pressure remains constant with further decrease in volume along an isotherm. This pressure relationship is characteristic of a vapor–liquid or vapor–solid phase transformation.

Even though there may be a question about the type of condensation phenomenon involved, the expression for $\ln \mathcal{Q}$ splits into two terms that can be of equal importance and the generating function becomes the product of generating functions $\mathcal{Q} = \mathcal{Q}_0 \mathcal{Q}_{\text{ex}}$. As noted throughout the text, this implies statistical independence of the two features, and in principle, there should exist experiments to determine each feature or filters to separate the features. Usually, one prefers to separate in physical space, but this is not necessary. Photons are separated in momentum space by gratings, and in principle, particles can be separated in momentum space. Figure 4.22c is an example of momentum selection for particles. A Josephson junction [*23*] can act as a "half-silvered" mirror for electrons in metallic conductors.

### *8.8.1 Helium-4*

Helium-4 forms a system of Bose particles. Equation (8.73) indicates that for the number density of particles occurring in liquid $^4$He, the critical temperature $T_c = 3.14°$K. This is remarkably close to the $\lambda$-transition temperature $T_\lambda = 2.18°$K at which the experimental specific heat is discontinuous in a logarithmically infinite manner and below which liquid $^4$He has peculiar flow properties. The discontinuity in the heat capacity indicates that the $\lambda$-transition is not a first-order transition and this limits the similarity. London [*21*] discussed $^4$He as a Bose–Einstein gas, and features that are characteristic of a Bose gas must be important since liquid $^3$He, which is a Fermi–Dirac system of particles, does not have this characteristic transition.

Some care is necessary in any analysis that is dependent on dimensional considerations. In examining the second virial coefficient by the phase-shift method, it was noted for hard spheres that for $\lambda > \pi r_0$, the hard-sphere repulsion could be ignored in the second virial coefficient. From another point of view, the second virial coefficient is almost independent of hard-core repulsion for $2^{-5/2}\lambda^3 > \frac{2}{3}\pi r_0^3$. Both of these conditions occur at a temperature of the same order of magnitude as the $\lambda$-transition in liquid $^4$He. Neither of these conditions is important for a Fermi gas, since very few particles can have this large value for the thermal de Broglie wavelength.

From elementary quantum mechanics, the energy of the lowest state is proportional to $V^{-2/3}$, and there remains at absolute zero a mechanical pressure which is proportional to $V^{-5/3}$. These are normally referred to as the zero-point energy and zero-point pressure, respectively, for a Bose–Einstein gas. They do not depend on the number density of particles and vanish as $V \to \infty$, even though $N/V$ remains finite. For small volumes, they can become important parameters.

## References


1. B. Kahn and G. E. Uhlenbeck, *Physica* (*Utrecht*) **5**, 399 (1938).
2. J. E. Mayer and M. G. Mayer, "Statistical Mechanics." Wiley, New York, 1940.
3. E. A. Mason and T. H. Spurling, The virial equation of state. *In* "International Encyclopedia of Physical Chemistry and Chemical Physics" (J. S. Rowlinson, ed.). Pergamon, Oxford, 1969.

3a. "American Institute of Physics Handbook." McGraw-Hill, New York, 1957.

4. R. B. Scott, "Cryogenic Engineering." Van Nostrand-Reinhold, Princeton, New Jersey, 1959.
5. E. S. Burnett, $N_2$ chart, U. S. Bur. Mines, Amarillo, Texas, 1949.
6. J. O. Hirschfelder, C. F. Curtiss, and R. B. Bird, "Molecular Theory of Gases and Liquids." Wiley, New York, 1954.
7. C. Block and C. E. Dominicus, *Nucl. Phys.* **10**, 509 (1959); R. Omnes, *Phys. Rev. Lett.* **23**, 38 (1969).
8. R. Dashen, S. Ma, and H. J. Bernstein, *Phys. Rev.* **187**, 345 (1969).
9. B. Kahn, On the theory of the equation of state. Thesis, Utrecht, 1938; see also Hirschfelder *et al.* [*6*].
10. J. de Boer, J. van Kranendonk, and K. Compaan, *Physica* (*Utrecht*) **16**, 545 (1950); J. E. Kilpatrick and M. F. Kilpatrick, *J. Chem. Phys.* **19**, 930 (1951).
11. J. de Boer, *Physica* (*Utrecht*) **14**, 139 (1948).
12. J. de Boer and R. J. Lunbeck, *Physica* (*Utrecht*) **14**, 510 (1948).

12a. B. M. Abraham, D. W. Osborne, and B. Weinstock, *Phys. Rev.* **80**, 366 (1950).

13. D. Pines, "Elementary Excitation in Solids." Benjamin, New York, 1964.

13a. N. M. Hugenholtz, *Rep. Progr. Phys.* **28**, 201 (1965).

14. F. S. Ham, *Phys. Rev.* **128**, 82, 2524 (1962); J. Callaway, *Solid State Phys.* **7**, 99 (1958).
15. C. Kittel, "Introduction to Solid State Physics." Wiley, New York, 1966.
16. L. Landau and E. Lifshitz, "Statistical Physics." Addison-Wesley, Reading, Massachusetts, 1958.
17. D. H. Menzel, P. Bhatnagar, and H. Sen, "Stellar Interiors." Wiley, New York, 1963.
18. E. E. Salpeter, Superdense equilibrium stars. *In* "Quasi-Stellar Sources and Gravitational Collapse" (I. Robinson, ed.), Chapter 32. Univ. of Chicago Press, Chicago, Illinois, 1965; see also H. Chiu, The formation of neutron stars and their surface properties. *In* "Quasi-Stellar Sources and Gravitational Collapse" (I. Robinson, ed.), Chapter 33. Univ. of Chicago Press, Chicago, Illinois, 1965.
19. M. Ruderman, *Comments Nucl. Particle Phys.* **3**, 37 (1969).
20. H. J. Lee, V. Canuto, H. Chiu, and C. Chiuderi, *Phys. Rev. Lett.* **23**, 390 (1969); R. Smoluchowski, *Ibid.* **24**, 923 (1970); M. Hoffberg, A. E. Glassgold, R. W. Richardson, and M. Ruderman, *Ibid.* **24**, 775 (1970); Chandrasekhar, S. *Ibid.* **24**, 611 (1970).
21. F. London "Superfluids," Vol. II. Wiley, New York, 1954.
22. K. Huang, "Statistical Mechanics." Wiley, New York, 1963.
23. B. D. Josephson, *Phys. Lett.* **1**, 251 (1962); *Advan. Phys.* **14**, 419 (1965); R. C. Jaklevic, J. Lambe, A. H. Silver, and J. E. Mercereau, *Phys. Rev. Lett.* **12**, 159 (1964); J. E. Zimmerman and J. E. Mercereau, *Ibid.* **13**, 125 (1964).

# Gases, Liquids, and Solids

# IX

## 9.1 Introduction

This chapter is primarily concerned with the equilibrium between phases. Some aspects of vapor-, liquid-, and solid-phase equilibria are considered for a single-component system and then the discussion is extended to binary mixtures or solutions. The osmotic pressure as a thermodynamic variable is introduced and then the liquid $^3He$–$^4He$ dilution refrigerator is used as an example which includes many of the conventional and some novel aspects of solutions. Since the counting of states plays a natural role in the quantum statistical mechanics approach, the Third Law of Thermodynamics appears as a natural aspect of the counting. Numerous experimental examples which support the third law are discussed. It always seems remarkable in such a discussion that the experimental entropy measured by heating a substance from zero temperature through all of its phase changes to the gas phase agrees with the entropy of the gas phase computed from spectroscopic data even though this agreement is expected by the third law.

The discussion then shifts to a study of the combinatorial problem of placing $N_A$ and $N_B$ particles of species A and B on an array of $N_A + N_B$ sites. The number of configurations of the A–B pairs is of primary interest, and then the effect of a constraint on the probable number of A–B pairs is studied. Recent techniques of counting the number of polygons and self-avoiding walks are used in the discussion of this problem. A remarkable aspect of this "three-dimensional checker game with a constraint" is the occurrence of a critical point or a singularity in the dispersion in $N_A$ and in the dispersion of the number of pairs $N_{AB}$. Even more remarkable is the realization during the past few years that as the theory became better and the experimental knowledge of the critical points improved, this combinatorial problem was in excellent agreement with the data near the critical point.

The chapter concludes with a discussion of the surface phase and of monolayers and multilayers on such surfaces.

## 9.2 Thermodynamic Aspects of Equilibrium between Phases

The thermodynamic considerations of Chapter VI apply to any physical system. Considerations in this section are limited to equilibrium between the vapor phase and either the solid or the liquid phase of the same substance. These phases are separated by surfaces and are regarded as distinct. In later sections of this chapter, the distinction between phases will be regarded as statistical independence with certain constraints. From the thermodynamic point of view, the additive variables for a vapor–solid system can be expressed as

$$U = U_v + U_s, \qquad S = S_v + S_s, \qquad V = V_v + V_s, \qquad N = N_v + N_s \quad (9.1)$$

and $A$ and $G$ follow from these. Gibbs considered equilibrium as occurring when the variation of the energy with respect to the virtual changes was zero. More recent developments require the entropy to be a maximum, or for a virtual change

$$\delta S = \sum_i \delta S_i = \sum_i \left( \frac{\delta U_i}{T_i} + \frac{P_i}{T_i}\,\delta V_i - \frac{\mu_i}{T_i}\,\delta N_i \right) = 0 \qquad (9.2)$$

subject to the constraints of fixed energy, volume, and number, or

$$\delta U = \sum_i \delta U_i = 0, \qquad \delta V = \sum_i \delta V_i = 0, \qquad \delta N = \sum_i \delta N_i = 0 \quad (9.3)$$

$i =$ v or s in the previous example, but can be more general. The variables in Eq. (9.2) are not independent, but can be made independent by the method of Lagrange multipliers. Introducing the Lagrange multipliers $\lambda_1, \lambda_2, \lambda_3$ into Eqs. (9.2) and (9.3) yields

$$\delta S = \sum_i \left[ \left( \frac{1}{T_i} - \lambda_1 \right) \delta U_i + \left( \frac{P_i}{T_i} - \lambda_2 \right) \delta V_i - \left( \frac{\mu_i}{T_i} - \lambda_3 \right) \delta N_i \right] = 0 \quad (9.4)$$

Now the $\delta U_i$, $\delta V_i$, and $\delta N_i$ can be regarded as independent and then the equality requires that

$$\frac{1}{T_i} = \lambda_i = \frac{1}{T}; \qquad \frac{P_i}{T_i} = \lambda_2 = \frac{P}{T}; \qquad \frac{\mu_i}{T_i} = \lambda_3 = \frac{\mu}{T}$$

or

$$T_i = T, \qquad P_i = P, \qquad \mu_i = \mu \tag{9.5}$$

Thus the constraint on the energy requires the temperature to be equal for all phases. In Chapter VI, this was shown to be the general condition for any system regardless of its composition. The constraint on the total volume of the phases requires the pressure to be a constant, or the force at an interface between phases to be equal and opposite. Finally, the constraint on the total number of particles of the same material requires the chemical potential for both phases to be equal. In general, these ideas are summarized by the equation

$$\mu_v(P, T) = \mu_s(P, T) \tag{9.6}$$

for equilibrium between the solid and vapor phases. Along the sublimation curve, this equality requires the vapor pressure to be a unique function of the temperature.

If three phases are present, then $i$ is summed over v, s, $l$ in the example of vapor–solid–liquid equilibrium. Then,

$$\mu_v(P, T) = \mu_s(P, T) = \mu_l(P, T) \tag{9.7}$$

and these three functions of two variables can be equal only at a single set of values of $P$ and $T$. This point is referred to as the "triple point." Equilibrium of more than three phases of a single substance is impossible. A phase diagram for water [1] with its various crystalline modifications is shown in Fig. 9.1.

Consider now the transformation of a fraction of one phase into another at a given point on the sublimation curve and let $N_s$ solid atoms be converted into $N_v = N_s$ atoms in the vapor. The Gibbs function $G = N\mu$ does not change in this process. Enthalpy does change, and the change in enthalpy at constant temperature and pressure is the heat added,

$$Q = H_v - H_s \tag{9.8a}$$

and is the *latent heat of sublimation in this case.* It is convenient to use the

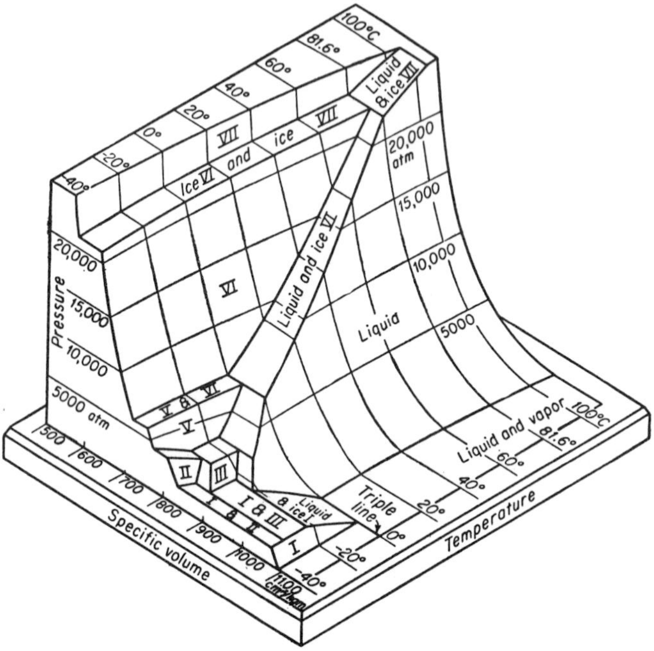


**Fig. 9.1** The *PVT* surface for $H_2O$ showing all triple points. (From "Heat and Thermodynamics," 5th ed., Fig. 11.10, p. 205, by M. W. Zemansky. Copyright 1968, McGraw-Hill, New York. Used with permission of McGraw-Hill Book Company.)

*latent heat per molecule*

$$q = h_v - h_s \qquad \text{(per molecule)} \tag{9.8b}$$

in many calculations.

### *9.2.1 Clausius–Clapeyron Equation*

A very powerful and useful relationship is now given, and for obvious reasons, the subscripts must be changed. From the condition for equilibrium between phases,

$$\mu_1(P, T) = \mu_2(P, T)$$

and from Eq. (6.31),

$$d\mu = -s\, dT + v\, dP$$

where $s$ and $v$ are entropy and volume per molecule, respectively. The relationship

$$-s_1\,dT + v_1\,dP = -s_2\,dT + v_2\,dP$$

follows from $d\mu_1 = d\mu_2$. This can be rewritten as

$$\frac{dP}{dT} = \frac{s_1 - s_2}{v_1 - v_2} \tag{9.9a}$$

or with the latent heat $q$ from Eq. (9.8b), as the *Clausius–Clapeyron equation*:

$$\frac{dP}{dT} = \frac{q}{T(v_1 - v_2)} \tag{9.9b}$$

Although these equations were derived by changing one molecule from phase 1 to phase 2, it remains correct when $N$ molecules are changed from phase 1 to phase 2, and

$$\frac{dP}{dT} = \frac{Q_{12}}{T(V_1 - V_2)} = \frac{S_1 - S_2}{(V_1 - V_2)} \tag{9.9c}$$

where $S_1$ and $S_2$ are the entropies of $N$ molecules in phases 1 and 2, respectively, and so on. The indices 1 and 2 can also refer to any two phases of the material and this equation remains valid along the sublimation curve, fusion curve, vaporization curve, and so forth. Freezing and boiling points and heats of fusion and sublimation at these points are given in Table 9.1 for some common substances.

Since the vapor is almost an ideal gas along the sublimation curve and the volume of the vapor is very much larger than that of the solid, the term $V_v - V_s$ is replaced by $NkT$. To a very good approximation,

$$\frac{dP}{P} \approx \frac{q\,dT}{kT^2} \tag{9.10}$$

relates the vapor pressure to the latent heat of sublimation per molecule. In many texts on thermodynamics, the latent heat of sublimation is related to the enthalpies, and Kirchhoff's equation is given as

$$q = \int_0^T C_P''\,dT - \int_0^T C_P'\,dT + q_0 \tag{9.11}$$

$C_P'$ is the heat capacity of the solid at constant pressure, $C_P''$ is the heat capacity of the vapor, and $q_0$ is the latent heat of sublimation at absolute zero. The vapor pressure equation is found by integration. A statistical approach is used in a subsequent section and this feature is not pursued.

**Table 9.1**

| Substance | Fusion temperature (°K) | Boiling point (°K) | Heat of fusion (in units of the gas constant $R$) | Heat of vaporization (in units of the gas constant $R$) |
|---|---|---|---|---|
| $^4He$ | — | 4.216 | — | — |
| $H_2$ | 13.96 | 20.39 | 14 | 108 |
| Ne | 24.57 | 27.1 | 40 | — |
| $O_2$ | 54.40 | 90.19 | 53 | 820 |
| $F_2$ | 55.20 | 85.24 | 61 | 785 |
| $N_2$ | 63.15 | 77.34 | 86 | 670 |
| CO | 68.10 | 81.66 | 100 | 727 |
| Ar | 83.85 | 87.29 | 140 | 780 |
| $CH_4$ | 90.68 | 111.67 | 112 | 1000 |
| NO | 109.51 | 121.39 | 276 | 1660 |
| Kr | 115.95 | 119.93 | 195 | 1080 |
| HCl | 158.91 | 188.11 | 240 | 1940 |
| Xe | 161.3 | 165.1 | 276 | 1520 |
| $SF_6$ | — | 209.5 | — | 2740 |
| $Cl_2$ | 172.16 | 239.10 | 770 | 2450 |
| $CO_2$ | — | 194.68 | — | 3030 |
| Hg | — | 629.73 | — | 7110 |
| Cs | 301.9 | 963 | 250 | 8160 |
| Na | 371 | 1156 | 311 | 12,700 |
| Pb | 600.6 | 2023 | 575 | 21,600 |
| $H_2O$ | 273.16 | 373.16 | 722 | 4885 |

### 9.2.2 *Statistical Aspects of Equilibrium between Phases*

Equilibrium between phases implies that the phases are statistically independent systems which are subject to constraints on the total volume, total number of particles, and total energy. These constraints were discussed in previous sections and are summarized by the equality of the chemical potentials of the phases. It was shown in Chapter VI that the chemical potential can be derived from either the canonical or grand canonical ensemble and is an important thermodynamic variable for processes in which the pressure and temperature are constant. Thus the basis of the statistical approach for the calculation of the sublimation curve is

$$\mathrm{V} \leftrightarrow \mathrm{S}$$

$$\mu_{\mathrm{V}}(P, T) = \mu_{\mathrm{S}}(P, T)$$

At low densities, the vapor is almost an ideal gas and the chemical potential for such a gas is given by Eqs. (7.5a) and (7.15),

$$\begin{aligned} \mu_{\mathrm{V}} &= -kT \ln \frac{Q_1}{N} \\ &= -kT \ln\left[\frac{V}{N}\left(\frac{2\pi MkT}{h^2}\right)^{3/2} Q_{\mathrm{vi}}\right] \\ &= kT \ln P - \frac{5}{2} kT \ln T - i_0 kT - kT \ln Q_{\mathrm{vi}} \end{aligned} \tag{9.12}$$

$Q_{\mathrm{vi}} = Q_{\mathrm{int}}$ is the internal generating function for the molecules in the vapor. The $N$ is replaced by $P/kT$ and $i_0$ involves the remaining constants:

$$i_0 = \ln\left[\left(\frac{2\pi M}{h^2}\right)^{3/2} k^{5/2}\right] \tag{9.13}$$

The chemical constant depends on the fundamental constants and the mass of the molecule. Other aspects of the vapor pressure equation are discussed in subsequent sections in this chapter.

### 9.2.3 *Grand Canonical Ensemble and Phases*

The very concept of phases suggests that the phases are statistically independent and the grand canonical ensemble should be a product of the generating functions for the phases. For solid–vapor phases,

$$\mathcal{Q} = \mathcal{Q}_{\mathrm{V}}(V_{\mathrm{V}}, z, \beta)\mathcal{Q}_{\mathrm{S}}(V_{\mathrm{S}}, z, \beta)$$

generates the conditions implied by Eqs. (9.1)–(9.3). Direct use of Eqs. (6.56)–(6.64) gives the equation

$$\frac{PV}{kT} = \ln \mathcal{Q} = \ln \mathcal{Q}_{\mathrm{V}} + \ln \mathcal{Q}_{\mathrm{s}} = \frac{P}{kT}(V_{\mathrm{V}} + V_{\mathrm{s}})$$

or

$$V = V_\text{v} + V_\text{s}$$

$$N = z\frac{\partial(\ln \mathcal{Q})}{\partial z} = z\frac{\partial(\ln \mathcal{Q}_\text{v})}{\partial z} + z\frac{\partial(\ln \mathcal{Q}_\text{s})}{\partial z} = N_\text{v} + N_\text{s}$$

$$U = -\frac{\partial(\ln \mathcal{Q})}{\partial \beta} = U_\text{v} + U_\text{s}$$

$$S = \frac{U + PV}{T} - Nk \ln z = S_\text{v} + S_\text{s}$$

These equations are the same as the set of equations given in Eq. (9.1) and the above procedure emphasizes that additive thermodynamic variables imply statistical independence of the additive quantities. This statistical independence requires a product of generating functions, or a product of generating functions implies additive thermodynamic variables. The selection of $P$, $z$, and $\beta$ fixes the volume $V$, average number of particles $N$, average energy $U$, and indirectly, the entropy $S$ and other thermodynamic functions. The author prefers the calculation of the properties of each phase separately. With this approach, superheated vapors and supercooled liquids can occur when equilibrium between the phases is not immediate. Equilibrium occurs when the same $P$, $z$, and $\beta$ are selected for each phase. It is not clear that a general proof exists to show that this is the correct form for $\mathcal{Q}$ and it may be that some features of fluctuations are missed by this simple and direct procedure. But as a first approximation, $\mathcal{Q}$ must have this form.

If all of the thermodynamics of a multiphase system must follow from

$$\mathcal{Q} = \sum_N z^N Q_N(V, \beta)$$

then a very formidable problem ensues. One must know the states of the canonical ensemble for $N_\text{v}$ particles in the vapor volume $V_\text{v}$ and $N_\text{s}$ particles in the solid volume $V_\text{s}$, subject to the constraint $N = N_\text{v} + N_\text{s}$ and $V = V_\text{v} + V_\text{s}$. But this statement of labeling implies that the canonical ensemble can be written as a product $Q_{N_\text{v}} Q_{N_\text{s}}$ and again the vapor and solid form independent systems. If one does not wish to treat them as independent, then somehow this observable information must be contained in the wavefunction describing the state and a many-body problem of this complexity is beyond the scope of this text. Whenever appropriate, an attempt is made to use simple models.

## 9.3 Phonon Gas

A different analysis is needed for the solid or condensed phase. $N$ atoms or molecules require $3N$ coordinates for a description of their position, or a system of $N$ particles has $3N$ degrees of freedom. The solid phase is an ordered array of these $N$ molecules with a crystalline structure which characterizes the phase. For a small displacement of an atom from its equilibrium position, the potential energy can be expanded in terms of the displacement. Equilibrium requires the first derivative of the potential energy be zero and the potential energy is proportional to the square of the displacement. The system of $N$ molecules in a solid crystalline phase is similar to a set of identical masses which are coupled together by springs. The potential energy depends on the pairwise displacements $(r_{ij} - r_{ij}^0)^2$ between all the nearest, next-nearest neighbors, and so on. In the system, the motions of the individual molecules are no longer independent. In classical mechanics, it can be shown that such a system of coupled particles can, by a suitable coordinate transformation, be described as a set of $3N$ independent normal modes. The Hamiltonian, and total energy of the system, is

$$H = \sum_{\alpha}^{3N} \left( \frac{p_\alpha^2}{2} + \frac{1}{2} \omega_\alpha^2 q_\alpha^2 \right) \tag{9.14}$$

and has the form of $3N$ independent harmonic oscillators. In principle, it is possible to excite a single normal mode of the system and the energy remains in this mode to the above degree of approximation. These normal modes can be quantized in the same manner as that used for a harmonic oscillator. Each normal mode has energy

$$\varepsilon_{\alpha n} = (n_\alpha + \tfrac{1}{2}) h\nu_\alpha \tag{9.15}$$

where $n_\alpha$ is the number of quanta associated with the normal mode with index $\alpha$ and frequency $\nu_\alpha$. The canonical generating function for this statistically independent mode is

$$Q_\alpha = \sum_n \exp\left[ -\beta\left(n + \frac{1}{2}\right) h\nu_\alpha \right] = \frac{\exp(-\beta h\nu_\alpha/2)}{1 - \exp(-\beta h\nu_\alpha)} \tag{9.16}$$

and the canonical generating function for all $3N$ statistically independent modes is

$$Q = \left( \prod_{\alpha}^{3N} Q_\alpha \right) (Q_s)^N \tag{9.17}$$

The internal coordinates of the molecule are included in $Q_s$ and are statistically independent from the normal modes $\alpha$. An expression for the free energy is

$$A(V, T, N) = -kT \ln Q = -kT \sum_{\alpha}^{3N} \ln Q_\alpha - NkT \ln Q_s \qquad (9.18)$$

There is no constraint on the number of quanta, and the chemical potential for the phase might be to be assumed zero; however, this is not true and the chemical potential is obtained from its definition,

$$\mu_s = A_{N+1}(V, T) - A_N(V, T) = \left(\frac{\partial A}{\partial N}\right)_{V,T}$$

The number of particles remains apparent in the calculation.

### 9.3.1 Debye Model of a Solid and Phonons

Debye regarded the excitations in a solid as very similar to the photons in a thermal enclosure. A plane-wave expansion is always possible for the description of the acoustic waves in a solid. As usual, these waves are of the form $\exp(i\mathbf{k} \cdot \mathbf{r})$, where $k_x = 2\pi n_x/L, \ldots$ for a discrete set of indices. The physics follows by regarding the solid as a continuous medium, with the waves traveling with the acoustic velocity $c$ in the medium. Three types of plane waves can propagate in a solid; two of these waves have polarization transverse to the direction of propagation $\hat{k}$, and the third has longitudinal polarization along $\hat{k}$.

In the Debye model, the quanta associated with the plane waves have energy $h\nu$ and momentum $h\nu/c$. They differ from the photon picture for thermal radiation by replacing the speed of light by the speed of sound. These quanta are called *phonons*. A second difference in the Debye model is that the number of normal modes is $3N$, or $N$ for each polarization. Thus the sum over indices $k$ or $n_x$, $n_y$, $n_z$ is limited to a maximum value,

$$\sum_{\alpha} \to \sum_{\Delta k} \to V \frac{4\pi k^2\, dk}{(2\pi)^3} \eta(k - k_m) \to V \frac{4\pi \nu^2\, d\nu}{c^3} \eta(\nu - \nu_m) \qquad (9.19)$$

where $\eta(x) = 1$ for $x < 0$ and $\eta(x) = 0$ for $x > 0$. The wave number $k$ and the frequency $\nu$ are related by

$$k = \frac{2\pi\nu}{c} \qquad (9.20)$$

and this describes the physics of acoustic waves. $k_m$ or $\nu_m$ is defined by

$$N = V \int_0^\infty \frac{4\pi\nu^2\, d\nu}{c^3} \eta(\nu - \nu_m) = \frac{4\pi V}{3}\left(\frac{\nu_m}{c}\right)^3 \tag{9.21a}$$

or

$$\nu_m = \left(\frac{3N}{4\pi V}\right)^{1/3} c \tag{9.21b}$$

A temperature defined by

$$k\Theta = h\nu_m \tag{9.22a}$$

or

$$\Theta = \frac{hc}{k}\left(\frac{3N}{4\pi V}\right)^{1/3} \tag{9.22b}$$

is referred to as the Debye $\Theta$. Transverse and longitudinal waves have different speeds $c$ in the solid and both $\Theta_t$ and $\Theta_l$ could be used. One can solve the problem for each polarization and then sum the results since they are independent. For the purposes of the discussion given here, only one average value of $\Theta$ is used and the sum is replaced by 3. With $Q_s = 1$, the free energy of the Debye solid is of the form

$$A(V, T, N) = A_0 + \frac{9Nk\Theta}{8} + \frac{9NkT^4}{\Theta^3} \int_0^{\Theta/T} x^2\, dx \ln(1 - e^{-x}) \tag{9.23}$$

where $x = h\nu/kT$ and $x_m = h\nu_m/kT = \Theta/T$. The term $9Nk\Theta/8$ is the zero-point contribution to the energy from the $\frac{1}{2}$ term in Eq. (9.15). Heat capacity at constant volume is given by

$$C = T\left(\frac{\partial S}{\partial T}\right)_{V,N} = -T\left(\frac{\partial^2 A}{\partial T^2}\right)_{V,N}$$

$$= -9NkT \frac{\partial^2}{\partial T^2}\left[\left(\frac{T^4}{\Theta^3}\right) \int_0^{\Theta/T} x^2\, dx \ln(1 - e^{-x})\right] \tag{9.24}$$

and this expression is a universal function of $T/\Theta$. A plot of $C_V$ against $T/\Theta$ is shown in Fig. 9.2 and very reasonable agreement is obtained for a large group of solids. Some values of $\Theta$ are given in Table 9.2. At high temperatures,

$$\frac{C_V}{N} \to 3k, \qquad T > \Theta \tag{9.25}$$

and this is the Dulong–Petit Law. It had been noted that most solids at

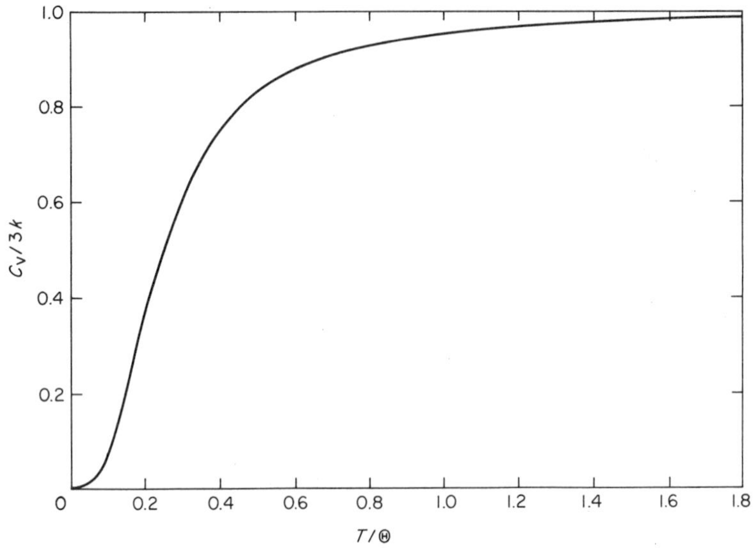


**Fig. 9.2** Heat capacity per mole as a function of the reduced temperature $T/\Theta$. Curve is for the Debye model.

room temperature had the same heat capacity per mole. This result was first explained by the equipartition theorem by associating $\frac{1}{2}kT$ with each squared term in the Hamiltonian. At low temperature, the integral can be taken to infinity and the well-known Debye $T^3$ Law for heat capacities follows,

$$\frac{C_V}{N} = \frac{12\pi^4}{5} k\left(\frac{T}{\Theta}\right)^3 \tag{9.26}$$

The Debye model cannot be used directly for the evaluation of the chemical potential. The phonon gas has been enclosed in volume $V$ and

**Table 9.2**

| Substance | Debye $\Theta$ | Substance | Debye $\Theta$ |
|---|---|---|---|
| Al | 426 | Na | 158 |
| Be | 1160 | Cs | 42 |
| Cd | 186 | Diamond | 2200 |
| Cu | 335 | Ice | 192 |
| Au | 165 | NaCl | 321 |
| Pb | 96 | KCl | 235 |
| Ni | 440 | | |

exerts a pressure on this surface. No such surface pressure exists in the true normal-mode analysis and the theory must be modified. Sublimation occurs from the bounding surface and in the absence of the vapor, a flow of momentum and energy should occur toward the surface.

### 9.3.2 *Chemical Potential of a Solid*

It was indicated at the end of the previous section that some care is necessary in estimating the chemical potential of a solid. First, it is assumed that it is possible to express the allowed frequency of oscillations of the solid phase as a function of a triple set of indices $n_x, n_y, n_z$ and use the label $k$ for this set of indices. A sum can be converted to an integral by means of a density-of-states function which is of the form $Ng(\nu, v)$, so that

$$\sum_{\alpha}^{3N} \rightarrow N \int_0^{\infty} d\nu \, g(\nu, v) \tag{9.27}$$

where $v$ is the volume per atom or unit cell. The Helmholtz function becomes

$$A(V, T, N) = N\varepsilon_0 - NkT \ln Q_s + NkT \int_0^{\infty} d\nu \, g(\nu, v) \ln(1 - e^{-\beta h\nu}) \tag{9.28}$$

The form $Ng(\nu, v)$ for the density of states ensures that the free energy is an additive function. In addition, $g(\nu, v)$ also limits the number of normal vibrational modes to three per atom or unit cell in the crystalline solid. Addition of a particle at constant volume $V$ requires $dV = 0 = N\,dv + v\,dN$, and the derivative of the density-of-state function is

$$\frac{\partial g}{\partial N} = \frac{\partial g}{\partial v}\frac{\partial v}{\partial N} = -\frac{v}{N}\frac{\partial g}{\partial v}$$

Denoting the integral in Eq. (9.28) as $\bar{g}$, the chemical potential is

$$\mu_s = \left(\frac{\partial A}{\partial N}\right)_{V,T} = \varepsilon_0 + Pv + kT\bar{g} - kT \ln Q_s$$

and the thermal pressure is

$$P = -\left(\frac{\partial A}{\partial V}\right)_{T,N} = -\frac{\partial \varepsilon_0}{\partial v} - kT\frac{\partial \bar{g}}{\partial v}$$

In this derivation,

$$\frac{\partial g}{\partial V} = \frac{\partial g}{\partial v}\frac{\partial v}{\partial V} = \frac{1}{N}\frac{\partial g}{\partial v}$$

is used. The change in the density of states with volume can be eliminated between these two equations and the chemical potential becomes

$$\mu_s = -q_0 - kT \ln Q_s + kT \int_0^\infty d\nu\, g(\nu, v) \ln(1 - e^{-\beta h\nu}) \qquad (9.29)$$

The zero-point energy is taken as the reference of energy and is not included. The term $q_0 = -(\varepsilon_0 + Pv)$ is the latent heat of sublimation at absolute zero.

For the Debye model, the density-of-states function can be taken as

$$g(\nu, v) \approx 3\left(\frac{4\pi\nu^2}{c^3}\right) v\eta(\nu - \nu_m) \qquad (9.30)$$

and this will now yield a good approximation to the chemical potential. More general density-of-state functions are given in a review article [*2*] and $g(\nu)$ is given in Fig. 9.3[1] for aluminum.

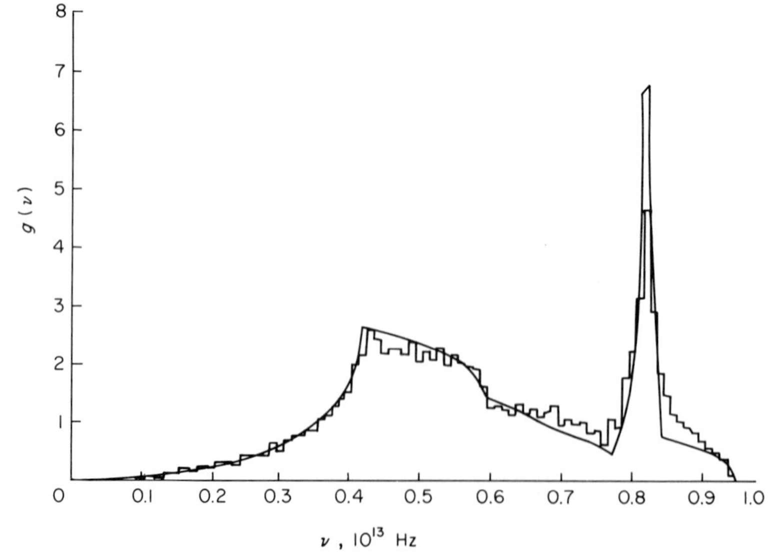


**Fig. 9.3** Approximate frequency spectrum for Al. [From J. C. Phillips, *Phys. Rev.* **104,** 1263 (1956), Fig. 9.]

[1] For original of Fig. 9.3, see Phillips [*2a*].

The physical properties of the substance are included in the latent heat of sublimation at absolute zero $q_0$, the velocity of sound $c$, and the density of states $g(\nu, v)$. This reduces to $q_0$ and $\Theta$ for the Debye model. Thermodynamic considerations which are included in Eq. (9.11) indicate that $q_0$ and the heat capacity at constant pressure $C_P''$ provide more information for the determination of the vapor pressure of the low-density vapor. A major contribution to the chemical potential of a solid is the latent heat of evaporation at absolute zero, or the change in enthalpy for sublimation at absolute zero. If $q_0$ is regarded as the binding energy per particle in the solid phase, then

$$\mu_s \approx -q_0$$

is a good initial approximation. It was noted in Chapter VIII, Exercise 8.18 that to a good approximation, the chemical potential of a nucleus was approximately equal to the binding energy $\mu(A, Z) = -\varepsilon(A, Z)$. Whenever a large binding energy is involved, the chemical potential is, in the first approximation, approximately equal to the binding energy. Conservation of energy in the transport of particles across a boundary is the dominant feature when the binding energy is large.

### *9.3.3 Sublimation Curve*

Equation (9.12) for $\mu_v$ and Eq. (9.29) for $\mu_s$ now yield the equation for the sublimation curve,

$$\mu_v = \mu_s$$

or

$$\ln P = \left(-\frac{q_0}{kT}\right) + \frac{5}{2}\ln T + kT\int_0^\infty d\nu\, g(\nu, v)\ln(1 - e^{-\beta h\nu}) + i_0 + \ln\frac{Q_{vi}}{Q_s} \tag{9.31}$$

This is in many respects similar to the integrated form of the Clapeyron equation (9.10), with Kirchhoff's equation (9.11) for the latent heat of sublimation. Here, $q_0$ is the latent heat of sublimation at absolute zero, $\frac{5}{2}\ln T$ is the contribution to $C_P''$ which would occur for a monatomic gas, and $Q_{vi}$ contains the remaining part of $C_P''$ and other statistical weight aspects. The heat capacity of the solid at constant pressure is contained in the integral over $\ln(1 - e^{-\beta h\nu})$. The term $i_0$ is the previously defined chemical constant. This equation is simpler in appearance than the integral of Clapeyron's equation. Since the latent heat of sublimation is obtained by integrating the heat capacity and then this quantity is integrated once more, double integrals appear. But the heat capacity follows from the free energy by two partial differentiations with respect to temperature. Thus the two integrations return the problem to the free energies and the rather simple form of Eq. (9.31) follows.

Since the latent heat is the dominant term in the equation for the sublimation curve and its value is taken from experiment rather than theory, Eq. (9.31) has only one feature which is not contained in the direct integration of the Clapeyron equation. This is the chemical constant $i_0$ and it is obtained from the gas rather than the solid analysis. It can be obtained from the experimental value of the vapor pressure as a function of temperature and is in good agreement with the value of $i_0$ determined from Eq. (9.13) for many gases. Deviations occur if the gas forms molecules in the vapor, and this occurs for sodium and potassium.

An interesting sublimation curve is the vapor pressure of electrons in equilibrium with a hot filament. Again, the chemical potentials are equal, and

$$\mu_e(\text{solid}) = \mu_e(\text{vapor})$$

Equation (7.15) can be used for the chemical potential in the vapor, and the chemical potential in the solid is approximately

$$\mu_e(\text{solid}) \approx -e\varphi$$

Equation (8.55a) can be used for the chemical potential, but the same zero of energy is needed for both phases and an electron at rest in the vapor is usually selected. If $e\psi$ is the total depth of the potential which holds the electrons in the metal, then the chemical potential of the electrons in the solid is $\mu_e(\text{solid}) = -e\psi + \mu_0 = -e\varphi$. The factor $e$ is the magnitude of the charge of the electron and $\mu_0$ is the Fermi energy. Since the term linear in temperature is usually small, this term is omitted and the chemical potential has a simple form. With these assumptions, the vapor pressure is given by

$$\ln P_e \approx -\frac{e\varphi}{kT} + \frac{5}{2}\ln T + i_0 + \ln 2 \tag{9.32}$$

Exercise 9.1 Use $\Gamma = \frac{1}{4}n\bar{c}$ as the number of electrons striking the metallic surface from the vapor. By detailed balance arguments, this is equal to the number leaving. Show that the thermionic emission current from a hot metallic surface is given by

$$j \approx AT^2 \exp\left(-\frac{e\varphi}{kT}\right) \quad \frac{\text{A}}{\text{m}^2}$$

where $A = 1.2\times 10^6$ A/m$^2$. For tungsten, $\varphi = 4.52$ V and $A = 0.6\times 10^6$ A/m$^2$. The difference in the values of $A$ is usually attributed to different values for $\varphi$ for different faces, and thermionic emission from definite crystals faces appear to have the theoretical value of $A$. The equation for thermionic emission is referred to as "Richardson's equation."

## 9.4 Thermodynamic Systems that Are Composed of Different Particles

In the earlier sections, a thermodynamic system of any composition was discussed in terms of the obvious variables $P$ and $V$ and the additional variables $T$, $U$, and $S$. Other thermodynamic variables were associated with special experiments or experimental arrangements. Thus a system composed of one type of particle or component can occur in many phases. These phases can be separated and the method of separation of the gas, liquid, or solid phase is so obvious that it is not usually discussed. From everyday experience, the liquid or solid is found at the bottom of the container and the liquid flows, and so on. In space or in the absence of a gravitational field, the separation of the phases or the condensation into one very large drop or solid is not quite so easily accomplished. Even so, a membrane of fine screen or a macroscopic membrane would be adequate for separating the phases. Additional variables were needed to describe the amount or number of particles in each phase of the single-component system which was discussed in the previous sections. Experiments were possible on each phase separately and the internal energy, entropy, and so forth of each phase became a measurable quantity.

If a system is composed of different particles, and "different" is used in the sense that there exists a microscopic membrane or semipermeable membrane which can be used to separate the system into various species, then the internal energy, entropy, and so forth become functions of the number of the various particles. The particles may be molecules formed from atoms, atoms, atoms in excited states, electrons, electrons in metals, and even elementary nuclear particles. Usually, the temperature and density determine the type of particle under consideration. The word "component" is used in chemical literature rather than particle.

The internal energy can be defined for a particular phase $i$ of the substance, and is a function of the entropy, volume, and number $N_i^\alpha$ of each component $\alpha$ in this phase, or, under a virtual change,

$$\delta U_i = T_i\,\delta S_i - P_i\,\delta V_i + \sum_\alpha \mu_i^\alpha\,\delta N_i^\alpha \tag{9.33a}$$

The chemical potential of each component in this phase is

$$\mu_i^\alpha = \left(\frac{\partial U_i}{\partial N_i^\alpha}\right)_{S_i, V_i, N_i^\beta} = \mu_i^\alpha(S_i, V_i, N_i^\beta, \ldots) \tag{9.33b}$$

or the change in internal energy at constant entropy, volume, and all other $N_i^\beta$ as the number $N_i^\alpha$ is changed for the $\alpha$th component. If all possible phases are placed in volume $V$ and the system is at equilibrium, then the requirement that the entropy be a maximum

$$\delta S = \sum_i \delta S_i = 0 \tag{9.34a}$$

subject to the constraints

$$\delta U = \sum_i \delta U_i = 0 \tag{9.34b}$$

and

$$\delta V = \sum_i \delta V_i = 0 \tag{9.34c}$$

yields the condition that the temperature and pressure are constant throughout the system.

The Gibbs function is defined for each phase and is a function of $P$, $T$, and $N_i^\alpha$:

$$G_i(P, T, N_i^\alpha, \ldots) = U_i + PV_i - TS_i \tag{9.35}$$

and the change in the Gibbs function is defined by

$$dG_i = -S_i\,dT + V_i\,dP + \sum_\alpha \mu_i^\alpha\,dN_i^\alpha \tag{9.36a}$$

The chemical potential is used whenever possible rather than the Gibbs function. The chemical potential derived from the internal energy is a function of the extensive variables $V$ and $S$ and the particle number $N^\beta$, but the chemical potential derived from the Gibbs function is given in terms of the convenient intensive variables $P$ and $T$ and the particle number

$$\mu_i^\alpha = \left(\frac{\partial G_i}{\partial N_i^\alpha}\right)_{T, P, N_i^\beta} = \mu_i^\alpha(P, T, N_i^\alpha, N_i^\beta, \ldots) \tag{9.36b}$$

The chemical potential for the simple gases discussed in Chapter VII gave the chemical potential in terms of either $V/N$ and $T$ or $P$ and $T$. The proper interpretation of the chemical potential as a thermodynamic function is often difficult, and it is useful to reconsider the problem in terms of $U$, $G$, $S$, .... If $G$ is integrated at constant temperature and pressure, the Gibbs function becomes

$$G_i = \sum_A N_i^A \mu_i^A \tag{9.37}$$

for each phase. The phases can be summed for the total $G$. If $dG_i$ is formed with Eqs. (9.37) and (9.36a) and then Eq. (9.33a) is used for $dU_i$, it follows that

$$-S_i\, dT + V_i\, dP - \sum_A N_i^A\, d\mu_i^A = 0 \tag{9.38}$$

This is one form of the Gibbs–Duhem relationship and for a single component this reduces to Eq. (6.31).

### 9.4.1 *Conservation of Particles and the Phase Rule*

In the statistical approach, the constraints on the particle number introduce the chemical potential. In this present problem, the general constraint is that the number of particles or the amount of a particular component is conserved, that is,

$$\delta N^\alpha = \sum_i \delta N_i^\alpha = 0 \tag{9.39a}$$

Again the method of Lagrangian multipliers can be used and at constant temperature and pressure, Eq. (9.34a) for $\delta S$ at constant $P$ and $T$ becomes

$$\sum_i (\mu_i^\alpha - \lambda^\alpha)\, \delta N_i^\alpha = 0 \tag{9.39b}$$

Now the $\delta N_i^\alpha$ can be treated as independent variables and,

$$\mu_i^\alpha = \lambda^\alpha, \qquad \alpha = 1, \ldots, \eta; \quad i = 1, \ldots, n \tag{9.39c}$$

The chemical potential of a component is the same for each phase. If there are $\eta$ components and $n$ phases, then there are $\eta n$ values of $N_i^\alpha$. Each $\mu_i^\alpha$ is a function of $P$, $T$, and the remaining values of $N_i^\beta$ in the phase, or $2 + (\eta - 1)$ variables. The set of the $\mu_i^\alpha$ depend on $2 + n(\eta - 1)$ variables, but Eq. (9.39c) represents a system of $\eta(n - 1)$ equations which relate these variables. For a nontrivial solution of these equations, the number of equations $\eta(n-1)$ must be less than or equal to the number of variables $2+n(\eta-1)$, or

$$\eta + 2 \geq n \tag{9.40}$$

This is known as the Gibbs Phase Rule and states that in a system which is composed of $\eta$ components, not more than $\eta+2$ phases can exist in equilibrium. In Section 9.2, a single-component system was considered and three phases could coexist at the triple point.

### *9.4.2 Chemical Equilibrium*

A chemical reaction between components is often written as

$$\mathrm{A} + \mathrm{B} \rightleftarrows 2\mathrm{C} + \mathrm{D} + \cdots$$

or more generally as

$$\sum_{\mathrm{A}} \nu_{\mathrm{A}} \mathrm{A} = 0 \tag{9.41a}$$

where $\nu_{\mathrm{A}}$ is a coefficient for species A, . . . , and is a positive or negative integer. In the example given, $\nu$ has values $+1, +1, -2, -1, \ldots$, respectively. From the previous discussion, equilibrium conditions require constant temperature and constant pressure. For convenience, the index $\alpha = \mathrm{A}$ denoting the component is used as a subscript, and

$$\sum \mu_{\mathrm{A}}\, \delta N_{\mathrm{A}} = 0 \tag{9.41b}$$

The constraint on the virtual change $\delta N_{\mathrm{A}}$ is given by the reaction Eq. (9.41a), or

$$\frac{\delta N_{\mathrm{A}}}{\nu_{\mathrm{A}}} = \frac{\delta N_{\mathrm{B}}}{\nu_{\mathrm{B}}} = \frac{\delta N_{\mathrm{C}}}{\nu_{\mathrm{C}}} = \cdots = \text{constant} \tag{9.41c}$$

Substituting these conditions into Eq. (9.41b) yields the equation for chemical equilibrium

$$\sum_{\mathrm{A}} \mu_{\mathrm{A}} \nu_{\mathrm{A}} = 0 \tag{9.42}$$

In Chapter VII, the chemical potential for the gaseous phase in the low-density region was used to discuss many simple reactions. The equation given here has greater generality, but the chemical potential as a function of all components is not known in general.

The reaction isochore which is used by chemists can be introduced by writing Eq. (7.15) as $\mu = \mu_0 + kT \ln P$ and then writing Eq. (9.42) as $\sum (\mu_{0A} + \ln P_A)\nu_A = 0$. The equilibrium constant $K_P$ at constant pressure is defined by

$$kT \ln K_P = \sum_A \nu_A \ln P_A = -\sum \nu_A \mu_{0A} \tag{9.43a}$$

This can be related to the heat of reaction $q_P = -\sum \nu_A h_A$ or enthalpies of the components by the relationship $[\partial(\mu/T)/\partial T]_P = h$ and then the reaction isochore follows:

$$\frac{\partial(\ln K_P)}{\partial T} = \frac{q_P}{kT^2} \tag{9.43b}$$

$q_P$ can be determined from experimental data.

## 9.5 The Third Law of Thermodynamics

In most specific problems discussed in this text, the statistical method is used for evaluation. Thermodynamics is used as a guide for both simple and complex systems. As absolute zero is approached, the quantum statistical system that is based on a set of discrete states tends toward a definite nondegenerate ground state. Since each state has weight unity, the statistical approach requires that the entropy vanish as the temperature tends toward zero for these simple systems.

The First and Second Laws of Thermodynamics do not give any information regarding the approach of a system toward absolute zero, although these two laws can be used to define the absolute zero of temperature. Nernst [*2b*], in order to predict the equilibrium conditions of chemical reactions, put forth in 1906 the Heat Theorem stating that the entropy change in a chemical reaction tends to vanish as the temperature approaches absolute zero. Ensuing research and discussions by Giauque [*2c*] and by Simon [*2d*] helped to clarify this statement and has given rise to the following statement of the Third Law of Thermodynamics.

*Third Law of Thermodynamics*

> The contribution to the entropy of a system by each aspect in internal thermodynamic equilibrium tends to zero at absolute zero.

A statement frequently attributed to Planck is that

$$\lim_{T \to 0} S = 0 \tag{9.44}$$

An immediate consequence of the Third Law is that the heat capacity, which is related to the entropy by the Second Law:

$$C_V = T\left(\frac{\partial S}{\partial T}\right)_V \xrightarrow[T\to 0]{} 0 \tag{9.45a}$$

must tend to zero as the temperature tends toward zero. Since $(\partial V/\partial T)_P = -(\partial S/\partial P)_T$, the coefficient of thermal expansion vanishes:

$$\lim_{T\to 0}\left(\frac{\partial V}{\partial T}\right)_P = 0 \tag{9.45b}$$

and from a similar relationship,

$$\lim_{T\to 0}\left(\frac{\partial P}{\partial T}\right)_V = 0 \tag{9.45c}$$

It is difficult to illustrate the importance of the Third Law of Thermodynamics in a discussion that uses the statistical approach. Most of the conclusions are inherent in the definition of a state and the weight of unity attached to the state. Thus the entropy of the ideal Fermi and Bose gases, the vibrational excitations of a solid, the magnetic systems, and so on, all tend toward zero as the temperature tends toward zero. The relationships for the heat capacity and thermal expansion are obeyed for these simple systems.

Some of the experimental evidence for the Third Law of Thermodynamics is now reviewed.

(a) The heat capacity of nonmagnetic solids decreases as $\alpha T + \beta T^3$ at low temperatures, where the linear term is the contribution of the electrons in metals and the $\beta T^3$ is characteristic of the vibrational modes in metallic and dielectric solids.

(b) The linear coefficient of expansion of solids tends toward zero as absolute zero is approached. An equation similar to Eq. (9.45c) occurs for the surface tension $\sigma$, $dW = -\sigma\, dA$,

$$\lim_{T\to 0}\left(\frac{\partial \sigma}{\partial T}\right)_A = 0 \tag{9.45d}$$

and $^3$He provides an interesting example. The surface tension [3] of liquid $^3$He as a function of temperature is shown in Fig. 9.4 and the slope approaches zero at the lowest temperatures.

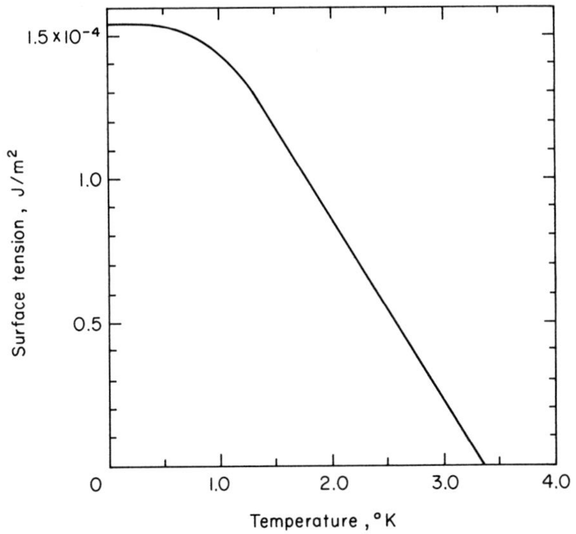


**Fig. 9.4** Surface tension [3] of liquid $^3$He.

(c) Other examples occur as other ultimate dynamic variables such as $B$ and $M$ are included with $P$ and $V$. For magnetic phenomena,

$$dG = -S\,dT + V\,dP - M\,dB$$

and from the relationship between coefficients,

$$\left(\frac{\partial M}{\partial T}\right)_B = \left(\frac{\partial S}{\partial B}\right)_T$$

one can show that the derivative of either the magnetization or the magnetic susceptibility with respect to temperature must tend toward zero as the temperature tends to zero. This is quite apparent as the magnetic materials undergo a phase transition to either ferromagnetic or antiferromagnetic states. Thermodynamics does not indicate the temperature at which ordering begins, but in paramagnetic salts, the Curie point occurs at temperatures of less than 0.01°K, while for pure iron, it occurs at 1000°K.

For superconducting materials, the difference in Gibbs function between the superconducting and normal phases is given by $G_n - G_s = \frac{1}{2}(H_c^2 - H^2)\mu_0$ and the entropy difference by $-(\partial G/\partial T)_B$, or

$$S_n - S_s = \mu_0 H_c \left(\frac{\partial H_c}{\partial T}\right)_P$$

where $H_c(T)$ is the critical field above which the metal is normal. Since the entropies tend toward zero by the Third Law, this implies that $\partial H_c/\partial T \to 0$ and this is in accord with experimental observations. The critical field as a function of temperature is shown for lead [4] in Fig. 9.5.

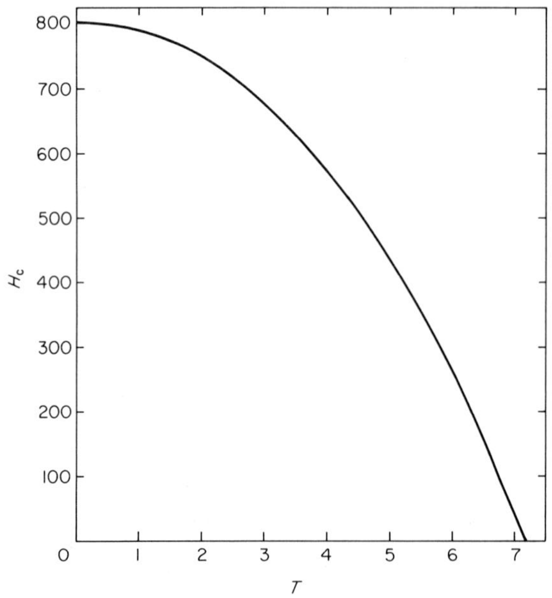


**Fig. 9.5** Critical magnetic field strength $H_c$ as a function of the temperature $T$ for lead. The units of $H_c$ are shown in Oersteds or $(10^3/4\pi)$ $H_c$ ampere-turns/meter. [Data for curve is taken from D. L. Decker, D. E. Mapother, and R. W. Shaw, *Phys. Rev.* **112**, 1888 (1958).]

Using $\xi\, dZ$ for the transport of charge, one can show that $d\xi/dT \to 0$ for a thermocouple.[2]

(d) The Third Law permits an experimental determination of the absolute value of the entropy as a pure substance is heated from absolute zero or a very low temperature to the boiling point. Statistical considerations permit a calculation of the entropy of the same gas from the Helmholtz function given by Eq. (7.4b) or Exercise 9.4. The experimental values for $N_2$ are [6] given in Table 9.3.

Calculations with Eq. (7.4b) yield $18.21R$ and the data are given in terms of the gas constant $R$. The calculations are made using Eq. (7.4b),

[2] See Wilks [5] for a very complete discussion of the Third Law of Thermodynamics.

**Table 9.3**

| | |
|---|---|
| 0–10°K (Debye) | $0.229R$ |
| 10–35.6°K | $3.017R$ |
| Transition at 35.61°K | $0.768R$ |
| 35.61–63.14°K | $2.794R$ |
| Melting | $1.365R$ |
| 63.14–77.32°K | $1.364R$ |
| Vaporization | $8.613R$ |
| | $18.15R$ |
| Correction for nonideality of gas | $0.11R$ |
| | $18.26R$ |

the mass of $N_2$, $k$ and $h$, and the rotational and vibrational constants which are given in Table 7.3. Even though it is expected, it still seems remarkable that the experimental entropy change which is determined from the Debye $\Theta$ from 0 to 10°K, the measured heat capacity of the solid from 10°K to the phase transition at 35.61°K, the latent heat of this phase transition, the heat capacity of the new solid phase from 35.61°K to the melting point at 63.14°K, the latent heat of melting, the heat capacity of the liquid, and the latent heat of vaporization at 77.32°K are in excellent agreement with the statistical calculation for the gas which seems to be composed of almost unrelated parameters. Data for substances like Ar, $O_2$, $Cl_2$, HCl, HD, and $CH_4$ provide the same excellent agreement between experiment and statistical calculations [5].

(e) Equation (9.42) provides the thermodynamic description for equilibrium among the chemical components. In the low-density gases, Eq. (7.15) can be used for the chemical potential of the gaseous components. This procedure was used for many different reactions in Chapter VII and the agreement with experimentally observed pressure data must be regarded as evidence for the validity of the Third Law. It was noted in these earlier discussions that even the statistical weight which was introduced by the electron spin or the small mass of the electron in the translational partition function could have considerable influence on the relative concentrations or pressures. In the reaction isochore given by Eq. (9.43b), the heat of reaction $q_P$ can be determined from experimental data and the temperature dependence of $K_P$ follows from the First and Second Laws. Nernst put forward the Heat Theorem in order to obtain an absolute value for $K_P$. It may be noted that the statistical approach yields the absolute value directly and the value of $\mu_{0A}$ and the gas entropy are taken from spectroscopic and other known physical data. This provides a more accurate value of $K_P$ than by calorimetric methods.

(f) The chemical constant of a vapor can be determined experimentally from the measurement of the vapor pressure as a function of temperature and the heat capacities of the solid and the vapor. Experimentally determined chemical constants agree with the $i_0$ given by Eq. (9.13), and again indicate the validity of the Third Law.

(g) Experimental values of the entropy of allotropic solids illustrate that if the entropy of one pure phase is zero, the entropy of the other pure phase is zero as the absolute zero of temperature is approached.

As an example, the entropy change of rhombic sulfur [7] as it is heated from 0 to 368.5°K is $4.40R$ and the entropy change for the phase transition from rhombic to monoclinic sulfur is $0.130R$. The total change in entropy as monoclinic sulfur is heated from 0 to 368.5°K is $4.52R$ and this is in excellent agreement with the sum of the values for rhombic sulfur. Similar experimental data [5] exist for tin, cyclohexanol, and phosphine.

The conversion of graphite to diamond is an exotic example of the use of the Third Law of Thermodynamics. The two phases of carbon are related by the reaction

$$\text{C(diamond)} \rightarrow \text{C(graphite)}$$

The heat of combustion of diamond to graphite, or the enthalpy difference, is $\Delta H = -1880$ J/g-mole at 298°K. The entropy of graphite at 298.15°K is 5.74 J/°K-g-mole and the entropy of diamond is 2.43. The difference in the Gibbs function is $\Delta G = \Delta H - T\Delta S = -2860$ J/g-mole. From this value and the relationship $(\partial G/\partial P)_T = V$, it follows from the difference in volume $\Delta V$ between diamond and graphite that $\Delta G$ becomes zero at a pressure of the order of 20,000 atm. It has been suggested [8] that diamond is the stable form at room temperature and a pressure of 15,000 atm. Since diamond is metastable at room temperature under atmospheric pressure and the conversion to graphite must exceed millions of years, the inverse process must be equally slow and diamonds cannot be made by mere application of pressure to graphite. Commercial diamonds are made by the General Electric Co. and it is suggested that temperatures as high as 3000°K and pressures as high as 100,000 atm are used in the preparation. Normally, the diamonds have been black, but during the past year, General Electric has produced clear, one-carat diamonds.

The condition of zero entropy at absolute zero may not be met unless the substance is a perfect crystalline material. The atomic arrangement in a glassy solid is similar to that of a liquid and a classic study of glycerine [*9*] indicated that the entropy of the supercooled liquid or glassy glycerine at absolute zero was $2.8R$ greater than crystalline glycerine. A disorder entropy is expected in the statistical approach of the order of $R \ln \eta$, where $\eta$ is a disorder parameter. Some pure substances which do not have zero entropy at absolute zero are [*5*] CO, $N_2O$, $H_2O$, $P_2O$, and NO. Spectroscopic data indicated that the entropy of CO at 298.15°K was $23.6R$, however the heat capacity measurements [*10*] indicated an entropy of $23.1R$. This excess entropy of $0.5R$ at absolute zero for solid CO is explained by observing that a CO molecule may be placed on a crystalline site with the orientation CO or OC relative to its neighbors,

CO OC OC CO

CO OC CO

OC CO CO OC

Random orientation on $N$ sites would indicate $2^N$ equivalent arrangements or an entropy contribution of $R \ln 2$. This value of $0.7R$ is comparable to the observed excess entropy. The explanation of the residual entropy of $0.4R$ for $H_2O$ molecules in ice is somewhat more complex [*11*].[3] The molecule $CH_3D$ has four equivalent positions for the D atom in the solid and this yields $4^N$ equivalent arrangements or a configurational entropy of $R \ln 4$. This is very close to the experimental difference [*12*] of $1.37R$.

These studies illustrate that the configurational entropy of the substance can be "frozen in" if the temperature is lowered more rapidly than reorientation of the molecules can occur. In a sense, the system becomes a large molecule with this particular configuration and with vibrational modes. During the cooling process, the energy constraint acts as a semipermeable membrane for the various configurations, but the rate at which the process of selection occurs may be too slow to permit the selection of the configuration of greatest order and the configuration of some higher temperature becomes the permanent arrangement at lower temperatures.

The atomic nuclei contribute by virtue of their nuclear spin states $I$ and their difference in isotopic mass. At sufficiently low temperatures, some state of order can be expected for the nuclear spin states, but above this temperature, their contribution to the solid-phase and to the vapor-phase entropies is $R \ln(2I + 1)$ and the effect of nuclear spin cancels. Isotopic mass differences contribute to the entropy since isotopes like $^{35}Cl$ and $^{37}Cl$

[3] For $D_2O$, see Long and Kemp [*11a*].

can be separated by a mass spectrometer. In solid sodium chloride, there is no tendency for the two isotopes to separate into different spatial regions and the entropy of mixing is the same as in the vapor phase of sodium chloride. For molecules, the formation of homonuclear molecules like $^{35}Cl_2$ and $^{37}Cl_2$ and heteronuclear molecules like $^{35}Cl^{37}Cl$ has an effect on the rotational partition function. The homonuclear molecules have only one-half as many rotational states as the heteronuclear molecules and there is an entropy difference of $R \ln 2$. Since the transition rate for

$$^{35}Cl_2 + {}^{37}Cl_2 \rightarrow 2\,{}^{35}Cl^{37}Cl$$

is negligible in both the vapor and solid phases, the relative concentration of each molecular species remains fixed and they can be treated as distinct species in both the solid and vapor phases. The mixing entropy is the same in both phases.

*Ortho*- and *para*-hydrogen were treated as distinct species in Chapter VII when the rate of temperature change was faster than the rate of *ortho–para* conversion. The solid and liquid phase have the "frozen-in" *ortho–para* concentration of the vapor phase. The lowest rotation state of *ortho*-hydrogen has $J = 1$ and a low-temperature residual entropy of $R \ln 3$. This aspect is discussed in Exercise 9.7. The heat capacity anomaly [*13*], which is introduced by the removal of this configurational aspect, is shown in Fig. 9.6 for two different concentrations of *ortho*-hydrogen.

**Exercise 9.2** Show that $C_P = 0$ if $C_V = 0$ as $T = 0$.

**Exercise 9.3** From what relationship does Eq. (9.45c) follow?

**Exercise 9.4** Find the expression for the entropy of a gas using Eq. (7.4b) and show that it is consistent with the value obtained from Eq. (9.12) and $(\partial\mu/\partial T)_P = -s$.

$$\text{Ans.}\quad \frac{S}{R} = \frac{5}{2} + i_0 + \ln\left(T^{5/2}\frac{Q_{\text{int}}}{P}\right) + T\frac{\partial(\ln Q_{\text{int}})}{\partial T} \tag{9.46}$$

**Exercise 9.5** The measured entropy of argon at the boiling point is $15.42R$. Determine the entropy of the vapor at the boiling point from Eq. (7.4b) and compare with the above value.

**Exercise 9.6** Compare the entropy of pure *para*-hydrogen gas at the boiling point (20.26°K) with the experimental value of $7.4R$.

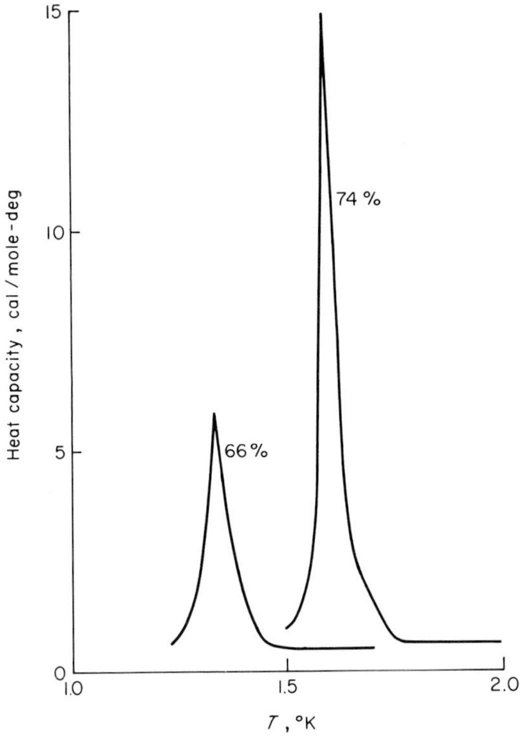


**Fig. 9.6** Heat capacity of solid hydrogen [*13*] with 66% and 74% *ortho*-hydrogen mixed with *para*-hydrogen. [From R. W. Hill and B. W. A. Ricketson, *Phil. Mag.* **45**, 277 (1954), Fig. 3.]

Exercise 9.7 Solid *ortho*-hydrogen has a sharp peak in the heat capacity near 2°K and this peak does not occur for *para*-hydrogen. The excess entropy from this peak is of the order of $R \ln 3$. The nuclear spin for *ortho*-hydrogen is $I = 1$ and at a lower temperature, another anomaly in the heat capacity should occur and contribute another $R \ln 3$. Show that the entropy of *ortho*-hydrogen at the boiling point at 20.26°K is consistent with the expected experimental value of $7.4R + 2R \ln 3$.

### *9.5.1 The Unattainability of Absolute Zero*

It was pointed out by Nernst [*14*] that the Third Law of Thermodynamics implies that it is impossible to cool a system to absolute zero. This implies that the ground state cannot be reached by a thermodynamic process. Some care is needed in the use of this statement, since nuclei with zero nuclear spin are in their ground state and atoms are in their lowest electronic configuration and this configuration can have zero total angular momentum. Thus the author prefers to regard the Third Law as applying to the ultimate dynamic variables, or the variables which appear in the definition of the work $dW$, that is, pressure and volume $(P, V)$, magnetization and magnetic field $(M, B)$, electric field and electric charge $(\xi, Z)$, and surface tension and surface area $(\sigma, A)$. Liquid mixtures of $^3$He and $^4$He require that osmotic pressure be included with this group. The entropy as a function of temperature and one of the intensive variables $P, M, \sigma, \ldots$ is shown in Fig. 9.7.

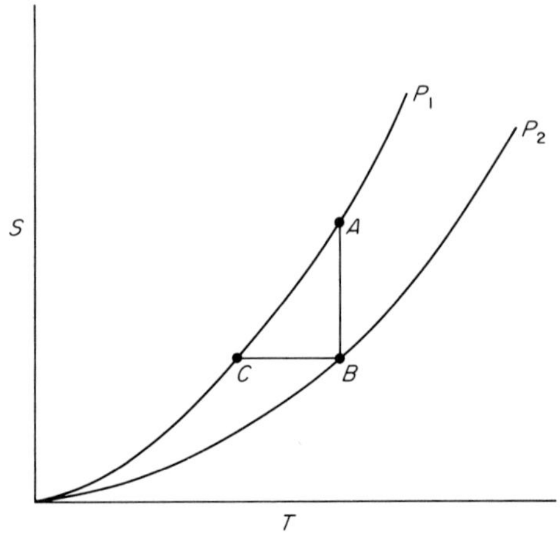


**Fig. 9.7** Entropy–temperature curves for two values of the parameter $P$. A sequence of constant-temperature, constant-entropy paths between these two curves cannot be used to reach absolute zero.

The Third Law requires that the entropy approach zero along both values of the intensive variable. The entropy is reduced along the isotherm $AB$ by changing $P_1$ to $P_2$. Then, the temperature is reduced at constant entropy along path $BC$ by changing the intensive variable from $P_2$ to $P_1'$. This process can be continued and it is quite apparent that the changes become smaller and smaller but the absolute value of zero is not reached.

### 9.5.2 *Negative Temperature*

In the discussion in Chapter III of the throwing of a die subject to a constraint, it became apparent that if the average number facing up exceeded $\frac{1}{2}(6 + 5 + 4 + 3 + 2 + 1)$, the problem required a negative value of $\beta$.

This is characteristic of any problem with a finite set of states if the average energy exceeds one-half the value without the constraint. In the counting of configurations in Chapter III, this was a natural conclusion and led to a canonical ensemble with $-\infty < \beta < \infty$. For an infinite set of states, only positive values of $\beta$ are permitted. In lasers, it is possible to place more atoms in an upper state than in a lower state by some external action and the occupation of the states is inverted from the Boltzmann distribution. A negative value of $\beta$ is often used to describe this inversion and the concept of "negative temperature" is often introduced.

An experiment by Pound, Ramsey, and Purcell is often regarded as having produced a negative temperature [*15*]. In pure lithium fluoride, it is found that hours are required for the nuclear spin system to come into thermal equilibrium with the lattice. The spin system can be isolated from the lattice, but thermal equilibrium can occur within the spin system itself. The lithium fluoride crystal, with a spin of $I = \frac{3}{2}$, is placed in a magnetic field. From the discussion in Chapter VII, the nuclear energy levels are separated into four states by the magnetic field and then take up a Boltzmann distribution as thermal equilibrium with the lattice occurs. The magnetic field is reversed in a time short compared with the frequency spacing between energy levels, $\Delta t \ll \nu = \gamma_I B/h$, and the spins are unable to follow the field reversal, and remain in their original orientation. This is the quantum mechanical problem of nonadiabatic passage. Since the axis of quantization is reversed by reversing the field, the process is equivalent to changing the magnetic quantum number $m \to -m$ and changing the level occupation $n(m)$ to $n(-m)$. Thus the level occupation is inverted and is described by changing $\beta \to -\beta$. This introduces the concept of negative temperature. In the subsequent transfer of energy to the lattice, the value of $\beta$ passes through zero or infinite temperature and then cools to the normal lattice temperature.

## 9.6 Binary Mixtures

From the discussion that leads to Eq. (9.39c), the properties of binary mixtures are described in terms of the chemical potentials $\mu_i^A(P, T, N^A, N^B)$ and $\mu_i^B(P, T, N^A, N^B)$ for a system composed of components A and B. The phase rule, which is stated in Eq. (9.40), permits as many as four phases for this binary system. In liquid or solid mixtures, the interactions are quite strong and thermodynamic expressions and experimental data provide most of the useful information. Statistical methods can be used for dilute systems, but for physical systems of interest, concentrations of greater than 1% usually occur and binary interactions become quite important. In Section 9.7, some aspects of a statistical problem for higher concentrations will be discussed. Since thermodynamics is so important in a binary mixture, some general features are introduced prior to the discussion of specific systems.

A readily measurable quantity is the partial molal volume. The volume of the unmixed components and the volume of the same components mixed are readily measurable quantities at constant pressure and temperature. The volume per particle $v^\alpha$ can be determined, and

$$v^{\mathrm{A}}(P, T, N^{\mathrm{A}}, N^{\mathrm{B}}) = \left(\frac{\partial V}{\partial N^{\mathrm{A}}}\right)_{P,T,N^{\mathrm{B}}} = \left(\frac{\partial \mu^{\mathrm{A}}}{\partial P}\right)_{T,N^{\mathrm{A}},N^{\mathrm{B}}} \tag{9.47a}$$

is the volume per particle in the solution. The deviation from the volume $v^{\mathrm{A}}(P, T)$ of the pure component is a measure of the interactions between particles and an important physical quantity. The volume of a sodium chloride–water solution as a function of the number of moles of water $n_1 = 55.51$, and of the number of moles $n_2$ of NaCl is given by

$$V = n_1 v_1 + n_2 v_2 = 55.51 v_1 + 16.4 n_2 + 2.5 n_2^2 - 1.2 n_2^3$$

The partial molar volume of the NaCl in water is given by $\partial V/\partial n_2$ and

$$v_2(n_1, n_2) = 16.4 + 5 n_2 - 3.6 n_2^2$$

Direct substitution yields $v_1(n_1, n_2)$ for water. These equations emphasize that the volume per particle or per mole is a function of $n_1$ and $n_2$ or of the concentration. If the total volume decreases with the addition of the solute, then $v^{\mathrm{A}}$ can be negative. This occurs for solutions of anhydrous magnesium sulfate and nickel sulfate in water and indicates some care is necessary in the interpretation of partial molal quantities. If hydrated salts are used and $n_{\mathrm{A}}$ refers to the amount of hydrated salt, then $v^{\mathrm{A}}$ is positive and indicates the formation of a hydrated complex in the solution. $v^{\mathrm{A}}$ is positive for most mixtures and the volume dependence of a $^3$He–$^4$He mixture is shown in Fig. 9.8 as an example. Other physically measurable quantities are the compressibility $\partial V/\partial P$ and the expansion coefficient $\partial V/\partial T$, where the volume is a function of the variables $V(P, T, N^{\mathrm{A}}, N^{\mathrm{B}})$. As A and B are mixed to form a single-phase solution at constant temperature and pressure, there is a heat of mixing, or the integral heat of solution, which can be

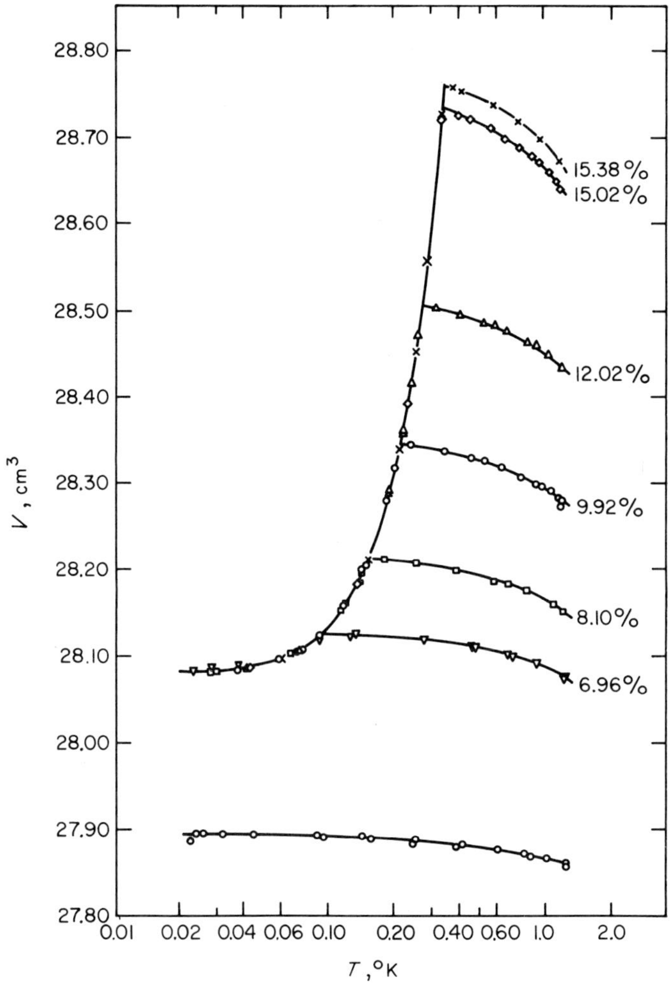


**Fig. 9.8** The gram-molar volume of dilute solutions of $^3He$ in $^4He$ at the saturated vapor pressure is shown. The lowest curve is the molar volume of pure $^4He$ as a function of temperature. If 0.12 moles of $^3He$ are mixed with 0.88 moles of $^4He$, the volume as a function of temperature is shown by the 12% curve. As the temperature is decreased, phase separation occurs near 0.3°K and this is indicated by the intersection with the two-phase curve. Further cooling results in a pure $^3He$ phase and a decreasing concentration of $^3He$ in the binary phase, and the two-phase curve gives the volume of these two phases. At the lowest temperature, the binary solution approaches a concentration of 6.4% of $^3He$. [From D. O. Edwards, E. M. Ifft, and R. E. Sarwinski, *Phys. Rev.* **177**, 380 (1969).]

measured in an elementary manner. The partial molal enthalpy is defined as the heat added to the mixture as $\Delta N^{\mathrm{A}}$ particles are added to a given phase,

$$h_i^{\mathrm{A}}(P, T, N^{\mathrm{A}}, N^{\mathrm{B}}) = \left(\frac{\partial H_i}{\partial N^{\mathrm{A}}}\right)_{\mathrm{P,T},N^{\mathrm{B}}} \tag{9.47b}$$

and can be obtained by adding increments of $\Delta N^{\mathrm{A}}$ to the solution or from a knowledge of the integral heat of mixing as a function of concentration. Extensive variables like volume and enthalpy can be expressed in terms of these partial quantities as

$$V = N^{\mathrm{A}}v^{\mathrm{A}} + N^{\mathrm{B}}v^{\mathrm{B}}$$

$$H = N^{\mathrm{A}}h^{\mathrm{A}} + N^{\mathrm{B}}h^{\mathrm{B}}$$

and the expressions are not as simple as they appear, since $v$ and $h$ are functions of $P$, $T$, $N^{\mathrm{A}}$, and $N^{\mathrm{B}}$.

The chemical potential can be expressed as

$$d\mu^{\mathrm{A}} = +v^{\mathrm{A}}\,dP - s^{\mathrm{A}}\,dT + \left(\frac{\partial \mu^{\mathrm{A}}}{\partial N^{\mathrm{B}}}\right)_{\mathrm{P,T},N^{\mathrm{A}}} dN^{\mathrm{B}} \tag{9.47c}$$

where $\mu^{\mathrm{A}}$, $v^{\mathrm{A}}$, and $s^{\mathrm{A}}$ are functions of $P$, $T$, $N^{\mathrm{A}}$, and $N^{\mathrm{B}}$ for the binary system. If $G$ is expressed as $G = H + T(\partial G/\partial T)$ and the partial derivative taken with respect to $N^{\mathrm{A}}$, it follows that the enthalpy per particle and the chemical potential are related by

$$h^{\mathrm{A}} = \mu^{\mathrm{A}} - T\left(\frac{\partial \mu^{\mathrm{A}}}{\partial T}\right)_{\mathrm{P},N^{\mathrm{A}},N^{\mathrm{B}}} \tag{9.47d}$$

Each of these equations is valid for each phase and for the index B interchanged with A. A partial entropy is implied in these equations and is defined in terms of the chemical potential as

$$s^{\mathrm{A}}(P, T, N^{\mathrm{A}}, N^{\mathrm{B}}) = \left(\frac{\partial \mu^{\mathrm{A}}}{\partial T}\right)_{\mathrm{P},N^{\mathrm{A}},N^{\mathrm{B}}} \tag{9.47e}$$

With this notation, the entropy of a phase is expressed as

$$S_i(P, T, N^{\mathrm{A}}, N^{\mathrm{B}}) = N_i^{\mathrm{A}}s_i^{\mathrm{A}} + N_i^{\mathrm{B}}s_i^{\mathrm{B}}$$

with similar expressions for other extensive variables.

### 9.6.1 Gas Mixtures

The chemical potential of a particular component in a low-density gas is given by Eq. (7.15) and many examples were discussed in Chapter VII for noninteracting and reacting components. If the two components do not interact, the system of particles denoted by the labels A and B form an ideal gas mixture. From the statistical point of view, the number of arrangements with the same energy is given by

$$\{\text{number of arrangements}\} = \frac{(N_A + N_B)!}{N_A! N_B!}$$

and using the statistical concepts introduced in Chapter III, the entropy of the AB mixture is approximately

$$\begin{aligned} S_{AB} &= k \ln \frac{(N_A + N_B)!}{N_A! N_B!} \\ &\approx k[(N_A + N_B)\ln(N_A + N_B) - N_A \ln N_A - N_B \ln N_B] \\ &\approx -k(N_A + N_B)[c_A \ln c_A + c_B \ln c_B] \end{aligned} \tag{9.48}$$

where

$$c_A = \frac{N_A}{N_A + N_B}$$

is the concentration. This is the "entropy-of-mixing" term.

It is now shown that the Gibbs function, which is given by

$$G = N_A \mu_A + N_B \mu_B$$

yields the same result when the chemical potential given by Eq. (7.15) in terms of $\mu(V, T, N_A)$ is expressed as $\mu(P, T, c_A)$. By replacing the volume $V$ with $V = (N_A + N_B)kT/P$ and by using direct substitution, the Gibbs function can be written as

$$\begin{aligned} G = &+kT(N_A + N_B) \\ &\times \left[ \{c_A \ln c_A + c_B \ln c_B\} - \ln \frac{P}{kT} - c_A \ln(\lambda_A^{-3} Q_{iA}) - c_B \ln(\lambda_B^{-3} Q_{iB}) \right] \end{aligned} \tag{9.49}$$

The entropy is given by

$$S = -\left(\frac{\partial G}{\partial T}\right)_{P,T,N_A,N_B}$$

and the term within the curly brackets is the same entropy-of-mixing per particle as was given by Eq. (9.48.) It should be noted that the mixing aspect is only apparent in the $P$, $T$ variables. Thus the chemical potential of each component can be written as $\mu^{A}(V, T, N_{A})$ or as $\mu^{A}(P, T, c_{A})$. The chemical potential of each component in an A–B mixture can be expressed as suggested in Eq. (7.15) or (9.12) as

$$\mu^{A}(P, T, c_{A}) = \mu_{0}{}^{A}(P, T) + kT \ln c_{A} \tag{9.50}$$

where $\mu_{0}{}^{A}$ is the chemical potential of the pure component. Most simple explanations of mixtures and solutions are based on this equation. It is used even when $\mu_{0}{}^{A}(P, T)$ for the pure component applies to liquids. It seems to be an excellent approximation whenever either $c_{A} \ll c_{B}$ or $c_{B} \ll c_{A}$ in the nonideal systems and this will become apparent in subsequent sections.

A mixture of an ideal Fermi gas and an ideal Bose gas requires special consideration as the temperature is reduced. Since the gases are regarded as noninteracting but occupying the same volume, the grand canonical generating function

$$\mathscr{Q}(V, \beta, z_3, z_4) = \mathscr{Q}_3(V, \beta, z_3)\, \mathscr{Q}_4(V, \beta, z_4)$$

is the product of the two generating functions. The indices 3 and 4 are used since $^{3}$He and $^{4}$He are the usual isotopes of interest. The degenerate form of $\mathscr{Q}$ was discussed in Sections 8.7 and 8.8 for the Fermi and Bose gases and the thermodynamic properties follow immediately from $\ln \mathscr{Q}$. The properties are additive and

$$P = P_3 + P_4, \qquad U = U_3 + U_4$$

and so forth. Only when the Gibbs function

$$G = \beta(N_3 \ln z_3 + N_4 \ln z_4) = N_3\mu_3 + N_4\mu_4$$

is expressed in terms of pressure $P$ and $T$ does a mixing term appear in $G(P, T, N_3, N_4)$ or the entropy $S(P, T, N_3, N_4)$. It is interesting to observe that the entropy of mixing tends toward zero as the temperature tends toward zero for this ideal system and is in accord with the Third Law of Thermodynamics. As absolute zero is approached, almost all of the contribution to the pressure and energy is due to the Fermi component.

Exercise 9.8 Show that

$$\mu(P_2, T) - \mu(P_1, T) = kT \ln \frac{P_2}{P_1} \tag{9.51}$$

for a low-density gas.

**Exercise 9.9** Since most of the problems of interest were treated in Chapter VII, an interesting and almost trivial problem is given. The rate of flow of He gas through the thin rubber of a child's balloon is 50 times greater than the flow of $N_2$. For convenience, assume there is no flow of $N_2$ and the balloon has some features of a semipermeable membrane. Let the balloon be filled with one liter of $N_2$ at 1 atm pressure and then placed in a very large container of He gas at 1 atm. Find the equilibrium pressure inside the balloon using very simple arguments and then by using the equality of the chemical potential of the He inside and outside the balloon, that is,

$$\mu_{\mathrm{He}}(\text{inside}) = \mu_{\mathrm{He}}(\text{outside})$$

Ans. $P \approx 2$ atm

**Exercise 9.10** Use $S = (U + PV - G)/T$ and the expressions from Sections 8.7 and 8.8 to show that the entropy of mixing at zero temperature is zero for a mixture of ideal Fermi and Bose gases.[4]

9.6.1.1 Generating Function for Low-Density Binary Gas Mixtures

The statistical extension which includes binary encounters or the second virial correction in the mixture is given by the generating function

$$\mathcal{Q} = \sum_{N_A, N_B} z_A^{N_A} z_B^{N_B} Q_{N_A N_B} \tag{9.52a}$$

A cumulant expansion of this function about $z_A = 0 = z_B$ in the manner which was used in Section 8.2 for the cluster expansion yields

$$\begin{aligned} \ln \mathcal{Q} = z_A Q_{1A} + z_B Q_{1B} + [z_A{}^2(Q_{2A} - \tfrac{1}{2}Q_{1A}^2) + z_A z_B (Q_{AB} - Q_{1A} Q_{1B}) \\ + z_B{}^2(Q_{2B} - \tfrac{1}{2}Q_{1B}^2) + \cdots \end{aligned} \tag{9.52b}$$

All of the arguments used in Chapter VIII can be used to develop expressions for the two-particle generating functions $Q_{2A}$, $Q_{AB}$, and $Q_{2B}$. The $Q$'s can be expressed in terms of $V$ and $T$ and then $N_A$ and $N_B$ can be related to the $z$'s. It follows that

$$n_{AB}(r)\,dr = \frac{N_A N_B}{V}\{[\exp -\beta U'(AB)] - 1\}4\pi r^2\,dr \tag{9.53}$$

is the number of AB pairs in volume $V$ within a distance of $r$ to $r + dr$

[4] This feature was discussed by Heer and Daunt [*16*].

of each other. This can be compared with Eq. (8.35) for the number of AA or BB pairs.

Since this is one of the few corrections which can be readily determined for the chemical potential, it is given in some detail. Solving for $z_A$ and $z_B$ in terms of $N_A$ and $N_B$, yields the expression for $\mu_A$:

$$\frac{\mu_A}{kT} \approx \ln\left(\frac{N_A}{Q_{1A}}\right) - \frac{2N_A[Q_{2A} - \frac{1}{2}Q_{1A}^2]}{Q_{1A}^2} - \frac{N_B[Q_{AB} - Q_{1A}Q_{1B}]}{Q_{1A}Q_{1B}} \tag{9.54}$$

Again the feature of the cumulant expansion is apparent; the quantities within the square brackets are zero for a classical gas for which the interaction potential $U' = 0$.

A quantum correction occurs in the more general case, but is small for heavy particles at reasonable temperatures. This chemical potential is given in the $V$, $T$, $N_A$, $N_B$ scheme and must be changed to the $P$, $T$, $N_A$, $N_B$ scheme for use with the Gibbs function.

### 9.6.2 *Osmotic Pressure*

Various materials act as semipermeable membranes. Thus stainless steel, platinum, iridium, and palladium are permeable to hydrogen, but not to other gases. A carrot is permeable to pure $H_2O$ but not to sugar. An ox bladder is permeable to $H_2O$ but not to ethyl alcohol. The stomach wall is permeable to $H_2O$ but not to salt. These are a few of many examples. A quite general system is shown in Fig. 9.9. The mixture of A and B in region I is separated from pure A in region II by the semipermeable membrane $M$,

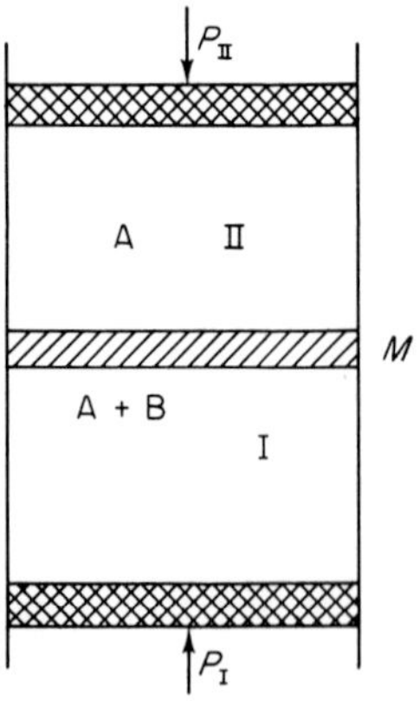


**Fig. 9.9** Components A and B form a mixture or a solution which is separated from the pure component A by a semipermeable membrane which is permeable to A. $P_I$ and $P_{II}$ are pressures applied to movable pistons.

through which B cannot pass. At a given temperature, the chemical potential must be the same on each side of the semipermeable membrane for component A and quite generally,

$$\mu_{\mathrm{I}}^{\mathrm{A}}(P_{\mathrm{I}}, T, c^{\mathrm{B}}) = \mu_{\mathrm{II}}^{\mathrm{A}}(P_{\mathrm{II}}, T) \tag{9.55}$$

This equation is valid for liquid, solid, or gaseous phases. At a given concentration $c^{\mathrm{B}}$, the pressure difference

$$P_{\mathrm{os}} = P_{\mathrm{I}} - P_{\mathrm{II}}$$

across the membrane is referred to as the osmotic pressure and is a function of the concentration. As the temperature is lowered, one of the regions solidifies and since this occurs first for the pure region, this is referred to as the *lowering of the melting point.* As the temperature is raised, one of the regions vaporizes and since the pure region boils first, this is referred to as the *raising of the boiling point.* All of these quantities can be measured, but the exact form of the chemical potential is difficult to obtain and Eq. (9.55) serves to illustrate the exact features of the problem.

#### 9.6.2.1 Statistical Treatment of Osmotic Pressure for a Dilute Solution

Assume that in principle it is possible to compute the necessary many-particle canonical generating functions $Q_{N_{\mathrm{A}}N_{\mathrm{B}}}$ and that the grand canonical generating function is given by

$$\mathscr{Q}_{\mathrm{I}} = \sum_{N_{\mathrm{A}}, N_{\mathrm{B}}} z_{\mathrm{A}}^{N_{\mathrm{A}}} z_{\mathrm{B}}^{N_{\mathrm{B}}} Q_{N_{\mathrm{A}}N_{\mathrm{B}}} \tag{9.56a}$$

If the solution is dilute, then $\ln \mathscr{Q}$ can be expanded about $z_{\mathrm{B}} = 0$ as

$$\ln \mathscr{Q}_{\mathrm{I}} \approx \ln \mathscr{Q}_{\mathrm{A}} + z_{\mathrm{B}}\left( \mathscr{Q}_{\mathrm{A}}^{-1} \sum_{N_{\mathrm{A}}} z_{\mathrm{A}}^{N_{\mathrm{A}}} Q_{N_{\mathrm{A}}1_{\mathrm{B}}} \right) \tag{9.56b}$$

where

$$\mathscr{Q}_{\mathrm{A}} = \sum_{N_{\mathrm{A}}} z_{\mathrm{A}}^{N_{\mathrm{A}}} Q_{N_{\mathrm{A}}}$$

is the generating function for pure A. The dilute solution approximation assumes that only $z_{\mathrm{B}}$ is needed. A term

$$z_{\mathrm{B}}^{2}\left[ \mathscr{Q}_{\mathrm{A}}^{-1} \sum_{N_{\mathrm{}}} z^{N_{\mathrm{A}}} Q_{N_{\mathrm{A}}2_{\mathrm{B}}} - \tfrac{1}{2}\mathscr{Q}_{\mathrm{A}}^{-2} \left(\sum z^{N_{\mathrm{A}}} Q_{N_{\mathrm{A}}1_{\mathrm{B}}}\right)^{2} \right]$$

is the term of next higher-order in the expansion. The coefficient of $z_{\mathrm{B}}^{2}$

is the pair correlation function for two B particles in medium A and is the equivalent of the second virial correction. It is assumed that this term is small and it is omitted in the subsequent discussion of dilute solutions.

Even in this approximation, the calculation of the states of a single B particle in a medium of A particles for the determination of $Q_{N_A 1_B}$ is a formidable problem. It is possible to determine the osmotic pressure without a knowledge of the form of $Q_{N_A 1_B}$.

Let $V_I = V_{II} = V$ so that $\ln \mathcal{Q}_A$ is the same in each volume. Since

$$z_I^A = z_{II}^A$$

this permits $\ln \mathcal{Q}_A$ to be replaced by $P_{II}V/kT$. The number of B particles is related to $z_B$ by

$$N_B = \frac{\partial(\ln \mathcal{Q})}{\partial(\ln z_B)} \approx z_B\left[\frac{1}{\mathcal{Q}_A} \sum_{N_A} z^{N_A} Q_{N_A 1_B}\right] \tag{9.57}$$

With these substitutions, the osmotic pressure follows from the equation of state $P_I V/kT = \ln \mathcal{Q}_I$, and

$$\frac{(P_I - P_{II})V}{kT} \approx N_B \tag{9.58a}$$

or

$$P_{os} \approx n_B kT \tag{9.58b}$$

where $n_B = N_B/V$. In a very clear and direct manner, the statistical approach yields the remarkable experimental result that a foreign particle in a solution transports the same momentum as a particle in a gas and the osmotic pressure in the first approximation is the same as the pressure exerted by a gas with the same number density. This conclusion follows for any particle and does not depend on the explicit form of $Q_{N_A 1_B}$ or the states of a single particle interacting with a medium. Since this number density in even dilute solutions approaches that of a high-density gas, the osmotic pressure can become quite large. In fact, pressures of up to 268 atm have been measured in sugar solutions. Equation (9.58) is usually referred to as van t'Hoff's equation for osmotic pressures.

The chemical potential of the solution can be obtained in the following manner:

$$\mu_I^A(P_I, T, c_I^B) = \mu_{II}^A(P_{II}, T) \approx \mu_{II}^A(P_I, T) + \frac{\partial \mu_{II}^A}{\partial P}(P_{II} - P_I)$$

But $(\partial\mu/\partial P)_T = v$ is the volume per particle and $P_{II} - P_I$ is given by Eq.

(9.58a), and with these substitutions, the chemical potential of the background medium A is given by

$$\mu_A(P, T, c_B) \approx \mu_A(P, T) - c_B kT \tag{9.59}$$

The chemical potential of the dominant medium is linear in the concentration of the dilute species. The chemical potential of the dilute species can be obtained by observing that the square bracket in Eq. (9.57) must be of the form $N_A f(P, T)$. This is not self-evident, but can be seen by observing that $Q_{N_A 1_B} \approx Q_{N_A} Q_{1_B}$ in the free-particle approximation. Then the square bracket reduces to $Q_{1_B}$, which is proportional to the volume $V$, and this volume is approximately $V \approx (N_A + N_B)v_A$. Equation (9.57) for $z_B$ is approximately

$$\mu_B \approx kT \ln c_B + \psi(P, T) \tag{9.60}$$

The explicit dependence on $c_B$ for $c_B \ll c_A$ is the same as derived earlier for a binary gas mixture and is again a manifestation of the entropy of mixing.

**Exercise 9.11** The concentration of sugar in water is $c_B = 10^{-4}$, or 0.01%. This is a rather dilute solution. Find the osmotic pressure at 300°K.

Ans. $P_{os} = 0.14$ atm

**Exercise 9.12** (a) Seawater is 3% NaCl by weight, or $c_B = 0.9\%$. Estimate the osmotic pressure of seawater. Assume complete dissociation and $n_B$ is the sum of the $Na^+$ and $Cl^-$ particles.

Ans. $P_{os} = 26$ atm

(b) Assume that a perfect membrane exists and is placed across the end of a vertical tube. The tube is filled with seawater and the membrane end is placed in normal water. At what height of seawater is the column in equilibrium? What happens if pressure is applied by a piston to the column of seawater? The stomach wall acts as a semipermeable membrane for $H_2O$. What happens to the water content of the blood when a person drinks either seawater or alcohol?

#### 9.6.2.2 Osmotic Pressure of a Degenerate Fermi–Bose Gas Mixture

It was shown in Section 8.7 that the pressure exerted by a degenerate Fermi gas at zero temperature is given by Eq. (8.55e) as

$$PV = \tfrac{2}{5} N\mu_0$$

where the chemical potential $\mu_0$ is given by Eq. (8.51). The pressure and number density are related by

$$P = \left[\frac{2}{5}\,\frac{h^2}{2M^*}\left(\frac{3}{8\pi}\right)^{2/3}\right]n^{5/3} \tag{9.61}$$

for spin-$\frac{1}{2}$ particles and the pressure is proportional to the number density to the $\frac{5}{3}$ power. The effective mass $M^*$ is introduced to approximate interactions with other particles.

Liquid $^3$He can be mixed with liquid $^4$He and superfluid flow acts as a semipermeable membrane for the separation of these two isotopes or components. Pure $^4$He can exist on one side of the membrane and the $^3$He–$^4$He mixture on the other. From the previous considerations, an osmotic pressure which is due to the $^3$He is expected, but from Eq. (9.58b), the osmotic pressure tends toward zero as the temperature tends toward zero. But the simple statistical analysis which leads to Eq. (9.58b) is no longer valid as the Fermi gas becomes degenerate. An elementary model is to return to Eq. (9.56a) and to regard the canonical generating function as a product of generating functions $Q_{N_A N_B} = Q(^3\text{He})Q(^4\text{He})$ with the constraint that the volume $V = N_3 v_3 + N_4 v_4$, where $v_3$ and $v_4$ are the atomic volumes in the liquid. With the two liquids treated as ideal gases, the osmotic pressure is given by Eq. (9.61) and

$$P_{\text{os}} \approx \text{constant} \times n_3^{5/3}$$

for the osmotic pressure at zero temperature.

The osmotic pressure of liquid $^3$He–$^4$He mixtures has been measured as a function of $^3$He concentration by either a direct measurement down to 0.027°K or by an extrapolation of other measurements to absolute zero [*17*]. The data are shown in Fig. 9.10. This curve has an approximate $n_3^{5/3}$ dependence on the $^3$He concentration, but the free $^3$He mass gives an osmotic pressure about three times larger than the observed value. Good agreement is obtained with an effective mass of $M^* = 2.4M_3$, but a theory which takes interactions into consideration is necessary for better agreement with the experimental data. Even so, the existence of an osmotic pressure at zero temperature is another manifestation of the importance of Fermi statistics. Equation (8.51) indicates that the chemical potential of $^3$He in a $^4$He mixture should vary as $n_3^{2/3}$. Measurements of the chemical potential have also been made at sufficiently low temperatures and show some features of this $n^{2/3}$ dependence on concentration [*17*]. Again, interaction between particles and the effective mass must be taken into consideration and the original reference is suggested.

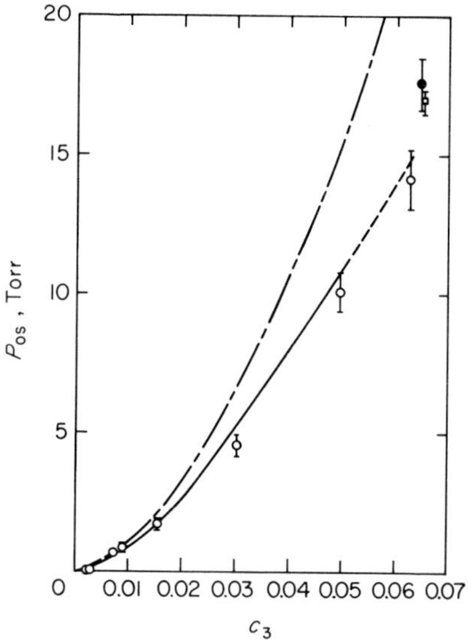


**Fig. 9.10** Osmotic pressure [*17*] at 0°K as a function of concentration $c_3$ for a liquid $^3He$ in $^4He$ solution. This osmotic pressure is characteristic of the pressure exerted by a degenerate Fermi–Dirac gas. [From P. Seligman, D. O. Edwards, R. E. Sarwinski, and J. Tough, *Phys. Rev.* **181**, 415 (1969), Fig. 5.]

It should be emphasized that the thermodynamic relationships remain correct and the approximations suggested here are used to infer the dependence of the chemical potential on concentration. If the necessary experimental data are available, then quantities like the osmotic pressure follow without the necessity for a model.

### 9.6.2.3 Liquid $^3He$–$^4He$ Dilution Refrigeration

The osmotic pressure, the unusual flow properties of pure $^4He$, and the two-phase equilibrium of pure $^3He$ and $^3He$–$^4He$ mixtures has made possible a dilution refrigerator [*18*] for the production of temperatures in the millidegree Kelvin region. A schematic diagram of such a dilution refrigerator [*19*][5] is shown in Fig. 9.11, Vessel A contains the pure phase of $^3He$ in equilibrium with the two-component mixed phase $^3He$–$^4He$. For a $^3He$ concentration $c_3 > 6.35\%$, the two-phase mixture coexists in equilibrium with the pure $^3He$ component. The chemical potentials are related by

$$\mu_3(P, T) = \mu_3(P, T, c_3) \tag{9.62a}$$

in vessel A. Vessel A is connected to a second vessel B which contains pure $^4He$ which is separated from a $^3He$–$^4He$ mixture by a membrane which is

[5] I am indebted to Professor D. O. Edwards for this model for the discussion of the dilution refrigerator.

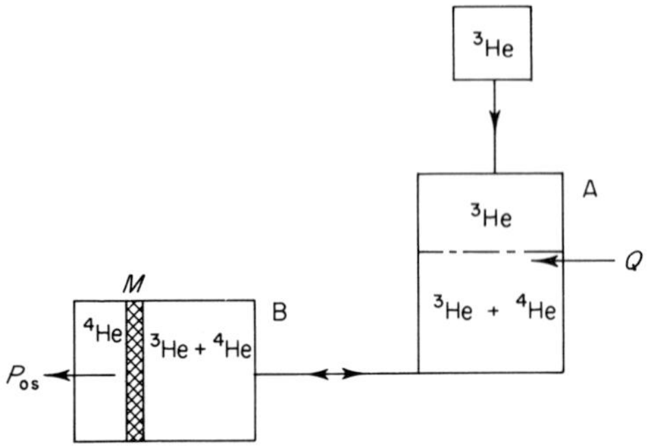


**Fig. 9.11** A $^3$He–$^4$He dilution refrigerator [19]. $M$ is a movable piston membrane permeable to liquid $^4$He and $P_{os}$ is the osmotic pressure exerted on the piston by the liquid $^3$He–$^4$He solution. $Q$ is the heat entering the vessel containing the two-phase system of pure liquid $^3$He and the 6.4% $^3$He–$^4$He mixture.

permeable to $^4$He. The superfluid flow properties of $^4$He make this membrane possible. The chemical potentials on each side of the semipermeable membrane in vessel **B** are related by

$$\mu_4(P_{\text{II}}, T, c_3) = \mu_4(P_{\text{I}}, T) \tag{9.62b}$$

and an osmotic pressure exists across the membrane. As $\Delta N_4$ atoms of $^4$He pass through the semipermeable membrane, the osmotic pressure does work

$$P_{os}(\Delta N_4)v_4$$

and the $^3$He concentration is reduced by $\Delta c_3$. This reduction in concentration is communicated to the two-phase mixture in vessel A, and, since at given $P$ and $T$ the concentration $c_3 = N_3/(N_3 + N_4)$ must remain fixed, some of the pure $^3$He must move from the pure phase into the mixture. This requires a heat $Q$ be extracted from the surroundings. As $\Delta N_4$ atoms pass from the pure $^4$He phase through the membrane,

$$\Delta N_3 = \frac{c_3}{1 - c_3} \Delta N_4$$

atoms of $^3$He pass from the pure $^3$He phase into the mixture. If this process of transfer is done in a reversible manner at constant $P$, $T$, and $c_3$, then the heat extracted is

$$\begin{aligned} Q &= TS(P, T, \Delta N_3, \Delta N_4) - T(\Delta N_3)s_3(P, T) - T(\Delta N_4)s_4(P, T) \\ &\approx \frac{\pi^2 k T^2}{2T_{\text{F}}} \Delta N_3 \end{aligned} \tag{9.63a}$$

As indicated earlier, almost all of the thermodynamic functions of pure $^4$He are zero at low temperatures. The osmotic pressure and the entropy are almost entirely due to the mixed phase and Eq. (8.55d) can be used with an effective mass of 2.4 for an excellent approximation [*17*]. Thus the chemical potential at zero temperature is

$$\mu_3 \approx \frac{h^2}{2M^*}\left(\frac{3}{8\pi}\right)^{2/3}\left(\frac{N_3}{N_3v_3+N_4v_4}\right)^{2/3} = kT_{\mathrm{F}} \tag{9.63b}$$

and defines the Fermi temperature of the $^3$He in the volume $V = N_3v_3 + N_4v_4$. Then the entropy at low temperatures is

$$S(P, T, \Delta N_3, \Delta N_4) \approx \frac{\pi^2 kT}{2T_{\mathrm{F}}}\Delta N_3$$

and the concentration $c_3$ is fixed by the two-phase equilibrium. At low temperatures, $s_3(P, T) \approx 0$ and $s_4(P, T) \approx 0$. The process just discussed is the latent heat of evaporation of the pure $^3$He phase into the phase with concentration $c_3$ at constant osmotic pressure.

Exercise 9.13 Estimate the Fermi temperature for $c_3 = 0.0132$ with $M^*/M = 2.4$. The molar volume of pure liquid $^4$He is 27.58 cm$^3$/g-mole and that of $^3$He is 36.83 cm$^3$/g-mole.

Ans. 0.141°K

Exercise 9.14 Estimate the heat extraction at $c_3 = 0.0635$ and a temperature of 0.003°K as $\Delta N_3 = 0.01$ g-mole of $^3$He is mixed at constant $c_3$, $P$, and $T$.

Exercise 9.15 Let the entropy of the mixture be given by $S = N_3s_3 + N_4s_4$, where Eq. (8.55d) is used for $s_3$ and Eq. (8.75c) for $s_4$ and where $V = N_3v_3 + N_4v_4$ is used for the volume of the mixture. Show that the contribution of $s_4$ is negligible in the millidegree temperature region for $c_3 = 0.064$.

#### 9.6.2.4 Lowering of Melting Point and Raising of Boiling Point

Consider a phase of the solution in equilibrium with another phase of the same substance. The semipermeable membrane is removed and the interface between the two phases is the distinguishing feature. In general,

$$\mu_{\mathrm{I}}^{\mathrm{A}}(P, T, c_{\mathrm{I}}^{\mathrm{B}}) = \mu_{\mathrm{II}}^{\mathrm{A}}(P, T, c_{\mathrm{II}}^{\mathrm{B}}) \tag{9.64a}$$

where A is the dominant component or solvent. Either Eq. (9.59) or Eqs. (9.38) and (9.60) can be used to form an equation similar to that used in deriving the Clausius–Clapeyron equation or $d\mu_{\rm I}{}^{\rm A} = d\mu_{\rm II}^{\rm A}$. Since $\mu_i{}^{\rm A}$ is linear in $c_{\rm B}$, $dc^{\rm B}$ can be replaced by $c^{\rm B}$. With these relationships and substitutions, it follows that

$$-s_{\rm I}{}^{\rm A}\,\Delta T + v_{\rm I}{}^{\rm A}\,\Delta P - c_{\rm I}{}^{\rm B}kT \approx -s_{\rm II}^{\rm B}\,\Delta T + v_{\rm II}^{\rm A}\,\Delta P - c_{\rm II}^{\rm B}\,kT \quad (9.64b)$$

$\Delta T$ and $\Delta P$ measure the change away from the equilibrium temperature and pressure of the two pure A phases. For $\Delta P = 0$, the change in transition temperature which is caused by the foreign B particles is

$$\Delta T \approx \frac{-(c_{\rm I}{}^{\rm B} - c_{\rm II}^{\rm B})}{(s_{\rm I}{}^{\rm A} - s_{\rm II}^{\rm B})}\,kT = \frac{(c_{\rm II}^{\rm B} - c_{\rm I}{}^{\rm B})}{q}\,kT^2 \quad (9.65)$$

where the latent heat of transformation is $q = T(s_{\rm I}{}^{\rm A} - s_{\rm II}^{\rm A})$. The latent heat is positive for I as the vapor phase and negative for I as the solid phase. If $c_{\rm I}{}^{\rm B} = 0$ or the solid phase is pure A, then the freezing point is lowered by the addition of B. If B is nonvolatile, or the vapor phase is pure A, then $c_{\rm I}{}^{\rm B} = 0$ and the boiling point is raised. This is in accord with the earlier comments.

Equation (9.65) implies that the raising of the boiling point and the lowering of the freezing point are dependent on only the heat of vaporization, or the heat of fusion of the solvent, or the dominant medium A. The change is independent of the nature of the foreign component B. The latent heat of fusion of water is $719R$ at 273.16°K and Eq. (9.65) yields

$$\Delta T = -103.5c_{\rm B}$$

for the lowering of the freezing point. As an example, consider ethylene glycol, which is one of the commercial antifreezes for automobile radiators. A 1% solution depresses the freezing point by 1.03° and is in excellent agreement with this simple equation. The depression of the freezing point by sugar or sucrose, with a molecular weight of 342, can be compared with that of ethylene glycol, with a molecular weight of 62. A 1% sugar solution has its freezing point reduced by 1.05°. Deviations from the linear dependence on concentration become appreciable in both examples at higher concentrations.

Since the osmotic pressure and the change in the boiling and freezing points depend only on the number density of the foreign constituents, $n_{\rm B}$ and $c_{\rm B}$ can be replaced by sums. This is quite apparent for NaCl, which dissociates when added to water, a pair of particles being formed. The experimental freezing point is lowered by $\Delta T = 200c_{\mathrm{NaCl}}$. This is twice the value expected for NaCl molecules and is in accord with the number of $Na^+$ and $Cl^-$ particles for complete dissociation [*20, 21*].

9.6.2.5 Vapor Pressure of Two-Phase, Two-Component Systems

If A and B are in solution in the liquid phase and A and Bare considered as an ideal gas mixture in the vapor phase, the vapor pressures of dilute systems follow from the previous considerations. If again A is the dominant component or solvent, Eq. (9.64a) can be used to find the shift in vapor pressure with the concentration of B. For $\Delta T = 0$, the shift in vapor pressure of the solvent A is given by

$$\Delta P_{\mathrm{A}} = kT \frac{c_{\mathrm{I}}^{\mathrm{B}} - c_{\mathrm{II}}^{\mathrm{B}}}{v_{\mathrm{I}}^{\mathrm{A}} - v_{\mathrm{II}}^{\mathrm{A}}} \tag{9.66a}$$

Let I denote the vapor phase and II the liquid phase and $v_{\mathrm{I}}^{\mathrm{A}} \gg v_{\mathrm{II}}^{\mathrm{A}}$. Then $kT/v_{\mathrm{I}}^{\mathrm{A}} = P_{\mathrm{A}}$ is the vapor pressure of the pure solvent A. For a nonvolatile *B* component or solute, $c_{\mathrm{I}}^{\mathrm{B}} = 0$. The reduction in vapor pressure due to the nonvolatile solute is

$$\frac{\Delta P_{\mathrm{A}}}{P_{\mathrm{A}}} = -c_l^{\mathrm{B}}$$

and is proportional to the concentration of the solute. This is Raoult's Law. A similar argument can be used if B is the dominant species. Again these equations require dilute solutions.

The vapor pressure of the solute or the minor component is related to concentration by Eq. (9.60), or

$$\mu_{\mathrm{II}}^{\mathrm{B}} = kT \ln c_{\mathrm{II}}^{\mathrm{B}} + \psi_{\mathrm{II}}^{\mathrm{B}}(P, T) = \mu_{\mathrm{I}}^{\mathrm{B}} = kT \ln P_{\mathrm{B}} + \psi_{\mathrm{I}}^{\mathrm{B}}(T)$$

where II is the liquid phase. This yields a vapor pressure at constant temperature of

$$P_{\mathrm{B}} = \text{constant} \times c_l^{\mathrm{B}} \tag{9.66b}$$

and when the solute dissolves in the liquid, the vapor pressure is proportional to the concentration $c_l^{\mathrm{B}}$ in the liquid phase. This is Henry's Law. For a "perfect" solution, the constant $\exp[(\psi_{\mathrm{II}}^{\mathrm{B}} - \psi_{\mathrm{I}}^{\mathrm{B}})/kT]$ is the vapor pressure of pure B, or the solute.

## 9.7 A Combinatorial Problem

An interesting combinatorial problem is the distribution of $N_A$ particles of type A and $N_B$ particles of type $B$ on an array of $N_A + N_B$ lattice sites. A two-dimensional array is shown in Fig. 9.12. It is assumed that the

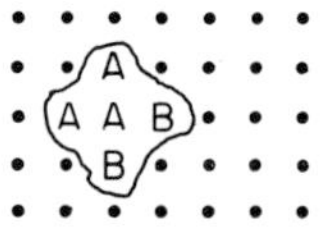


**Fig. 9.12** An ordered array of sites filled with A and B.

probability of a particular configuration is proportional to the number of A–B pairs, or

$$p(N_A, N_B, N_{AB}) = Q_{N_A N_B}^{-1} g(N_A, N_B, N_{AB}) e^{-\beta N_{AB} \varepsilon} \tag{9.67}$$

$g(N_A, N_B, N_{AB})$ is the number of configurations with $N_{AB}$ pairs and $N_{AB}\varepsilon$ is the energy of this configuration. Only potential energy occurs in this combinatorial problem and no dynamic aspects, which cause the change in configurations, are included. The canonical ensemble describing the array is

$$Q_{N_A N_B} = \sum_{N_{AB}} g(N_A, N_B, N_{AB}) x^{N_{AB}} \tag{9.68}$$

where for convenience,

$$x = e^{-\beta\varepsilon}$$

This problem is solved as soon as the number of configurations is counted for each number of pairs $N_{AB}$. Since the enumeration of the number of configurations $g(N_{AB})$ may not be possible, some general aspects are noted and then approximate methods are used. It should also be noted that a knowledge of $g(N_A, N_B, N_{AB})$ determines all aspects of the problem and Eq. (9.68) can be regarded as the generating function for the distribution. If $x$ is made variable, this corresponds to a single constraint on the system.

If $x = 1$, then it follows that

$$\sum_{N_{AB}} g(N_A, N_B, N_{AB}) = \frac{(N_A + N_B)!}{N_A! N_B!} \tag{9.69a}$$

is the total number of configurations. This is the total number of ways of distributing the $N_A$ and $N_B$ particles on $N_A + N_B$ sites. The average number of pairs can also be obtained and

$$\frac{\sum N_{AB}g}{\sum g} = \frac{\gamma N_A N_B}{N_A + N_B} \tag{9.69b}$$

where $\gamma$ is the number of nearest neighbors for the set of lattice points, and is determined by the crystal lattice; some values of $\gamma$ are given in Table 9.4.

**Table 9.4**

| $\gamma$ | Lattice |
|---|---|
| 4 | Two-dimensional square |
| 6 | Simple cubic |
| 8 | Body-centered cubic |
| 12 | Face-centered cubic |

If $N_A = N_B = N/2$, then two configurations of particular interest follow. For $x = 0$, only the array with the minimum value of $N_{AB}$ occurs and this is shown in Fig. 9.13. The system forms two distinct lattices,

**Fig. 9.13** A highly ordered array of A's in I and B's in II.

```
A A A B B B
A A A B B B
A A A B B B
A A A B B B
A A A B B B
```

I of A particles and II of B particles, and shows some aspects of phase separation or ferromagnetism. For $x \to \infty$, only the array with the maximum value of $N_{AB}$ occurs and this is shown in Fig. 9.14. The system forms two interpenetrating lattices I and II and has some aspects of antiferro-

**Fig. 9.14** A highly ordered array of two sublattices I and II where sublattice I has all A's and sublattice II all B's.

```
A B A B A B
B A B A B A
A B A B A B
B A B A B A
A B A B A B
```

magnetism. From this elementary discussion, it is apparent that some features of physical systems are included in this configurational problem. It may be further noted that the total number of configurations is

$$\sum_{N_{AB}} g = (N!)/[(N/2)!]^2 = C_N \cong 2^N$$

and the maximum number of pairs is

$$(N_{\mathrm{AB}})_{\max} = \frac{\gamma N}{2}$$

The average number of pairs for $x = 1$ is

$$\frac{\sum N_{\mathrm{AB}} g}{\sum g} = \frac{\gamma N}{4}$$

and it follows that the number of configurations $g(N_{\mathrm{AB}})$ is symmetric about the average value for $N_{\mathrm{A}} = N_{\mathrm{B}}$. The $C_N$ configurations are distributed between $N_{\mathrm{AB}} = 0$ and $N_{\mathrm{AB}} = \gamma N/2$, and $g(N_{\mathrm{AB}})$ is the number of configurations for each value of $N_{\mathrm{AB}}$. Some difficulty is encountered in the analysis since there is not much difference between a $\delta$ function with height $2^N$ and a function with an average height $2^N/\gamma N$.

**Exercise 9.16** Use the observation that the probability of finding an A on a site is $N_{\mathrm{A}}/(N_{\mathrm{A}} + N_{\mathrm{B}})$ to show that Eq. (9.69b) follows for a random distribution.

### *9.7.1 Low-Temperature Expansion*

The separation into two distinct lattices I and II indicates that the generating function can be written as a product

$$Q = Q_{\mathrm{I}} Q_{\mathrm{II}}$$

when the boundary effect is neglected. Interchange of A and B particles between I and II and the generating function of either I or II are of the same form as Eq. (9.68) with $N_{\mathrm{A}}$ and $N_{\mathrm{B}}$ variable. Thus the generating function for $Q_{\mathrm{II}}$ is

$$Q_{\mathrm{II}} = \sum_{N_{\mathrm{A}}=0}^{N/2} z^{N_{\mathrm{A}}} Q_{N_{\mathrm{A}} N_{\mathrm{B}}} = \sum_{N_{\mathrm{A}}}^{N/2} \sum_{N_{\mathrm{AB}}} g(N_{\mathrm{A}}, N_{\mathrm{B}}, N_{\mathrm{AB}}) z^{N_{\mathrm{A}}} x^{N_{\mathrm{AB}}} \tag{9.70a}$$

Since $N$ is fixed and $N_{\mathrm{B}} = \frac{1}{2}N - N_{\mathrm{A}}$, the term $N_{\mathrm{B}}$ is omitted in subsequent expressions. The generating function for I has the subscripts A and B interchanged and I can be regarded as the dual of II or the reflection of II. Since

$$g_{\mathrm{II}}(N_{\mathrm{A}}, N_{\mathrm{B}}, N_{\mathrm{AB}}) = g_{\mathrm{I}}(N_{\mathrm{B}}, N_{\mathrm{A}}, N_{\mathrm{AB}})$$

and

$$z^{N_B} = z^{(N/2)-N_A}$$

we have

$$Q_I = z^{N/2} Q_{II}(z^{-1}) \tag{9.70b}$$

These generating functions have the appearance of the grand canonical ensemble and for this reason, the grand canonical ensemble is often introduced for the discussion of this configurational problem. It was actually introduced in this discussion to select the configuration with $N_A$ particles in II. Following the procedure in Chapter III, the coefficient $Q_{N_A N_B}$ is selected by taking the $N_A$th derivative and then setting $z$ equal to zero. The concept of the grand canonical ensemble has advantages since it permits the selection of the average value of $\bar{N}_A$ by the choice of $z$. Hence the choice of $z$ permits the introduction of an external field in magnetic analogs and the vapor phase in gas–solid analogs.

The previous discussion applies to either the ferromagnetic or antiferromagnetic lattices by the choice of $x$. If $0 < x < 1$, then the discussion applies to Fig. 9.13, or the ferromagnetic model. For $1 < x < \infty$, the discussion applies to the antiferromagnetic model. It is convenient to start the counting at $N_{AB} = \gamma N/2$ in this latter case and observe that $g(N_A, N_{AB})$ for Fig. 9.13 is equal to $g(N_A, \frac{1}{2}\gamma N - N_{AB})$ for Fig. 9.14.

A series expansion is now used to obtain $Q_{II}$ for a simple cubic lattice, and for convenience, it is assumed there are $N$ sites in II. For $N_{AB} = 0$, there is only one configuration. If one A is interchanged with one B in region II, six pairs are formed and $N_{AB} = 6$ and there are $N$ ways for the interchange. If two adjacent A's are selected and interchanged with two adjacent B's, then $N_{AB} = 10$ and there are $3N$ ways of selecting the pairs. If two nonadjacent pairs are selected, then $N_{AB} = 12$ and there are $\frac{1}{2}N \times (N - 7)$ pairs. The results for $g(N_A, N_{AB})$ are given in Table 9.5.

**Table 9.5**

| $N_A$ | $N_{AB}$ | $g(N_A, N_{AB})$ |
|---|---|---|
| 0 | 0 | 1 |
| 1 | 6 | $N$ |
| 2 | 10 | $3N$ |
| 2 | 12 | $\frac{1}{2}N(N-7)$ |

The generating function $Q_{\mathrm{II}}$ is

$$Q_{\mathrm{II}} = 1 + Nzx^6 + 3Nz^2x^{10} + \tfrac{1}{2}N(N-7)z^2x^{12} + \ldots \tag{9.71a}$$

and this expansion is given up to $x^{28}$ in the literature [*22*]. Since $\ln Q_{\mathrm{II}}$ must be linear in $N$ in order that the problem be independent of the size selected for large $N$, the normal procedure is to extract the $N$th root with the binomial expansion and neglect terms of order $N^{-1}$.[6] The cumulant expansion provides a simpler procedure and the expansion of $\ln Q_{\mathrm{II}}$ about $z = 0$ yields

$$\ln Q_{\mathrm{II}} = \sum a_n z^n$$

with

$$\begin{aligned} a_0 &= \ln g(0, 0) = 0 \\ a_1 &= Q_1 = \sum g(1, N_{\mathrm{AB}})x^{N_{\mathrm{AB}}} = g(1, 6)x^6 \\ a_2 &= Q_2 - \tfrac{1}{2}Q_1{}^2 = \sum g(2, N_{\mathrm{AB}})x^{N_{\mathrm{AB}}} - \tfrac{1}{2}Q_1{}^2 \end{aligned}$$

and higher $a_n$ given from Eqs. (8.4). Then, the one-particle generating function is

$$\ln q \approx N^{-1} \ln Q_{\mathrm{II}} \approx (zx^6 + 3z^2x^{10} - \tfrac{7}{2}z^2x^{12} + \cdots)_{\mathrm{sc}} \tag{9.71b}$$

where the subscript sc indicates "simple cubic." Coefficients through $a_5$ or $Q_5$ are needed for the expansion up to $z^{28}$. In order to avoid interchanging region I with region II, the value of $z$ is limited to the range $0 \leq z \leq 1$. All interesting properties follow from $\ln q$, and $Q_{\mathrm{I}}$ is not needed since it is the same as II with A and B interchanged. Thus the various quantities of interest are

$$p_{\mathrm{A}} = \left[\frac{\partial(\ln q)}{\partial(\ln z)}\right]_x = (zx^6 + 6z^2x^{10} - 7z^2x^{12} + \cdots)_{\mathrm{sc}} = \frac{\bar{N}_{\mathrm{A}}}{N} \tag{9.72a}$$

$$\frac{\langle (N_{\mathrm{A}} - \bar{N}_{\mathrm{A}})^2 \rangle}{N} = \left\{\frac{\partial^2(\ln q)}{[\partial(\ln z)]^2}\right\}_x = (zx^6 + 12z^2x^{10} - 14z^2x^{12} + \cdots)_{\mathrm{sc}} \tag{9.72b}$$

$$\left(\frac{\gamma}{2}\right)p_{\mathrm{AB}} = \left[\frac{\partial(\ln q)}{\partial(\ln x)}\right]_z = (6zx^6 + 30z^2x^{10} - 42z^2x^{12} + \cdots)_{\mathrm{sc}} = \frac{\bar{N}_{\mathrm{AB}}}{N} \tag{9.72c}$$

$$\frac{\langle (N_{\mathrm{AB}} - \bar{N}_{\mathrm{AB}})^2 \rangle}{N} = \left\{\frac{\partial^2(\ln q)}{[\partial(\ln x)]^2}\right\}_z = (36zx^6 + 300z^2x^{10} - 504z^2x^{12} + \cdots)_{\mathrm{sc}} \tag{9.72d}$$

[6] The articles by Domb [*23*] review much of the important research up to 1960 and give an excellent list of references.

where $p_A$ is the probability of an A in region II or among the B's, and $p_{AB}$ is the probability of an AB pair. It should be noted that $p_A$ does not make much sense unless I and II are defined and for this reason, $p_A$ is a measure of the *long-range order*. In general, $p_{AB}$ has a meaning and is referred to as the short-range order parameter. The coefficient $\langle (N_{AB} - \bar{N}_{AB})^2 \rangle$ is a direct measure of the deviation from a random distribution. This term also contributes to the heat capacity and indicates the existence of thermal fluctuations. A long-range order parameter is usually defined as

$$\{\text{long-range order}\} = 1 - 2p_A \tag{9.73a}$$

where $p_A \leq \frac{1}{2}$. The short-range order parameter is defined as

$$\{\text{short-range order}\} = 1 - 2p_{AB} \tag{9.73b}$$

where $0 < p_{AB} < \frac{1}{2}$ is appropriate for Fig. 9.13 and $\frac{1}{2} < p_{AB} < 1$ is appropriate for Fig. 9.14. Long-range order disappears for $p_A = \frac{1}{2}$ and short-range order disappears for $p_{AB} = \frac{1}{2}$. The region of $z \approx 1$ is of particular interest in the low-temperature expansion and corresponds to spontaneous effects. The range of convergence for $z = 1$ appears to be limited to $x < 0.58$ and unfortunately this direct expansion cannot be used in the vicinity of the critical point [*23*].

Exercise 9.17 Show for a square lattice that $g(0, 0) = 1$, $g(1, 4) = N$, $g(2, 6) = 2N$, $g(2, 8) = \frac{1}{2}N(N - 5)$, $g(3, 8) = 6N$, and that $\ln q = zx^4 + z^2(2x^6 - \frac{5}{2}x^8) + z^3(6x^8) + z^4x^8 + \cdots$ or

$$q(x, z) \cong [Q(x, z; N)]_{N=1} \cong 1 + zx^4 + z^2(2x^6 - 2x^8) + z^3(6x^8 - 14x^{10} + 12x^{12}) + \cdots \tag{9.74a}$$

For $z = 1$,

$$q = 1 + x^4 + 2x^6 + 5x^8 + 14x^{10} + 44x^{12} + 152x^{14} + \cdots \tag{9.74b}$$

Exercise 9.18 Let $z = 1$ and plot the quantities in Eqs. (9.72a–d) and (9.73a,b) up to $x = 0.5$.

### *9.7.2 High-Temperature Expansion*

Many special techniques have been developed for the high-temperature expansion. The exact series expansion is a beautiful technique and is discussed in some detail for the special case of $z = 1$, or $p_A = \frac{1}{2}$. No external constraints are on the system. The expansion is made on the full set of points shown in Fig. 9.12 and as noted earlier, an equal number of A and B particles are placed on $N$ sites. The general generating function for the distribution is

$$Q_N = \sum_{N_{AB}} g(N_{AB})x^{N_{AB}} \tag{9.75a}$$

and the average number of AB pairs is given by

$$\bar{N}_{AB} = \frac{d(\ln Q_N)}{d(\ln x)} \tag{9.75b}$$

and the dispersion by

$$\langle(N_{AB} - \bar{N}_{AB})^2\rangle = \frac{d^2(\ln Q_N)}{[d(\ln x)]^2} \tag{9.75c}$$

and the heat capacity is related to the dispersion by

$$\frac{C}{k} = (\ln x)^2\langle(N_{AB} - \bar{N}_{AB})^2\rangle = (\beta\varepsilon)^2\langle(N_{AB} - \bar{N}_{AB})^2\rangle \tag{9.75d}$$

If $x^{N_{AB}}/Q$ is the probability of a particular configuration, then the entropy, $S/k = -\sum p \ln p$, becomes

$$\frac{S}{k} = \ln Q_N - \bar{N}_{AB} \ln x \tag{9.75e}$$

The counting of the configurations $g(N_{AB})$ is the primary problem and it is approached in the following manner.

The quantity $\sigma = \pm 1$ is introduced to aid in the counting procedure. If an A is placed on a site in Fig. 9.12, then by definition, $\sigma = +1$ and if a B is placed on the site, then by definition, $\sigma = -1$:

$$\sigma = \begin{cases} +1, & \text{A on a site} \\ -1, & \text{B on a site} \end{cases}$$

With this definition,

$$N_A - N_B = \sum_{i}^{N} \sigma_i \tag{9.76}$$

where the sum is over all sites. For pair interactions,

$$N_{AB} = \frac{1}{2}\left(\frac{1}{2}\gamma N - \sum_{ij}^{\gamma N/2} \sigma_i\sigma_j\right) \tag{9.77}$$

where the sum is restricted to nearest neighbors. $g(N_{AB})$ is the number of ways the double sum can be performed for a given value of $N_{AB}$, and

$$Q_N = \sum_{N_{AB}} g(N_{AB})x^{N_{AB}} = x^{\gamma N/4} \sum_{\text{all}} \prod_{ij} x^{-\sigma_i\sigma_j/2} \tag{9.78}$$

The sum is over all $C_N$ arrangements of an equal number of A and B particles placed on $N$ sites. The product is over the $\gamma N/2$ nearest neighbors. This equation relates the combinatorial problem to the Ising model.[7] Further reduction is possible by observing that $\sigma_i\sigma_j = \pm 1$, and

$$x^{-\sigma_i\sigma_j/2} = \tfrac{1}{2}(x^{+1/2} + x^{-1/2}) - \tfrac{1}{2}\sigma_i\sigma_j(x^{1/2} - x^{-1/2}) \tag{9.79a}$$

forms an identity. With the substitution

$$u = \frac{1 - x}{1 + x} \quad \text{or} \quad -\tanh\left(\frac{1}{2}\ln x\right) \tag{9.79b}$$

the generating function can be written as

$$Q(x) = \left(\frac{1 + x}{2}\right)^{\gamma N/2}\left[\sum_{\text{all}} \prod_{ij} (1 + \sigma_i\sigma_j u)\right] \tag{9.80}$$

The term within the square brackets is of primary importance and can be expanded as a finite polynomial,

$$\begin{aligned} Z_N &= \sum_{\text{all}} \prod_{ij} (1 + \sigma_i\sigma_j u) = \sum_{\text{all}} [1 + (\sigma_i\sigma_j)u + (\sigma_i\sigma_j)(\sigma_k\sigma_l)u^2 + \cdots] \\ &= 2^N\left[1 + \sum_{r=1}^{\gamma N/2} G(r)u^r\right] \end{aligned} \tag{9.81}$$

The simplicity of the final expression is truly remarkable. In forming the sum of the $C_N$ possible arrangements, it is quite apparent that

$$\sum_{\text{all}} 1 = C_N \cong 2^N$$

and that

$$\sum_{\text{all}} \sigma_i\sigma_j = 0$$

This sum is over all $C_N$ ways of selecting the $ij$ pairs. Since there is no constraint in this sum, A occurs as often as B. In a similar manner,

$$\sum_{\text{all}} (\sigma_i\sigma_j)(\sigma_k\sigma_l) = 0$$

[7] A history of the Ising model during the past 50 years with important references is given by Brush [*24*]. See also Newell and Montroll [*24a*] for an early review.

Not until the sequence $ij \ldots i$ forms a *closed polygon* is there a contribution to the sum. For the two-dimensional square lattice and the three-dimensional cubic lattice, this occurs first for

$$\sum_{\text{all}} (\sigma_i\sigma_j)(\sigma_k\sigma_l)(\sigma_m\sigma_n)(\sigma_p\sigma_q) \tag{9.82}$$

and the closed polygon is the square (see Fig. 9.15), or the sequence $(\sigma_1\sigma_2) \times(\sigma_2\sigma_3)(\sigma_3\sigma_4)(\sigma_4\sigma_1)$. All other contributions add up to zero and this can be checked by drawing some open configurations. Also, any configuration

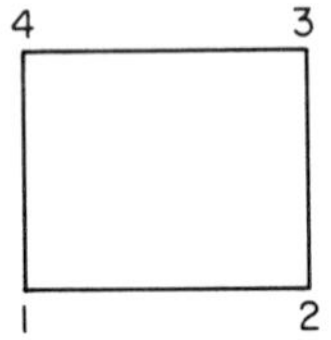


**Fig. 9.15** Closed polygon configuration with four sides for $(\sigma_1\sigma_2)(\sigma_2\sigma_3)(\sigma_3\sigma_4)(\sigma_4\sigma_1)$.

with an odd number of lines meeting at a site or a vertex will have an odd $\sigma_i$ left in the summation and will yield a zero contribution. Only nonzero contributions arise from closed polygons with 2, 4, 6, ... lines meeting at the vertices. From another point of view, each subscript must occur twice for a nonzero value. $G(r)$ is the number of independent closed polygons with $r$ connections or sides which can be drawn in the lattice, and

$$\begin{aligned} G(4) &= 1N, &&\text{two-dimensional square}\\ G(6) &= 2N, &&\text{two-dimensional square}\\ G(4) &= 3N, &&\text{simple cubic}\\ G(6) &= 22N, &&\text{simple cubic}\end{aligned}$$

This counting procedure has been systematized by Wakefield, and Domb summarizes much of the work up to 1960 in a set of review articles [*23*]. $G(r)$ is given for all of the common lattices. The logarithm of the generating function $Z_N$ for the square and simple cubic lattices are

$$\ln Z_N = N(u^4 + 2u^6 + 4u^8 + \cdots)_{\text{sq}} \tag{9.83a}$$

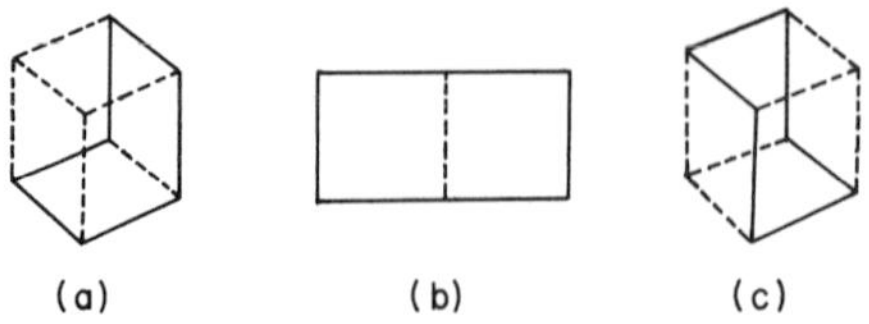


**Fig. 9.16** Closed polygons with six sides for a cubic lattice.

and

$$\ln Z_N = N(3u^4 + 22u^6 + 187u^8 + 1980u^{10} + 24044u^{12} + \cdots)_{\text{sc}} \quad (9.83\text{b})$$

Whenever the quantity within the parentheses is comparable to unity, the expansion fails.

The generating function $Z$ can be used to find the average-size polygon and the dispersion in the average-size polygon,

$$\bar{r} = \frac{d(\ln Z)}{d(\ln u)} \quad (9.84\text{a})$$

$$\langle (r - \bar{r})^2 \rangle = \frac{d^2(\ln Z)}{[d(\ln u)]^2} \quad (9.84\text{b})$$

The results are summarized by restating the *exact* generating function in terms of the number of closed polygons,

$$Q_N(x) = \left(\frac{1+x}{2}\right)^{\gamma N/2} Z_N \quad (9.85)$$

The number of AB pairs is given by

$$\bar{N}_{\text{AB}} = \frac{d(\ln Q)}{d(\ln x)} = \frac{\gamma N}{2}\left(\frac{x}{1+x}\right) - \frac{2\bar{r}x}{(1-x^2)} \quad (9.86\text{a})$$

and the dispersion in AB pairs is

$$\langle (N_{\text{AB}} - \bar{N}_{\text{AB}})^2 \rangle = \frac{\gamma N}{2}\,\frac{x}{(1+x)^2} + \langle (r-\bar{r})^2 \rangle \frac{4x^2}{(1-x^2)^2} - 2\bar{r}\,\frac{x(1+x^2)}{(1-x^2)^2} \quad (9.86\text{b})$$

Thus as $\beta \to 0$ or $x \to 1$, the average value tends to $\gamma N/4$, the dispersion tends to $\gamma N/8$, and the heat capacity to zero. These expressions are given in order to relate the average number of pairs with the average number of closed polygons.

**Exercise 9.19** Show that $G(5) = 0$ and $G(6) = 2N$ for a two-dimensional square lattice.

**Exercise 9.20** Using Fig. 9.16, show that $G(6) = 22N$ for a simple cubic lattice.

### 9.7.2.1 One-Dimensional or Linear Array Aspects

Whenever the generating function can be written as a product of generating functions, it is a manifestation of mutually independent aspects. Equation (9.85) is such a product. The first term $(1 + x)$ is the generating function of a pair and when taken to the $\gamma N/2$ power, is the generating function for $\gamma N/2$ mutually independent pairs. Since

$$1 + x = 1 + e^{-\beta\varepsilon}$$

this is the generating function for a two-energy-level unit and gives rise to a Schottky anomaly of the type shown in Fig. 9.17 in the heat capacity. In the high-temperature region, in which only terms proportional to $1/T^2$ are kept, the entire contribution to $Q$ comes from this term.

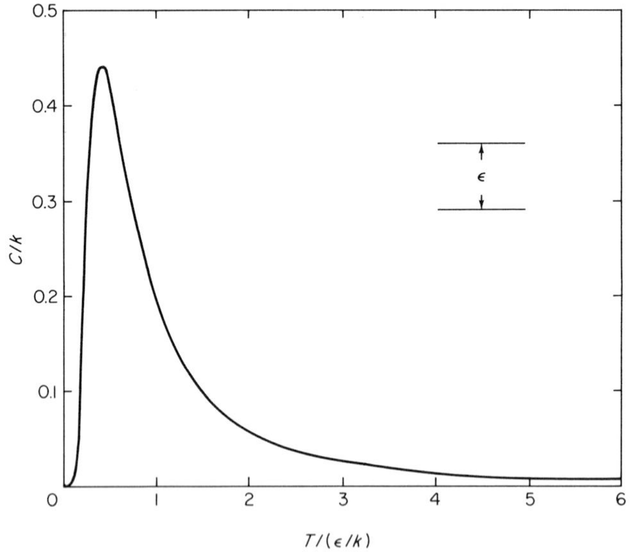


**Fig. 9.17** Schottky anomaly in the heat capacity for two states separated by $\varepsilon$.

**Exercise 9.21** Find the Schottky heat capacity for a system with two energy levels separated by $\varepsilon$.

Ans. $$\frac{C}{k} = (\beta\varepsilon)^2 \frac{e^{-\beta\varepsilon}}{(1 + e^{-\beta\varepsilon})^2} = (\ln x)^2 \frac{x}{(1 + x)^2} \qquad (9.87a)$$

This curve is shown in Fig. 9.17.

Exercise 9.22 (a) Show that $Z_N = C_N$ for a linear chain and therefore the statistical aspects of the linear chain are completely described by the pair distribution function

$$Q_N = C_N\left(\frac{1+x}{2}\right)^N = \sum g(N_{AB})x^{N_{AB}} \tag{9.87b}$$

Show that Eq. (9.87a) gives the heat capacity per pair. Why are there no connected diagrams?

(b) Equate equal powers of $x$ in Eq. (9.87b) to find $g(N_{AB})$.

$$\text{Ans.} \quad g(N_{AB}) = \frac{N!}{N_{AB}!(N - N_{AB})!}(C_N/2^N) \tag{9.87c}$$

Show that expression follows from combinatorial arguments. Where does the maximum value of $N_{AB}$ occur? $\{C_N = (N!)/[(N/2)!]^2\}$.

9.7.2.2 Dispersion in the Number of Pairs and Polygons

The high-temperature approximation for $Q_N$ in terms of pairs and closed polygons and the low-temperature expansion can be used to obtain expressions for the dispersion in $N_{AB}$. Dispersion and heat capacity are related by Eq. (9.75d) and a curve for the heat capacity of a simple cubic lattice [22] is shown in Fig. 9.18. A distinguishing feature of this curve is

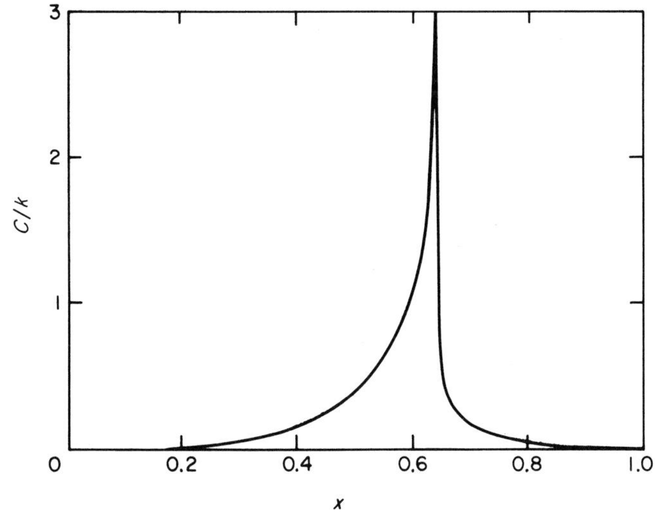


**Fig. 9.18** Heat capacity [22] per site for a cubic lattice with nearest-neighbor interaction. Discontinuity occurs at $x_c = 0.64$. [From A. J. Wakefield, *Proc. Cambridge Phil. Soc.* **47**, 799 (1951), Fig. 4.]

the very rapid rise at $x = 0.64$, which indicates that the dispersion may become extremely large at this value of $x$. The series expansions have been extended to a large number of terms in order to establish the nature of this critical region and this will be pursued in greater detail in a subsequent section. Since the shape of this curve is characteristic of the heat capacity anomalies that occur during configurational ordering in solids, this configurational problem has been of great interest in the physics of solids.

From the preceding discussion, it is quite clear that the knowledge of the pair distribution function $g(N_{AB})$ or the polygon distribution function $G(r)$ provides a complete description of the problem. This is apparent if the generating function is expressed in the alternate forms

$$Q_N(x) = \sum_{N_{AB}=0}^{\gamma N/2} g(N_{AB})x^{N_{AB}} = [\tfrac{1}{2}(1+x)]^{\gamma N/2} Z_N(u) \tag{9.88}$$

where

$$Z_N(u) = 2^N \sum_{r=0}^{\gamma N/2} G(r)u^r$$

and $G(0) = 1$ is introduced. The first expression for the generating function emphasizes that $Q_N(x)$ is a polynomial of degree $\gamma N/2$ in $x$ and by equating terms of equal powers of $x$, the pair distribution function $g(N_{AB})$ can be related to the polygon distribution function $G(r)$. Both distribution functions are distributed over $\gamma N/2$ values. The function $g(N_{AB})$ is limited to peak values characteristic of $2^N$, but $G(r)$ has peak values characteristic of $2^{[(\gamma/2)-1]N}$ and there are more polygon configurations than pair configurations in a three-dimensional lattice. This is apparent for $x = 0$, where lattices $Q_N(x) \approx 1$. This requires

$$\sum_{r=0}^{\gamma N/2} G(r) = 2^{[(\gamma/2)-1]N}$$

for the total number of polygons and some individual terms must be of this order of magnitude. Since the next term in $x$ for $g(N_{AB})$ is $x^\gamma$, the form of $G(r)$ must yield zero for the first few powers of $x$. This yields the moment expansion of $G(r)$ up to $r^{\gamma-1}$ and yields

$$\bar{r}(0) = \frac{\sum rG(r)}{\sum G(r)} = \frac{\gamma N}{4}$$

and

$$\langle (r - \bar{r})^2 \rangle = \tfrac{1}{8}\gamma N$$

for the average number of polygons and the dispersion in the average number of polygons, respectively, at $x = 0$, or for the low-temperature expansion. It is interesting to note that the average-size polygon at zero temperature is the same as the average number of pairs at infinite temperature. This is also true for the dispersion, and the moment expansion illustrates that $G(r)$ must cancel all terms of the pair generating function $(1 + x)^{\gamma-1}$ at low temperatures. If the heat capacity curve shown in Fig. 9.17 is placed in Fig. 9.18, its peak occurs near $x = 0.1$ and falls to small values by $x = 0.2$. This peak is canceled by the first terms in the moment expansion of $G(r)$. If only $Z_N$ is examined, the term in the dispersion $d^2(\ln Z_N)/[d(\ln x)]^2$ becomes negative in this region, but the dispersion

$$\langle (r - \bar{r})^2 \rangle = \frac{d^2(\ln Z_N)}{[d(\ln u)]^2}$$

remains positive, as it must. In order to show features of the observed peak at the critical temperature $x_c$, a very high-order expansion in powers of $x$ is needed. For this reason, the high-temperature expansion is regarded as more useful for finding $x_c$.

In the generating function at high temperatures, the term $(1 + x)^{\gamma N/2}$ provides the $1/T^2$, high-temperature tail and becomes of less importance as the temperature is reduced and the first polygon becomes important. In the expansion of the generating function $Z_N(u)$, the term $u^r/Z_N$ can be regarded as the probability of a polygon with $r$ sides and $G(r)$ as the number of such polygons. Then the discussion of $Z_N(u)$ becomes an independent problem and a cumulant expansion can be made about $u = 0$. This is the high-temperature expansion given by Eq. (9.81) and shows the features of the small polygons. As $u$ increases, higher-order polygons become more important and the expression diverges as $u$ approaches $u_c = (1 - x_c)/(1 + x_c)$. This divergence has been studied for the various lattices [*23, 25*] and occurs at

$$\begin{aligned} x_c &= \sqrt{2} - 1, && \text{square lattice} \\ x_c &\approx 0.64, && \text{simple cubic lattice} \end{aligned} \tag{9.89}$$

for these two simple lattices.

Exercise 9.23 Find the average-size polygon at $x = 0.9$ for a simple cubic lattice.

### 9.7.3 Bragg–Williams and Bethe Approximations

The Bragg–Williams and the Bethe models are included purely as historical examples, and this section can be neglected without impairing the continuity of the argument in this chapter.

(a) The simplest method for approximating $Q_{\mathrm{II}}$, which is given by Eq. (9.70a), is to replace the sum by the largest term and to regard the largest term as occurring at the average values for the random distribution given by Eq. (9.69b). Then,

$$Q_{\mathrm{II}} \approx \frac{(N_{\mathrm{A}} + N_{\mathrm{B}})!}{\bar{N}_{\mathrm{A}}!\bar{N}_{\mathrm{B}}!} x^{\bar{N}_{\mathrm{AB}}} z^{\bar{N}_{\mathrm{AB}}} \tag{9.90}$$

and this is the Bragg–Williams approximation. It is convenient to introduce the concentration as a variable and with $c_{\mathrm{A}} = N_{\mathrm{A}}/(N_{\mathrm{A}} + N_{\mathrm{B}})$, the average values become

$$p_{\mathrm{A}} = \frac{\bar{N}_{\mathrm{A}}}{N_{\mathrm{A}} + N_{\mathrm{B}}} = c_{\mathrm{A}} \tag{9.91a}$$

$$p_{\mathrm{AB}} = \frac{\bar{N}_{\mathrm{AB}}}{\gamma N/2} = 2c_{\mathrm{A}}(1 - c_{\mathrm{A}}) \tag{9.91b}$$

In the Bragg–Williams approximation, the long-range order

$$L = 1 - 2p_{\mathrm{A}} = 1 - 2c_{\mathrm{A}} \tag{9.92a}$$

and the short range order

$$1 - 2p_{\mathrm{AB}} = (1 - 2c_{\mathrm{A}})^2 \tag{9.92b}$$

are simply related to the concentration. Since $z$ is no longer a constraint on the system, it is chosen as 1. Then, the $N$th root of the generating function is

$$\ln q_{\mathrm{II}} = -[c_{\mathrm{A}} \ln c_{\mathrm{A}} + (1 - c_{\mathrm{A}}) \ln(1 - c_{\mathrm{A}})] + \gamma c_{\mathrm{A}}(1 - c_{\mathrm{A}}) \ln x \tag{9.93}$$

and the first term is the entropy of mixing and is similar to Eq. (9.48). For fixed $x$, the value of $c_{\mathrm{A}}$ that gives the maximum value of $q_{\mathrm{II}}$ is needed, and this occurs when $\partial(\ln q)/\partial c_{\mathrm{A}} = 0$, or

$$x = \left(\frac{c_{\mathrm{A}}}{1 - c_{\mathrm{A}}}\right)^{1/\gamma(1-2c_{\mathrm{A}})} \tag{9.94a}$$

This equation can be written in terms of a long-range order parameter as

$$L = -\tanh\left(\frac{\gamma L}{2} \ln x\right) = -\tanh\left(\frac{\gamma \beta \varepsilon L}{2}\right) \tag{9.94b}$$

and this is the conventional form. While $c_A = \frac{1}{2}$ or $L = 0$ is always a solution, it is unstable. If $x$ is less than

$$x_c = e^{-2/\gamma}$$

there is a stable solution. The solution can be shown to be stable by showing that $\partial^2(\ln q)/\partial c_A{}^2 = 0$. The number of AB pairs is given by Eq. (9.91b) and the dispersion or heat capacity per particle is proportional to $\varepsilon \, d\bar{N}_{AB}/dT$, or

$$C = \varepsilon \frac{d[\gamma c_A(1 - c_A)]}{dT} = -\frac{\gamma\varepsilon}{4}\frac{dL^2}{dT} \tag{9.95}$$

This contribution to the heat capacity is shown in Fig. 9.19(a). Above the

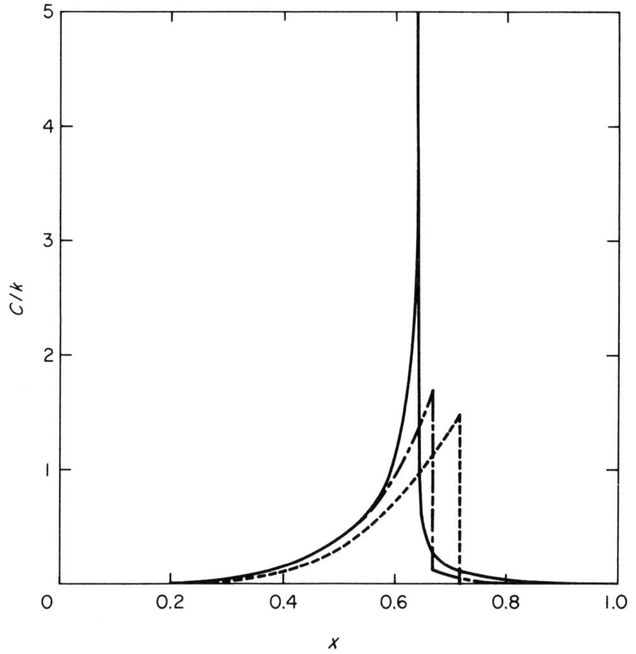


**Fig. 9.19a** The heat capacity $C/k$ or the dispersion in pairs $C/k = (\ln x)^2\langle(N_{AB} - \bar{N}_{AB})^2\rangle$ as a function of $x = \exp -\varepsilon/kT$ for a simple cubic lattice. The solid curve is the exact series expansion, the dot-dash curve is the Bethe approximation, and the dashed curve is the Bragg–Williams approximation.

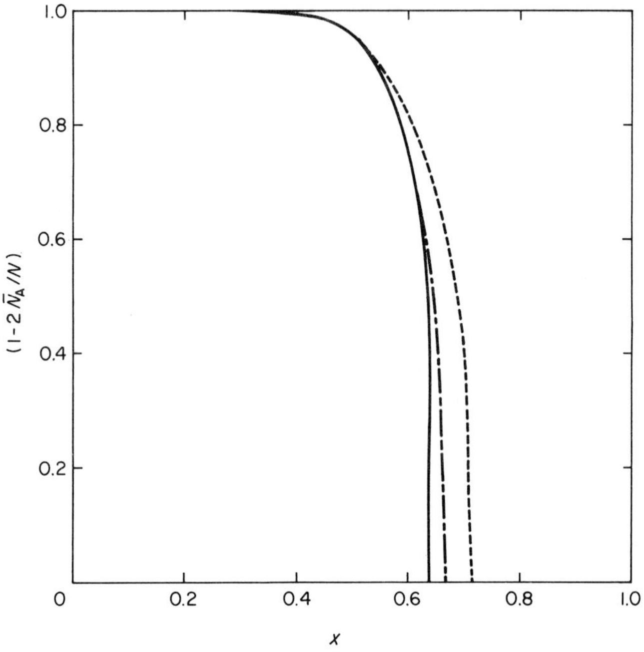


**Fig. 9.19b** The long range order parameter $(1 - 2\bar{N}_A/N)$ as a function of $x$ for a simple cubic lattice. The solid curve is the exact series expansion, the dot-dash curve is the Bethe approximation, and the dashed curve is the Bragg–Williams approximation.

critical temperature, or $x > x_c$, the system has a random distribution and $c_A = \frac{1}{2}$. At the critical temperature, $c_A = \frac{1}{2}$ and begins to decrease as $x$ decreases. The long-range order parameter changes from $L = 0$ to 1 as $x$ decreases from $x_c$ to 0 and is shown in Fig. 9.19(b).

(b) The Bethe method can be regarded as a first order approximation method. Again, lattice II is considered in Fig. 9.13 or 9.14 and $P(A, n)$ is regarded as the probability that a central site has an A particle on it and is surrounded by $n$ neighbors of the B type. $n$ is the number of AB pairs and the probability of the configuration is

$$P(\mathrm{A}, n) = \left[\frac{\gamma!}{n!(\gamma - n)!} \; \frac{z^{\gamma-n}x^n}{q}\right] \tag{9.96a}$$

$z^{\gamma-n}x^n/q$ is the probability of finding $n$ neighbors of the B type, and $\gamma!/n! \times(\gamma - n)!$ is the number of configurations with $n$ pairs. $z$ is a constraint other than the energy which controls the probability of an A in region II.

Then the sum over $n$ yields

$$\sum_{n=0}^{\gamma} P(\mathrm{A}, n) = \frac{(z + x)^\gamma}{q} = p_\mathrm{A} \tag{9.96b}$$

and is the probability of finding an A on the site. The alternate choice is a B on a central site and the probability of a B on the central site with $n$ neighbors of A type is

$$P(\mathrm{B}, n) = \frac{\gamma!}{n!(\gamma - n)!} z^n x^n \tag{9.97a}$$

The sum over $n$ yields

$$\sum_n P(\mathrm{B}, n) = \frac{(1 + zx)^\gamma}{q} = p_\mathrm{B} \tag{9.97b}$$

for the probability of a B on the central site. Since

$$p_\mathrm{A} + p_\mathrm{B} = 1$$

the generating function for the site is

$$q = (z + x)^\gamma + (1 + zx)^\gamma \tag{9.98}$$

In order for the solution to be self-consistent, the probability of finding an A on the central site must equal the average number of A's on the neighboring sites. The number of A neighbors is $\gamma p_\mathrm{A}$, and

$$\gamma p_\mathrm{A} = \left[\frac{\partial(\ln q)}{\partial(\ln z)}\right]_x \tag{9.99a}$$

Direct evaluation yields

$$z = \left(\frac{z + x}{1 + zx}\right)^{\gamma - 1} \tag{9.99b}$$

for the solution to be self-consistent. Then,

$$p_\mathrm{A} = [1 + z^{\gamma/(1-\gamma)}]^{-1} = \tfrac{1}{2}(1 + L) \tag{9.100}$$

is the probability of finding an A among the B's and is a measure of the long-range order.

The average number of pairs is given by

$$\bar{n} = \left[\frac{\partial(\ln q)}{\partial(\ln x)}\right]_z = \frac{\gamma}{2} p_\mathrm{BA} \tag{9.101a}$$

and is a measure of the short-range order. Dispersion in the number of pairs is given by

$$\langle(n - \bar{n})^2\rangle = \left\{\frac{\partial^2(\ln q)}{[\partial(\ln x)]^2}\right\}_z \tag{9.101b}$$

After differentiation, $z$ is replaced by $x$ in these expressions. The range of $p_A$ is $0 \leq p_A \leq \frac{1}{2}$ and this restricts the value of $z$ to $0 \leq z \leq 1$. The constraint relationship between $z$ and $x$ requires that

$$x_c > 1 - \frac{2}{\gamma}, \qquad z = 1 \quad \text{and} \quad p_A = \tfrac{1}{2}$$

$$x_c < 1 - \frac{2}{\gamma}, \qquad z < 1 \quad \text{and} \quad p_A < \tfrac{1}{2}$$

The existence of a solution [*26*] other than $z = 1$ requires that the slope of $[(z + x)/(1 + zx)]^{\gamma-1}$ at $z = 1$ exceed 1 and the critical value of $x$ is determined at the slope of 1. This model has spontaneous long-range order below $x_c$ and a discontinuity in the dispersion at $x = x_c$. Curves for the dispersion or heat capacity and the long-range order are shown in Fig. 9.19(a) and in Fig. 9.19(b), respectively, and it is quite apparent that the Bethe model has some features of the exact solution.

### 9.7.4 Nature of the Singularity

The generating function $Q_{II}$ can also be written in terms of the $\sigma_i$. Using the identity

$$z^{\sigma_i/2} = \tfrac{1}{2}(z^{1/2} + z^{-1/2}) + \tfrac{1}{2}\sigma_i(z^{1/2} - z^{-1/2}) \tag{9.102}$$

and Eq. (9.76) as well as the expressions that yielded Eq. (9.80), the generating function $Q_{II} = Q$ can be written as

$$\begin{aligned} Q &= \sum_{N_A, N_{AB}} g(N_A, N_{AB}) z^{N_A} x^{N_{AB}} \\ &= \left(\frac{1+z}{2}\right)^N \left(\frac{1+x}{2}\right)^{\gamma N/2} \left\{\sum_{\text{all}} \left[\prod_{ij} (1+\sigma_i\sigma_j u) \prod_m (1+\sigma_m v)\right]\right\} \end{aligned} \tag{9.103}$$

where

$$u = \frac{1-x}{1+x} \qquad \text{and} \qquad v = \frac{z-1}{z+1}$$

The notation uses $2N$ sites and $N$ sites occur in II. The observable aspects are

$$\bar{N}_{\mathrm{A}} = \left[\frac{\partial(\ln Q)}{\partial(\ln z)}\right]_x = Np_{\mathrm{A}} \tag{9.104a}$$

$$\langle(N_A - \bar{N}_{\mathrm{A}})^2\rangle = \left\{\frac{\partial^2(\ln Q)}{[\partial(\ln z)]^2}\right\}_x \tag{9.104b}$$

$$\bar{N}_{\mathrm{AB}} = \left[\frac{\partial(\ln Q)}{\partial(\ln x)}\right]_z = \left(\frac{\gamma N}{2}\right)p_{\mathrm{AB}} \tag{9.104c}$$

$$\langle(N_{\mathrm{AB}} - \bar{N}_{\mathrm{AB}})^2\rangle = \left\{\frac{\partial^2(\ln Q)}{[\partial(\ln x)]^2}\right\}_z \tag{9.104d}$$

The average number of A's in region II among the B's is a measure of the long-range order, and with $p_{\mathrm{A}} = \bar{N}_{\mathrm{A}}/N$,

$$(1 - 2p_{\mathrm{A}}) = -\langle\sigma_i\rangle \tag{9.105a}$$

The short-range order is measured by $p_{\mathrm{AB}} = \bar{N}_{\mathrm{AB}}/(\gamma N/2)$, and

$$(1 - 2p_{\mathrm{AB}}) = \langle\sigma_i\sigma_j\rangle \tag{9.105b}$$

where angular brackets as well as bars are used to denote averages. Thus $\langle\sigma_i\rangle$ and $\langle\sigma_i\sigma_j\rangle$ become measures of the long-range and short-range order, respectively.

In physical measurements in which there is an external control parameter, the fluctuation in the particle number $\langle(N_{\mathrm{A}} - \bar{N}_{\mathrm{A}})^2\rangle$ is an observable parameter and corresponds to the susceptibility in magnetic measurements and compressibility in vapor–liquid models. It is normally assumed that this dispersion is linear in $N$ and marked deviations are regarded as anomalous. The explicit relationship between $N_{\mathrm{A}}$ and $\sigma_i$ in the dispersion is given by

$$\langle(N_{\mathrm{A}} - \bar{N}_{\mathrm{A}})^2\rangle = \tfrac{1}{4}\left\langle\left[\sum_i^N (\sigma_i - \bar{\sigma}_i)\right]^2\right\rangle \tag{9.106a}$$

Correlation between different lattice sites is apparent in this expansion and by invariance to choice of origin Eq. (9.106a) becomes equal to

$$\tfrac{1}{4}N\sum_{\varkappa}[\langle\sigma_i\sigma_\varkappa\rangle - \langle\sigma_i\rangle\langle\sigma_\varkappa\rangle] \tag{9.106b}$$

where the index $\varkappa$ is used to emphasize that these quantities need no longer be nearest neighbors and is independent of the index $i$.

For small values of $v$, the last term within the curly brackets in Eq. (103) can be expanded as

$$1 + v\sigma_m + v^2\sigma_m\sigma_\varkappa$$

and then the expansion completed in the manner of Eq. (9.81) as a polynomial in $u^n$. If higher-order terms in $v$ are neglected, the dispersion $N_A$ is independent of $v$ or $z$ and emphasizes another aspect of the distribution. Since terms like

$$[(\sigma_i\sigma_j)](\sigma_\varkappa\sigma_m)$$
$$[(\sigma_i\sigma_j)(\sigma_j\sigma_k)](\sigma_\varkappa\sigma_m)$$
$$\leftarrow \text{ walk } \rightarrow$$

occur and since $m$ and $\varkappa$ need not be nearest neighbors, all self-avoiding walks as well as closed polygons must be counted. Since $m$ and $\varkappa$ are no longer restricted to nearest neighbors, terms like $(\sigma_1\sigma_2)(\sigma_2\sigma_1)$, $(\sigma_1\sigma_2)\times(\sigma_2\sigma_3)(\sigma_3\sigma_1)$, ... are positive and have nonzero average values.

Typical chains and polygons which must be counted are shown in Fig. 9.20. It is apparent from these simple diagrams that if the pairs are joined in a manner which is referred to as a self-avoiding chain or walk and $\varkappa$

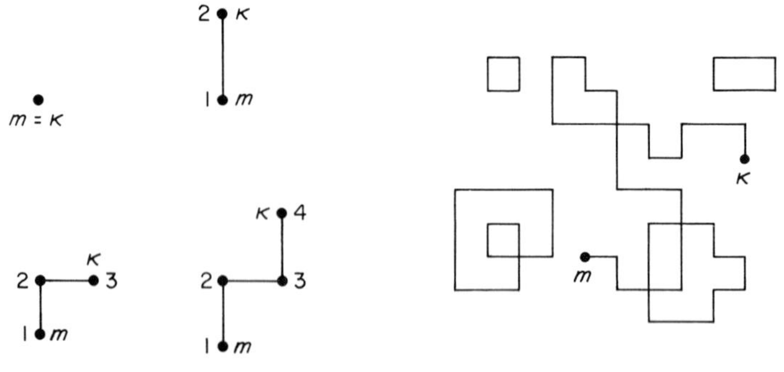


**Fig. 9.20** Walks or chains and polygons contributing to $a_n$ in Eq. (9.107). The term $(\sigma_m\sigma_\varkappa)$ has the same subscripts at the beginning and the end of the walk and these subscripts occur twice.

has the same subscript as the last position of the walk and $m$ the same subscript as the start of the walk, then every subscript occurs twice and the product is positive. A concept of correlation distance is introduced by $(\sigma_m\sigma_\varkappa)$ since the index $m$ corresponds to the beginning of the walk and $\varkappa$ to the end of the walk. Self-avoiding walks occur in the study of polymers.

If the average value and dispersion of a system described by a characteristic function $\varphi(a, N) \sum P(m, N)e^{ima}$ are linear in $N$, then

$$\varphi(a, 1) = [\varphi(a, N)]^{1/N} = \left[\sum_m P(m, N)e^{ima}\right]_{N=1}$$

Invariance to choice of origin implies linearity in $N$ and the one-particle generating function follows from

$$q(x, z) = [Q(x, z; N)^{1/N} = Q(x, z; N)]_{N=1} \qquad (9.107\text{a})$$

This permits the dispersion in $N_A$ to be written in terms of the coefficient $a_n$, which is given in Table 9.6.

**Table 9.6**

Number of Self-Avoiding Walks $c_n$ and the Coefficient $a_n$ in Eq. (9.107)[a]

| | Square lattice | | Simple cubic lattice | |
|---|---|---|---|---|
| $n$ | $a_n$ | $c_n$ | $a_n$ | $c_n$ |
| 1 | 4 | 4 | 6 | 6 |
| 2 | 12 | 12 | 30 | 30 |
| 3 | 36 | 36 | 150 | 150 |
| 4 | 100 | 100 | 726 | 726 |
| 5 | 276 | 284 | 3510 | 3534 |
| 10 | 34,876 | 44,100 | $8.31\times10^6$ | $8.81\times10^6$ |
| 15 | $3.76\times10^6$ | $6.42\times10^6$ | — | $21.2\times10^9$ |

[a] From M. E. Fisher, *Rep. Prog. Phys.* **30**, 680 (1967). Reproduced by permission of The Institute of Physics and The Physical Society, London.

$$\langle(N_A - \bar{N}_A')^2\rangle \approx \tfrac{1}{4}N\left(1 + \sum_{n=1} a_n u^n\right) \qquad (9.107\text{b})$$

The form of $a_n$ for large $n$ has been studied in great detail.

Finally, the pair dispersion, or heat capacity, is related to the $\sigma_i\sigma_j$ variable by

$$\langle(N_{AB} - \bar{N}_{AB})^2\rangle = \tfrac{1}{4}\left\langle\left[\sum_{i,j}(\sigma_i\sigma_j - \langle\sigma_i\sigma_j\rangle)\right]^2\right\rangle \qquad (9.108\text{a})$$

[8] The article by Fisher [*25*] gives an excellent review of the research up to 1967 and an excellent list of references.

and by using invariance to choice of origin becomes equal to

$$\frac{\gamma N}{8} \sum_{\varkappa,\varepsilon} [\langle\sigma_i\sigma_j\sigma_\varkappa\sigma_\varepsilon\rangle - \langle\sigma_i\sigma_j\rangle\langle\sigma_\varkappa\sigma_\varepsilon\rangle] \tag{9.108b}$$

where $\varkappa\varepsilon$ can form a pair at some distance from the $ij$ pair. This introduces the concept of a pair correlation distance. Since the correlation with distant pairs would seem intuitively to be small, the decrease in correlation with distance has been another area of considerable interest. The decrease in correlation has the form [*28*]

$$r_{i\varkappa}^{-m} e^{-br_{i\varkappa}} \tag{9.109}$$

where $r_{i\varkappa}$ is the distance from the $ij$ to $\varkappa\varepsilon$ pair, and $m$ and $b$ depend on whether the correlation is above or below the transition point and whether the pair or four correlation functions are being considered.

These are the primary observables for the combinatorial problem. Fluctuations are normally linear in $N$ and are extensive variables. The maximum possible fluctuation is of the order of $N^2$ and infinite fluctuations do not occur in finite systems. As noted earlier, the first two terms in Eq. (9.103) can give rise to a small anomaly, but they cannot give rise to a very large effect. Any large fluctuation must come from the term within the curly brackets, and this can be regarded as the primary combinatorial problem. Onsager [*24, 24a*] was able to find an exact solution for the two-dimensional square lattice and found a logarithmic singularity of the type

$$\frac{C}{k} \approx A \left| \ln \left| \ln \frac{x}{x_c} \right| \right| \approx B \left| \ln | T - T_c | \right| \tag{9.110}$$

in the heat capacity of an infinite lattice.

The fluctuations of these two quantities in the three-dimensional lattices have been extensively studied by the exact series approximation methods. The method of Padé approximants [*25, 29*] has been used extensively in recent studies and has proven very useful in studying the nature of these singularities. It has been shown that these singularities are of the type [*25*]

$$\bar{N}_A \sim (T_c - T)^\beta, \qquad T < T_c \tag{9.111a}$$

$$\langle (N_A - \bar{N}_A)^2 \rangle \sim \frac{1}{(T - T_c)^\gamma} \qquad T > T_c \tag{9.111b}$$

$$\langle (N_{AB} - \bar{N}_{AB})^2 \rangle \sim \begin{cases} (T - T_c)^{-\alpha}, & T > T_c \\ (T_c - T_c)^{-\alpha'}, & T < T_c \end{cases} \tag{9.111c}$$

These detailed studies for a square and a simple cubic lattice indicate that

$$\gamma \approx \begin{cases} \frac{7}{4}, & \text{square} \\ \frac{5}{4}, & \text{simple cubic} \end{cases} \tag{9.112a}$$

Estimates for other three-dimensional lattices indicate that $\gamma = \frac{5}{4}$ to within an accuracy of $\frac{1}{2}\%$. Estimates for $\beta$ for the three-dimensional lattices give

$$\beta \approx 0.31 \tag{9.112b}$$

Heat capacity calculations converge more slowly and the study of three-dimensional lattices suggest values of

$$\begin{aligned} \alpha &\approx 0.12 \quad \text{or} \quad \tfrac{1}{8} \\ \alpha' &\approx 0.066 \quad \text{or} \quad \tfrac{1}{16} \end{aligned} \tag{9.112c}$$

Values of $\alpha \approx 0.1$ are almost indistinguishable from logarithmic singularities.

These quantities must be regarded as general features of the term within the curly brackets in Eq. (9.103). These singularities are a very sensitive measure of the form of $G(r)$ for closed polygons or other density functions $a_n$ used in the counting process.

The problem which has been discussed in this section is a combinatorial problem and would correspond to the "moves" in a three-dimensional checkers game. $z$ and $x$ infer constraints that give preference to certain configurations in this game, and then the nonconstant dispersion or correlation functions indicate that the system of "checkers," or A and B, is no longer randomly distributed. *The constraints introduce correlations into the distribution of walks or* A–B *pairs.* By varying $z$ and $x$, the character of the random distribution could be inferred and these become the typical thermodynamic experiments discussed in the next section. If the scattering of the A and B is different for x rays or neutrons, then the spherical scattering by each center and the intensity distribution of the scattered wave can be used as a measure of these dispersive quantities.

### *9.7.5 Elastic Scattering of Plane Waves by the A–B Lattice*

Assume that a plane wave $\exp(i\mathbf{k}_0 \cdot \mathbf{r})$ is incident on this array and that the scattering cross sections for A and B are different. At a distant position, the sum of all the spherically scattered wavelets is proportional to

$$\sum_m e^{i\varphi_m} y^{\sigma_m/2}$$

where $\varphi_m$ is the phase difference $2\pi\lambda^{-1}\times$ distance from scatterer to detector, and is discussed in greater detail in Sections 10.11 and 11.8; $y^{\sigma_m/2}$ is the scattering amplitude and is $y^{+1/2}$ for scattering from an A and $y^{-1/2}$ for the scattering from B. Equation (9.102) can be used to write $y$ in product form. The scattered intensity is proportional to the average square of the superposition of the scattered wavelets and is divided into a coherent part and an incoherent part. The coherent part is proportional to $\langle\ \rangle^2$ and the incoherent part to $\langle(\ )^2\rangle - \langle\ \rangle^2$. Thus the incoherent scattered intensity is assumed proportional to

$$\left\langle \sum_{m,\varkappa} e^{i(\varphi_m-\varphi_\varkappa)} y^{\sigma_m/2} y^{\sigma_\varkappa/2} \right\rangle - \left\langle \sum_m e^{i\varphi_m} y^{\sigma_m/2} \right\rangle^2$$

$$= \left(\frac{y^{1/2} - y^{-1/2}}{2}\right)^2 \sum_{m,\varkappa} e^{i(\varphi_m-\varphi_\varkappa)} [\langle\sigma_m\sigma_\varkappa\rangle - \langle\sigma_m\rangle\langle\sigma_\varkappa\rangle] \quad (9.113a)$$

where the average is over the $2^N$ possible configurations subject to the constraint imposed earlier by $z$ and $x$. If $I_0$ is the scattered intensity at $x = 1$ or infinite temperature, then the distribution is random and only for $m = \varkappa$ does a term occur. The scattered intensity as a function of $\mathbf{K}$ or angle $\theta$ becomes

$$\frac{I(\mathbf{K})}{I_0} = 1 + \frac{1}{N} \sum_{m\neq\varkappa} [\langle\sigma_m\sigma_\varkappa\rangle - \langle\sigma_m\rangle\langle\sigma_\varkappa\rangle] \exp i\mathbf{K}\cdot(\mathbf{r}_m - \mathbf{r}_\varkappa) \quad (9.113b)$$

Here $\mathbf{r}_m - \mathbf{r}_\varkappa$ is the distance between lattice points, and $\mathbf{K} = \mathbf{k}_0 - \mathbf{k}$ is the change in direction of the elastically scattered wave. Further details are given in Section 10.11.1. If $\theta$ is the direction of the scattered radiation $\hat{k}$ relative to that of the incident radiation $\hat{k}_0$, then $K = (4\pi/\lambda)\sin(\theta/2)$. If the scattering is in the forward direction, then $K = 0$, or $\varphi_m = \varphi_\varkappa$, and the forward-scattered intensity is proportional to the fluctuation in $N_A$,

$$\frac{I(0)}{I_0} = \frac{\langle(N_A - \bar{N}_A)^2\rangle}{N} \quad (9.113c)$$

This expression can be related to $a_n$ by Eq. (9.107). The scattering is expected to become intense at the critical point for small values of $K$. The dependence on $K$ is of considerable interest [*25*, *26*] since it is a measure of the correlation distance and is already incorporated in Eq. (9.113b) in terms of the correlation $\langle\sigma_m\sigma_\varkappa\rangle$. If $\mathbf{K}$ is a reciprocal lattice vector, then $\mathbf{K}\cdot\mathbf{r}_m = 2\pi(\text{integer})$ and $\exp i\mathbf{K}\cdot(\mathbf{r}_m - \mathbf{r}_\varkappa) = 1$. Equation (9.113c) indicates that the incoherent scattering is proportional to the dispersion in $N_A$ at each Bragg angle.

Below the critical point, $\langle\sigma_m\rangle$ is a measure of the long-range order and has a nonzero average value. At the Bragg angles, the coherent scattering is important and is given by the $\langle\ \rangle^2$ term in Eq. (9.113a). The coherent scattered intensity is

$$\frac{I_c}{I_0} = \frac{1}{N} \sum_{m,\varkappa} \langle\sigma_m\rangle\langle\sigma_\varkappa\rangle \exp i\mathbf{K} \cdot (\mathbf{r}_m - \mathbf{r}_\varkappa)$$

and at the Bragg angles,

$$\frac{I_c}{I_0} = \frac{1}{N} \sum_{m,\varkappa} \langle\sigma_m\rangle\langle\sigma_\varkappa\rangle \propto \left(1 - \frac{2\bar{N}_A}{N}\right)^2 \propto (T_c - T)^{2\beta} \qquad (9.113d)$$

This provides a second method for measuring the parameter $\beta$ below the critical point. In magnetic materials, the coherent scattering is proportional to the square of the magnetization.

## 9.8 Experimental Critical Points

Change of phase remains as one of the striking aspects of the physical world. Macroscopic phases are usually apparent to visual observation and are separable by macroscopic techniques. But as the thermodynamic properties are varied, the properties of the two phases can be made more similar until at the *critical point*, the differences vanish. The most familiar examples of such critical points are [*25*, *30*][9]: (a) the temperature at which a vapor and liquid cease to coexist; (b) binary fluid mixtures; (c) binary metallic alloys for which the components mix homogeneously and in all proportions above the critical point; (d) the Curie point of a ferromagnetic crystal, at which spontaneous magnetization occurs; (e) the Néel point, at which the alternating spin order of an antiferromagnet goes to zero and the two regions become indistinguishable; (f) the ordering temperature in a binary crystal, such as $\beta$-brass (Cu–Zn), above which there is no order and below which the Cu occupies one sublattice and Zn the other; (g) the transition from normal to superconducting state for a metal; (h) the lambda point transition of liquid helium.

At a first-order phase transition, the extensive variables of the Gibbs function, such as the entropy $S$, the volume $V$, or the magnetization $M$,

[9] Heller [*30*] presents a review of the experimental research on critical points and includes references to a large body of original research.

are discontinuous. Transitions for which these variables remain continuous and for which variables like the heat capacity and the susceptibility are discontinuous can be conveniently referred to as *continuous transitions.* It is for transitions of this type that the term *critical point* may be used. Critical points for liquid–vapor systems are shown in Fig. 9.21.

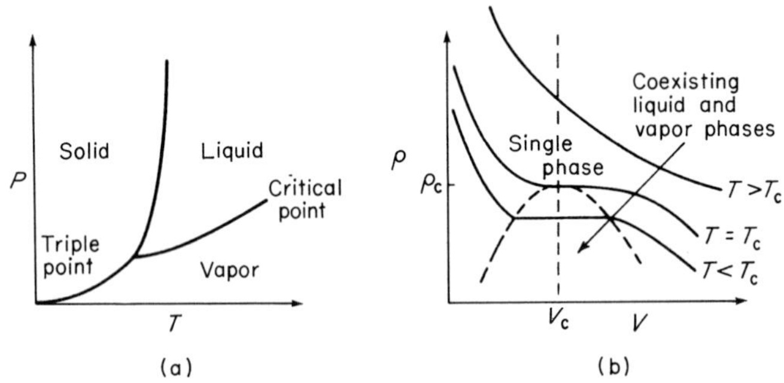


**Fig. 9.21** (a) $P$–$T$ diagram for a typical substance. (b) $P$–$V$ diagram for a typical substance. [From P. Heller, *Rep. Prog. Phys.* **30**, 372 (1967). Reproduced by permission of the Institute of Physics and The Physical Society, London.]

For a gas–liquid critical point, the difference in density between the vapor and the liquid is

$$(\varrho_l - \varrho_g) \sim (T_c - T)^{\beta} \sim \left(\ln \frac{x}{x_c}\right)^{\beta} \tag{9.114a}$$

where $T < T_c$. In these expressions, $\ln(x/x_c) \sim (T_c - T)$. The experimental evidence [*31*] for $CO_2$ suggests $\beta \approx \frac{1}{3}$ and data for $CO_2$ are shown in Fig. 9.22. From above the critical point, the isothermal compressibility characterizes the discontinuity and

$$K_T = -\frac{1}{v}\left(\frac{\partial v}{\partial P}\right)_T \sim \frac{1}{(T - T_c)^{\gamma}} \tag{9.114b}$$

for $T > T_c$. A value of $\gamma \approx 1.2$ occurs for xenon. The heat capacity of the simple gases diverge in an almost logarithmic manner near the critical point. Near the critical point,

$$C_V \sim \begin{cases} (T - T_c)^{-\alpha}, & T > T_c \\ (T_c - T)^{-\alpha'}, & T < T_c \end{cases} \tag{9.114c}$$

$\alpha$ and $\alpha'$ are probably of the order of 0.1 or less and a small negative exponent has the appearance of a logarithmic discontinuity.

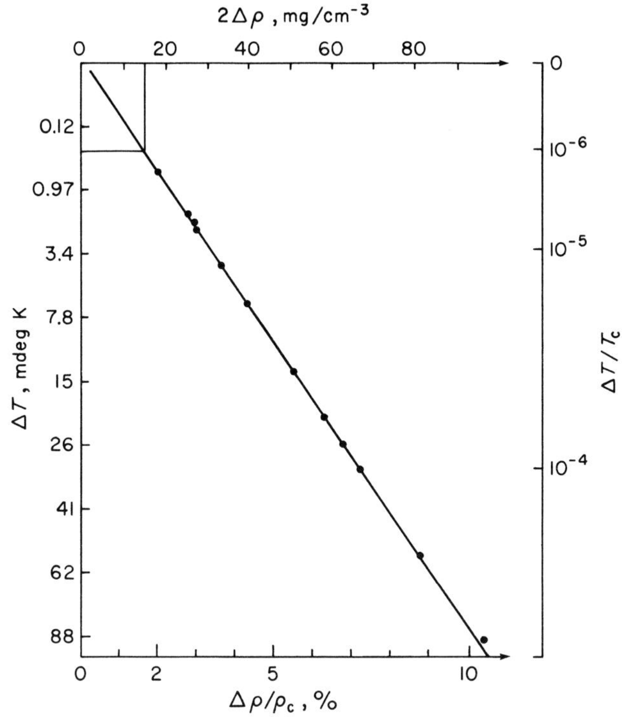


**Fig. 9.22** Plot of the cube root of $\Delta T = T_c - T$ against $\Delta \varrho = \varrho_l < \varrho_g$ for $CO_2$. The data demonstrate the validity of the $\frac{1}{3}$ law to high accuracy over three decades of temperature [*30*]. [From M. E. Fisher, *Rep. Prog. Phys.* **30**, 616 (1967), Fig. 1. Reproduced by permission of The Institute of Physics and The Physical Society, London.]

For ferromagnetics, the isothermal susceptibility is a measure of the fluctuations, and

$$\chi_T = \left(\frac{\partial M}{\partial B}\right)_T \sim \frac{1}{(T - T_c)^\gamma} \tag{9.115a}$$

for $T > T_c$. Experimental values for Ni, Fe, Gd, $YtFeO_3$, and so forth are of the order of 1.35 or $\frac{4}{3}$. Heat capacity measurements yield

$$c_{B=0} \sim \begin{cases} (T - T_c)^\alpha, & T > T_c \\ (T_c - T)^{\alpha'}, & T < T_c \end{cases} \tag{9.115b}$$

The singularity in EuO is roughly logarithmic [*32*] and is shown in Fig. 9.23.

Scattering by fluids and magnetic materials [*30*] can be examined by scattering of electromagnetic radiation or by neutrons. These methods

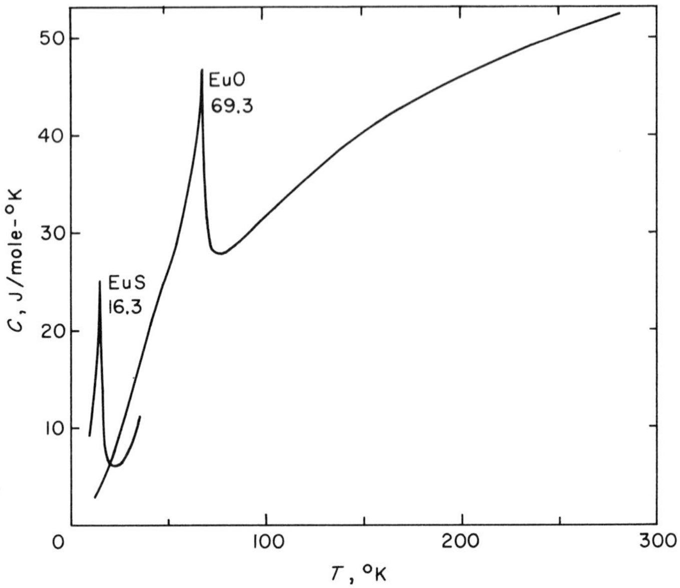


**Fig. 9.23** Heat capacity [*31*] of EuO and EuS showing discontinuity at magnetic critical point superimposed on lattice heat capacity. [From M. E. Fisher, *Rep. Prog. Phys.* **30**, 616 (1967), Fig. 5. Reproduced by permission of The Institute of Physics and The Physical Society, London.]

measure the fluctuation in pair correlations and are discussed in greater detail in the next two chapters.

In binary fluids and alloys, the concentration difference between the two phases vanishes as $(T_c - T)^\beta$ and $\beta \approx \frac{1}{3}$ in the detailed study on $CCl_4 + C_7F_{16}$. For a discussion of liquid helium, superconductors, ferroelectrics, and so on, reference to recent review articles is suggested [*30*]. These references are also suggested for the exponent inequalities [*25, 30*] $\alpha' + 2\beta + \gamma' \geq 2$ and $\alpha' + \beta(1 + \delta) \geq 2$.

In order to make these experimentally observed quantities analogous to the combinatorial problem of placing $N_A$ and $N_B$ particles on $N_A + N_B$ lattice sites, the following models are given.

(a) *Lattice gas.* Regard B as a site without a particle and A as a site with a particle. The vapor phase is in region II in Fig. 9.13 and has a few A particles among the B's or vacant sites. The B's in region I are vacant sites in the liquid phase. Thus the density of the liquid phase is proportional to $1 - (N_A/N)$ and that of the gas phase to $N_A/N$,

$$\varrho_l \propto 1 - \frac{\bar{N}_A}{N} \qquad \text{and} \qquad \varrho_v \propto \frac{\bar{N}_A}{N} \tag{9.116a}$$

so that the difference in density is proportional to

$$\varrho_l - \varrho_v \propto 1 - \frac{2\bar{N}_A}{N} \tag{9.116b}$$

and has the temperature dependence of $\bar{N}_A/N$ and should be compared with Eq. (9.111a). The compressibility given by Eq. (9.114b) should be compared with Eq. (9.111b), or

$$K_T \propto \langle (N_A - \bar{N}_A)^2 \rangle \tag{9.116c}$$

(b) *Ferromagnet or the Ising model for a ferromagnet.* Regard $N_A$ as the number of spins with $m = +\frac{1}{2}$, or which point in the $+z$ direction, and $N_B$ as the number of spins with $m = -\frac{1}{2}$, or which point in the $-z$ direction. Then the magnetization of region II in Fig. 9.13 is

$$M \propto 1 - \frac{\bar{N}_A}{N} \tag{9.117a}$$

and its temperature dependence near the critical point is the same as $\bar{N}_A$. For this model, the susceptibility is proportional to the dispersion in $N_A$, or

$$\chi \propto \langle (N_A - \bar{N}_A)^2 \rangle \tag{9.117b}$$

(c) *Antiferromagnet.* Below the critical or Néel temperature, the spins are the same as in (b), but are located on the lattice sites shown in Fig. 9.14. It is necessary to define the magnetization of a sublattice and the deviation of this sublattice from maximum magnetization is given by

$$M_{II} \propto 1 - \frac{\bar{N}_A}{N} \tag{9.118}$$

where $\bar{N}_A$ is the number of A's, or number of spins with $m = +\frac{1}{2}$, among the B's on the II sublattice. Nuclear magnetic resonance has been used to measure the sublattice magnetization [*30*] of an antiferromagnetic crystal of $MnF_2$ and the variation near $T_c$ is in good agreement with $\beta = 0.335$. Neutron scattering can be used to measure the fluctuation in $N_A$ and agreement with value expected for the combinatorial model occurs.

(d) *Binary liquid mixtures.* Here, we use Fig. 9.13 and regard regions I and II as two distinct phases with concentrations $c_{II}^A = \bar{N}_A/N$ and $c_I^A = 1 - (\bar{N}_A/N)$, respectively. Then, near the critical point, we have

$$c_I^A - c_{II}^A \propto 1 - \frac{2\bar{N}_A}{N} \tag{9.119a}$$

and a temperature dependence of $\beta \approx \frac{1}{3}$ has been found [*25*]. In problems of this type, it is the osmotic compressibility that is proportional to the dispersion in $N_{\mathrm{A}}$,

$$K_{\mathrm{os}} = \frac{1}{c_{\mathrm{A}}} \frac{\partial c_{\mathrm{A}}}{\partial P_{\mathrm{os}}} \propto \langle (N_{\mathrm{A}} - \bar{N}_{\mathrm{A}})^2 \rangle \tag{9.119b}$$

and near the critical point, critical opalescence has been observed.

(e) *Binary alloys.* These are composed of A and B particles on sublattices of the I–II type as shown in Fig. 9.14. Neutron diffraction, x-ray studies, and heat capacity measurements have permitted comparison with the combinatorial problem.

Ferroelectrics, superconductors, and the lambda point in liquid helium have also been compared with the combinatorial problem [*30*].

In closing this section on the temperature dependence of the critical points, it must be emphasized that experimental techniques have provided a very important part of the improved data. Since even gravitational effects are of extreme importance, it is necessary to refer to the original papers for a proper appreciation of the experimental research. Fisher [*25*] and Heller [*30*] provide a very complete list of references to this experimental research. During this same period, the theoretical development has been equally important and again reference to the original papers is necessary for an appreciation of the theoretical developments [*25*]. Rechten *et al.* [*33*], Griffin [*33a*], and Wu [*33b*] include references to current papers on many of these topics.[10]

In comparing experimental data with the combinatorial problem, it must again be emphasized that no dynamic aspects are included in the model. Thus agreement is not expected in the low-temperature region, where quantum effects become important and the problem must be analyzed in terms of phonons, magnons, and so on. The potential and kinetic aspects are no longer approximately statistically independent and a new set of modes is needed for the analysis.

## 9.9 Surface–Gas-Phase Equilibrium

The interaction between a molecule and a surface was discussed briefly in Section 4.8. As in the case of intermolecular collisions, the wall presents

[10] Rechten *et al.* [*33*] discuss neutron scattering in CoO, Griffin [*33a*] discusses critical indices, and Wu [*33b*] discusses a ferroelectric model.

a short-range repulsive force and a long-range attractive force for all molecules. The molecules adsorbed on the surface form a surface phase and the chemical potential of the surface phase must be equal to the chemical potential of the vapor phase in equilibrium with the surface,

$$\mu_a(P, T) = \mu_g(P, T) \tag{9.120}$$

The terminology gas phase rather than vapor phase is used in the literature. As in the example of vapor–solid equilibrium, a Clausius–Clapeyron equation follows from $d\mu_a = d\mu_g$ and the vapor pressure which exists above the surface is related to the latent heat of vaporization by Eq. (9.9b), or

$$\frac{\delta P}{\delta T} = \frac{s_g - s_a}{v_g - v_a} = \frac{q}{T(v_g - v_a)} \tag{9.121a}$$

The gas phase can be treated as almost ideal, and then

$$\frac{\delta(\ln P)}{\delta(\ln T)} \approx \frac{q}{kT} \tag{9.121b}$$

relates the change in pressure to the change in temperature along the vapor–surface equilibrium curve, where $q$ is the latent heat of vaporization. $q$ is a function of surface coverage and changes to the latent heat of vaporization of the bulk phase as the surface layer becomes thick. Since $q$ depends on the surface coverage, the change $\delta(\ln P)/\delta(\ln T)$ is evaluated at constant coverage and is called the isosteric heat.

Since gases are usually confined in vessels, equilibrium with the walls occurs and in this sense, two phases always coexist in equilibrium with each other. The number of molecules in the surface phase is normally small and is ignored in most experiments. Yet at 1 Torr, the number of wall atoms is comparable to the gas atoms in a tube of less than 1 cm diameter. The surface densities which occur in a single layer of a liquid are of the order of $0.5–1.0 \times 10^{15}$ molecules/cm$^2$ for most common gases. Some specific values for the number of molecules per cm$^2$ are given in Table 9.7 for some common liquids. Just as the vapor pressure as a function of temperature provides information regarding the vapor–liquid phases, the adsorption

**Table 9.7**

| He | $H_2$ | Ne | $N_2$ | Ar | $O_2$ | $CO_2$ | $H_2O$ |
|---|---|---|---|---|---|---|---|
| 0.72 | 0.69 | 1.0 | 0.61 | 0.69 | 0.71 | 0.59 | $0.95 \times 10^{15}$ |

isotherms provide information regarding the surface–vapor phases. Since the attractive forces differ for various surfaces, there exists an adsorption isotherm for each type of surface. The number of atoms associated with the surface phase and the vapor phase will depend on the type of surface, the amount of surface area, and pressure $P$ for a given temperature $T$. The heat of vaporization will depend on the number of layers in the surface phase and add to the complexity of the problem. In order to explore this problem further, some simple models are considered.

### 9.9.1 Low-Pressure Adsorption and Cryopumping

Adsorption at very low pressures, or in the pressure range between $10^{-10}$ and $10^{-1}$ Torr, has been investigated extensively and is of considerable interest for cryopumping for the production of ultrahigh vacuums [*34*]. At these low pressures, the surface coverage is much less than one monolayer. Experimental data for surface coverage as a function of pressure and chemical potential are shown in Fig. 9.24 for nitrogen and argon [*35*]. The data are in remarkably good agreement with the straight lines shown in the figure. If $\bar{N}_a/N_0$ is the fraction of a monolayer, then the data are in good agreement with

$$\frac{\bar{N}_a}{N_0} = \exp[-B(\mu - \mu_0)^2] \tag{9.122}$$

where $\mu_0$ is a reference chemical potential and $\mu$ the chemical potential at the experimental gas pressure.

The form of the adsorption isotherm suggests the direction of approach for a statistical model [*36*]. If the grand canonical generating function for a surface of area $A$ is

$$\mathcal{Q}_a = \sum_{N_a} z^{N_a} Q_{N_a} \tag{9.123a}$$

where $Q_{N_a}$ is the canonical generating function for $N_a$ molecules adsorbed on the surface of area $A$, then the average number of atoms on the surface is

$$\bar{N}_a = \left(\frac{\partial(\ln \mathcal{Q}_a)}{\partial(\ln z)}\right)_T \tag{9.123b}$$

At small surface coverages, it is expected that $\bar{N}_a/N_0$ will be a very small number, and that an expansion of $\ln \bar{N}_a$ as a function of $\ln z$ would provide a better approximation than the expansion of $\bar{N}_a$ as a function of $z$. This

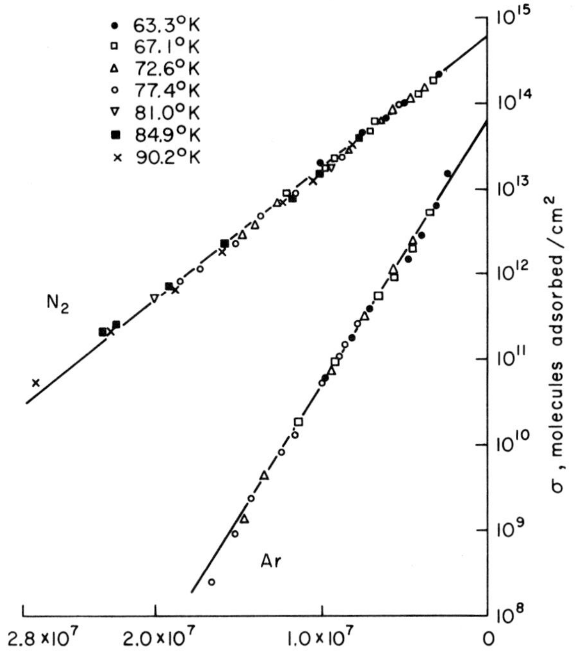


**Fig. 9.24** Physical adsorption isotherm [35] of $N_2$ and Ar on Pyrex plate as a function of $(\mu - \mu_0)^2 = [RT\ln(P/P_0)]^2$. [From J. P. Hobson and R. A. Armstrong, *J. Phys. Chem.* **67**, 2000 (1963), Fig. 6. Copyright 1963 by the American Chemical Society. Reproduced by permission of the copyright owner.]

has the appearance of a cumulant expansion for the surface coverage $N_a$. Direct expansion about $\ln z_0$ at constant $T$ yields the expansion for $\ln \bar{N}_a$ as

$$\ln \bar{N}_a(z) = \ln \bar{N}_a(z_0) + \left[\frac{\partial(\ln \bar{N}_a)}{\partial(\ln z)}\right]_{z_0}\left(\ln\frac{z}{z_0}\right) + \cdots$$

$$= \sum_n b_n\left(\ln\frac{z}{z_0}\right)^n \tag{9.124}$$

Under the assumption that the coefficients of $[\ln(z/z_0)]^3, \ldots$ are small, the approximate equation for the surface coverage is

$$\bar{N}_a(T, z) \approx \bar{N}_a(T, z_0)\exp\left(b_1 \ln\frac{z}{z_0}\right)\exp\left[b_2\left(\ln\frac{z}{z_0}\right)^2\right] \tag{9.125}$$

For a surface phase in equilibrium with the gas phase, the fugacity $z$ and the gas-phase pressure are related by

$$\ln \frac{z}{z_0} = \beta(\mu - \mu_0) = \ln \frac{P}{P_0}$$

The chemical potential $\mu$ of the gas phase is known from Eq. (9.12).

For sufficiently small $\ln(z/z_0)$, this is an exact expansion and will always yield a good approximation to the surface coverage. Experimental data [*36*] seem to indicate that it is an excellent expansion for large negative values of $\ln(z/z_0) = \ln(P/P_0)$ and this requires that the expansion coefficients $b_3$, $b_4, \ldots$ be negligibly small. This feature requires detailed consideration.

The coefficient $b_1$ is directly related to the fluctuation in the number of surface molecules and direct evaluation yields (see Section 10.10)

$$b_1(T, z_0) = \left[\frac{\partial(\ln \bar{N}_a)}{\partial(\ln z)}\right]_T = \frac{1}{\bar{N}_a}\langle(N_a - \bar{N}_a)^2\rangle \tag{9.126a}$$

and all quantities are evaluated at the expansion origin $T$, $z_0$. Evaluating $b_2$ in a similar manner yields

$$2b_2 = \frac{1}{\bar{N}_a}\left\{\frac{\partial^3(\ln \mathcal{Q}_a)}{[\partial(\ln z)]^3}\right\}_{T,z_0} - b_1^2 \tag{9.126b}$$

$b_2$ depends on higher-order fluctuations and its physical interpretation is not as apparent as that for $b_1$.

If the surface particles are mobile and behave as an ideal gas, then the $N$-particle generating function is given by $Q_N = Q_1^N/N!$ and the grand canonical generating function can be written as $\mathcal{Q} = e^{zQ_1}$. The fluctuation in number of surface particles is $\langle(N_a - \bar{N}_a)^2\rangle = \bar{N}_a$ and the expansion coefficients are

$$b_1 = 1 \qquad \text{and} \qquad b_n = 0 \qquad \text{for} \quad n > 1$$

This yields Henry's Law for the surface coverage:

$$\frac{\bar{N}_a(T, P)}{\bar{N}_a(T, P_0)} = \frac{P}{P_0} \tag{9.127}$$

and the ratio of the surface coverage at two pressures is given by the ratio of the pressures.

This same result follows from the localized adsorbed particle model, which has a generating function for a particular site of

$$\mathcal{Q}_\alpha = 1 + zQ_1 \tag{9.128a}$$

$1/\mathcal{Q}_\alpha$ is the probability that there is no particle adsorbed on the site and $Q_1/\mathcal{Q}_\alpha$ is the probability that there is one particle adsorbed on the site. If there are $N_0$ adsorbing sites, then the generating function for the surface is

$$\mathcal{Q}_{\mathrm{a}} = \prod \mathcal{Q}_\alpha = (1 + zQ_1)^{N_0} \tag{9.128b}$$

The surface coverage follows from Eq. (9.123b), $\ln \mathcal{Q}_{\mathrm{a}}$, and

$$\frac{\bar{N}_{\mathrm{a}}}{N_0} = \frac{1}{(1 + z^{-1}Q_1^{-1})} \tag{9.129}$$

If $z_0$ is defined so that $z_0Q_1 = 1$, then the expression for surface coverage can be written as

$$\frac{\bar{N}_{\mathrm{a}}(T, P)}{N_0} = \frac{P}{P + P_0} \approx \frac{P}{P_0}, \qquad P \ll P_0 \tag{9.130}$$

and again yields Henry's Law for small surface coverages with the proper choice of $z_0$. Equation (9.130) is the Langmuir isotherm and can be derived by the kinetic method of Section 4.8. If $\varepsilon_0$ is the binding energy to the surface site, then $Q_1 = e^{-\beta\varepsilon_0}$. This yields $\beta^{-1} \ln z_0 = \mu_0 = \varepsilon_0$ and for this choice of $z_0$, $b_1 = 1$ and all other coefficients $b_n = 0$ for $n > 1$. For $z = z_0$, the probability of a site being occupied is one-half and suggests that the proper choice of $P_0$ is the gas pressure at one-half coverage.

If the temperature is reasonably high and the surface coverage is sufficiently small that interaction between nearest neighbors is small, then a homogeneous surface should obey Henry's Law for either model. Adsorbent P33(2700), a graphitized carbon black with a highly homogeneous surface, is found experimentally to obey Henry's Law along an isotherm.

In both of the above models, there is no $b_2$ term and the results are not in agreement with the surface coverage on the average laboratory surfaces shown in Fig. 9.24. For large $[\ln(z/z_0)]^2$, the second term is dominant and Eq. (9.125) is an excellent approximation over 10 orders of magnitude in surface coverage.

It has been suggested that the surface conditions of average laboratory surfaces are not homogeneous and that a nonhomogeneous surface can be used to explain the coverage. This is not possible for the gas model having different surfaces with different values of binding energy. Henry's Law always follows. A localized model and a heterogeneous surface does permit the introduction of a $b_2$ coefficient and small $b_n$ coefficients and this model is now examined in greater detail.

Assume the surface of the sample is heterogeneous and the binding energy of a site is $\varepsilon_\alpha$. Again, the generating function for a site is

$$\mathcal{Q}_\alpha = 1 + zQ_{1\alpha} = 1 + ze^{-\beta\varepsilon_\alpha} \tag{9.131a}$$

where 1 is proportional to the probability of no particle on the site and $e^{-\beta\varepsilon_\alpha}$ is proportional to the probability of one particle on the site. The generating function for $N_0$ sites is

$$\mathcal{Q}_\mathrm{a} = \prod_\alpha^{N_0} (1 + ze^{-\beta\varepsilon_\alpha}) \tag{9.131b}$$

and

$$\ln \mathcal{Q}_\mathrm{a} = \sum_\alpha \ln(1 + ze^{-\beta\varepsilon_\alpha}) = N_0 \int_{-\infty}^{0} d\varepsilon\, g(\varepsilon) \ln(1 + ze^{-\beta\varepsilon}) \tag{9.131c}$$

$g(\varepsilon)$ is the probability of a site with a characteristic binding energy $\varepsilon$. This expression has the mathematical features of a Fermi–Dirac distribution. Surface coverage is given by

$$\bar{N}_\mathrm{a} = N_0 \int_{-\infty}^{0} d\varepsilon\, g(\varepsilon) n(\varepsilon) \tag{9.132a}$$

where

$$n(\varepsilon) = \frac{1}{1 + \exp \beta(\varepsilon - \mu)} \tag{9.132b}$$

From physical considerations, it is expected that the probability $g(\varepsilon)$ of a site having binding energy $\varepsilon$ is limited in energy spread near some value $\mu_0$. The area under the curve is normalized so that

$$\int_{-\infty}^{0} d\varepsilon\, g(\varepsilon) = 1$$

Let $g(\varepsilon)$ be the second derivative of a function $\psi$,

$$g(\varepsilon) = \frac{d^2\psi}{d\varepsilon^2} = \psi''(\varepsilon)$$

then the procedure that was used for Eq. (8.53a) yields the grand canonical ensemble for surface coverage for a heterogeneous surface as

$$\ln \mathcal{Q}_\mathrm{a} = N_0\beta \int_{-\infty}^{0} d\varepsilon\, \psi(\varepsilon) n(1 - n) \tag{9.133a}$$

The function $n(1 - n)$ is an exponentially sharp, symmetric function about

$\varepsilon = \mu$ with a width of the order of $\beta^{-1} = kT$. If all sites have the same binding energy $\varepsilon_0$ and $g(\varepsilon) = \delta(\varepsilon - \varepsilon_0)$, this equation reduces to the generating function for the Langmuir isotherm. If $g(\varepsilon)$ has a spread in energy which is large compared to $kT$, then $\psi(\varepsilon)$ is a slowly changing function and the sharpness of $n(1 - n)$ at $\varepsilon = \mu$ can be used to approximate the integral as $\psi(\mu)$. Using Eq. (8.53a) for Fermi–Dirac integrals yields

$$\ln \mathcal{Q}_a = N_0\beta\left[\psi(\mu) + \frac{\pi^2}{6\beta^2} g(\mu) + \cdots\right] \tag{9.133b}$$

Surface coverage follows from this expression and is given by

$$\bar{N}_a = N_0\left[\psi'(\mu) + \frac{\pi^2}{6\beta^2} g'(\mu) + \cdots\right] \tag{9.134a}$$

The other thermodynamic variables $U_a, S_a, \ldots$ follow from $\ln \mathcal{Q}_a$. If the second term is neglected, the surface coverage is approximately

$$\bar{N}_a \approx N_0\psi'(\mu) = N_0 \int_{-\infty}^{\mu} d\varepsilon\, g(\varepsilon) \tag{9.134b}$$

and is a measure of the number of surface sites with energy less than $\mu$. $g(\varepsilon)$ can be selected as a Gaussian distribution about $\varepsilon_0$ such that $\bar{N}_a/N_0$ has the form of Eq. (9.125). Values of $N_0$, $b_1$, and $b_2$ must be taken from the experimental surface coverage data.

The heterogeneous surface models seem appropriate on a material like glass of Pyrex, for which the surface is made up of different kinds of atoms and these atoms can be arranged in a very large number of local arrays. A Gaussian distribution of binding energies would seem to be appropriate. Various defects, steps, fissures, cracks, and so on can also provide a random distribution of binding energies. It would seem that in a certain sense, all surfaces are heterogeneous and the dependence of the coverage isotherms on gas pressure will depend on whether the width of the energy distribution $g(\varepsilon)$ is large or small compared to $kT$. If the width is large compared to $kT$, the coverage at low pressures will depend primarily on the number of sites with binding energy $\varepsilon < \mu$. If the width is less than $kT$, the simpler Henry's Law expansion given by Eq. (9.127) follows. If the average binding energy is $\varepsilon_0$, then it is not surprising that the width of $g(\varepsilon)$ is comparable to from one-third to one-fifth of $\varepsilon_0$.

The amount of material absorbed at various pressures follows from the curves of the type shown in Fig. 9.24 and indicates the amount of cryopumping which can be expected at a given temperature. Effective areas of some surface absorbents are given in Table 9.8. The physical adsorption

**Table 9.8**

| Absorbent | Effective area ($m^2$/g) |
|---|---|
| Silica gel type R granule for $N_2$ | 784 |
| Porous glass for $N_2$ | 117 |
| Sanan charcoal-S85 for $N_2$ | 1170 |
| Tungsten for Kr | 40 |

isotherm represents the lower limit of cryopumping and the fraction of surface area covered represents the maximum amount of material which can be pumped at this pressure. The area per gram of sorbent can be combined with the surface coverage for a quantitative estimate of the pumping capacity of the material.

In concluding this section on monolayers, some experiments are selected for discussion. Adsorbed films [*37*] of $^4$He and of $^3$He on Vycor have been studied in some detail between 1 and 3°K. The heat capacity of the monolayer follows a $T^2$ law with a Debye $\Theta \approx 30$. A $T^2$ temperature dependence is expected for a two-dimensional model since the density of states $g(\nu)$ is proportional to $\nu$ rather $\nu^2$ in Eq. (9.29). Helium-4 forms a very thick film below the $\lambda$ point and this film has been studied extensively.

Experiments [*38*] on the electronic and lattice structure of Cs films absorbed on a W surface illustrate other features of adsorption. The first monolayer of Cs on the W has a nearest-neighbor distance of $4.46 \times 10^{-10}$ m and reaches a surface density of $5 \times 10^{14}/\text{cm}^2$. This layer reduces the work function of the surface by 2 eV. Each Cs atom is partially ionized and there is one $Cs^+$ ion for two W atoms. The second layer starts to form with a similar structure as the first and is assumed to form in the holes of the first layer. After $7.5 \times 10^{14}/\text{cm}^2$ is reached, continuous rearrangement occurs and the second layer tends toward a hexagonal-close-packed layer with half the atoms over the holes and the other half over the sites of the monolayer. The density of this layer is near that of bulk Cs.

## References

1. M. W. Zemansky, "Heat and Thermodynamics," 5th ed. McGraw-Hill, New York, 1968.
2. A. A. Maradudin, I. P. Ipatova, E. W. Montroll, and G. H. Weiss, eds., *Solid State Phys. Suppl.* **3**, *2nd Ed.* (1971).

2a. J. C. Phillips, *Phys. Rev.* **104**, 1263 (1956).
2b. W. Nernst, "The New Heat Theorem." Methuen, London, 1926.
2c. W. F. Giauque and H. L. Johnston, *J. Amer. Chem. Soc.* **50**, 3221 (1928).
2d. F. Simon, *Ergeb. Exakt. Naturw.* **9**, 222 (1930).
3. K. W. Zinov'eva, *Zh. Eksp. Teor. Fiz.* **29**, 899 (1955); *JETP* **2**, 774 (1956); H. M. Guo, D. O. Edwards, R. E. Sarwinski, and J. T. Tough, *Phys. Rev. Lett.* **27**, 1259 (1971).
4. D. L. Decker, D. E. Mapother, and R. W. Shaw, *Phys. Rev.* **112**, 1888 (1958).
5. J. Wilks, "The Third Law of Thermodynamics." Oxford Univ. Press, London and New York, 1961.
6. W. F. Giauque and J. O. Clayton, *J. Amer. Chem. Soc.* **55**, 4875 (1933).
7. E. D. Eastman and W. C. McGavock, *J. Amer. Chem. Soc.* **59**, 145 (1937).
8. R. Berman and F. E. Simon, *Z. Elektrochem.* **59**, 333 (1955).
9. G. E. Gibson and W. F. Giauque, *J. Amer. Chem. Soc.* **45**, 93 (1923).
10. J. O. Clayton and W. F. Gaiuque, *J. Amer. Chem. Soc.* **54**, 2610 (1932).
11. L. Pauling, *J. Amer. Chem. Soc.* **57**, 2680 (1935); W. F. Giauque and W. Stout, *J. Amer. Chem. Soc.* **58**, 1144 (1936).
11a. E. A. Long and J. D. Kemp, *J. Amer. Chem. Soc.* **58**, 1829 (1936).
12. K. Clusius, L. Popp, and A. Frank, *Physica* (*Utrecht*) **4**, 1105 (1937).
13. R. W. Hill and B. W. A. Ricketson, *Phil. Mag.* **45**, 277 (1954).
14. W. Nernst, *Ber. Ko. Preuss. Acad.* (February 1, 1912).
15. R. V. Pound, *Phys. Rev.* **81**, 156 (1951); E. M. Purcell and R. V. Pound, *Phys. Rev.* **81**, 279 (1951); N. F. Ramsey, *Phys. Rev.* **103**, 20 (1956).
16. C. V. Heer and J. G. Daunt, *Phys. Rev.* **81**, 447 (1951).
17. P. Seligmann, D. O. Edwards, R. E. Sarwinski, and J. Tough, *Phys. Rev.* **181**, 415 (1969); J. Landau, J. T. Tough, N. R. Brubaker, and D. O. Edwards, *Phys. Rev. Lett.* **23**, 283 (1969). (References to other research is given in these papers.)
18. H. London, G. R. Clarke, and E. Mendoza, *Phys. Rev.* **128**, 1992 (1962); O. E. Vilches and J. C. Wheatley, *Phys. Lett. A* **24**, 740 (1967); **25**, 344 (1967).
19. D. O. Edwards, Private communication, 1971.
20. M. Tribus, "Thermostatics and Thermodynamics." Van Nostrand-Reinhold, Princeton, New Jersey, 1961.
21. F. H. Getman and F. Daniels, "Outlines of Theoretical Chemistry." Wiley, New York, 1937.
22. A. J. Wakefield, *Proc. Cambridge Phil. Soc.* **47**, 419, 799 (1951).
23. C. Domb, *Advan. Phys.* **9**, 150–361 (1960).
24. S. G. Brush, *Rev. Mod. Phys.* **39**, 883 (1967).
24a. G. F. Newell and E. W. Montroll, *Rev. Mod. Phys.* **25**, 352 (1953).
25. M. E. Fisher, *Rep. Progr. Phys.* **30**, 616 (1967).
26. K. Huang, "Statistical Mechanics." Wiley, New York, 1963.
27. M. F. Sykes and M. E. Fisher, *Physica* (*Utrecht*) **28**, 919, 939 (1962).
28. L. P. Kadanoff, W. Gotze, D. Hamblen, R. Hecht, E. A. S. Lewis, V. V. Palciauskas, M. Raye, J. Swift, D. Aspnes, and J. Kane, *Rev. Mod. Phys.* **39**, 395 (1967); J. Stephenson, *J. Math. Phys.* **7**, 1123 (1966).
29. G. A. Baker, Jr. *Phys. Rev.* **124**, 768 (1961).
30. P. Heller, *Rep. Progr. Phys.* **30**, 731 (1967).
31. H. L. Lorentzen, Statistical mechanics of equilibrium and non-equilibrium. *Proc. Int. Symp., Aachen, 1964*, p. 262.

32. D. T. Teaney, Critical phenomena. *Nat. Bur. Stand.* (*U.S.*) *Spec. Publ.* **No. 273**, 50 (1966).
33. M. D. Rechten, S. C. Moss, and B. L. Averback, *Phys. Rev. Lett.* **24**, 1485 (1970).
33a. R. B. Griffith, *Phys. Rev. Lett.* **24**, 1479 (1970).
33b. F. Y. Wu, *Phys. Rev. Lett.* **24**, 1476 (1970).
34. P. A. Redhead, J. P. Hobson, and E. V. Kornelsen, "The Physical Basis of Ultrahigh Vacuum." Chapman & Hall, London, 1968.
35. J. P. Hobson and R. A. Armstrong, *J. Phys. Chem.* **67**, 2000 (1963).
36. C. V. Heer, *J. Chem. Phys.* **55**, 4066 (1971).
37. D. Brewer, *J. Low Temp. Phys.* (1970); see also J. G. Dash, *Ibid.* (1970).
38. A. V. MacRae, K. Muller, J. J. Lander, J. Morrison, and J. C. Phillips, *Phys. Rev. Lett.* **22**, 1048 (1969).

# Stochastic Processes, Noise, and Fluctuations

# X

## 10.1 Introduction

This chapter introduces the time domain into the statistical problem and considers a number of simple problems. The response of a damped simple harmonic oscillator driven by a series of random pulses or by thermal noise is studied in considerable detail. This detail seems warranted since a thorough knowledge of this simple system permits a better understanding of systems with more than one degree of freedom. Brownian motion, random pulses which drive mechanical and electrical systems, and the response of these systems to thermal noise have a natural place in these discussions. The correlation function and spectral density are introduced as the parameters which describe the response of these systems.

The general theory of fluctuations in the canonical and grand canonical ensembles is included in this chapter. Measurement of the fluctuation in number density by using waves is introduced as a method for measuring the fluctuations. Fluctuations in energy are measured by means of heat capacity measurements and have been included in earlier chapters.

Nonlinear systems and systems with gain rather than loss are included for consideration. A pair of coupled oscillators, which are self-sustained and therefore nonlinear, is discussed as an example of a pair of systems which tends to a singularity in the motion. With suitable coupling parameters, one oscillator can entrain the other and this phenomenon occurs in lasers, mechanical systems, cosmic systems, and perhaps even in the human heart.

In the latter part of the chapter, the characteristic function for continuous variables is introduced and placed in such a form that it can be used in the study of line shapes. This procedure is generalized and the chapter concludes with a formal discussion of random processes.

## 10.2 Thermal Motion of a One-Dimensional System

A simple harmonic oscillator, such as a mass–spring or an inductance–capacitance system, is a basic and simple system and is considered first. In the sense in which statistical mechanics has been used in the previous chapters, for very short periods of time, either system can be regarded as a simple isolated system with energy

$$E = \frac{1}{2} m\dot{q}^2 + \frac{1}{2} Kq^2 \quad \text{or} \quad E = \frac{1}{2} L\dot{q}^2 + \frac{q^2}{2C} \tag{10.1}$$

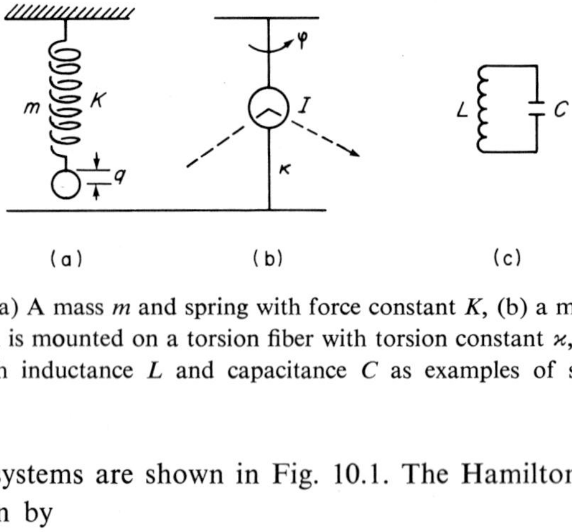


**Fig. 10.1** (a) A mass $m$ and spring with force constant $K$, (b) a mirror of moment of inertia $I$ which is mounted on a torsion fiber with torsion constant $\varkappa$, and (c) an electrical circuit with inductance $L$ and capacitance $C$ as examples of simple harmonic systems.

These simple systems are shown in Fig. 10.1. The Hamiltonian for either system is given by

$$H = \frac{p^2}{2m} + \frac{1}{2} m\omega_0^2 q^2 \quad \text{or} \quad H = \frac{p^2}{2L} + \frac{1}{2} L\omega_0^2 q^2$$
$$\omega_0^2 = \frac{K}{m} \quad \text{or} \quad \omega_0^2 = \frac{1}{LC} \tag{10.2}$$

where $p = m\dot{q}$ or $p = L\dot{q}$. By the equipartition theorem, the average energy is

$$\langle H \rangle = kT \tag{10.3a}$$

for either system. Quantum mechanical considerations require the average thermal energy of an oscillator to be

$$\langle H \rangle = \frac{1}{2} h\nu_0 + \frac{h\nu_0}{e^{h\nu_0/kT} - 1} \tag{10.3b}$$

and this reduces to Eq. (10.3a) for $kT > h\nu_0$. This is the problem considered first. The average value of the coordinates or the momenta is $\frac{1}{2}kT$ and in an experiment that measures the spring displacement,

$$\langle \tfrac{1}{2} m\omega_0{}^2 q^2 \rangle = \tfrac{1}{2}kT \tag{10.4}$$

In the same manner, an experiment that measures the momentum or velocity yields

$$\left\langle \frac{p^2}{2m} \right\rangle = \frac{1}{2} kT \tag{10.5}$$

Similar relationships hold for the electrical current. Neither of these are easy to observe or measure, but the twisting motion of a mirror mounted on a suspension wire can be measured. Figure 10.1(b) shows such a suspension. The kinetic energy is $\frac{1}{2}I\dot{\varphi}^2$ and the potential energy is $\frac{1}{2}\varkappa\varphi^2$, where $\varphi$ is the displacement, $I$ the moment of inertia, and $\varkappa$ the torsional constant. The average angular displacement of the mirror is

$$\langle \varphi^2 \rangle = \frac{kT}{\varkappa} \tag{10.6}$$

Results of an experimental measurement [*1*] are shown in Fig. 10.2 and

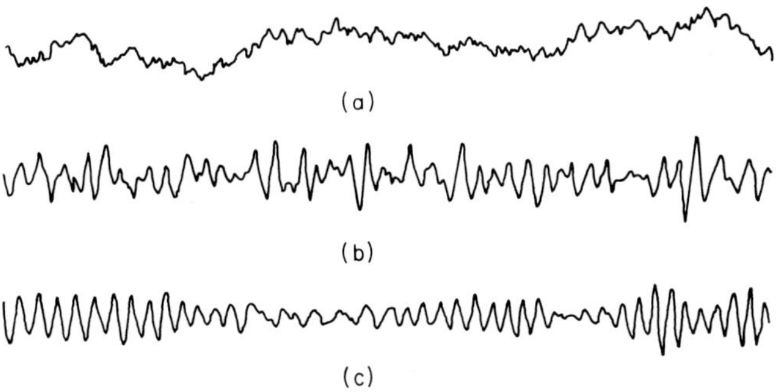


**Fig. 10.2** Record of Brownian motion [*1*] of a sensitive torsion balance: (a) atmospheric pressure, (b) $10^{-3}$ Torr, (c) $10^{-4}$ Torr. [From D. K. C. MacDonald, *Rep. Progr. Phys.* **12**, 56 (1949), Fig. 1. Reproduced by permission of The Institute of Physics and The Physical Society, London.]

will be discussed in greater detail in a later section. It is adequate to note that the temperature can be measured from these curves.

Macroscopic objects always have energy $\frac{1}{2}kT$ associated with each squared term in the appropriate expression for the energy or Hamiltonian. An interesting example is the mounting of two laser mirrors on a massive granite block. The mirrors are separated by a distance $L$. Young's modulus $E_Y$ relates the elongation and elastic energy by

$$\{\text{elastic energy}\} = \frac{1}{2} E_Y \left(\frac{\Delta L}{L}\right)^2 V = \frac{1}{2} kT$$

and the average squared change in length between the mirrors is apparent. Here, $V$ is the volume of the block. The resonant frequency of the light waves between the mirrors follows from $L = (\text{integer})(\lambda/2)$ and $\nu\lambda = c$, and

$$\frac{\Delta\nu}{\nu} = \frac{\Delta L}{L}$$

The mean-square fluctuation in frequency is of the order of

$$\left\langle \left(\frac{\Delta\nu}{\nu}\right)^2 \right\rangle = \frac{kT}{E_Y V} \tag{10.7}$$

Optical lasers operate at a frequency of the order of $5 \times 10^{14}$ Hz, and a $\Delta\nu$ of the order of 1 Hz can occur on a small mounting structure. Unless great care is taken, other sources of noise, such as the air conditioning units vibrating the floor or people walking, can be much more important. Blocks mounted on air cushions allow one to approach the thermal limit of $\frac{1}{2}kT$ for the block.

**Exercise 10.1** Find the average square voltage across a capacitor in an $LC$ circuit in thermal equilibrium with its surroundings. Let $L = 1 \times 10^{-6}$ H, $C = 100 \times 10^{-12}$ F, and $T = 300$°K.

**Exercise 10.2** A laser is supported on a black granite block with a length of 1 m between the mirrors. The block cross section is $10 \times 50$ cm. Find the thermal line broadening for a helium–neon laser operating at 0.63 μm.

Ans. $\langle \Delta\nu^2 \rangle^{1/2} = 1$ to 2 Hz

### *10.2.1 Brownian Motion*

Next, consider the motion of a macroscopic particle in a liquid or gaseous medium. The average kinetic energy of the particle is

$$\langle \tfrac{1}{2} m v_x{}^2 \rangle = \tfrac{1}{2} kT \tag{10.8}$$

by the arguments given in the previous section. For three degrees of freedom, the average energy is $\frac{3}{2}kT$ and the same as for a gas particle. Again, this is all the information given by statistical thermodynamics.

Further analysis of the problem requires an appeal to our knowledge of macroscopic mechanical systems. It is experimentally observed that the driven motion of an object through a medium is accompanied by damping and the equation of motion is of the form

$$m\ddot{x} + \gamma \dot{x} = f(t) \tag{10.9}$$

where $\gamma$ is the damping coefficient; $f(t)$ is the force exerted by the surrounding medium and by external agencies on the mass. Usually, $f(t)$ is the random thermal force of the surrounding medium. The average square displacement of the macroscopic particle has been of first concern in the theory of Brownian motion. This problem was first solved by Einstein [*2*]. A very elementary analysis due to Langevin indicates the nature of the problem. Equation (10.9) can be written as an energy equation by the substitution

$$x\ddot{x} = \frac{1}{2} \frac{d^2x^2}{dt^2} - \dot{x}^2$$

and then

$$\frac{m}{2} \frac{d^2x^2}{dt^2} - m\dot{x}^2 + \frac{\gamma}{2} \frac{dx^2}{dt} = xf(t) \tag{10.10a}$$

This equation is integrated over time to yield

$$x^2 = \frac{2}{\gamma} \int_0^t (m\dot{x}^2)\, dt + \frac{2}{\gamma} \int_0^t xf\, dt - 2\frac{m}{\gamma} x\dot{x} \Big|_0^t \tag{10.10b}$$

Either many systems are averaged as in the ensemble averages, or many measurements are made on the same system. The latter is more like the experimental observations in which the path of one particle is followed. Equation (10.10) is correct if $x^2$ is the displacement during interval $t$ in each case. If a series of such squared displacements $x^2$ is measured, the average value of $\langle \frac{1}{2} m\dot{x}^2 \rangle = \frac{1}{2}kT$ and is replaced by this value before integration. Since the integral is a sum, the order of integration and averaging can be interchanged. The averages $\langle xf \rangle$ and $\langle x\dot{x} \rangle$ can be argued to be zero since $f$ and $\dot{x}$ are random variables with zero average values and are statistically independent of $x$ at the same instant of time $t$. With these assumptions, the average squared displacement during time interval $t$ is

$$\langle x^2 \rangle = \frac{2kT}{\gamma} t \tag{10.11}$$

and *the squared displacement is linear in time t*.

The term $\gamma$ is the damping constant for the particle. If the particle is spherical and the average path of a molecule in the gas is small compared to the size of the object or if the particle is in a liquid, Stokes's Law can be used for the viscous drag force, and

$$\gamma = 6\pi\eta a \tag{10.12}$$

where $\eta$ is the coefficient of viscosity and $a$ the radius. In his experiments, Perrin [3] observed the motion of a single particle with the aid of a microscope. The microscope had in its field of view a series of mutually perpendicular lines as on graph paper with $3.4 \times 10^{-6}$ m between divisions. The distance $x$ traveled in 30 sec was measured. This value of $x$ was squared and then the average over a set of $\langle x^2 \rangle$ was taken. Figure 10.3 shows the characteristic random motion. In another set of experiments, Perrin used a sphere with a spot on it and observed the rotational Brownian motion.

In Chapter III, the random walk continuous in space and time was examined in some detail. It was shown in Eq. (3.103a) that the average square distance traveled was

$$\langle x^2 \rangle = 2Dt \tag{10.13}$$

Einstein used the concept of osmotic pressure acting on a particle to relate the viscous force acting on a particle to the diffusion coefficient. This is equivalent to noting that the probability of being at position $x$ at time $t$ is governed by the diffusion equation (3.101):

$$\frac{\partial p(x, t)}{\partial t} = D \frac{\partial^2 p(x, t)}{\partial x^2} \tag{10.14}$$

and as noted in Chapter III is the Fokker–Planck equation. Einstein further derived the famous formula

$$\langle x^2 \rangle = \frac{2kT}{\gamma} t = 2Dt \tag{10.15}$$

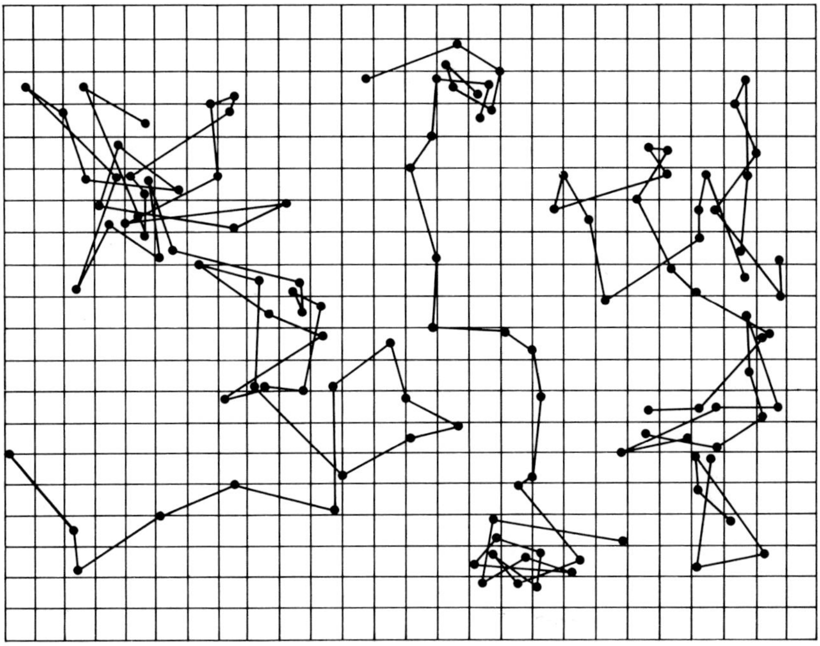

**Fig. 10.3** The position of three different granules of mastic at intervals of 30 sec. The granules have a radius of $0.52 \times 10^{-6}$ m and the spacing between divisions is $3.4 \times 10^{-6}$ m.

which relates the average square displacement, viscous drag, and diffusion coefficient. Equation (10.15) permits the measurement of the Boltzmann constant $k$ and hence Avogadro's number. Perrin was awarded the Nobel prize in 1926 for this measurement.

## 10.3 Random Pulses

The next simplest problem is the examination of a set of random pulses considered as a driving force.[1] Suppose there are a series of $K$ random pulses during the interval $T$; then the force as a function of time is

$$f_K(t) = \sum_{k=1}^{K} b(t - t_k) \tag{10.16}$$

[1] The discussion of random noise uses many of the methods discussed by Rice [*4*]. These papers are collected, along with many other excellent papers, by Wax [*4a*].

where each pulse starts at time $t_k$ and is zero after $t_k + \Delta$. During $t_k$ to $t_k+\Delta$, its shape is described by the function $b(t - t_k)$. Such a function has a Fourier transform

$$b(\nu) = \int_{t_k}^{t_k+\Delta} dt\, b(t - t_k) e^{i2\pi\nu t} \tag{10.17}$$

The Fourier integral representation is used throughout this chapter, with

$$y(t) = \int_{-\infty}^{\infty} d\nu\, y(\nu) e^{-i2\pi\nu t} \tag{10.18a}$$

$$y(\nu) = \int_{-\infty}^{\infty} dt\, y(t) e^{i2\pi\nu t} \tag{10.18b}$$

Inherent in the use of these transformations is the definition of the $\delta$ function,

$$\int_{-\infty}^{\infty} d\xi\, e^{i2\pi\xi\eta} = \delta(\eta) \tag{10.18c}$$

which is zero unless $\eta = 0$. The Fourier transform of $f_K(t)$ is

$$f_K(\nu) = \int_0^T dt \sum_k b(t - t_k) e^{i2\pi\nu t} \approx b(\nu) \sum_{k=1} e^{i\omega t_k} \tag{10.19}$$

Here $\omega = 2\pi\nu$ and is used whenever convenient. Since the $t_k$ are chosen at random, the terms in the sum tend to cancel and this is similar to adding a number of sine and cosine waves with the same frequency and random phase. The addition of a set of random vectors in two dimensions where $\cos \omega t_k$ is along $\hat{x}$ and $\sin \omega t_k$ is along $\hat{y}$ is a similar problem. For $K$ terms, the sum is of the order of $\sqrt{K}$. There is a term which is independent of the random phase and we have the product

$$f_K(\nu) f_K{}^*(\nu') = b(\nu) b^*(\nu') \sum_{\alpha,\beta} \exp i(\omega t_\alpha - \omega' t_\beta) \approx K \,|\, b(\nu) \,|^2\, \delta_{\nu\nu'}$$

The double sum is almost zero unless $\alpha = \beta$ and $\omega = \omega'$. A term of this type incorporates almost all of the observable phenomena for a series of random pulses.

If an ensemble of intervals of duration $T$ is averaged and if the average number of pulses per second is $\lambda$, the probability of $K$ pulses in interval $T$ is given by the Poisson distribution (3.50), and

$$P_K = e^{-\lambda T} \frac{(\lambda T)^K}{K!}$$

The spectral distribution of this random source is

$$\langle | f(\nu) |^2 \rangle = \sum_K P_K | f_K(\nu) |^2 = \lambda T | b(\nu) |^2 \tag{10.20}$$

This expression is linear in the time interval $T$ selected for the measurement, and since the interval $T$ is arbitrary, this quantity can be related to the spectral density for the source by the definition given in the next section:

$$\psi(\nu) = \lim_{T\to\infty} \frac{1}{T} \lambda T | b(\nu) |^2 = \lambda | b(\nu) |^2 \tag{10.21}$$

As will be discussed in subsequent sections, this contains most of the known information for this random source. The spectral distribution is that of a single pulse and the average rate of arrival is $\lambda$. All phase information between pulses is lost.

A second quantity of interest is the average value of the square of the difference of the force at $t + \tau$ and the force at $t$, or $\langle [f(t + \tau) - f(t)]^2 \rangle$. It is sufficient to determine the average value of the quantity

$$\psi(\tau) = \langle f(t)f(t + \tau) \rangle = \sum_K P_K \overline{f_K(t) f_K(t + \tau)} \tag{10.22}$$

where the bar indicates a time average. The quantity

$$f_K(t)f_K(t + \tau) = \sum_{k,m} b(t - t_k)b(t + \tau - t_m)$$

is completely determined for given values of $t_k$. But $t_k$ is to be treated as a random variable and the probability that $t_k$ occurs in interval $dt_k$ is $dt_k/T$. An average over $t_k$ for $m = k$ yields

$$-\int_0^T dt_k \frac{1}{T} b(t - t_k)b(t + \tau - t_k) = \frac{1}{T} \int_{-\infty}^{\infty} d\xi \, b(\xi)b(\xi + \tau)$$

A change of variable to $\xi = t - t_k$ and the knowledge that $b(\xi)$ is zero outside $0 < \xi < \Delta$ permits the simplification. There are $K$ of these terms and an average over $P_K$ for the Poisson distribution yields $\lambda T$. For $m \neq k$, the product of two independent integrals occurs and reduces to

$$\left(\frac{\bar{b}}{T}\right)^2 = \left[\frac{1}{T} \int d\xi \, b(\xi)\right]^2$$

or the square of the average value of the pulse. There are $K(K - 1)$ of these terms and an average over $P_K$ yields a coefficient of $(\lambda T)^2$. Both terms are independent of the interval $T$ as long as $T$ is large compared to the pulse duration $\Delta$. With these, the average value becomes

$$\psi(\tau) = \langle f(t)f(t+\tau)\rangle = \lambda \int_{-\infty}^{\infty} d\xi\, b(\xi)b(\xi+\tau) + (\lambda\bar{b})^2 \qquad (10.23)$$

and this is the correlation function defined in the next section. This expression is independent of the selection of the measuring time $t$ and depends only on the interval $\tau$. Random processes that are invariant under time translation are referred to as stationary random processes, and for these processes,

$$\langle f(t+\tau)^2\rangle = \langle f(t)^2\rangle \qquad (10.24)$$

The intuitively apparent quantity in a measurement can be written as

$$\langle [f(t+\tau) - f(t)]^2\rangle = 2[\psi(0) - \psi(\tau)] \qquad (10.25)$$

for these random pulses.

### 10.3.1 *Radiation as an Example of a Series of Random Pulses*

A specific form is now selected for the random pulse $b(t - t_k)$. In the classical wave train theory of emission by atoms, the pulse train emitted by an atom is assumed to have the form

$$b(\xi) = 2Ae^{-\Gamma\xi/2}\cos 2\pi\nu_0\xi, \qquad \xi = t - t_k > 0 \qquad (10.26a)$$

The Fourier transform of this function is

$$b(\nu) \approx A[i(\omega_0 - \omega) + \tfrac{1}{2}\Gamma]^{-1}$$

and

$$|\, b(\nu)\,|^2 \approx \frac{A^2}{(\omega_0 - \omega)^2 + (\Gamma/2)^2} \qquad (10.26b)$$

and is usually referred to as the Lorentz line shape. The approximate sign indicates that the high-frequency term $(\omega_0 + \omega)$ is omitted. Here $b(\xi)$ has an average $\bar{b}$ which is small for $\omega_0 \gg \Gamma$ and is neglected in further considerations. The correlation function for the pulse follows from

$$\overline{b(\xi)b(\xi+\tau)} = \int_0^{\infty} d\xi\, b(\xi)b(\xi+\tau) = \frac{2A^2}{\Gamma} e^{-\Gamma|\tau|/2}\cos 2\pi\nu_0\tau \qquad (10.26c)$$

Now assume that the source is composed of atoms which emit an average number $\lambda$ of quanta per second and these quanta are described by $b(\xi)$ at the position of observation. The electric field is a superposition of all pulses, and for observation through a linear polarizer,

$$E_x(t) = \sum_k b(t - t_k)$$

is the electric field at the observer. The emission time $t_k$ is unknown and the phase is not a measurable quantity. A measurable quantity is

$$\psi(\tau) = \langle E_x(t)E_x(t+\tau)\rangle = \lambda \int_{-\infty}^{\infty} d\xi\, b(\xi)b(\xi+\tau) \tag{10.27a}$$

or the spectral density of the radiation is

$$\psi(\nu) = \lambda \,|\, b(\nu)\,|^2 \tag{10.27b}$$

Intensity and spectral density are intimately related. The intensity of radiation is given by the Poynting vector $\mathbf{E} \times \mathbf{H}$. For plane waves in free space, a traveling wave $E_0 \cos(kx - \omega t)$ has an intensity of $\frac{1}{2}c\varepsilon_0 E_0{}^2$, or $2c\varepsilon_0 E_\nu{}^2$, where $E_\nu$ is the coefficient of $e^{-i\omega t}$, and $\frac{1}{2}c\varepsilon_0\psi(\tau = 0)$ is the energy crossing a surface. Intensity and spectral density are related by

$$I(\nu)\, d\nu = 2c\varepsilon_0\psi(\nu)\, d\nu \quad \frac{\text{W}}{\text{m}^2} \tag{10.28}$$

which is the energy per second crossing 1 m$^2$. This is the energy which would pass through a narrowband filter of width $d\nu$ and stimulate a detector.

Exercise 10.3 Show that the spectral density and correlation function for a pulse rate of $\lambda$ per second for pulses with the shape given by Eq. (10.26a) are

$$\psi(\nu) = \lambda A^2 \frac{1}{(\omega - \omega_0)^2 + (\Gamma/2)^2} \tag{10.29a}$$

and

$$\psi(\tau) = 2A^2\left(\frac{\lambda}{\Gamma}\right)e^{-\Gamma|\tau|/2} \cos 2\pi\nu_0\tau \tag{10.29b}$$

## 10.4 Correlation Function and Spectral Density

If a physical process is described by $x(t)$ and if the ensemble average of the quantity $x(t)x(t+\tau)$ is independent of the choice of the instant of time which is selected for the average, the ensuing function

$$\psi(\tau) = \langle x(t)x(t+\tau)\rangle \tag{10.30}$$

is defined as the correlation function of a stationary random process. A more general correlation function and an extended discussion are given in Section 10.14. Inherent in this definition is that the average value of $x(t)$ is independent of the choice of $t$ and is constant:

$$\langle x(t) \rangle = \text{constant}$$

and also the average square is independent of the instant of time or

$$\langle x(t)^2 \rangle = \langle x(t+\tau)^2 \rangle$$

As $\tau$ becomes large, it is expected from physical considerations that the correlation between $x(t)$ and $x(t+\tau)$ decreases and the two quantities become statistically independent, or

$$\langle x(t)x(t+\tau) \rangle \underset{\tau \to \infty}{\longrightarrow} \langle x \rangle \langle x \rangle$$

A consideration of the average value of the nonnegative quantity $\langle [x(t) \pm x(t+\tau)]^2 \rangle = 0$ permits one to show that $\psi(0)$ is the maximum value attained by the correlation function, that is,

$$\psi(0) \geq |\psi(\tau)| \tag{10.31a}$$

Since the function is stationary and cannot depend on $t$, it follows that $\psi(\tau)$ is an even function of $\tau$,

$$\psi(-\tau) = \psi(\tau) \tag{10.31b}$$

This is readily shown by replacing $t$ by $t - \tau$ in $\langle x(t)x(t+\tau) \rangle$. These are some general features of the correlation function for a stationary random process. These features are apparent for the correlation function of a series of random pulses which was discussed in Section 10.3.

Let $x(t)$ be a particular realization of the physical process which is described by a correlation function for a stationary random process. The measurement interval $T$ can be chosen arbitrarily long and the Fourier integral transform of this realization is given by

$$x(\nu) = \int_0^\infty dt\, x(t) e^{i\omega t}$$

where $x(t)$ is regarded as zero before the start of the measurement. Since $x(t)$ is a real function, the Fourier coefficients are related by

$$x(-\nu) = x(\nu)^*$$

Next, form the quantity

$$\lim_{T\to\infty} \frac{|x(\nu)|^2}{T} = \lim_{T\to\infty} \frac{1}{T} \int_0^T dt' \int_0^T dt''\, x(t')x(t'') \exp i\omega(t' - t'')$$

for a particular realization. If an ensemble of realizations of this form is summed, this average can be written as the power spectrum of this stationary random process:

$$\psi(\nu) = \lim_{T\to\infty} \frac{\langle |x(\nu)|^2\rangle}{T} = \lim_{T\to\infty} \frac{1}{T} \int_0^T dt' \int_0^T dt''\, \langle x(t')x(t'')\rangle \exp i\omega(t'-t'') \tag{10.32}$$

Since the process of integration is a sum, it is permissible to interchange the order of integration and the ensemble average. With the substitution $t' - t'' = \tau$, the integrand can be written in terms of the correlation function

$$\langle x(t')x(t'')\rangle = \langle x(t)x(t+\tau)\rangle = \psi(\tau)$$

and it is apparent that the integrand is independent of the variable $t$. Completing the integration yields

$$\psi(\nu) = 2\int_0^\infty d\tau\, \psi(\tau) \cos \omega\tau \tag{10.33a}$$

and the transformation of the integral is discussed in Exercise 10.5. From its definition or from the above equation, it is apparent that $\psi(\nu)$ is an even function of $\nu$, and

$$\psi(-\nu) = \psi(\nu)$$

The cosine transform of the even function of $\psi(\nu)$ is (see Exercise 10.4)

$$\psi(\tau) = 2\int_0^\infty d\nu\, \psi(\nu) \cos \omega\tau \tag{10.33b}$$

These equations relating $\psi(\tau)$ and $\psi(\nu)$ are known as the Wiener–Khinchin relations.

If $x(t)$ is the coordinate $q(t)$ for a particle, then it is expected that for a particular realization of the coordinate, $q(t)$ is a continuous function and has a derivative $\dot{q}(t)$. If $q(\nu)$ is the Fourier component of the coordinate, then

$$\dot{q}(\nu) = -i\omega q(\nu)$$

is the Fourier component $\dot{q}(\nu)$ of the velocity. Let $\theta(\nu)$ be the spectral density of the coordinate $q$ and let $\dot{\theta}(\nu)$ be the spectral density of the velocity. Then the spectral densities are related by

$$\dot{\theta}(\nu) = \omega^2\theta(\nu) \tag{10.34a}$$

Using the Fourier transform of the spectral density, Eq. (10.33b), the correlation function of the velocity $\dot{q}(t)$ can be written as

$$\dot{\theta}(\tau) = \langle \dot{q}(t)\dot{q}(t+\tau)\rangle = 2\int_0^\infty d\nu\, [\omega^2\theta(\nu)] \cos \omega\tau \tag{10.34b}$$

Direct differentiation of the coordinate correlation function yields this same relationship,

$$\langle \dot{q}(t)\dot{q}(t+\tau)\rangle = -\frac{d^2}{d\tau^2}\langle q(t)q(t+\tau)\rangle \tag{10.34c}$$

One can show in a similar manner that the correlation between the coordinate and the velocity is given by

$$\langle q(t)\dot{q}(t+\tau)\rangle = -2\int_0^\infty d\nu\, [\omega\theta(\nu)] \sin \omega\tau \tag{10.35a}$$

or

$$\langle q(t)\dot{q}(t+\tau)\rangle = \frac{d}{d\tau}\langle q(t)q(t+\tau)\rangle \tag{10.35b}$$

A more general discussion of these relationships is given by Beckman [5]. Equation (10.35a) is zero at $\tau = 0$ and there is no correlation between the coordinate and the velocity at the same instant, that is,

$$\langle q(t)\dot{q}(t)\rangle = 0 \tag{10.35c}$$

This is expected from physical considerations. The velocity $\dot{q}$ depends on the prior history, but is not correlated with $q(t)$ at the same instant.

**Exercise 10.4** Show from the definition of the Fourier integral transform that an even function $f(\xi) = f(-\xi)$ has a cosine transform

$$f(\eta) = 2\int_0^\infty d\xi\, f(\xi) \cos 2\pi\xi\eta \tag{10.36}$$

**Exercise 10.5** This problem considers the steps necessary to reduce the double integral given by Eq. (10.32), which occurs in problems on transition rates in Chapter XI.

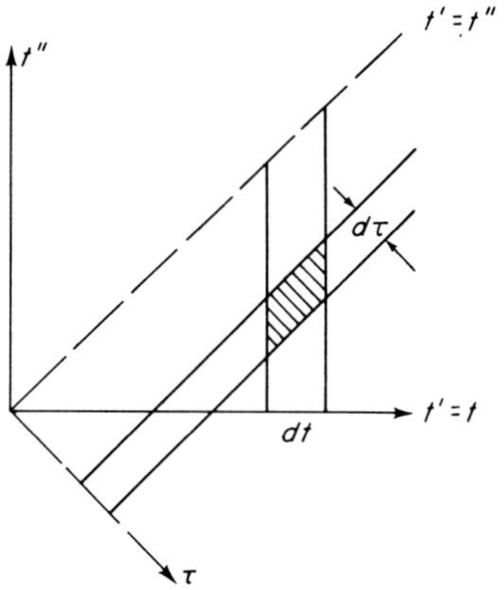


**Fig. 10.4** Coordinate transformation from the variables $t'$ and $t''$ to $t' - t'' = \tau$ and $t' = t$, which is discussed in Exercise 10.5.

(a) Show that

$$\lim_{T\to\infty} \frac{1}{T} \int_0^T dt' \int_0^T dt'' \, g(t' - t'') = \int_0^\infty d\tau \, [g(\tau) + g(-\tau)] \tag{10.37}$$

when the integrand falls to zero for large $\tau$.

(b) Show that integration procedures over the area of the rectangle permit

$$\int_0^T dt' \int_0^T dt'' \, g(t', t'') = \int_0^T dt' \int_0^{t'} dt'' \, [g(t', t'') + g(t'', t')]$$

(c) Assume that $g(t', t'') = g(t' - t'')$ and is a function of the difference only. Let $t' - t'' = \tau$ and $t' = t$. Show that $dt' \, dt'' = dt \, d\tau$ from the Jacobian of the transformation given by Eq. (4.69). Examine the region of integration shown in Fig. 10.4 and show that the integral reduces to

$$\int_0^T d\tau \, [g(\tau) + g(-\tau)] \int_\tau^T dt$$

The last integral reduces to $T - \tau$ and in the limit of large $T$, is linear in $T$ if $g(\tau)$ falls to zero for large $\tau$.

## 10.5 Response of a Simple Harmonic Oscillator to Random Pulses and Thermal Noise

In this section, the equation of motion of a damped driven simple harmonic oscillator is assumed to have the form

$$m\ddot{q} + \gamma\dot{q} + m\omega_0^2 q = f(t) \tag{10.38}$$

$\gamma\dot{q}$ is the viscous drag. If $m$ is replaced by $L$ and $\gamma$ by $R$, this is the equation for an $RLC$ circuit. Multiplying Eq. (10.38) by $\dot{q}$ and rearranging terms yields an expression for power of the form

$$\frac{dH}{dt} = \frac{d}{dt}\left(\frac{m\dot{q}^2}{2} + \frac{m\omega_0{}^2q^2}{2}\right) = \dot{q}f - \gamma\dot{q}^2 \tag{10.39}$$

and relates the change in stored energy to the power furnished by the source $\dot{q}f$ and the power dissipated by frictional losses $\gamma\dot{q}^2$.

This simple harmonic oscillator has a linear response to the driving force $f(t)$. If $f(t)$ is a sinusoidal signal $f_\nu e^{-i\omega t}$ and the response is $q_\nu e^{-i\omega t}$, Eq. (10.38) can be written as $q_\nu = G(\nu)f_\nu$. If $f(t)$ has the more general form

$$f(t) = \int_{-\infty}^{\infty} d\nu\, f(\nu)e^{-i2\pi\nu t} \tag{10.40a}$$

then the solution is

$$q(t) = \int_{-\infty}^{\infty} d\nu\, q(\nu)e^{-i2\pi\nu t} \tag{10.40b}$$

where the Fourier components are simply related by the linear response function

$$q(\nu) = G(\nu)f(\nu) \tag{10.41a}$$

The linear response function for Eq. (10.38), or the damped simple harmonic oscillator, is

$$G(\nu) = \frac{1}{m(\omega_0{}^2 - \omega^2) - i\gamma\omega} \tag{10.41b}$$

The squared values of the Fourier coefficients are related by

$$|\, q(\nu)\,|^2 = |\, G(\nu)\,|^2\,|\, f(\nu)\,|^2 \tag{10.42}$$

for a particular realization of $f(t)$.

The response of the simple harmonic oscillator to a stationary random perturbation is now considered. Following our previous discussion in Section 10.4, an ensemble average is taken of each side of Eq. (10.42), and then the limit, as the measuring interval $T$ becomes large, can be written as

$$\lim_{T\to\infty} \frac{1}{T}\langle|\, q(\nu)\,|^2\rangle = |\, G(\nu)\,|^2 \lim_{T\to\infty} \frac{1}{T}\langle|\, f(\nu)\,|^2\rangle \tag{10.43}$$

The character of $|\, G(\nu)\,|^2$ is fixed by the response function of the system and is independent of the average. The driving function is described by the spectral density for the source or $\psi(\nu)$. For a stationary random driving force, that is, a force for which $\langle f(t)f(t+\tau)\rangle$ is independent of $t$, the spectral density $\theta(\nu)$ of the coordinate $q$ is

$$\theta(\nu) = |G(\nu)|^2\psi(\nu) \qquad (10.44a)$$

Equation (10.33b) can be used to obtain the correlation function for coordinate $q$, and

$$\theta(\tau) = \langle q(t)q(t+\tau)\rangle = 2\int_0^\infty d\nu\, |G(\nu)|^2\psi(\nu)\cos 2\pi\nu\tau \qquad (10.44b)$$

### *10.5.1 Random Pulses*

If the driving force is a series of random pulses, the spectral density of the source is given by Eq. (10.27b) as $\psi(\nu) = \lambda|b(\nu)|^2$. Again, $\lambda$ is the average number of pulses per second and $|b(\nu)|^2$ is the Fourier coefficient that describes the pulse. Thus the correlation function for the coordinate $q$ is given by

$$\langle q(t)q(t+\tau)\rangle = 2\int_0^\infty d\nu\, [\lambda|G(\nu)|^2|b(\nu)|^2]\cos 2\pi\nu\tau \qquad (10.45a)$$

Features characteristic of the source $|b(\nu)|^2$ and features characteristic of the detector $|G(\nu)|^2$ are apparent in this correlation function. A quantity which is somewhat easier to visualize is the coordinate difference between a measurement at $t$ and $t+\tau$, or

$$\langle[q(t+\tau) - q(t)]^2\rangle = 2[\theta(0) - \theta(\tau)] \qquad (10.45b)$$

for the coordinate $q$. If $q = \varphi$ is the angular displacement of the mirror shown in Fig. 10.1(b) or the displacement shown by the traces in Fig. 10.2, this measurement is easily visualized.

Exercise 10.6 (a) Assume that the force on an electron or a charged particle is due to an electric field $f(t) = eE(t)$, which is due to a sequence of random pulses with a spectral density given by Eq. (10.29a). Then assume that the equation of motion of the electron is given by Eq. (10.38). With these assumptions, show that the spectral density for the electron coordinate $q$ is given by

$$\theta(\nu) = \frac{1}{m^2(\omega_0^2 - \omega^2)^2 + \gamma^2\omega^2}\,\frac{e^2\lambda A^2}{(\omega - \omega_s)^2 + (\Gamma/2)^2} \qquad (10.46)$$

Let $\nu_s$ be the basic frequency of the source and $\nu_o$ the resonant frequency of the harmonic oscillator.

(b) Show that $\Gamma \gg \gamma$ permits a measurement of $\gamma$ for the detector and $\gamma \gg \Gamma$ permits a measurement of $\Gamma$ for the source. Why does large $\Gamma$ imply a pulse of short duration and why is the response of the oscillator similar to natural decay for a pulse of short duration? Find an approximate value for Eq. (10.45b) for $\Gamma \gg \gamma$.

### *10.5.2 Thermal Source*

If the simple harmonic oscillator is in thermal equilibrium with its surroundings, for short periods of time, the system can be regarded as isolated and the energy of the system is

$$H = \tfrac{1}{2}m\dot{q}^2 + \tfrac{1}{2}m\omega_0{}^2q^2 \tag{10.47a}$$

If $q(t)$ is expanded in terms of the Fourier integral coefficients $q(\nu)$, then

$$H = \int_{-\infty}^{\infty} d\nu \int_{-\infty}^{\infty} d\nu' \; [\tfrac{1}{2}m(\omega\omega' + \omega_0{}^2)][q(\nu)q(\nu')^*] \exp i(\omega' - \omega)t \tag{10.47b}$$

From the equipartition theorem, the average thermal value of $H$ is $\langle H\rangle = kT = 1/\beta$, where $\beta$ is used to avoid confusion between the period $T$ and the temperature $T$. Thus the ensemble average of Eq. (10.47b) is independent of the time variable $t$ and this is only possible with $\nu = \nu'$. Performing the following operation on Eq. (10.47b):

$$\lim_{T\to\infty} \frac{1}{T} \int_0^T dt \langle \; \rangle$$

permits the proper introduction of $\delta(\nu - \nu')$ by the time integration and the introduction of the spectral density $\theta(\nu)$ for the coordinate $q$. Equation (10.42), $\theta(\nu) = |\, G(\nu)\,|^2\psi(\nu)$, can be used to replace $\theta(\nu)$ by the spectral density of the source. With these operations, the expression reduces to

$$\frac{1}{\beta} = m \int_0^{\infty} d\nu \; (\omega^2 + \omega_0{}^2)\,|\, G(\nu)\,|^2\, \psi_{\mathrm{T}}(\nu) \tag{10.48}$$

Since this integral is invariant to various choices of $\nu_0$ and $\gamma$, the spectral density of the source $\psi_{\mathrm{T}}(\nu)$ must be independent of frequency. $\psi_{\mathrm{T}}$ can be taken outside the integral and then by integrating around the poles of the response function $| G(\nu) |^2$, the following relationship for the spectral density of the thermal source is obtained:

$$\psi_T(\nu) = \frac{2\gamma}{\beta} = 2\gamma kT \tag{10.49}$$

The equipartition theorem can be used only when $h\nu < kT$ and is limited to values of $\nu < kT/h$, or frequencies less than $10^{14}$ Hz at 300°K. The spectral density $\psi_T(\nu)$ must decrease toward zero for large $\nu$. This spectral density implies a correlation function for a thermal source or thermal noise source of

$$\psi_T(\tau) = \langle f(t)f(t+\tau)\rangle_T = 2\gamma kT\,\delta(\tau) \tag{10.50}$$

This follows since twice the integral of the cosine function, integrated from zero to infinity, is another definition of the $\delta$ function. Again, this is overly restrictive since $\psi(\nu)$ must cutoff at large $\nu$ and this implies some correlation for very short times.

The time correlation between the positions of the coordinate $q$ is given by

$$\langle q(t)q(t+\tau)\rangle = (2\gamma kT)2\int_0^\infty d\nu\,| G(\nu) |^2 \cos 2\pi\nu\tau \tag{10.51}$$

For $\tau = 0$, this yields the expected value

$$\tfrac{1}{2}m\omega_0^2\langle q^2\rangle = \tfrac{1}{2}kT$$

from equipartition and is in agreement with the average thermal value of the coordinate given by statistical mechanics.

Exercise 10.7 Show that the spectral density for the motion of the mass in Eq. (10.38) is

$$\theta(\nu) = | G(\nu) |^2\,\psi_T(\nu) = 2\gamma kT\frac{1}{m^2(\omega_0^2 - \omega^2)^2 + \gamma^2\omega^2} \tag{10.52}$$

for a thermal source.

#### 10.5.2.1 Narrow Resonance or Small Damping

If $\gamma \ll \omega_0$ and the resonance is very sharp, the spectral density describing the motion of the coordinate $q$ is approximately

$$\theta(\nu) \approx \frac{kT}{m\omega_0^2}\,\frac{\gamma/2m}{(\omega_0 - \omega)^2 + (\gamma/2m)^2} \tag{10.53a}$$

This expression can be integrated to obtain the correlation function

$$\theta(\tau) = \langle q(t)q(t+\tau)\rangle \approx \frac{kT}{m\omega_0^2} e^{-(\gamma/2m)\tau} \cos 2\pi\nu_0\tau \qquad (10.53b)$$

and for small $\gamma/2$, the motion of this system is almost sinusoidal about the resonant frequency $\nu_0$. From the development in Section 10.13, the probability of a particular value of the coordinate $q$ at time $t$ is

$$P[q] = \frac{1}{[2\pi\theta(0)]^{1/2}} \exp\left(-\frac{q^2}{2\theta(0)}\right) = \left(\frac{m\omega_0^2}{2\pi kT}\right)^{1/2} \exp\left(-\frac{\frac{1}{2}m\omega_0^2q^2}{kT}\right)$$

where $\theta(0) = \langle q^2\rangle$. This is in accord with the canonical distribution, in which the probability of a particular energy is given by $e^{-E/kT}$. A question of interest is the expected rate of fluctuation in amplitude. Since the time for decay of any particular value of $q$ is, from Eq. (10.53b), of the order of $\tau = (\gamma/2m)^{-1}$, the expected time between fluctuations in amplitude is of the order of

$$\Delta\omega \approx \frac{\gamma}{m} = 2\pi\,\Delta\nu$$

where $\Delta\nu$ is the bandwidth between half-power points:

$$\{\text{number of large fluctuations per second}\} \approx \Delta\nu \approx \frac{\gamma}{2\pi m}$$

This is a very elementary derivation. Rice [*4*] shows in a very much more involved analysis that

$$\left\{\begin{matrix}\text{number of zeros in amplitude} \\ \text{of a narrow band Gaussian system}\end{matrix}\right\} \approx \{\text{bandwidth}\} \qquad (10.54)$$

The random motion of a galvanometer mirror is described by Eq. (10.38) and Eq. (10.55) (see Exercise 10.8) and this is readily apparent in Fig. 10.2. Narrowband noise has the appearance of a sinusoid. With the very narrowband amplifiers now available, considerable care is necessary when the output is observed on an oscilloscope. Noise and signal give quite similar traces for short time intervals.

Exercise 10.8 Use the equation of motion for a torsional oscillator $I\ddot{\varphi} + \gamma\dot{\varphi} + \varkappa\varphi = f(t)$ and the spectral density for a thermal source to find the spectral density of the response $\theta(\nu)$. Assume the damping is small, or $\gamma \ll \omega$, and show that

$$\langle[\varphi(t) - \varphi(t+\tau)]^2\rangle \approx 2\frac{kT}{I\omega_0^2}[1 - e^{-(\gamma/2I)\tau}\cos 2\pi\nu_0\tau] \quad (10.55)$$

Sketch the type of motion you might observe for $\varphi(t)$ and compare with Fig. 10.2. Explain the effect of varying the gas pressure.

## 10.6 Correlation Function and Spectral Density for Brownian Motion

Let the equation of motion of a particle in a viscous medium be given by

$$m\ddot{q} + \gamma\dot{q} = f(t) \quad (10.56a)$$

or by the equivalent response equation

$$q(\nu) = G(\nu)f(\nu) \quad (10.56b)$$

where the response function is

$$G(\nu) = \frac{1}{-m\omega^2 - i\omega\gamma} \quad (10.57)$$

The correlation for velocity $\dot{q}$ is examined first and the symbolic notation for the Fourier coefficients of $\dot{q}(t)$ is used, that is, $\dot{q}(\nu) = -i\omega q(\nu)$. With this notation, the correlation function for the velocity for a thermal source is given by

$$\langle\dot{q}(t)\dot{q}(t+\tau)\rangle = (2\gamma kT)2\int_0^\infty d\nu\,[\omega^2 \mid G(\nu)\mid^2]\cos 2\pi\nu\tau$$

$$= \frac{kT}{m}e^{-(\gamma/m)|\tau|} \quad (10.58)$$

The velocity is correlated for a time of the order of $(\gamma/m)^{-1}$. This is in accord with the expected average velocity for a free particle, and at $\tau = 0$,

$$\tfrac{1}{2}m\langle\dot{q}^2\rangle = \tfrac{1}{2}kT$$

The correlation function for position requires greater care. The appropriate expression is the same as Eq. (10.58) without the $\omega^2$ term under the integral. The response function $\mid G(\nu)\mid^2$ contains a singularity at $\nu = 0$ of the order of $\nu^{-2}$. The term $\nu^2 \mid G(\nu)\mid^2$ occurred in the correlation function for velocity and the singularity was removed and of no concern. Since $\psi_T(\nu)$ does not contain a dc term with $\nu = 0$, this singularity must be treated as a principal value in the correlation function for position. This procedure is avoided by examining the quantity that is more nearly related to measurements:

$$\begin{aligned}\langle [q(t) - q(t+\tau)]^2 \rangle &= 2[\theta(0) - \theta(\tau)] \\ &= 4\gamma kT \int_{-\infty}^{\infty} d\nu \, \frac{1}{m^2\omega^2 + \gamma^2} \, \frac{1 - \cos \omega\tau}{\omega^2} \\ &\approx \left(\frac{2kT}{\gamma}\right) |\,\tau\,| \quad \text{or} \quad 2D\tau \end{aligned} \tag{10.59}$$

This expression is evaluated by placing both terms under the integral before performing the integration. The term $(1 - \cos y)/y^2$ is only large near $y = 0$ and has the properties of a $\delta$ function and in general,

$$\int_{-\infty}^{\infty} dy \, f(y) \, \frac{1 - \cos y}{y^2} = \pi f(0) \tag{10.60}$$

can be used. The result is in accord with the earlier development for Brownian motion and helps clarify the definition of $\langle \Delta x^2 \rangle$, that is,

$$\langle (\Delta x)^2 \rangle = \langle [q(t) - q(t+\tau)]^2 \rangle \approx \frac{2kT}{\gamma} \tau$$

A measurement of the coordinate at time $t$ and a subsequent measurement at time $t + \tau$ have this as an average squared difference.

The procedure used in developing Eq. (10.59) is only valid for $\tau > (\gamma/m)^{-1}$. Since the coordinate $q(t)$ is a continuous function of $t$, the derivative $\dot{q}(t)$ exists and the development which leads to Eqs. (10.34) and (10.35) can be used. Direct application yields the expression

$$\langle [q(t) - q(t+\tau)]^2 \rangle = \frac{2kT}{\gamma} |\,\tau\,| + \frac{2mkT}{\gamma^2} (e^{-(\gamma/m)|\tau|} - 1) \tag{10.61}$$

and gives the correlation for small $\tau$, that is, $\tau < (\gamma/m)^{-1}$. As expected from elementary considerations, for small values of $\tau$, the change in position is proportional to the velocity and time interval, or

$$|\, q(t) - q(t+\tau) \,| \approx \left(\frac{kT}{m}\right)^{1/2} \tau$$

and this is in agreement with the expression just given. For large $\tau$, Eq. (10.61) is linear in $\tau$ and is characteristic of a random walk.

## 10.7 Response of Electrical Circuits to Random Pulses and Thermal Noise

If the analysis of the simple harmonic oscillator is applied to an electrical *RLC* circuit, the equation of motion is

$$L\ddot{q} + R\dot{q} + \frac{q}{C} = e(t) \tag{10.62a}$$

and this can be written in terms of the Fourier coefficients

$$q(\nu) = G(\nu)e(\nu) \tag{10.62b}$$

where the response function is

$$G(\nu) = [L(\omega_0^2 - \omega^2) - i\omega R]^{-1} = [-i\omega Z(\nu)]^{-1} \tag{10.63}$$

The natural frequency of oscillation $\omega_0^2 = 1/LC$, and $Z(\nu)$ is the circuit impedance, which is usually used for electrical circuits. The argument used in Section 10.5.2 can be repeated to show that the spectral density of the thermal source or thermal noise is

$$\psi_{\mathrm{T}}(\nu) = 2RkT \tag{10.64}$$

This also follows by changing $\gamma$ to $R$ in Eq. (10.49). The spectral density describing the charge $q$ on the capacitor is

$$\theta(\nu) = |\, G(\nu)\,|^2 \psi_{\mathrm{T}}(\nu) \tag{10.65a}$$

and the correlation function for the charge is

$$\theta(\tau) = \langle q(t)q(t+\tau)\rangle_{\mathrm{T}} = (2RkT)2\int_0^\infty d\nu\, |\, G(\nu)\,|^2 \cos 2\pi\nu\tau \tag{10.65b}$$

In a similar manner, the current flowing through the inductance and the resistance is described by the correlation function

$$\langle \dot{q}(t)\dot{q}(t+\tau)\rangle_{\mathrm{T}} = 4RkT \int_0^\infty d\nu\, [\omega^2\, |\, G(\nu)\,|^2] \cos 2\pi\nu\tau \tag{10.66}$$

at $\tau = 0$, these expressions reduce to the equipartition values of

$$\frac{1}{2}kT = \frac{1}{2}\left\langle \frac{q^2}{C} \right\rangle = \frac{1}{2}\langle L\dot{q}^2\rangle$$

It is apparent in Eq. (10.66) that the same current will flow in the circuit with a resistor as a noise source having a thermal output power of

$$2\psi_T(\nu)\, d\nu = 4RkT\, d\nu \tag{10.67}$$

into bandwidth $d\nu$. Figure 10.5(a) shows an equivalent representation for a resistor placed into a circuit. A noise voltage with a spectral density $2\psi_T$,

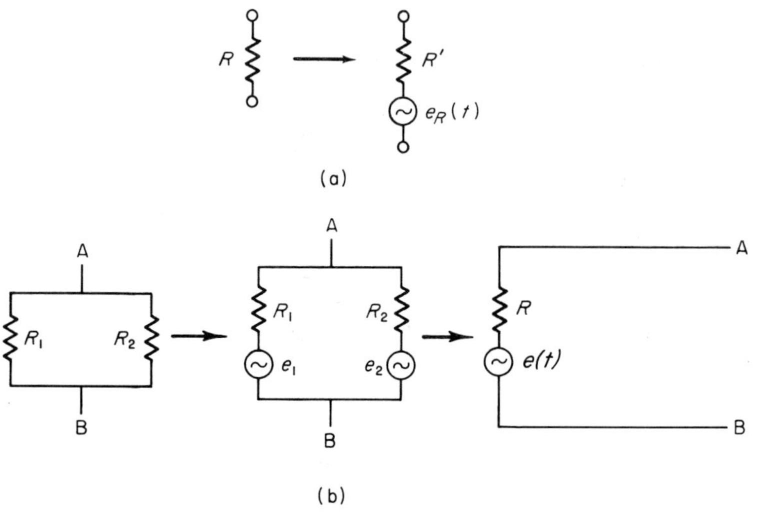


**Fig. 10.5** (a) Replacement of a resistor $R$ in an electrical circuit by a "noiseless" resistor $R'$ and a thermal noise voltage $e_R(t)$. (b) Replacement of resistor $R_1$ at temperature $T_1$ and resistor $R_2$ at temperature $T_2$ by a single "noiseless" resistor $R = [(1/R_1) + (1/R_2)]^{-1}$ and a noise voltage $e(t) = (e_1R_2 + e_2R_1)/(R_1 + R_2)$.

or a thermal noise power $4RkT\, d\nu$, is associated with a resistor. If the circuit is completed with an $L$ and a $C$ or with a circuit with a bandwidth $\Delta\nu_B$, the average square voltage across the resistor is approximately

$$\langle e^2 \rangle \approx 4RkT\, \Delta\nu_B \tag{10.68}$$

Nyquist and Williams [6] were responsible for early developments in resistor noise, and the thermal noise in a circuit is often referred to as "Nyquist noise."

Regarding the resistor as a noise source has proved to be a very useful concept. Equipartition indicated that $\frac{1}{2}kT$ should be associated with each squared term in the Hamiltonian. Inherent in the subsequent discussion was the assumption that the driving force of the surroundings was independent of the damping mechanism, but at equilibrium, these two terms were equal. It is now assumed that under nonequilibrium conditions, the resistor is a source of noise equivalent to its damping action under equilibrium conditions at the same temperature $T$. This is the significance of Fig. 10.5(a) and Eqs. (10.66) and (10.67). This is reminiscent of the procedure used to discuss the rate of evaporation and condensation of a vapor. A similar procedure was used in Chapter II to find the energy radiated by a surface. With this concept, the thermal noise due to resistors at different temperatures can be determined and is the basis of Williams' Theorem. Each resistor $R_a$ in the circuit is replaced by a "noiseless resistor" and a noise voltage with an amplitude $e_a(\nu)$ at frequency $\nu$. Thevenin's Theorem can be used to replace the noise generators by a single generator and the impedances by a single impedance. This impedance is the impedance with all generators short-circuited and the voltage is the open-circuit voltage. As an example, consider the noise across terminals $AB$ in Fig. 10.5(b). The squared voltage for frequency $\nu$ is

$$| e(\nu) |^2 = \left(\frac{R_1 R_2}{R_1 + R_2}\right)^2 \left| \frac{e_1(\nu)}{R_1} + \frac{e_2(\nu)}{R_2} \right|^2$$

and the spectral density across terminals $AB$ is given by

$$2\psi(\nu) = 2 \lim_{T\to\infty} \frac{1}{T} \langle | e(\nu) |^2 \rangle = \left(\frac{R_1 R_2}{R_1 + R_2}\right)^2 4k\left(\frac{T_1}{R_1} + \frac{T_2}{R_2}\right) \qquad (10.69)$$

According to the earlier discussion,

$$2\psi_{\mathrm{T}} = 2 \lim_{T\to\infty} \frac{1}{T} \langle | e_1(\nu) |^2 \rangle = 4R_1 k T_1$$

for each source independently. Since the sources $e_1$ and $e_2$ are assumed independent, there is no correlation between the sources and $\langle e_1(t) e_2(t+\tau) \rangle = 0$ or $\langle e_1(\nu) e_2^*(\nu) \rangle = 0$. The cross terms do not contribute to the spectral density. $\psi_{\mathrm{T}}(\nu)$ can be used as the spectral description of this noise source, which is composed of two resistors at different temperatures. Circuit impedances $Z_a$ can be treated in exactly the same manner. Each $R_a$ is replaced by an $R_a$ and an $e_a(\nu)$. Thevenin's Theorem is used to find the impedance $Z$ in series with $e(\nu)$. The term $| e(\nu) |^2$ is formed and terms in $\langle | e(\nu) |^2 \rangle$ of the type $\langle e_a(\nu) e_b^*(\nu) \rangle$ are zero; terms of the type $2\langle | e_a(\nu) |^2 \rangle$ yield $4R_a k T_a$ in the spectral density; and so on. The correlation function for the current flow follows from Eqs. (10.65)–(10.66), where $G(\nu)$ is for the final simplified circuit and $\psi(\nu)$ is the spectral density for this circuit and is as given earlier.

The output of a device or an electrical circuit is either the charge $q(t)$ or the current $\dot{q}(t)$. In the absence of a signal, there is a fluctuating output current with a spectral density $\langle\dot{q}(t)\dot{q}(t+\tau)\rangle_{\mathrm{N}}$. The response to a signal is $\langle\dot{q}(t)\dot{q}(t+\tau)\rangle_{\mathrm{S}}$ and the spectral density of the signal is usually regarded as a sharp function of frequency. Signal-to-noise ratio is defined as the ratio of the signal power to the noise power in this output. If $\theta_{\mathrm{S}}$ and $\theta_{\mathrm{N}}$ are spectral densities for these correlation functions, the signal-to-noise ratio is

$$\frac{S}{N} = \frac{\int_0^\infty d\nu\,\theta_{\mathrm{S}}(\nu)}{\int_0^\infty d\nu\,\theta_{\mathrm{N}}(\nu)} \tag{10.70}$$

The limit of the sensitivity of the instrument is usually defined as $S/N = 1$.

Exercise 10.9 (a) Show that the correlation function for

$$q(t) = A + B\cos(2\pi\nu_{\mathrm{e}}t - \varphi)$$

is

$$\langle q(t)q(t+\tau)\rangle = A^2 + \tfrac{1}{2}B^2\cos 2\pi\nu_{\mathrm{e}}\tau$$

Then show that the spectral density for this dc component and a component at frequency $\nu_{\mathrm{e}}$ is

$$\theta(\nu) = A^2\,\delta(\nu) + \tfrac{1}{4}B^2\,[\delta(\nu-\nu_{\mathrm{e}}) + \delta(\nu+\nu_{\mathrm{e}})]$$

(b) The spectral density of a sequence of random pulses plus a sinusoidal signal is

$$\psi(\nu) = \psi_{\mathrm{R}}(\nu) + \tfrac{1}{4}B^2\,[\delta(\nu-\nu_{\mathrm{e}}) + \delta(\nu+\nu_{\mathrm{e}})]$$

where $\psi_{\mathrm{R}}(\nu) = \lambda\,|\,b(\nu)\,|^2$ for identical pulses. Find the correlation function for the coordinate $q$ for an *RLC* circuit.

$$\text{Ans.}\quad \langle q(t)q(t+\tau)\rangle = \tfrac{1}{2}\,|\,G(\nu_{\mathrm{e}})\,|^2\,B^2(\cos 2\pi\nu_{\mathrm{e}}\tau) + 2\int_0^\infty d\nu\,|\,G(\nu)\,|^2\,\psi_{\mathrm{R}}(\nu)\cos 2\pi\nu\tau$$

(c) Find the signal-to-noise ratio $S/N$ for part (b). What is the smallest detectable signal?

Exercise 10.10 (a) Find the signal-to-noise ratio in an *RLC* electrical circuit for a spectral density $\psi(\nu) = 2RkT + \tfrac{1}{4}B^2\,[\delta(\nu-\nu_{\mathrm{e}}) + \delta(\nu+\nu_{\mathrm{e}})]$.

$$\text{Ans.}\quad \frac{\tfrac{1}{4}B^2\,|\,G(\nu_{\mathrm{e}})\,|^2}{\tfrac{1}{2}kTC}$$

$$\text{or at resonance,}\quad \frac{\tfrac{1}{4}B^2}{\tfrac{1}{2}RkT}\,\frac{L}{R}$$

Note that the bandwidth is $\Delta\omega \approx R/L$ and $N \approx \pi RkT\,\Delta\nu$. Show that the signal-to-noise ratio can be improved by using smaller bandwidths $\Delta\nu$. How does this effect the measuring time?

(b) What is the smallest signal that can be detected for a bandwidth of 1, $10^2$, $10^4$, $10^5$ Hz at a resonant frequency of $10^6$ Hz for a value of $R = 50$ and for $R = 10^6$ ohm?

### 10.7.1 Shot Noise

Current in electrical circuits is carried by electrons and the charge crossing a surface is composed of discrete events in some experiments. This is apparent in vacuum tubes, in which the electrons leave a hot filament and the current is sufficiently small that space-charge effects can be ignored. This effect was first studied by Schottky. The effect also occurs in solid-state devices such as transistors and diodes. Since the electrons can be treated independently, the current can be considered as a set of random pulses,

$$I(t) = e \sum b(t - t_k) \tag{10.71a}$$

In a vacuum tube, the $t_k$ are the random emission times and the pulse shape depends on the geometry. The average current flow is given by

$$\langle I \rangle = \lambda e \int_0^\infty d\xi\, b(\xi) = \lambda e \tag{10.71b}$$

where the area under the $b(\xi)$ curve for the pulse shape is unity and the average current is the pulse rate $\lambda$ multiplied by the electron charge $e$. The correlation function is given by Eq. (10.23), and

$$\psi(\tau) = \langle I(t) I(t+\tau) \rangle = \lambda e^2 \int_{-\infty}^{\infty} d\xi\, b(\xi) b(\xi + \tau) + (\lambda e)^2 \tag{10.72}$$

Since there is a dc component, the power spectral density has a $\nu = 0$ term, or

$$\psi(\nu) = \lambda e^2 \,|\, b(\nu) \,|^2 + (\lambda e)^2\, \delta(\nu) \tag{10.73}$$

As a specific example, assume that the probability that an electron leaves a filament or crosses a surface is given by $b(\xi) = \tau_c^{-1} e^{-\xi/\tau_c}$. Then the spectral density of the "shot noise" is given by

$$\psi_S(\nu) = \frac{e\langle I \rangle}{1 + \omega^2 \tau_c^{\,2}} \tag{10.74}$$

In a noise-generating diode, $\tau_c \approx 3\times10^{-10}$ and for frequencies less than 500 MHz, this temperature-limited diode generates a noise power of $e\langle I\rangle$.

## 10.8 Fluctuation Dissipation Theorem

Both the mechanical system and the electrical circuit were described by an equation of motion for $q(t)$ when driven by $f(t)$ or $e(t)$. The spectral density for the coordinate $q$ for a thermal source is given by Eq. (10.49) or Eq. (10.67). Either of these expressions is an expression of the fluctuation dissipation for a very simple physical system. An alternate procedure is to use an integral expression for the linear response of the physical system,

$$q(t) = \int_{-\infty}^{\infty} \varkappa(\tau)f(t-\tau)\,d\tau, \qquad \varkappa(\tau) = 0, \quad \tau < 0 \tag{10.75}$$

Expanding each quantity as a Fourier integral yields the response for each frequency

$$q(\nu) = \varkappa(\nu)f(\nu) \tag{10.76}$$

and corresponds to the previous development with

$$\varkappa(\nu) = G(\nu)$$

A linear amplifier has a response equation similar to Eq. (10.75). The function $\varkappa(\nu)$ may be called the generalized susceptibility of the system and can be written in terms of its real and imaginary parts as

$$\varkappa(\nu) = \varkappa' + i\varkappa'' = \int_{-\infty}^{\infty} d\tau\, \varkappa(\tau)e^{i\omega\tau} \tag{10.77}$$

Here $\varkappa(\tau)$ is the response function which relates the generalized coordinate $q$ with the generalized force $f$. The function $\varkappa(\tau)$ is real and therefore it follows that

$$\varkappa(-\nu) = \varkappa^*(\nu)$$

Expanding Eq. (10.77) as a sine and cosine function, it is apparent that $\varkappa'(\nu)$ is an even function of $\nu$ and $\varkappa''(\nu)$ an odd function of $\nu$. The form of the integral equation ensures that the response occurs after the signal and the events are causal. This causality is manifested by the absence of singularities for real $\nu$, except possibly at the origin. If $\nu$ is regarded as complex, $\nu' + i\nu''$, then there are no singularities for $\varkappa(\nu)$ in the upper half-plane. This discussion is not pursued here, but it may be noted that the Kramers–Kronig relations are a direct consequence of this causality [7].

In analogy with the discussion used for the simple systems and which yielded Eq. (10.49) or (10.67), it follows that for this more general response, the fluctuation in the spectral density for the coordinate $q$ with $\varkappa''(\nu)/\omega = \gamma \mid G(\nu) \mid^2$ is given by

$$\theta(\nu) = 2kT \frac{\varkappa''(\nu)}{\omega} \tag{10.78}$$

and the correlation function for the coordinate by

$$\begin{aligned} \langle q(t)q(t+\tau)\rangle &= 2 \int_0^\infty d\nu\, \theta(\nu) \cos \omega\tau \\ &= 4kT\mathscr{P} \int_0^\infty d\nu \frac{\varkappa''(\omega)}{\omega} \cos \omega\tau \end{aligned} \tag{10.79}$$

This is the Fluctuation Dissipation Theorem of Callen and Greene [8]. For real $\nu$, $\varkappa(\nu)$ has no pole at $\nu = 0$ and $\varkappa(0)$ is finite. In this development, $\varkappa(\nu)$ is assumed to have no poles for real $\nu$ and is finite for all values of $\nu$. In the earlier discussion in this chapter, the dc or $\nu = 0$ term was excluded. This term is excluded in the expression for the correlation function by taking the principal value. The analogy given here does not constitute a proof and the work of Callen and Greene [8] and de Groot and Mazur [9] is suggested for a detailed analysis.

The principal value can be avoided by discussing the average value of

$$\begin{aligned} \langle [q(t) - q(t+\tau)]^2 \rangle &= 4 \int_0^\infty d\nu\, \theta(\nu)(1 - \cos \omega\tau) \\ &\approx 4kT \int_{-\infty}^{\infty} d\nu\, \omega\varkappa''(\omega) \frac{1 - \cos \omega\tau}{\omega^2} \end{aligned} \tag{10.80}$$

This integral can be taken from $-\infty$ to $+\infty$ since the integrand is an even function of $\nu$. The principal value is of no concern in the problem of the damped simple harmonic oscillator, but for a system with a pole on the imaginary axis in the sense of the equation for Brownian motion, the expression reduces to

$$\langle [q(t) - q(t+\tau)]^2 \rangle = 2kT\tau[\omega\varkappa''(\omega)]_{\omega=0} \tag{10.81}$$

Equation (10.60) is used in this reduction and as noted in the discussion of

Eqs. (10.59) and (10.61), this implies $\tau$ sufficiently large that effects near $\tau = 0$ are smoothed out. For a singularity at $\nu = 0$, this yields a growth of the difference squared which is linear in $\tau$.

De Groot and Mazur [9] give an extended discussion of the fluctuation dissipation for systems in which more than one coordinate is considered and in systems in which magnetic fields are included. In the discussion of transport processes, a relationship between transport coefficients and correlation function can be obtained which are similar to the relationships given here. These are referred to as the Kubo relations [9] and are discussed in the appropriate references. In a system of two or more variables, the Onsager reciprocal relations between the variables occur and these systems are discussed by de Groot and Mazur [9].

In order to appreciate properly the integral form of the linear response of a physical system, consider a rather complex electrical system with two connecting wires. This system could be a large array of $R$, $L$, and $C$ components connected in any manner, an inductance coil which surrounds a large number of paramagnetic nuclei or ions, or molecules contained between the parallel plates of a capacitor. Although many individual systems are involved and their interaction with their surroundings may be complex, the measurable response of the system depends on a single coordinate $q(t)$ and its derivative $\dot{q}(t)$ at the two terminals. If a sinusoidal signal $f_\nu e^{-i\omega t}$ + complex conjugate is applied to the terminals, the response is $G_\nu$ or $\varkappa_\nu$ and in the continuous index notation, yields Eq. (10.76), or $q(\nu) = \varkappa(\nu)f(\nu)$. The Fourier integral transform is Eq. (10.75). The measurable quantity in many experiments in physics is a single coordinate and the complex system responds as if it had only one degree of freedom.

## 10.9 Fluctuations in Thermal Radiation

Thermal radiation in a cavity is described by a set of normal modes which in Chapters II and VII were denoted by the index $a$. Either the free energy $A_a$ or the generating function $Q_a$ of the mode can be used to discuss fluctuations in energy, and by using Eqs. (7.18) and (7.20), either of these functions is given by

$$A_a = -kT \ln Q_a = kT \ln(1 - e^{-\beta h\nu_a})$$

Since modes are independent, they can be treated separately. Fluctuations are given by the average value and the dispersion of this distribution. The average value of the energy of the mode is

$$E_a = -\frac{\partial(\ln Q_a)}{\partial\beta} = \bar{n}_a h\nu_a \tag{10.82}$$

where

$$\bar{n}_a = \frac{1}{e^{\beta h\nu_a} - 1}$$

Dispersion in energy is given by

$$\begin{aligned}\langle(E_a - \bar{E}_a)^2\rangle &= \left[\frac{\partial^2(\ln Q_a)}{\partial\beta^2}\right]_V = -\left(\frac{\partial\bar{E}_a}{\partial\beta}\right)_V = kT^2C_{aV} \\ &= \bar{n}_a(\bar{n}_a + 1)(h\nu_a)^2 \end{aligned}\tag{10.83}$$

Dispersion in photon number is usually of greater interest and is given by

$$\langle(n_a - \bar{n}_a)^2\rangle = \bar{n}_a(\bar{n}_a + 1) \tag{10.84}$$

This indicates that the thermal fluctuation in the number of photons in a normal mode is of the order of the number of photons in the normal mode. This large thermal fluctuation is characteristic of systems which obey Bose–Einstein statistics. It can also be noted that the dispersion in energy is proportional to the heat capacity of the mode.

If the modes in range $\Delta\nu$ are averaged together and treated as plane waves as in Chapter II,

$$\sum_a \to 2V\frac{4\pi\nu^2\,\Delta\nu}{c^3}$$

then the fluctuation in energy for the radiation in this frequency interval is

$$\langle(E - \bar{E})^2_{\Delta\nu}\rangle = \sum_{\Delta\nu} \bar{n}_\nu(\bar{n}_\nu + 1)(h\nu)^2 = h\nu\bar{E}_{\Delta\nu} + \frac{(\bar{E}_{\Delta\nu})^2}{V8\pi\nu^2\,\Delta\nu/c^3} \tag{10.85}$$

Average energy $\bar{E}_{\Delta\nu} = \bar{n}_\nu h\nu(V8\pi\nu^2\,\Delta\nu/c^3)$ is used in the second equation, which was derived by Einstein in 1909 for thermal radiation. The fluctuation term proportional to $n$ or $E_{\Delta\nu}$ was interpreted as the corpuscular aspect of radiation. This is the fluctuation characteristic of the Poisson distribution or of particles in a cell. The second term, proportional to $n^2$ or $E^2$, was interpreted as the fluctuation due to the wave aspects of the thermal radiation.

## 10.10 General Theory of Fluctuations in Statistical Thermodynamics

Fluctuations in the canonical and grand canonical ensembles can be derived in a very direct manner as the cumulants of the generating function for the ensemble. The fluctuation in energy of an $N$-particle system in volume $V$ in thermal equilibrium with its surroundings at temperature $T$ is given by dispersion of the distribution $\ln Q_N$,

$$\langle (E-\bar{E})^2\rangle = \frac{\partial^2(\ln Q_N)}{\partial\beta^2} = -\left(\frac{\partial \bar{E}}{\partial\beta}\right)_{\mathrm{V}} = kT^2C_{\mathrm{V}} \tag{10.86}$$

The dispersion in energy of a physical system is proportional to the heat capacity of the system and is proportional to $N$ for an $N$-particle system. Fluctuations in energy are directly related to the experimentally measurable heat capacity.

Fluctuations which occur in a volume $V$ can be discussed in terms of the cumulants of the generating function of the grand canonical ensemble. Temperature $T$ and chemical potential $\mu$ or fugacity $z$ are constant for the physical system in volume $V$ in equilibrium with the surroundings. Only a single chemical species is considered. The dispersion in energy and particle number are given by the expressions

$$\langle (E-\bar{E})^2\rangle = \left[\frac{\partial^2(\ln \mathcal{Q})}{\partial\beta^2}\right]_{z,\mathrm{V}} = -\left[\frac{\partial \bar{E}}{\partial\beta}\right]_{z\mathrm{V}} = kT^2C_{\mathrm{V}z}$$

$$\langle (N-\bar{N})^2\rangle = \left[\frac{\partial^2(\ln \mathcal{Q})}{\partial(\ln z)^2}\right]_{\mathrm{T,V}} = \frac{1}{\beta}\left(\frac{\partial \bar{N}}{\partial\mu}\right)_{\mathrm{T,V}} = \frac{N^2}{\beta V}\varkappa_{\mathrm{T}} \tag{10.87}$$

and

$$\langle (E-\bar{E})(N-\bar{N})\rangle = -\frac{\partial^2(\ln \mathcal{Q})}{\partial(\ln z)\,\partial\beta} = \frac{\partial \bar{E}}{\partial(\ln z)} = -\frac{\partial \bar{N}}{\partial\beta}$$

Notation causes some difficulty since $U = \bar{E} = \langle E\rangle$ and $N$ rather than $\bar{N}$ is used in most of the earlier calculations. The term $(\partial\bar{N}/\partial\mu)_{\mathrm{T,V}}$ is not an apparently measurable quantity, but with some manipulation of thermodynamic variables, one can show that

$$\left(\frac{\partial N}{\partial\mu}\right)_{\mathrm{T,V}} = -\left(\frac{N}{V}\right)^2\left(\frac{\partial V}{\partial P}\right)_{\mathrm{T},N} = \frac{N^2}{V}\varkappa_{\mathrm{T}} \tag{10.88}$$

Thus the fluctuation in the number of particles in volume $V$ is directly related to the isothermal compressibility $\varkappa_{\mathrm{T}}$. These relationships are of general validity and apply to physical systems in which either classical or quantum effects are important. If the number of particles is held constant.

$$0 = \delta N = \frac{\partial \bar{N}}{\partial(\ln z)}\,\delta(\ln z) + \frac{\partial \bar{N}}{\partial \beta}\,\delta\beta$$

then

$$-q = \left[\frac{\Delta(\ln z)}{\Delta\beta}\right]_N = \frac{\partial \bar{E}/\partial(\ln z)}{\partial \bar{N}/\partial(\ln z)} = \left(\frac{\partial \bar{E}}{\partial \bar{N}}\right)_{\beta V} = \frac{\langle \Delta E\,\Delta N\rangle}{\langle(\Delta N)^2\rangle} \tag{10.89}$$

is the energy per particle at constant temperature and volume. In experiments on surface absorption, the surface coverage as a function of pressure for the various isotherms can be used to obtain this term.

The arguments that were used in discussing Eq. (3.85) are now repeated for the canonical ensemble. The probability that the state $a$ occurs in the canonical ensemble is given by

$$p_a(\beta) = \frac{e^{-\beta E_a}}{Q}$$

for a system with $N$ particles and fixed volume $V$. The average energy $E$ and the entropy $S$ of this canonical distribution are discussed in Section 6.4.2, and are

$$\bar{E} = -\left[\frac{\partial(\ln Q)}{\partial\beta}\right]_V$$

and

$$\frac{S}{k} = -\sum p_a \ln p_a = \beta\bar{E} + \ln Q$$

In the microcanonical ensemble, the energy is fixed at $\mathscr{U}$ and this constraint introduces the parameter $\beta_0$. If a sequence of measurements is made on the canonical ensemble, then the energy will have an average value $\bar{E}$ and a dispersion about this average value of $\langle(E - \bar{E})^2\rangle$. A measurement of the entropy will also show a variation in this sequence of measurements. Since $\beta$ is the only parameter in the probability law, the fluctuation appears in $\beta$. Let

$$\Delta\beta = \beta - \beta_0 \qquad \text{and} \qquad \Delta E = E - \bar{E}$$

then $\ln Q$ can be expanded about $\beta_0$ as

$$\ln Q = \ln Q(\beta) = \ln Q(\beta_0) - \bar{E}\,\Delta\beta + \tfrac{1}{2}\langle(\Delta E)^2\rangle(\Delta\beta)^2$$

The probability that a measurement finds the canonical ensemble in state $a$ with temperature $\beta$ is

$$p_a(\beta) \simeq p_a(\beta_0)\exp[-(E_a - \bar{E})\,\Delta\beta]\exp[-\tfrac{1}{2}\langle(\Delta E)^2\rangle(\Delta\beta)^2]$$

At $E_a = \bar{E}$, the probability of $\Delta\beta$ is given by

$$p(\Delta\beta) = \text{constant} \times \exp[-\tfrac{1}{2}\langle(\Delta E)^2\rangle(\Delta\beta)^2]$$
$$= \text{constant} \times \exp(\Delta_2 \ln Q) \qquad (10.90a)$$

where $\Delta_2 \ln Q$ indicates second-order terms. The average fluctuation in temperature can be obtained from this Gaussian distribution in the variable $\Delta\beta$, and

$$\langle(\Delta E)^2\rangle\langle(\Delta\beta)^2\rangle = 1 \qquad (10.90b)$$

The earlier discussion can be repeated for the grand canonical ensemble and the probability of deviations $\Delta\beta$ and $\Delta(\ln z)$ is given by

$$p(\Delta\beta, \Delta[\ln z]) = \text{constant} \times \exp \Delta_2(\ln \mathcal{Q}) \qquad (10.91a)$$

where

$$\Delta_2(\ln \mathcal{Q}) = -\frac{1}{2}\{\langle(\Delta E)^2\rangle(\Delta\beta)^2 - 2\langle\Delta E\,\Delta N\rangle\,\Delta\beta\,\Delta(\ln z)$$
$$+\langle(\Delta N)^2\rangle[\Delta(\ln z)]^2\}_V$$
$$= \frac{1}{2}[\Delta\bar{E}\,\Delta\beta - \Delta\bar{N}\,\Delta(\ln z)]_V \qquad (10.91b)$$

are the second-order terms in the expansion of $\ln \mathcal{Q}$. The shorthand notation

$$\Delta\bar{E} = \frac{\partial\bar{E}}{\partial\beta}\Delta\beta + \frac{\partial\bar{E}}{\partial(\ln z)}\Delta(\ln z)$$

and so on for $\Delta\bar{N}$ is introduced. Cross-correlation terms occur in the Gaussian distribution in the variables $\Delta\beta$ and $\Delta(\ln z)$. Average values can be obtained for this Gaussian by placing the terms of (10.91b) in diagonal form. Exercise 10.11 uses $\Delta\beta$ and $\Delta N$ as independent fluctuating quantities and these variables are usually more convenient. The average fluctuations in temperature and fugacity $z$ are

$$\langle(\Delta\beta)^2\rangle = \frac{\langle(\Delta N)^2\rangle}{D}$$
$$\langle\Delta\beta\,\Delta(\ln z)\rangle = \frac{\langle\Delta E\,\Delta N\rangle}{D} \qquad (10.92)$$
$$\langle[\Delta(\ln z)]^2\rangle = \frac{\langle(\Delta E)^2\rangle}{D}$$

where

$$D = \langle(\Delta E)^2\rangle\langle(\Delta N)^2\rangle - \langle \Delta E\, \Delta N\rangle$$

and these quantities can be written in terms of thermodynamic variables by using Eqs. (10.87).

It is possible to reconsider the microcanonical ensemble and use walls for the canonical ensemble that can change in a slow manner. Then, the constraint $\delta V_b = 0$ for each canonical ensemble is removed and the single constraint $\delta V = \sum \delta V_b = 0$ is introduced. This feature was discussed in Section 9.2 and the pressure is now constant throughout the microcanonical ensemble. For any given system, both fluctuations in volume and temperature can occur. Rather than repeat the earlier discussion for two variables, the method used by Landau and Lifshitz [7] is introduced. As the system exchanges energy and changes its volume, the microcanonical ensemble is assumed sufficiently large that the pressure and temperature remain unchanged at $P_0$ and $T_0$. The minimum work required to cause this change is given by

$$R_{\mathrm{m}} = \Delta \bar{E} - T_0\, \Delta S + P_0\, \Delta V$$

where $\Delta \bar{E}$, $\Delta S$, and $\Delta V$ are the changes in energy (or adiabatic work), entropy, and volume of the small system, respectively. Expanding $\Delta \bar{E}$ as a function of $S$ and $V$ through second order yields

$$\begin{aligned} R_{\mathrm{m}} &= \frac{1}{2}\left[\frac{\partial^2 \bar{E}}{\partial S^2}(\Delta S)^2 + 2\frac{\partial^2 \bar{E}}{\partial S\, \partial V}\Delta S\, \Delta V + \frac{\partial^2 \bar{E}}{\partial V^2}(\Delta V)^2\right] \\ &= \frac{1}{2}(\Delta S\, \Delta T - \Delta P\, \Delta V)_N \end{aligned}$$

for a fixed number of particles. The probability of a fluctuation is proportional to the fluctuation in entropy of the microcanonical ensemble $\Delta S_{\mathrm{t}} = -R_{\mathrm{m}}/T_0$, or

$$p(R_{\mathrm{m}}) = \text{constant} \times \exp(-\beta R_{\mathrm{m}}) = \text{constant} \times \exp[-\tfrac{1}{2}\beta(\Delta S\, \Delta T - \Delta P\, \Delta V)] \tag{10.93}$$

If $\Delta\beta$ and $\Delta V$ are selected as the independent variables, the expansion of $\Delta S$ and $\Delta P$ in berms of $\Delta\beta$ and $\Delta V$ permits the probability to be written as [7]

$$p(\Delta\beta, \Delta V) = \text{constant} \times \exp\left\{-\frac{1}{2}\left[\langle(\Delta E)^2\rangle(\Delta\beta)^2 - \beta\left(\frac{\partial P}{\partial V}\right)_{\mathrm{T}}(\Delta V)^2\right]\right\} \tag{10.94a}$$

No cross term occurs and for a constant number of particles, the fluctuations in volume and temperature are independent,

$$\langle \Delta\beta \, \Delta V \rangle = 0$$

It is much easier to work with a constant number of particles than with constant volume, since most well-known thermodynamic relationships are for fixed number.

Exercise 10.11 (a) The independent variables are $\Delta E$ and $\Delta(\ln z)$ or $\Delta\beta$ and $\Delta N$ in Eq. (10.91b). Show that an expansion in $\Delta\beta$ and $\Delta N$ yields

$$p(\Delta\beta, \Delta N) = \text{constant} \times \exp\left\{\frac{1}{2}\left[\frac{\partial E}{\partial \beta}(\Delta\beta)^2 - \beta\frac{\partial \mu}{\partial N}(\Delta N)^2\right]\right\} \tag{10.94b}$$

Use the relationship $(\partial\mu/\partial N)_{T,V} = (V/N^2)\varkappa_T^{-1}$ to show that

$$\langle(\Delta\beta)^2\rangle = \frac{1}{C_V kT^2} \quad \text{and} \quad \langle(\Delta N)^2\rangle = \left(\frac{N^2}{\beta V}\right)\varkappa_T$$

for the fluctuation in temperature at constant number and the fluctuation in number at constant temperature, respectively. Then show that

$$\langle \Delta\beta \, \Delta N \rangle = 0$$

(b) Show that pressure and entropy can be used as the independent fluctuating variables and that

$$p(\Delta P, \Delta S) = \text{constant} \times \exp\left\{\frac{1}{2}\left[\beta\left(\frac{\partial V}{\partial P}\right)_S(\Delta P)^2 - \frac{1}{kC_P}(\Delta S)^2\right]\right\} \tag{10.94c}$$

and then show that

$$\langle(\Delta S)^2\rangle = kC_P \quad \text{and} \quad \langle(\Delta P)^2\rangle = -kT\left(\frac{\partial P}{\partial V}\right)_S$$

### *10.10.1 Bose–Einstein and Fermi–Dirac Systems*

Bose and Fermi distributions were discussed in Section 7.3 and it was noted that the one-particle states of the system in volume $V$ with label are statistically independent and can be treated as distinct physical systems. The generating function for the entire physical system is given by Eq. (7.25):

$$\mathcal{Q} = \prod_a \mathcal{Q}_a \tag{10.95}$$

and for a particular state $a$, the grand canonical generating function is

$$\ln \mathcal{Q}_a = \pm\ln[1 \pm e^{\beta(\mu-\varepsilon_a)}] \qquad \begin{Bmatrix} \text{FD} \\ \text{BE} \end{Bmatrix} \tag{10.96}$$

Equation (10.88) can be used and the fluctuation in number in state $a$ in volume $V$ in equilibrium with its surroundings is

$$\langle(n_a - \bar{n}_a)^2\rangle = \bar{n}_a(1 \mp \bar{n}_a) \qquad \begin{Bmatrix} \text{FD} \\ \text{BE} \end{Bmatrix} \tag{10.97}$$

$\bar{n}_a$ is given by Eq. (7.29) or Eq. (7.31). The fluctuation in energy is given by Eq. (10.87) or by multiplying the above by $\varepsilon_a^2$. For a Fermi–Dirac gas, the average occupation number $n_a$ is in the range $0 \leq n_a \leq 1$ and the fluctuations are small. Since the size of $n_a$ is unlimited for Bose–Einstein gases, the fluctuation is of the order of the square of the average number of particles in the state. In most simple problems, the index $a$ refers to plane-wave states with the index $k$ and can be treated in a manner similar to that used in the previous section for photons.

A gas at low densities, or a Boltzmann gas, has all occupation numbers $\bar{n}_a \ll 1$, and the dispersion is given by

$$\langle(n_a - \bar{n}_a)^2\rangle \approx \bar{n}_a \tag{10.98a}$$

where $\bar{n}_a \approx \exp(\mu - \varepsilon_a)/kT$ by Eq. (7.33). This expression can be summed over the states of the system to yield

$$\langle(N - \bar{N})^2\rangle \approx \bar{N} \tag{10.98b}$$

for the fluctuation in the total number of particles in $V$. This is in accord with the general equation (10.88) with $PV = NkT$.

**Exercise 10.12** Show that the fluctuation in volume $V$ for a gas obeying $PV = NkT$ is given by Eq. (10.98b).

### *10.10.2 Poisson Distribution*

It was shown in the previous section that the average number of particles in volume $V$ is $N$ and the fluctuation is given by Eq. (10.98b). The development in Section 3.5 for the Poisson distribution suggests that the probability of finding $N$ particles in volume $V$ is

$$p(N) = \frac{\bar{N}^N \exp -\bar{N}}{N!} \tag{10.99a}$$

This function has the same dispersion as that given by Eq. (10.88) for an ideal gas. Even so, this equation contains more information. It is interesting to test the grand canonical or Gibbs distribution to see if it yields the same information as the Poisson distribution. Equation (6.62a) gives the probability of $N$ particles in volume $V$, that is,

$$p(N) = \mathcal{Q}^{-1}(z^N Q_N)$$

For a gas at low density, $Q_N = Q_1{}^N/N!$, $z = \bar{N}/Q_1$, and $\mathcal{Q} = \exp \bar{N}$, so that

$$p(N) = \frac{\bar{N}^N \exp -\bar{N}}{N!} \tag{10.99b}$$

and this is the same as Eq. (10.99a) for the Poisson distribution. The grand canonical distribution function yields the expected result for $\bar{N}$'s and $N$'s comparable to unity as well as the very large values for a macroscopic system.

## 10.11 Correlation of Fluctuations and Measurement with Waves

Fluctuations in a gas at low density or a gas in which there are no intermolecular potentials or quantum effects are given by the Poisson distribution. The choice of position is completely random. In a gas of moderate density, the intermolecular potential is important and the second virial coefficient is a manifestation of these potentials. Equation (10.88) is of general validity and yields a fluctuation in volume $V$ for a gas of

$$\langle (N - \bar{N})^2 \rangle = \bar{N}\left[1 - 2\left(\frac{\bar{N}}{V}\right)B\right] \tag{10.100a}$$

where $\bar{N}$ is the average number in $V$. Depending on the sign, the second virial coefficient $B$ either increases or decreases the fluctuation in number density from the Poisson distribution. In this same degree of approximation, the pair distribution function given by Eq. (8.35) yields a similar result.

The grand canonical ensemble can be interpreted in the following manner. The probability of one particle in volume $V$ is $\mathcal{Q}^{-1}zQ_1$, the probability of two particles in $V$ is $\mathcal{Q}^{-1}z^2Q_2$, or, by Eq. (6.62a),

$$p(N) = \mathcal{Q}^{-1}z^N Q_N$$

This moment expansion does not illustrate in a direct manner the effect of correlations, and the cumulant expansion of $\ln \mathcal{Q}$ is needed. Then the coefficient $a_1$ of $z$ is related to the probability of independent particles, $a_2$ of $z^2$ to the probability of particle pairs, and so on. In the absence of correlation, $Q_N \neq 0$, but $a_N = 0$ for $N > 1$. This feature was discussed in greater detail in Chapter VIII. Suppose, for convenience, that $a_3 z^3 \approx 0$; then

$$\ln \mathcal{Q} \approx a_1 z + a_2 z^2$$

Direct application of Eq. (10.88) yields, with $\bar{N} = a_1 z + 2a_2 z^2$, a fluctuation of

$$\langle (N - \bar{N})^2 \rangle = \bar{N} + 2a_2 z^2 = \bar{N}\left[1 + 2\bar{N}\left(\frac{a_2}{a_1}\right)\right] \tag{10.100b}$$

and is the same as Eq. (10.100a). The argument leading to Eq. (8.35) can be repeated and the "pair correlation" is the integrand of $(Q_2 - \frac{1}{2}Q_1^2)$. If the phase-space approximation is used, then this integrand yields

$$\begin{Bmatrix}\text{probability of a pair molecule} \\ \text{between} \quad r \quad \text{and} \quad r + dr\end{Bmatrix} = 2\pi \bar{n}(e^{-\beta U} - 1) r^2 \, dr \tag{10.100c}$$

where $\bar{n} = N/V$ is the average number of molecules per cubic meter. It should be noted that Eq. (10.100c) is the probability of a pair molecule and implies a correlation aspect. This same term is important in the scattering of particles. For a perfect Boltzmann gas, this term is zero, but there is the usual Poisson distribution for finding another particle within $r$.

Equation (10.88) is of general validity and relates the compressibility of gases, liquids, and solids to the fluctuation in the number density. If the extensive variable is the magnetic field $B$, then the fluctuation in number density can be related to the magnetic susceptibility. A fluctuation in number density also occurred in the combinatorial problem discussed in Section 9.7 and was found to be characteristic of many materials at their critical points. Experimental measurements can be made on the compressibility, magnetic susceptibility, osmotic compressibility in binary mixtures, and so on. Some experimental aspects of the measurement of fluctuations using waves are now discussed.

### 10.11.1 Measurement of Fluctuations and Correlation of Fluctuations

Fluctuations in the number of particles can become appreciable in small volumes. An observable measurement would require an average on an ensemble of small volumes and this is not possible to obtain in a direct manner. Elastic scattering of electromagnetic waves or of neutrons provides an indirect method of measuring fluctuations and correlations of fluctuations. The interaction of plane electromagnetic waves with an atom was considered in Section 2.8 and of a plane matter or de Broglie wave with another molecule in Section 4.3. Further details are considered in Chapter XI. Both of these phenomena can be considered at the same time by considering the general scattering problem discussed in Chapter IV.

If correlations are unimportant, the scattering by $N$ molecules is just $N$ times the scattering by one molecule. For long-wavelength electromagnetic waves, the scattering amplitude is proportional to $\lambda^{-2}$ and the intensity to $\lambda^{-4}$. This is the simple Rayleigh scattering and explains the blue sky and red sunset. The simplicity of this problem depends on the random position of the scattering atom in the volume, and the decrease in forward intensity is a direct measure of the number of molecules per cubic meter and can be used to determine Avogadro's number. In a system with a correlation between the molecular positions, there are additional effects, and both the previous effect and these other effects are discussed.

Elastic scattering occurs when the wavelength $\lambda$ or the wave vector $\mathbf{k}$ is a function of position in the scalar Helmholtz equation,

$$\nabla^2\psi + \varkappa^2\psi = 0$$

This equation can be written as

$$\nabla^2\psi + k_0{}^2\psi = [k_0{}^2 - \varkappa^2(\mathbf{r})]\psi = k_0{}^2[\Delta\eta(\mathbf{r})]\psi$$

and the discussion that leads to Eq. (4.30) can be used to show that the asymptotic solution at large $r$ is given by the integral equation

$$\psi_k = (\exp i\mathbf{k}_a \cdot \mathbf{r}) - k^2 \frac{\exp ikr}{4\pi r} \int d\mathbf{r}\, (\exp -i\mathbf{k} \cdot \mathbf{r})[\Delta\eta(\mathbf{r})]\psi_k(\mathbf{r})$$

With this notation, $\exp i\mathbf{k}_a \cdot \mathbf{r}$ is the incident plane wave in direction $\hat{k}_a$ and the integral is proportional to the amplitude of the scattered wave in direction $\hat{k}$. If the scattering is not too large, the Born approximation can be used, and under the integral sign, $\psi_k$ can be replaced by $\exp i\mathbf{k}_a \cdot \mathbf{r}$.

With this approximation, the scattering cross section is proportional to

$$\int d\mathbf{r}'\, d\mathbf{r}''\, [\exp i(\mathbf{k}_a - \mathbf{k}) \cdot (\mathbf{r}' - \mathbf{r}'')]\, \Delta\eta(\mathbf{r}')\, \Delta\eta(\mathbf{r}'')$$

An ensemble average is taken of this quantity and it is assumed that the ensemble average depends only on the coordinate difference $\mathbf{r} = \mathbf{r}' - \mathbf{r}''$. The correlation function $\langle \Delta\eta(\mathbf{r}')\, \Delta\eta(\mathbf{r}'')\rangle$ is a stationary random variable in the space coordinates, and the scattering cross section under quite general conditions for all systems that obey a wave equation becomes

$$\sigma(\mathbf{K}) = \left(\frac{\pi^2}{\lambda^4}\right) V \int d\mathbf{r}\, (\exp -i\mathbf{K} \cdot \mathbf{r}) \langle \Delta\eta(0)\, \Delta\eta(\mathbf{r})\rangle \qquad (10.101a)$$

where the change in the wave vector is

$$\mathbf{K} = \mathbf{k} - \mathbf{k}_a \qquad \text{or} \qquad K = 2k_0 \sin\frac{\theta}{2} = \frac{4\pi}{\lambda} \sin\frac{\theta}{2} \qquad (10.101b)$$

$\theta$ is the direction between the incident wave $\hat{k}_a$ and the scattered wave $\hat{k}$. Since the space correlation function is real and since

$$\langle \Delta\eta(0)\, \Delta\eta(\mathbf{r})\rangle = \langle \Delta\eta(0)\, \Delta\eta(-\mathbf{r})\rangle \qquad (10.101c)$$

the scattering cross section, or the spectral density, has the properties

$$\sigma(\mathbf{K}) = \sigma(-\mathbf{K}) = \sigma^*(\mathbf{K}) \qquad (10.101d)$$

and cosine transforms can be used to relate the correlation function and the spectral density. If the intensity of the incident wave is $I_0$, then the intensity of the scattered wave is

$$I(\mathbf{K}) = I_0 \sigma(\mathbf{K}) \qquad (10.101e)$$

Einstein (see Landau and Lifshitz [7]) used scattering relationships of this type to discuss the scattering of light by the fluctuation in the dielectric constant, or $\Delta\eta = \Delta\varepsilon$. As noted earlier, the wave approach has general validity for all phenomena that obey the scalar Helmholtz equation.

The correlation function for scattering can be related to the correlation function for the number density and this aspect of the problem has been emphasized by van Hove [*10*]. A convenient approach is to use the $\delta$ function for the location of the individual scatterers for the wave, and then to write the number density as

$$n(\mathbf{r}) = \sum_m \delta(\mathbf{r} - \mathbf{r}_m)$$

The average number of particles in volume $V$ is given by the grand canonical ensemble average and

$$\bar{N} = \int d\mathbf{r} \left\langle \sum_m \delta(r - r_m) \right\rangle$$

It is assumed that this average is not dependent on the location of $V$ in the sample and the bracketed quantity, or $\langle n(\mathbf{r}) \rangle$, is a stationary random variable and therefore constant. The joint probability of finding a particle at $\mathbf{r}'$ and a particle at $\mathbf{r}''$ is given by the ensemble average or the correlation function,

$$\langle n(\mathbf{r}')n(\mathbf{r}'') \rangle = \left\langle \sum_{m,n} \delta(\mathbf{r}' - \mathbf{r}_m)\, \delta(\mathbf{r}'' - \mathbf{r}_n) \right\rangle$$

It is assumed that this correlation function does not depend on the location in the sample, is a function only of the distance, and is a stationary random process in the position coordinate. The correlation function with $\mathbf{r} = \mathbf{r}' - \mathbf{r}''$ becomes

$$\begin{aligned} \langle n(0)n(\mathbf{r}) \rangle &= \bar{n}\, \delta(\mathbf{r}) + \left\langle \sum_{m \neq n} \delta(\mathbf{r} + \mathbf{r}_m - \mathbf{r}_n) \right\rangle \\ &= \bar{n} G(\mathbf{r}) = \bar{n}[\delta(\mathbf{r}) + \bar{n} g(\mathbf{r})] \end{aligned} \tag{10.102a}$$

Van Hove introduced the correlation function $G(\mathbf{r})$ to describe the important aspects of the correlation in thermodynamic systems. The fluctuation of the number of particles in a volume $V$ is related to the isothermal compressibility by Eq. (10.88),

$$\begin{aligned} \langle (N - \bar{N})^2 \rangle &= \int d\mathbf{r}'\, d\mathbf{r}''\, \langle n(\mathbf{r}')n(\mathbf{r}'') \rangle - \bar{N}^2 \\ &= \bar{N}\left\{1 + \bar{n} \int d\mathbf{r}\, [g(\mathbf{r}) - 1)]\right\} = \left(\frac{N^2}{\beta V}\right) \varkappa_{\mathrm{T}} \end{aligned} \tag{10.102b}$$

and is a measurable quantity.

For comparison with the scattering cross section, it is convenient to write

$$\Delta n(\mathbf{r}) = n(\mathbf{r}) - \bar{n}$$

and then the fluctuation in the number density becomes

$$\langle (N - \bar{N})^2 \rangle = V \int d\mathbf{r}\, \langle \Delta n(0)\, \Delta n(r) \rangle \tag{10.102c}$$

Thus the measurable quantity in an experiment is the correlation function,

$$\langle \Delta n(0)\, \Delta n(\mathbf{r}) = \bar{n}\{\delta(\mathbf{r}) + \bar{n}[g(\mathbf{r}) - 1]\} \qquad (10.102d)$$

The Fourier transform of this correlation function, or the space spectral density, contains the measurable properties of the system, which depend on number density,

$$\int d\mathbf{r}\, (\exp -i\mathbf{K} \cdot \mathbf{r})\langle \Delta n(0)\, \Delta n(\mathbf{r})\rangle \qquad (10.102e)$$

If the change in $\eta$ in any scattering system is related to the change in number density by

$$\Delta\eta = \frac{d\eta}{dn}\, \Delta n$$

then the correlation functions are related by

$$\langle \Delta\eta(0)\, \Delta\eta(\mathbf{r})\rangle = \left(\frac{d\eta}{dn}\right)^2 \langle \Delta n(0)\, \Delta n(\mathbf{r})\rangle \qquad (10.103a)$$

Direct substitution into Eq. (10.101a) yields a general scattering cross section or spectral density for the sample of

$$\sigma(\mathbf{K}) = \bar{N}\,|f|^2 \left\{1 + \bar{n} \int d\mathbf{r}\, [g(\mathbf{r}) - 1] \exp -i\mathbf{K} \cdot \mathbf{r}\right\} \qquad (10.103b)$$

with a scattering amplitude per scatterer of

$$|f|^2 = \frac{\pi^2}{\lambda^4}\left(\frac{d\eta}{dn}\right)^2 \qquad (10.103c)$$

For a one-component system, the cross section is directly proportional to the quantity within the curly brackets in (10.103b), which is the spectral density of the space correlation function for the number density.

For gases and liquids, the orientation of the sample is not important in the measured cross section and the ensemble average becomes

$$\left\langle \sum_{m \neq n} \delta(\mathbf{r} + \mathbf{r}_m - \mathbf{r}_n) \right\rangle \rightarrow \left\langle \sum_{m \neq n} \delta(r - r_{mn}) \right\rangle$$

or

$$g(\mathbf{r}) \rightarrow g(r) \qquad (10.104a)$$

Only the scalar distance between particles is important. Even so, structure is apparent. If there is a tendency toward cubic structure, then the distances $a$, $\sqrt{2}\,a$, and $\sqrt{3}\,a$ will appear as peaks at angles $\theta_{100}$, $\theta_{110}$, and $\theta_{111}$ when $[(4\pi/\lambda)\sin(\theta/2)]r = 2\pi$ for each characteristic distance between particles in the (100), (110), and (111) directions. Further simplification is possible when only scalar distances are important. Let $\gamma$ be the angle between $\hat{K}$ and $\hat{r}$, and then integration over $d\mathbf{r} = 2\pi r^2 \sin\gamma\, d\gamma\, dr$ yields a scattering cross section of

$$\sigma(K) = \bar{N}\,|f|^2 \left\{1 + 4\pi\bar{n} \int_0^\infty r^2\,dr\,[g(r) - 1]\frac{\sin Kr}{Kr}\right\} \qquad (10.104b)$$

The integral is taken to infinity since $g(r) - 1$ must tend toward zero for large $r$. The Fourier transform of Eq. (10.104b) is $V\,|f|^2$ times Eq. (10.102d) and an experimental measurement of $\sigma(K)$ or of the intensity of the scattered wave $I(K)$ can provide a determination of the correlation $g(r) - 1$. Scattering of x rays by high-density argon gas [*10b*] illustrates some features of Eq. (10.104b). The scattering intensity $I(K)$ as a function of $K = (4\pi/\lambda)\sin(\theta/2)$ is shown in Fig. 10.6(a). The correlation function $g(r) - 1$ determined from these data is shown in Fig. 10.6(b). The data are for small angles, or angles of less than 10°. For small $r$, the pair correlation is negative and indicates the hard-core repulsion between the two particles. The probability of finding two particles at these small distances is less than that for a random, or Poisson, distribution. As $r$ increases, the correlation becomes positive and indicates the presence of attractive forces and a potential minimum. In these regions, the probability of finding a pair is greater than for a random distribution.

A grand canonical ensemble average is implied by all of the brackets in these equations. In detail, the correlation term is

$$\left\langle \sum_{m\neq n} \delta(\mathbf{r} + \mathbf{r}_m - \mathbf{r}_n)\right\rangle = \mathcal{Q}^{-1} \sum_N z^N \operatorname{tr}\left[e^{-\beta H_N} \sum_{m\neq n} \delta(\mathbf{r} + \mathbf{r}_m - \mathbf{r}_n)\right]$$

and requires a knowledge of the solution of the $N$-body problem for its evaluation. If $z \ll 1$ and only the $z^2$ term is appreciable in the sum, an approximate solution is given in Section 8.5.1 and yields

$$2z^2 \operatorname{tr}[e^{-\beta H_2}\,\delta(\mathbf{r} + \mathbf{r}_1 - \mathbf{r}_2)] = \bar{N}\bar{n}e^{-\beta U(r)}$$

This is the approximation in which binary interactions are important and the atoms or molecules are treated as spherical scatterers with coordinates $\mathbf{r}_1$ and $\mathbf{r}_2$. If the scattering is by the atomic electrons, then $\mathbf{r}_1$ and $\mathbf{r}_2$ are the coordinates of the center of mass of the atoms and the double sum must

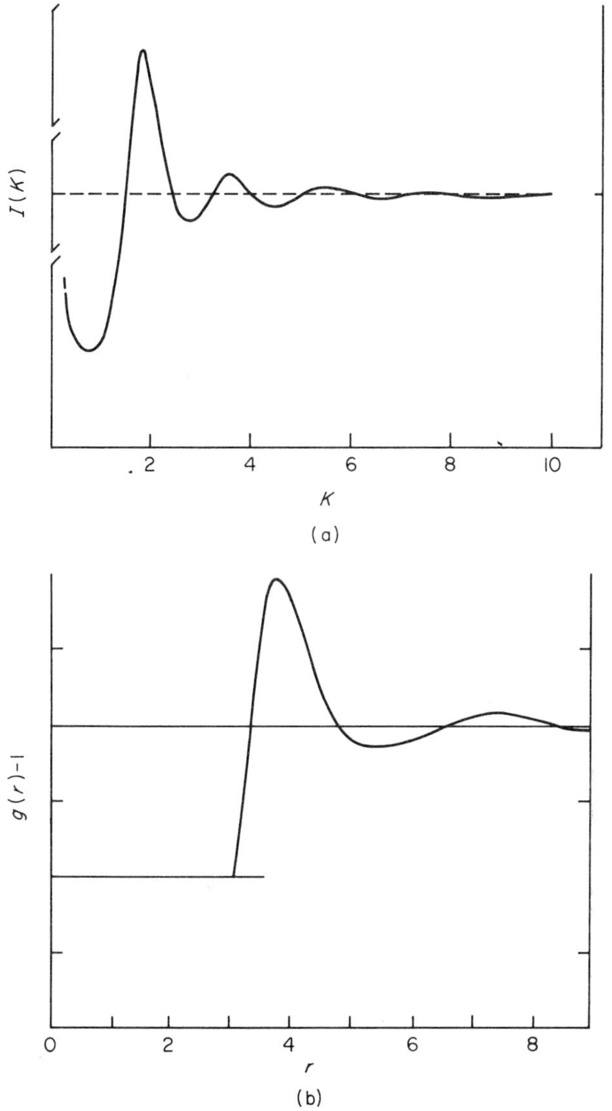


**Fig. 10.6** (a) X-ray scattering intensity $I(K)$ as a function of $K = (4\pi/\lambda)\sin(\theta/2)$ for argon gas at a density of $\varrho = 0.780$ g/cm$^3$ and a temperature of 148°K. $K$ is in units of $10^{10}$ m$^{-1}$. (b) Pair correlation function $[g(r) - 1]$ derived from $I(K)$ as a function of the pair distance $r$, which is in units of $10^{-10}$ m. [From C. J. Pings, *in* "Physics of Simple Liquids" (H. N. Temperley, ed.), Figs. 6 and 10; Wiley, New York, 1968. Reproduced by permission of copyright owner, North-Holland Publishing Co., Amsterdam.]

include the electronic coordinates. A random distribution occurs as $\beta \to 0$ or $T \to \infty$, and this corresponds to the $z \ll 1$, or single-particle, solution.

Scattering by low-number-density gases is an example of a random distribution of scatterers, or a Poisson distribution. By Eq. (10.89b), the fluctuation in number in a Poisson distribution is $\langle (N - \bar{N})^2 \rangle = \bar{N}$ and this requires $g - 1 = 0$. Since $\delta(r - r_{mn})$ is equal to unity for any choice of $r$ for a random distribution, this also implies $g - 1 = 0$. Hence for uncorrelated scatterers, the scattered wave is spherically symmetric and is a measure of the number of scatterers,

$$\sigma(K) = N\,|\,f\,|^2$$

This is the Rayleigh scattering term and is proportional to $\lambda^{-4}$ for constant $d\eta/dn$. As the binary interaction term becomes important, there is a correlation between the particle positions, and comparison of the pair correlation function given by Eq. (10.100c) with Eq. (10.102b) suggests that

$$g(r) - 1 \approx e^{-\beta U(r)} - 1$$

This also follows from the discussion in the previous paragraph in which only the $z^2$ term was retained in the grand canonical average. Again, $e^{-\beta U(r)} - 1$ is the deviation from a random, or Poisson, distribution in a gas in which three-particle collisions can be ignored. The effect of this term on the scattering is comparable to the size of the second virial term and will become important when deviations from the Ideal Gas Law are observable. If

$$U(r) = 4\varepsilon\left[\left(\frac{r_0}{r}\right)^{12} - \left(\frac{r_0}{r}\right)^{6}\right]$$

for the Lennard–Jones interaction between two molecules, then for small $r$, the hard-core repulsion is dominant and the exponential is zero, or the correlation is $-1$. As $r$ increases, a minimum is reached at $r = r_0$ and the correlation is zero. With a further increase in $r$, the attractive force becomes dominant and pairs are favored, or $g - 1 > 0$. For large $r$, the potential and the correlation tend to zero, $g - 1 \to 0$. Oscillations about zero do not occur for this type of potential. The term $g - 1$ starts at $-1$ for small $r$, reaches zero at $r_0$, has a peak as $r$ increases, and then falls to zero for large $r$.

Since the calculation of $g(r) - 1$ is a formidable problem for most experimental samples, it is convenient to have an estimate of its magnitude.

Forward scattering provides such an estimate,

$$\sigma(0) = |f|^2\langle(N - \bar{N})^2\rangle = |f|^2 (\bar{N}\bar{n}kT)\varkappa_T$$

when the isothermal compressibility is known.

It is illuminating to consider the problem as a set of scatterers at positions $\mathbf{r}_m$ interacting with the incident wave with strength $f_m$ and generating new spherical waves $f_m(k, \theta)r^{-1}e^{ikr}$. If some position $\mathbf{r}_0$ is chosen as the origin, the scattering amplitude is proportional to

$$F = \sum_m f_m \exp[i\mathbf{K} \cdot (\mathbf{r}_m - \mathbf{r}_0)]$$

where $\mathbf{K} \cdot (\mathbf{r}_m - \mathbf{r}_0)$ is the difference in phase between the two paths shown in Fig. 10.7(a). An experiment which measures the amplitude of a wave will measure the average value of $F$,

$$\langle F\rangle = \left\langle \sum_m f_m \exp[i\mathbf{K} \cdot (\mathbf{r}_m - \mathbf{r}_0)]\right\rangle \qquad (10.105\text{a})$$

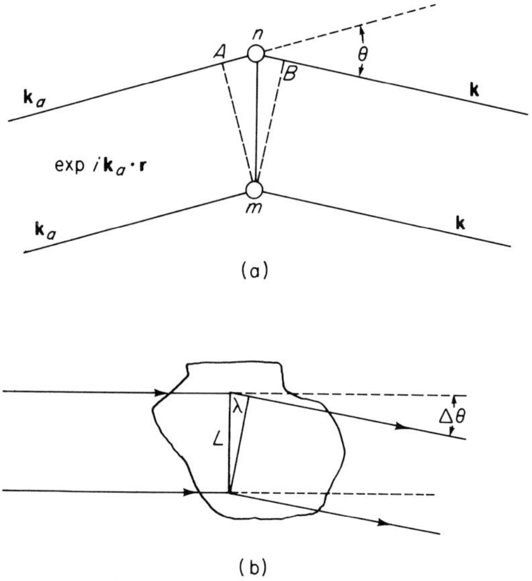


**Fig. 10.7** (a) Scattering by two particles $m$ and $n$, which may be in a gas, liquid, or solid. Here, $\exp i\mathbf{k}_a \cdot \mathbf{r}$ is the incident plane wave and $\exp i\mathbf{k} \cdot \mathbf{r}$ a part of the scattered wave. $AB$ is the path difference which gives rise to the phase difference between the two scatterers. $K = |\mathbf{k}_a - \mathbf{k}| = (4\pi/\lambda)\sin(\theta/2)$. (b) Homogeneous scattering, or the diffraction effect in the forward direction due to shape.

Typical values of $\mathbf{r}_m - \mathbf{r}_0$ are of the order of the sample dimension $L$ and for values of $KL \gg 1$, the sum becomes a sum of terms with random phases and is small. For $K = 0$, this sum reduces to

$$\begin{aligned} \bar{F} &= \left\langle \sum_m f_m \right\rangle, & K &= 0 \\ &\approx 0, & \delta\theta &> \frac{\lambda}{L} \end{aligned}$$

Figure 10.7(b) indicates the salient features of this term. All of the scattered waves in the forward direction add coherently and yield this maximum value for the average. As $K$ is increased, the waves will interfere at $\delta\theta \approx \lambda/L$ and yield the first diffraction minimum. The angular width of the amplitude measures the sample size and $\delta\theta$ is a very small angle in most experiments. It should be noted that this small-angle scattering is independent of the location of the individual scatterers in the sample and the scattering is the same as that for a homogeneous sample with the average scattering amplitude. Since there is an arbitrary phase in $\bar{F}$, its measurement requires an interference type of experiment with the incident radiation or a measurement of the index of refraction. This term is almost insensitive to the internal coordinates of the system and can be written in the approximate form

$$\bar{F} \approx \left\langle \sum_m f_m \right\rangle \delta(\mathbf{K})$$

Since the average amplitude $\bar{F}$ can be made to interfere with the incident wave, it cannot be regarded as scattered, and the intensity measured in scattering experiments is proportional to

$$\begin{aligned} S(\mathbf{K}) &= \langle | F - \bar{F} |^2 \rangle = \langle | F |^2 \rangle - | \bar{F} |^2 \\ &= \left\langle \sum_m | f_m |^2 \right\rangle + \left\langle \sum_{m \neq n} f_m f_n^* \exp[i\mathbf{K} \cdot (\mathbf{r}_m - \mathbf{r}_n)] \right\rangle - | \bar{F} |^2 \delta(\mathbf{K}) \end{aligned} \quad (10.105b)$$

The first term is the self-scattering, or single-particle scattering, term and is spherically symmetric. The second term is the pair correlation function and depends on the space vector between the pair and on the scattering amplitude of each member of the pair. This term is highly sensitive to the internal coordinates of the system. The average of this pair correlation cannot be sensitive to the choice of $K$ and this becomes apparent by writing this expression in integral form as

$$\left\langle \sum_{m \neq n} f_m f_n^* \exp[i\mathbf{K} \cdot (\mathbf{r}_m - \mathbf{r}_n)] \right\rangle = \int d\mathbf{r}\, (\exp - i\mathbf{K} \cdot \mathbf{r}) \left\langle \sum_{m \neq n} f_m f_n^* \, \delta(\mathbf{r} + \mathbf{r}_m - \mathbf{r}_n) \right\rangle \quad (10.105c)$$

For $f_m = f$, this becomes the previous spectral density of the space correlations. Since the integral is an average, the processes of integration and averaging can be interchanged and this permits the above definition of the correlation. For this correlation function, $f_m$ can change with position and $f_m$ can be a function of $k$ and $\theta$, $f_m(k, \theta)$. This permits the scattering atoms to have an internal form factor and the sum is over atomic or molecular coordinates.

Incoherent elastic scattering can occur if the internal state of the scattering molecule or atom changes during the scattering process. If a photon is scattered off an atom in the $m$ state and after the scattering event the atom is in the state $m'$, the event is incoherent and the spherical scattered wave has a random phase factor $\delta_m$,

$$f_m \to f_m e^{i\delta_m}.$$

All double sums average to zero,

$$\sum_{m \neq n} e^{i(\delta_m - \delta_n)} = 0$$

and only the self-scattering term remains.

### *10.11.2 Scattering of Visible Radiation by Gases and Liquids*

This theory can be applied to the scattering of visible radiation by gases and liquids. Since $g - 1$ falls to zero in distances of the order of the collision or correlation radius, the wavelength is much larger than the range of $g - 1$ and $(Kr)^{-1} \sin Kr \approx 1$. Equations (10.102a–e) and (10.104a, b) simplify in this approximation. All directional aspects of scattering are contained in the scattering amplitude $f(k, \theta)$ or in the polarization of the light waves. Scattering by gases and by dilute binary liquid mixtures are considered first. Since the second virial coefficient is small in gases, and the interaction between the dilute species is small in liquid mixtures, the correlation term can be omitted. The intensity of the scattered radiation is directly proportional to the number of scattering centers, or $\bar{N} \mid f(k, \theta) \mid^2$.

The scattering of radiation by an atom or molecule is a two-photon process in the quantum mechanical calculation and is directly related to the dispersion and the index of refraction. Classical scattering by bound electrons illustrates many features of the problem. The equation of motion of a bound electron is given by Eq. (10.38) with a driving force

$$f(t) = \hat{x} e E_x \cos(k_0 z - \omega t)$$

Linear polarization along $\hat{x}$ is assumed and the incident radiation is along $\hat{k}_0 = \hat{z}$. The polarization induced by the electric field is $p_x = eq$ and the amplitude of the scattered wave is given as

$$E_s \propto \frac{\ddot{p}_x}{R}$$

For radiation scattered in direction $\hat{k}$ making angle $\theta$ with $\hat{k}_0$, the polarization of the wave is the component of $p_x$ transverse to $\hat{k}$, or along $\hat{\theta}$ and $\hat{\varphi}$. For induced polarization along $\hat{x}$, the polarization of the scattered wave is

$$\hat{u} = \hat{k} \times (\hat{x} \times \hat{k})$$

The scattering cross section per particle is proportional to $|f(k, \theta)|^2$ or

$$\sigma(k, \theta, \psi) = |f(k, \theta)|^2 = \frac{\pi^2 \alpha^2}{\lambda^4} \sin^2 \psi$$

where $\psi$ is the angle between $\hat{x}$ and $\hat{k}$ and is related to $\theta$ and $\varphi$ by

$$\cos \psi = \sin \theta \cos \varphi$$

Rayleigh scattering occurs for $\nu \ll \nu_0$ and the scattering cross section is proportional to

$$\sigma_p(k, \theta, \psi) = b\lambda^{-4} \sin^2 \psi \tag{10.106a}$$

Blue light is scattered more than red, and the polarization of the scattered light depends on $\psi$. If the incident light is unpolarized, then it is necessary to average over the angle $\varphi$, and the cross section for unpolarized light is

$$\sigma_u(k, \theta) = \frac{b}{2\lambda^4} (1 + \cos^2 \theta) \tag{10.106b}$$

For unpolarized incident radiation, the ratio of the intensities of the polarization of the scattered radiation with linear polarizations $\hat{\theta}$ and $\hat{\varphi}$ is given by

$$\frac{I_\theta}{I_\varphi} = \cos^2 \theta$$

Thus at $\theta = 90°$, or transverse to $\hat{k}_0 = \hat{z}$, the radiation is linearly polarized. The blue sky and the linear polarization of the radiation from the blue sky are in accord with this simple explanation.

Scattering from the incident radiation gives rise to an attenuation of the radiation in the forward direction of $\bar{n}\sigma$, that is,

$$\frac{I}{I_0} = \exp(-\bar{n}\sigma d)$$

where $d$ is the sample thickness. The total scattering cross section requires an average over $\psi$ for polarized radiation and over $\theta$ for unpolarized radiation. Thus the attenuation is

$$\bar{n}\sigma_{\mathrm{u}} = 2\bar{n}\sigma_{\mathrm{p}} = \frac{16\pi^3}{3}\,\frac{\alpha^2\bar{n}}{\lambda^4} \tag{10.106c}$$

The constant $b$ follows from the equation for the classical scattering cross section, and

$$b = \alpha^2 = \left(\frac{e^2}{4\pi\varepsilon_0 mc^2}\right)^2 \frac{(\lambda^4/\pi^2)}{[1-(\omega_0/\omega)^2]^2 + (\gamma/m\omega)^2}$$

but for atoms and molecules, a more detailed calculation is necessary. A more general quantum mechanical calculation will yield

$$p_x = \varepsilon_0\alpha E_x$$

where $\varepsilon_0\alpha$ is the polarizability of the atom or molecule. For free electrons $\omega_0 = 0$ and $r \simeq 0$ and the Thomson cross section is

$$\sigma = \frac{8\pi}{3}\,r_e^{\,2} = \frac{8\pi}{3}\left(\frac{e^2}{4\pi\varepsilon_0 mc^2}\right)^2 \tag{10.106d}$$

If the Lorentz local field model of the type introduced in Section 7.10 for the local magnetic field is used for $E_x$, then the index of refraction and the polarizability are related by

$$\frac{\varkappa - 1}{\varkappa + 2} = \frac{\mu^2 - 1}{\mu^2 + 2} = \frac{\bar{n}\alpha}{3} \tag{10.106e}$$

and permits a determination of $\alpha$ from experimental data for the index of refraction. The value of $\varkappa = \varepsilon/\varepsilon_0 = \mu^2$ is determined at the scattering frequency $\nu$.

Since the scattering by the molecules in liquids may be sensitive to the molecular volume, it may be necessary to determine $(\partial\varepsilon/\partial n)_{\mathrm{T}}$ for the liquid and use Eq. (10.101a) for the scattering. In the absence of correlation,

$$\langle \Delta n(0)\,\Delta n(\mathbf{r})\rangle = \bar{n}\,\delta(\mathbf{r})$$

**Table 10.1**

Intensity of Light Scattered by Gases and Liquids[a]

| | | | |
|---|---|---|---|
| Air (*g*) | 1 | Xe (*g*) | 5.5 |
| $N_2$ (*g*) | 1 | $C_2H_5Cl$ (*g*) | 16 |
| $H_2$ (*g*) | 0.23 | Benzene (*l*) | 1 |
| Ar (*g*) | 0.8 | Water (*l*) | 0.055 |
| $O_2$ (*g*) | 0.9 | Acetone (*l*) | 0.26 |
| Kr (*g*) | 2.0 | $CS_2$ (*l*) | 4.1 |
| $CO_2$ (*g*) | 2.6 | | |

[a] Values for gases (*g*) are relative to an attenuation coefficient $\alpha = \bar{n}\sigma$ for unpolarized light at $\lambda = 0.4358\ \mu$m for air of $\alpha_{\text{air}} \approx 3.0 \times 10^{-9}$/m, and values for liquids (*l*) are relative to $\alpha_{\text{benzene}} \approx 6.6 \times 10^{-6}$/m.

and

$$\frac{I(\mathbf{K}, \psi)}{I_0} = \left(\frac{\pi^2}{\lambda^4}\right)\bar{n}V\left(\frac{\partial \varkappa}{\partial n}\right)_{\mathrm{T}}^2 \sin^2 \psi \tag{10.106f}$$

for the scattering by a sample of volume $V$.

Table 10.1 indicates the intensity of light scattered by some common gases and liquids [*10c, d*].

For scattering by dilute gas or liquid mixtures, the fluctuation in the dielectric constant with concentration can cause appreciable scattering. If $c = N_{\mathrm{B}}/(N_{\mathrm{A}} + N_{\mathrm{B}})$ is the concentration of the minor species, then in principle, the fluctuation in dielectric constant with concentration at constant temperature and density is given by

$$\Delta\varepsilon = \left(\frac{\partial \varepsilon}{\partial c}\right)_{N,\mathrm{T}} \Delta c = \left(\frac{\partial \varepsilon}{\partial n_{\mathrm{B}}}\right)_{N,\mathrm{T}} \Delta n_{\mathrm{B}}$$

Following the previous development,

$$\langle \Delta\varepsilon(0)\, \Delta\varepsilon(\mathbf{r})\rangle_{N,\mathrm{T}} = \left(\frac{\partial \varepsilon}{\partial n_{\mathrm{B}}}\right)_{N,\mathrm{T}}^2 \langle \Delta n_{\mathrm{B}}(0)\, \Delta n_{\mathrm{B}}(\mathbf{r})\rangle$$

The correlation in number density can be related to the osmotic compressibility in the manner used for Eq. (10.88). For dilute solutions, the correlation is small between B particles and the fluctuation in number density is given by

$$\langle \Delta n_{\mathrm{B}}(0)\, \Delta n_{\mathrm{B}}(\mathbf{r})\rangle \approx \bar{n}_{\mathrm{B}}\, \delta(\mathbf{r})$$

Equation (9.58b) relates $\bar{n}_B$ to the osmotic pressure. The scattering intensity is the same as Eq. (10.106f) with $\bar{n}$ replaced by $\bar{n}_B$. Binary mixtures were discussed in relationship to the three-dimensional Ising model and can be compared with Eq. (9.113b).

Anisotropic molecules, for which $P_a = \varepsilon_0 \alpha_{ab} E_b$, will be consider in greater detail in the discussion of time-dependent aspects of molecular scattering in Chapter XI.

### *10.11.3 Scattering of Light by Particles*

If the size of a dielectric object is small compared to the wavelength, then the induced polarization in the dielectric medium is almost the same as that in a constant electric field. The induced polarization in a sphere of dielectric constant $\varkappa_1$ in a surrounding medium of dielectric constant $\varkappa_2$ is

$$p_x = 3\varepsilon_0 \left( \frac{\varkappa - 1}{\varkappa + 2} \right) V E_x = \varepsilon_0 \alpha E_x$$

where $\varkappa = \varkappa_1 / \varkappa_2$ is the relative dielectric constant and $V$ is the volume of the sphere or object. The induced polarization oscillates with frequency $\omega$ and scatters proportional to $\ddot{p}_x$ or $\omega^2 p_x$. If the incident wave is along $\hat{k}_0 = \hat{z}$ and polarized along $\hat{x}$, the scattered wave in direction $\hat{k}$ and with polarization $\hat{u} = \hat{k} \times (\hat{x} \times \hat{k})$ has a scattering cross section of $| f(k, \theta) |^2$, or

$$\sigma_p(k, \theta, \psi) = 9\pi^2 V^2 \left( \frac{\varkappa - 1}{\varkappa + 2} \right)^2 \frac{1}{\lambda^4} \sin^2 \psi \tag{10.107a}$$

Integration over $\psi$ yields a total cross section for scattering $\sigma_p(k)$, or an attenuation coefficient $\bar{n}\sigma_p(k)$, for linearly polarized light of

$$\bar{n}\sigma_p = 24\pi^3 V^2 \left( \frac{\varkappa - 1}{\varkappa + 2} \right)^2 \frac{\bar{n}}{\lambda^4} \tag{10.107b}$$

This equation can be used to estimate the average number density $\bar{n}$ and volume $V$ of scattering objects which are much less than one wavelength in diameter. This attenuation of the forward wave is often called the turbidity. Scattering of light by colloidal particles was studied by Tyndall in 1869 and he observed the bluish light of linear polarization scattered at 90°. Dust particles were suggested as the cause of the blue sky, but Rayleigh suggested that molecules as well as macroscopic particles can scatter. Molecular scattering is the dominant scattering in the atmosphere.

Sulfur hydrosols are used as an example of colloidal scattering of light. On the mixing of dilute solutions of hydrochloric acid and sodium thiosulfate, the molecularly dispersed sulfur forms condensed droplets of supercooled sulfur and these droplets grow as the sol ages. The initial radii are of the order of 0.01 $\mu$m and these droplets scatter in the domain of Rayleigh scattering. This forms the basis for the experiment often shown in elementary physics to illustrate the cause of the blue sky and the red sunset.

Another interesting problem is the scattering of a polymer molecule, which is analogous to a disordered ball of thread. Such a thread is shown in Fig. 10.8(a), and the phase difference between the scattering of points $m$ and $n$ on the thread is given by

$$\delta = \mu 2\pi \frac{Bm - An}{\lambda} = \mu \frac{2\pi L}{\lambda} [\cos\varphi - \cos(\varphi - \theta)]$$

For $\theta = 0$ the phase difference $\delta = 0$ and for $\theta = \pi$, the phase difference reaches a maximum. The scattering intensity $I(\theta)/I(0)$ has been examined for balls of thread with various thread concentrations as a function of the radius or a characteristic length $L$, and the shape dependence is shown in Fig. 10.8(b). The scattering by a thin rod and a sphere are shown as limiting examples. Scattering coefficients of proteins in water can become quite large and protein molecules with a molecular weight of $5 \times 10^5$ and a concentration of $10^{-3}$ g/cm$^3$ scatter 100 times more strongly than pure water.

If the atom or molecule is near resonance, the cross section for absorption is dependent on other factors and can approach $\lambda^2/4\pi$. If small particles are conducting, their cross section is modified, and more detailed references are suggested [*10e, f*]. These aspects were not included in the discussion here. Scattering due to temperature fluctuations of the index of refraction is neglected and is usually much smaller than density fluctuations.

Exercise 10.13 (a) Use Eq. (10.107b) to develop a method for estimating the number of sulfur droplets and their size.

(b) Estimate the number density of scatterers of 0.01 $\mu$m diameter when $\alpha = 0.5$/m for light with a wavelength of 0.5 $\mu$m. Estimate the osmotic pressure of this same solution of water and sulfur droplets.

Exercise 10.14 The index of refraction of air is 1.00029. Estimate the attenuation per meter and compare with $3.0 \times 10^{-9}$/m for unpolarized incident radiation.

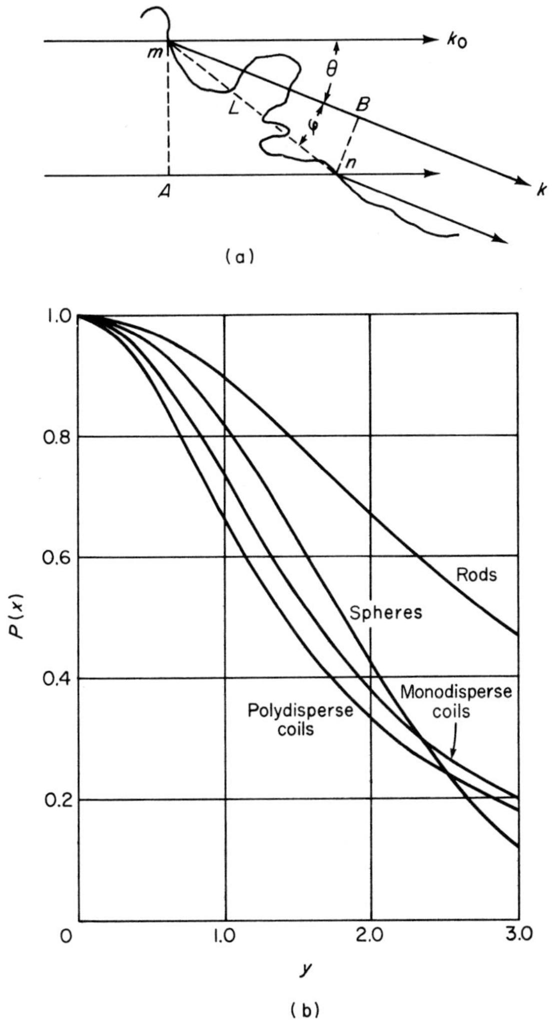


**Fig. 10.8** (a) Scattering of light by a chain or thread. Scattering centers $m$ and $n$ on the thread are separated by distance $L$. Here, $\theta$ is the angle between the incident wave $k_0$ and the scattered wave $k$. A phase difference $\delta = \mu(2\pi L/\lambda)[\cos\varphi - \cos(\varphi - \theta)]$ occurs between the two centers separated by a distance $L$. (b) The scattering of unpolarized light is given by $I(\theta)/I(0) = [(1 + \cos^2\theta)/2]P(x)$, where $P(x)$ is shown for: (a) spheres of radius $R$ with $x = (4\pi\mu/\lambda)R\sin\theta/2$; (b) rods of length $L$ with $x = (4\pi\mu/\lambda) \times (L/2)\sin\theta/2$; (c) balls or coils of thread or systems of similar construction, where $L$ is the root mean square distance between the ends of the coil and $x^{1/2} = (4\pi\mu/\lambda)(L/6^{1/2}) \times \sin\theta/2$. The curves for $P(x)$ for rods, spheres, and monodisperse and polydisperse randomly kinked coils are taken from the tabulated data. $y = x$ for rods and spheres and $y = x^{1/2}$ for coils. [From P. Doty and R. F. Steiner, *J. Chem. Phys.* **18**, 1211 (1950), Fig. 1. Reproduced by permission of the *Journal of Chemical Physics*.]

### *10.11.4 Critical Opalescence*

The correlation of fluctuation term in Eq. (10.106a), or $g(r) - 1$, has been small in most of the previously discussed experiments. In a certain sense, this term contributed appreciably to the scattering by the large molecules in the previous section. The individual scattering centers were highly correlated and gave rise to diffraction or shape effects for small rods, spheres, chains, and so on. It has been noted throughout this section that for $\mathbf{K} = 0$, the quantity within the curly brackets in Eq. (10.104b) is proportional to the compressibility. Near the critical point in vapor–liquid systems, the compressibility becomes large and in binary liquid mixtures, the osmotic compressibility becomes large at the critical concentration. This implies that the correlation term is important and long-range aspects must be considered. Scattering becomes large and the phenomenon is referred to as critical opalescence.

Some theories suggest[2]

$$g(r) - 1 = \frac{Ae^{-br}}{r^{1+\eta}} \tag{10.108a}$$

as the long-range correlation function. This becomes the Ornstein–Zernike theory for $\eta = 0$. Recent developments of the three-dimensional Ising model which were discussed in Section 9.7 suggest a value of $\eta = 0.06$. This form of the correlation function was apparent in the combinatorial problem and was referred to in Eq. (9.109). The coefficients $A$ and $b$ are related to the compressibility by Eq. (10.103e). Omitting the constant term $\bar{N}$ and with $\eta = 0$, the coefficients are related by

$$\frac{4\pi A}{b^2} \approx \beta \varkappa_{\mathrm{T}} \tag{10.108b}$$

Near the critical point, the correlation length $b$ increases as $(\varkappa_{\mathrm{T}})^{1/2}$ and the scattering at angle $\theta$ is related to the forward scattering by

$$\frac{I(\mathbf{K})}{I(0)} \approx \left(1 + \frac{K^2}{b^2}\right)^{-1+(\eta/2)} \tag{10.108c}$$

For binary liquid mixtures, the osmotic compressibility replaces $\varkappa_T$ for the determination of the coefficients.

[2] See Fisher [*25*] in Chapter IX, this volume, for the $\eta$ coefficient.

Reliable data near the gas–liquid critical point are not as yet available. Critical opalescence in binary liquids provides a better test of the theory. The scattering of 0.546-$\mu$m light near the critical point of a nitroethane–3-methylpentane system [*10c*] yields an angular dependence consistent with this scattering equation with $\eta \approx 0.2 \pm 0.1$.

## 10.12 Nonlinear Physical Systems

The physical systems discussed in the earlier part of this chapter were damped systems which were driven by a signal, by random pulses, or by thermal noise. Physical systems with "negative damping" can also occur. Damping under normal conditions brings the system into thermal equilibrium with its surroundings. "Negative damping" requires a special arrangement for each physical system. In a watch, the stored energy of a spring is used to drive a resonant balance wheel via an escapement mechanism. In laboratory oscillators, the stored energy of a battery or dc power supply is converted into electrical oscillations in a resonant $LC$ circuit via a triode or transistor. Some recent systems use an Esaki diode in the megahertz region and a Gunn diode in the microwave region. Lasers use dc electrical discharges, rf electrical discharges, flash lamps, or chemical reactions to place a greater number of molecules in an upper state $b$ than in a lower state $a$, and then stimulated emission of radiation in a resonant cavity, which is formed by two mirrors, provides an oscillator in the optical, infrared, or far-infrared spectral region. The first radio transmitters used a resonant $LC$ circuit and the negative resistance of an arc to provide oscillatory energy for the antenna. If in each of these systems, the coordinate $q$ describes the angular motion of the balance wheel, the charge on the capacitor in the $LC$ circuit, the electrical field in the microwave waveguide, or the electrical field in the laser, the power supplied by the external reservoir is proportional to a gain term of the type $a''\dot{q}$. In the laser, this is the polarization of the laser medium, in the Gunn diode, this is a characteristic of the current–voltage relationship $dV/dI$ for a particular value of the dc current, and so on. By its very definition, every oscillator is nonlinear and its ultimate amplitude is limited by a nonlinear process which can often be described by a loss term of the type $-b''\dot{q}^3$. In the remainder of this section, a simple oscillator is discussed, and then the very unusual aspects of coupled oscillators are considered.

### *10.12.1 Simple Self-Sustained Oscillator*

A simple harmonic oscillator, which has damping $\gamma\dot{q}$ in the absence of the source and which is supplied energy through terms of the type $a''\dot{q} - b''\dot{q}^3$, obeys a nonlinear equation of the form

$$m\ddot{q} + m\omega_0^2 q + \gamma\dot{q} = ma'\omega_0 q - mb'\omega_0 q^3 + a''\dot{q} - b''\dot{q}^3 \quad (10.109)$$

Here $m$ is a generalized coefficient, so that $m\omega_0^2 q^2$ has the units of energy. This problem is not analyzed in detail here, but some salient features become apparent by assuming that the resonance is narrow, or $\omega_0 \gg \gamma$. Then, an approximate solution is of the form

$$q(t) = \xi(t)e^{-i\omega_0 t} + \text{complex conjugate}$$

Terms like $\ddot{\xi}$, $\gamma\dot{\xi}$, ... are regarded as small when compared with $\omega_0\dot{\xi}$, $\omega_0\gamma\xi$, .... With these approximations, Eq. (10.109) reduces to

$$2m\dot{\xi} = \xi[a'' - \gamma - 3b\omega_0^2\xi\xi^* + im(a' - 3b'\omega_0^2\xi\xi^*)] \quad (10.110)$$

A steady-state solution is readily obtained with the substitution $\xi = (\xi\xi^*)^{1/2} \times e^{iu}$. Then,

$$m\frac{d(\xi\xi^*)}{dt} = \xi\xi^*(a'' - \gamma - 3b''\omega_0^2\xi\xi^*) \quad (10.111a)$$

and at steady-state operation, the stored energy of the oscillation is limited to

$$\omega_0^2\xi\xi^* = \frac{a'' - \gamma}{3b''} \quad (10.111b)$$

and the frequency of oscillation is shifted off resonance by $\dot{u}$, or

$$\omega_{\text{osc}} = \omega_0 - \tfrac{1}{2}(a' - 3b'\omega_0^2\xi\xi^*) \quad (10.112)$$

The system operates at a stored energy and frequency such that the energy furnished by the external source is equal to the loss term $\gamma\dot{q}^2$. Note that the oscillator is pulled off resonance by a term which is proportional to the energy stored in the resonant circuit. For many problems, the power output is measured and this is proportional to $\gamma\dot{q}^2 = 2\gamma\omega_0^2\xi\xi^*$.

The effect of thermal noise on the oscillator can be estimated by adding a noise source $f(t)$ to the equation for $q(t)$. This adds a term $-fe^{i\omega_0 t}/i\omega_0$ to Eq. (10.110) and

$$f\omega_0^{-1}(\xi\xi^*)^{1/2}\sin(\omega_0 t - u)$$

to Eq. (10.111a) for the stored energy. Since $f$ and $u$ are not correlated, the average value of $\langle \xi\xi^* \rangle$ is the same as for the oscillator without noise. The stored energy in the oscillator is $2m\omega_0^2\xi\xi^*$ and the thermal energy is $kT$. This suggests that the fluctuation in the output power is of the order of $kT/2m\omega_0^2\xi\xi^*$ and is quite small well away from threshold. The probability of the signal power being reduced to zero is quite small. Fluctuation in frequency is more important. Replacing $u$ by

$$u = v + \tfrac{1}{2}(a' - 3b'\omega_0^2\xi\xi^*)$$

the phase equation becomes

$$\dot{v} = \frac{f(t)}{2m\omega_0(\xi\xi^*)^{1/2}} \cos(\omega_{\text{osc}}t - v) \tag{10.113a}$$

This equation cannot be solved since $f(t)$ is a thermal noise function, but the correlation function for the frequency

$$\dot{v} = \Delta\omega(t) = \omega(t) - \omega_{\text{osc}}$$

can be obtained and is

$$\langle \Delta\omega(t)\, \Delta\omega(t+\tau) \rangle = \langle \dot{v}(t)\dot{v}(t+\tau) \rangle = \frac{\gamma kT}{2m^2\omega_0^2\xi\xi^*}\, \delta(\tau) \tag{10.113b}$$

From earlier considerations, the correlation function of the noise source is

$$\langle f(t)f(t+\tau) \rangle = 2\gamma kT\, \delta(\tau)$$

for the resonant circuit alone.

A correlation function and spectral density can now be obtained for the oscillator. Noting that the phase

$$v(t+\tau) - v(t) = \int_t^{t+\tau} \dot{v}\, dt'$$

the correlation function for the coordinate $q$ can be written as

$$\langle q(t)q(t+\tau) \rangle = \left\langle \xi\xi^* \exp\left( i\omega_{\text{osc}}\,\tau - i\int_0^\tau \dot{v}\, dt' \right) + \text{complex conjugate} \right\rangle$$

$$\approx 2\langle \xi\xi^* \rangle (\exp -\alpha\tau) \cos \omega_{\text{osc}}\tau \tag{10.114a}$$

where

$$\alpha = \frac{\gamma kT}{2m^2\omega_0^2\xi\xi^*} \tag{10.114b}$$

Using a technique discussed in greater detail in a later section in this chapter, the average value of the exponential is obtained from

$$\left\langle \exp -i \int_0^\tau \dot{v}\, dt \right\rangle \approx \exp\left[-\tfrac{1}{2}\int_0^\tau dt' \int_0^\tau dt''\, \langle \dot{v}(t')\dot{v}(t'')\rangle\right]$$
$$\approx \exp -\alpha\tau \tag{10.114c}$$

The double integral is evaluated by the procedure used for Eq. (10.37), and Eq. (10.113b) is used for the average value of the term inside the angular brackets.

Equation (10.114a) is the same as the correlation function for a Lorentz line shape with a spectral width between half-power points of $2\alpha$, or

$$\Delta\omega_{1/2} = \frac{\gamma kT}{m^2\omega_0^{\,2}\xi\xi^*} \tag{10.115a}$$

The line width without gain is $2\gamma/m = \Delta\omega_0$, and

$$\frac{\Delta\nu_{1/2}}{\Delta\nu_0} = \left\{\frac{\text{stored thermal energy}}{\text{stored oscillator energy}}\right\} \tag{10.115b}$$

This expression is not quite correct, since the saturation effect of the gain medium should be included in the noise. Original references are suggested for this aspect [*11*].

Note that Eq. (10.109) is similar to the equation used for a nonquantized maser or laser oscillator [*12*]. The generalized coordinate $q$ refers to a cavity mode and is related to the electric field by

$$m\omega_0^{\,2}q^2 \to \varepsilon_0 E^2$$

$a\dot{q}$ and $b\dot{q}^3$ are related to the electric polarization of the medium of the molecules. The cavity losses are described by a cavity $Q$, and $2\gamma/m \to \omega/Q = \Delta\omega_0$. For optical lasers, $h\nu \gg kT$ and quantum noise replaces thermal noise. The spectral line width can be estimated by replacing $kT \to \frac{1}{2}h\nu$ and $2m\omega_0^{\,2}\xi\xi^* \to \bar{n}h\nu$, where $\bar{n}$ is the average number of quanta in the cavity. Then

$$\frac{\Delta\nu_{1/2}}{\Delta\nu_0} \approx \frac{1}{2\bar{n}} \tag{10.116}$$

is the ratio of the spectral width of the power output to the empty cavity width. The phase of the wave shifts 1 rad in a time of the order of $1/\Delta\nu_{1/2}$. Since $\bar{n} = 10^{10}$ and $\Delta\nu_0 = 10^6$ in a simple optical laser, a rather long correlation time is expected. Usually, other disturbances are more important and the correlation time is much shorter. Even so, beats between two laser outputs can be measured. The fluctuation in power output of the laser is quite small and is of the order of $1/\bar{n}$ in the ideal system. Saturation also increases the width, and original references are suggested for this contribution [*13*].

### *10.12.2 Coupled Self-Sustained Oscillators*

A problem which occurs in many diverse areas in physics is the coupling of two self-sustained oscillators. Very interesting anomalous effects occur. Van der Pol considered the problem of entrainment of an oscillator by a sinusoidal signal. More recently, the coupling of two laser modes with different frequencies, the coupling of the two states of polarization in lasers, a tentative explanation for the circular polarization of the 1700-MHz radiation from OH clouds in our galaxy, and the effect of magnetic fields on the coupled modes in lasers have been discussed.[3] Another interesting example is the human heart and its entrainment by a pacemaker or external impulse generator.

Features common to all of these problems occur in the generalized equation for a pair of coupled oscillators,

$$\ddot{q}_1 + \omega_1{}^2 q_1 + \gamma_1 \dot{q}_1 = a_1 \dot{q}_1 - b_1 \dot{q}_1{}^3 - c_{12} \dot{q}_1 \dot{q}_2{}^2 \tag{10.117a}$$

and a similar equation with subscripts 1 and 2 interchanged. Frequency-shifting terms like $q_1$, $q_1{}^3$, ... and the generalized mass $m$ are omitted for convenience. It is assumed that the resonance of both oscillators is narrow and the difference in frequency between $\omega_1$ and $\omega_2$ is much less than either of these frequencies. With these assumptions, the solution is of the form

$$q_1(t) = \xi_1(t) e^{-i\omega_1 t} + \text{complex conjugate}$$

and Eq. (10.117a) yields the following equation for $\xi_1$:

$$2\dot{\xi}_1 = \xi_1(a_1 - \gamma_1 - 3b_1\omega^2 \xi_1 \xi_1{}^* - 2c_{12}\omega^2 \xi_2 \xi_2{}^*) - c_{12}\omega^2 \xi_1{}^* \xi_2{}^2 e^{i2(\omega_1 - \omega_2)t} \tag{10.117b}$$

[3] See Lamb [*12*] for coupling between modes separated in frequency; Heer and Graft [*14*] for coupling between two states of polarization, magnetic effects, and coupling of clockwise and counterclockwise traveling waves in a laser gyroscope; Heer [*14a*] for the anomalous polarization of the cosmic OH radiation; and Sargent *et al.* [*14b*] and Settles and Heer [*14c*] for other magnetic effects.

A similar equation follows for $\xi_2$ by interchanging subscripts 1 and 2. Considerable insight into the solution can be obtained by the following procedure. Let

$$\xi_1 = Ue^{iu} \qquad \text{and} \qquad \xi_2 = Ve^{iv}$$

and then direct substitution into Eq. (10.117b) yields a set of four equations. These may be reduced to three by noting that only the relative phase

$$\varphi = 2(v - u) + 2(\omega_1 - \omega_2)t$$

is important. Then, in detail,

$$\dot{\varphi} = 2(\omega_1 - \omega_2) + (c_{12}\omega^2 V^2 + c_{12}\omega^2 U^2) \sin \varphi \tag{10.118a}$$

$$\frac{dU^2}{dt} = U^2(a_1 - \gamma_1 - 3b_1\omega^2 U^2 - 2c_{12}\omega^2 V^2 - c_{12}\omega^2 V^2 \cos \varphi) \tag{10.118b}$$

$$\frac{dV^2}{dt} = V^2(a_2 - \gamma_2 - 3b_2\omega^2 V^2 - 2c_{21}\omega^2 U^2 - c_{21}\omega^2 U^2 \cos \varphi) \tag{10.118c}$$

are the approximate equations of change. The phase equation has a standard form

$$\dot{\varphi} = a + b \sin \varphi$$

and can be integrated to yield

$$t(\varphi) = \int_{\varphi_0}^{\varphi} \frac{d\varphi'}{a + b \sin \varphi'} \tag{10.119a}$$

This is a standard integral and has the following properties:

$$t \to \infty, \qquad b^2 < a^2, \qquad \varphi \to at$$

$$t \to \infty, \qquad b^2 > a^2, \qquad \tan \frac{\varphi}{2} \to \frac{-b - (b^2 - a^2)^{1/2}}{a}$$

and these two cases are considered.

For $b^2 > a^2$, the relative phase $\varphi$ tends to a fixed value, or $\dot{\varphi} \to 0$ as $t \to \infty$. This implies that both oscillators will oscillate at the same frequency $\omega$,

$$\omega = \omega_1 - \dot{u} = \omega_2 - \dot{v}$$

and the two oscillators are "entrained." Two oscillators become entrained whenever the nonlinear coupling term and frequency separation are related by

$$(\omega_1 - \omega_2)^2 < \omega^2(c_{12}V^2 + c_{21}U^2)^2 \tag{10.119b}$$

The power furnished by each oscillator is proportional to $U^2$ and $V^2$. The evolution of the amplitude equations in time can be analyzed for stable points by a technique used in the analysis of nonlinear differential equations [*15*]. A criterion for stability, examines the ratio $dV^2/dU^2$ for singular points. Equations (10.118b) and (10.118c) can be placed in the form

$$\frac{dV^2}{dU^2} = \frac{V^2}{U^2}\,\frac{A_2 - B_2V^2 - c_2U^2}{A_1 - B_1U^2 - c_1V^2} \tag{10.120}$$

where $A$, $B$, and $C$ are immediately apparent coefficients. Singular points occur whenever the numerator and denominator are simultaneously zero. Four singular points occur and are given by

$$U^2 = 0 \quad \text{and} \quad V^2 = 0 \tag{10.121a}$$

$$U^2 = 0 \quad \text{and} \quad V^2 = \frac{A_2}{B_2} \tag{10.121b}$$

$$U^2 = \frac{A_1}{B_1} \quad \text{and} \quad V^2 = 0 \tag{10.121c}$$

$$U^2 = \frac{A_1B_2 - A_2C_1}{B_1B_2 - C_1C_2} \quad \text{and} \quad V^2 = \frac{B_1A_2 - C_2A_1}{B_1B_2 - C_1C_2} \tag{10.121d}$$

The first point, $U^2 = V^2 = 0$, is not stable for gain, or $A > 0$. The stability of the remaining singular points can be examined by drawing the curves of $dV^2/dU^2$ in the $U^2$, $V^2$ plane. Only the slope is needed, as the trajectory crosses the straight lines that are given by the numerator and denominator; that is,

$$A_1 - B_1U^2 - C_1V^2 = 0 \quad \text{and} \quad A_2 - B_2V^2 - C_2U^2 = 0.$$

Some trajectories [*12*] are shown in Fig. 10.9. If the two lines intersect, then the singular point at their intersection is stable for

$$\frac{B_1}{C_2} > \frac{A_1}{A_2} > \frac{C_1}{B_2}$$

and the two amplitudes are given by Eq. (10.121d). If the inequality is reversed, the singular point is not stable and either $U^2 = 0$, or $V^2 = 0$, with amplitudes given by Eq. (10.121b) or (10.121c). The choice between $U^2$ or $V^2$ depends on the trajectory near the origin. If the lines do not intersect, the choice is fixed on one of the two amplitudes.

For $b^2 < a^2$, entrainment does not occur and the phase $\varphi$ is linear in $t$,

$$\dot{\varphi} = 2(\omega_1 - \omega_2) = 2(\nu - u) \tag{10.122}$$

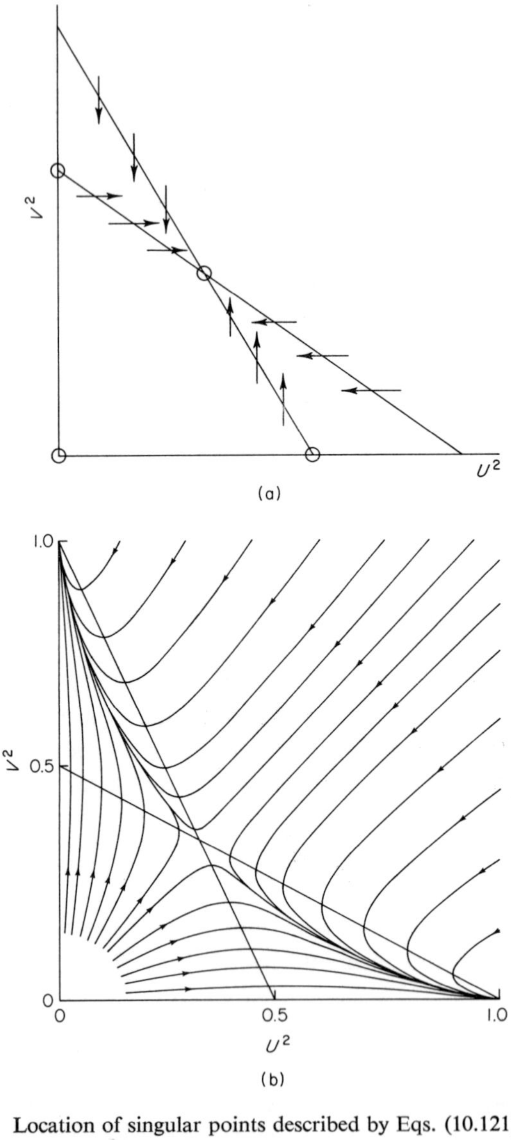


**Fig. 10.9** Location of singular points described by Eqs. (10.121). Straight lines are for the equations $A_1 - B_1U^2 - C_1V^2 = 0$ and $A_2 - B_2V^2 - C_2U^2 = 0$. (a) The slope $dV^2/dU^2$ of the trajectory as it crosses these lines is shown and the direction of the slope indicates that the stable singular point occurs at the intersection and the amplitudes are given by Eq. (10.121d). (b) Trajectories from various initial configurations with stable singular points at $(U^2 = 0,\ V^2 = 1)$ and at $(U^2 = 1,\ V^2 = 0)$ [see Eqs. (10.121b) and (10.121c)]. [Part (b) from W. E. Lamb, Jr., *Phys. Rev. A* **134**, 1429 (1964), Fig. 5.]

Since terms like $a'q$ and $b'q^3$ have been omitted, each oscillator oscillates at its characteristic frequency $\omega_1$ or $\omega_2$. The amplitude equations contain a time-dependent term and a computer analysis is necessary for detailed amplitude curves. If the beat frequency $\omega_1 - \omega_2$ is observed, the beat is periodic, but is not sinusoidal. As $\omega_1 - \omega_2$ is increased, the oscillatory term $\cos[2(\omega_1 - \omega_2)t]$ can be ignored in the response. Equations (10.118b) and (10.118c) can be analyzed in the same manner and the condition for stability is again given by Eq. (10.120) with the appropriate coefficients.

All of these phenomena can be observed in laser systems [*12*, *14*, *14a–c*]. If the laser is oscillating in two well-separated modes $\omega_1$ and $\omega_2$ and the coupling term $c$ is sufficiently large, the laser will select one or the other mode. As $c$ decreases, the oscillator oscillates in both modes. A solution of a more complex problem indicates that $\omega_1$ and $\omega_2$ have opposite states of polarization. A laser oscillating at a single frequency or two closely spaced frequencies $\omega_1$ and $\omega_2$ "locks" on a single frequency $\omega$. For a $J_a = 2$–$J_b = 2$ transition, the sum of the Clebsch–Gordon coefficients is such that $c > b$ and the system operates in a state of right or left circular polarization. For a $J_a = 1$–$J_b = 2$, transition, $c < b$, and the laser operates in a state of linear polarization. The photon rate gyroscope is a system in which a resonant cavity is formed by placing mirrors at the vertices of a triangle. Light rays travel around the closed triangular path in clockwise and counterclockwise directions. A laser provides gain and the system forms an oscillator. As the entire system rotates about an axis perpendicular to the triangle, the cw and ccw modes have different resonant frequencies and the frequency difference $\omega_{\mathrm{cw}} - \omega_{\mathrm{ccw}}$ is proportional to the angular rotation rate $\Omega$, or $\omega_{\mathrm{cw}} - \omega_{\mathrm{ccw}} = g\Omega$, where $g$ is a constant depending on the cavity geometry. At low rotation rates, no beats are observed and the cw and ccw oscillators are entrained. As the rate is increased, a periodic beat is observed, and finally, at larger rates, the beats approach a sinusoidal signal. This sequence of events is readily observed on an oscilloscope trace.

An estimate of whether or not entrainment or locking of two similar oscillators occurs can be made with Eq. (10.119b), and suggests for strong coupling $b = c$ that entrainment occurs when

$$|\omega_1 - \omega_2| < a$$

This is almost the Rayleigh criterion for resolution, since $a$ is the line width of the atom, molecule, or medium supplying the gain. In the example of an atom, $a$ is the natural width or the width $\frac{1}{2}(\Gamma_a + \Gamma_b)$ of the discussion in Chapter II.

If two very stable atomic clocks are placed in the vicinity of each other, the isolation between the clocks must be increased as the line width decreases. If the isolation is not adequate, the two clocks will entrain each other and oscillate at the same frequency.

If one of the oscillators has a very large amplitude, or if by suitable arrangement the coupling constant $c_{21} = 0$, the problem reduces to the entrainment of an oscillator by an external signal. For the model given by Eq. (10.109), a term $A \cos \omega t$ is added as a source. The entrainment of the human heart by a pacemaker is an example of entrainment in a living system. It should be noted that entrainment is not possible in a system with a linear response, and a nonlinear element is necessary.

In his outline of present problems in statistical mechanics, Uhlenbeck [*16*] points out that an understanding of nonlinear problems of this type can be of assistance in a description of the approach of a macroscopic system to equilibrium. These systems approach distinct singular points whatever the initial conditions.

## 10.13 Characteristic Functions for Random Pulses

In a manner similar to that used in the discussion on the characteristic functions of discrete quantities used in the development in Chapter III, the characteristic function of $x(t)$ is given by

$$\varphi(a) = \langle e^{iax(t)} \rangle = \int_{-\infty}^{\infty} e^{iax} P(x)\, dx \tag{10.123}$$

where $P(x)\, dx$ is the probability of a particular value of $x$ at a time $t$. Before discussing its general features, the example of random pulses discussed in Section 10.3 is examined by this method.

The characteristic function of a single pulse [*4*, *4a*] in Section 10.3 is given by

$$\begin{aligned} \varphi(a) &= \left(1 - \frac{\Delta}{T}\right) + \int_{t_k}^{t_k+\Delta} \frac{dt}{T}\, e^{iab(t-t_k)} \\ &= 1 + \frac{1}{T} \int_{-\infty}^{\infty} d\xi\, (e^{iab} - 1) \end{aligned} \tag{10.124}$$

This last form follows since $b(\xi) = 0$ when $\xi = t - t_k$ lies outside the range $\xi < 0$ or $\xi > \Delta$. This expression is very similar to that for the discrete example in Chapter III. The factor $1 - (1/g)$ is replaced by $1 - (\Delta/T)$, or the probability that the pulse is not received in interval $\Delta$. The integral term is analogous to $g^{-1}e^{ia}$.

The characteristic function for $K$ mutually independent random pulses is

$$\Phi_K = \prod_{k=1}^{K} \varphi_k(a) \tag{10.125a}$$

where

$$f_k(t) = \sum_{k}^{K} b(t - t_k) \tag{10.125b}$$

Expanding $\ln \Phi_K$ yields

$$\ln \Phi_K = K \ln \varphi(a) = KT^{-1} \int d\xi \, (e^{iab} - 1) \tag{10.126}$$

In the limit that $K/T \to \lambda$, or the average number of pulses per second, the characteristic function of $f(t)$ is

$$\Phi(a) = \langle \exp iaf(t) \rangle = \exp\left\{\lambda \int_{-\infty}^{\infty} d\xi \, [(\exp iab) - 1]\right\} \tag{10.127}$$

This is similar in form to the discrete Poisson characteristic function. The probability that $f$ has a particular value is given by the Fourier transform,

$$P(f) = \int_{-\infty}^{\infty} \frac{da}{2\pi} \, e^{-iaf}\Phi(a) \tag{10.128}$$

The expectation value of $f$ is given by

$$\mathscr{E}f = -i\left[\frac{d(\ln \Phi)}{da}\right]_{a=0} = \lambda \int_{-\infty}^{\infty} d\xi \, b(\xi) = \lambda \bar{b} \tag{10.129a}$$

and the dispersion is given by

$$\mathscr{D}f = -\left[\frac{d^2(\ln \Phi)}{da^2}\right]_{a=0} = \lambda \int_{-\infty}^{\infty} d\xi \, b^2(\xi) = \lambda \langle b^2 \rangle \tag{10.129b}$$

and is the spectral density. If $\lambda$ is large, the characteristic function is limited to small values of $a$ and as in Chapter III tends to the Gaussian form

$$\begin{aligned}\Phi(a) &= \exp\left[\lambda \int_{-\infty}^{\infty} d\xi \, (iab - \tfrac{1}{2}a^2b^2 - \cdots)\right] \\ &\approx \exp[-\tfrac{1}{2}a^2\lambda\langle b^2 \rangle + ia\langle b \rangle]\end{aligned} \tag{10.130}$$

Although it is not necessary, $\langle b \rangle \approx 0$ is assumed in this section. The probability that $f$ has a particular value can readily be evaluated for this characteristic function, and

$$P(f) = \frac{1}{(2\pi\lambda\langle b^2\rangle)^{1/2}} \exp\left(-\frac{f^2}{2\lambda\langle b^2\rangle}\right) \tag{10.131}$$

is the expected Gaussian distribution with dispersion $\lambda\langle b^2\rangle$. This illustrates some of the techniques used with continuous variables.

A joint distribution characteristic function can be defined in a similar manner [*4*, *4a*]:

$$\Phi(a_1, a_2) = \langle\exp[ia_1 f(t) + ia_2 f(t+\tau)]\rangle \tag{10.132}$$

and for the random pulses,

$$\Phi(a_1, a_2) = \exp\left(\lambda \int_{-\infty}^{\infty} d\xi \,\{\exp[ia_1 b(\xi) + ia_2 b(\xi+\tau)] - 1\}\right) \tag{10.133}$$

If this is expanded as a MacLaurin series expansion in $a_1$ and $a_2$, the second moments are generated as the coefficients of $a_1{}^2$, $a_1 a_2$, and $a_2{}^2$. The coefficient of $a_1{}^2$ or $a_2{}^2$ is

$$\psi(0) = \langle f^2(t)\rangle = \lambda \int_{-\infty}^{\infty} d\xi \, b^2(\xi) = \lambda\langle b^2\rangle = \lambda \int_{-\infty}^{\infty} d\nu \, |\, b(\nu)\,|^2 \tag{10.134}$$

Parseval's theorem can be used to write

$$\int_{-\infty}^{\infty} d\xi \, b^2(\xi) = \int_{-\infty}^{\infty} d\nu \, |\, b(\nu)\,|^2$$

and an average is not needed since the form of the pulse is assumed known. The coefficient of $a_1 a_2$ is the correlation function

$$\psi(\tau) = \langle f(t)f(t+\tau)\rangle = \lambda \int_{-\infty}^{\infty} d\xi \, b(\xi)b(\xi+\tau) = \lambda\langle b(\xi)b(\xi+\tau)\rangle$$

$$= 2\lambda \int_{-\infty}^{\infty} d\nu \, |\, b(\nu)\,|^2 \cos 2\pi\nu\tau \tag{10.135}$$

This procedure can be extended to a series of random pulses with different pulse functions $b_\alpha(\xi)$. Since the pulses are mutually independent random variables, the characteristic function is a product of characteristic functions with an extra index $\alpha$,

$$\Phi(a_1, a_2) = \langle\exp[ia_1 f(t) + ia_2 f(t+\tau)]\rangle$$

$$= \exp\left(\sum_\alpha \lambda_\alpha \int_{-\infty}^{\infty} d\xi \,\{\exp[ia_1 b_\alpha(\xi) + ia_2 b_\alpha(\xi+\tau)] - 1\}\right) \tag{10.136}$$

As the number of pulses becomes large, the higher-order terms in the cumulant expansion can be neglected and the characteristic function and joint characteristic function have the Gaussian form

$$\Phi(a) \approx \exp[-\tfrac{1}{2}a^2\psi(0)] \tag{10.137a}$$

$$\Phi(a_1, a_2) \approx \exp[-\tfrac{1}{2}(a_1^2 + a_2^2)\psi(0) - a_1a_2\psi(\tau)] \tag{10.137b}$$

where again

$$\psi(\tau) = \langle f(t)f(t+\tau)\rangle = \sum_\alpha \lambda_\alpha \overline{b_\alpha(\xi)b_\alpha(\xi+\tau)} \tag{10.138}$$

A Gaussian distribution has modes which are independent and this is apparent by replacing $a_1$ and $a_2$ with $(u \pm v)/\sqrt{2}$ and

$$F_\pm = \frac{f(t) \pm f(t+\tau)}{\sqrt{2}}$$

It then follows that the Gaussian distribution can be written as

$$\Phi_G(a_1, a_2) = \Phi(u)\Phi(v)$$

and the $F_\pm$ are statistically independent variables in this approximation. This is expected since $\langle F_+F_-\rangle = 0$. With this idea, the characteristic function is

$$\Phi_G(v) = \left\langle \exp \frac{iv[f(t) - f(t+\tau)]}{\sqrt{2}} \right\rangle \approx \exp\left\{-\frac{v^2[\psi(0) - \psi(\tau)]}{2}\right\}$$

$$= \exp\left\{\frac{v^2[\langle f(0)f(\tau)\rangle - \langle f(0)\rangle^2]}{2}\right\} \tag{10.139}$$

and this Gaussian distribution describes many physical processes.

### *10.13.1 Line Shapes*

In the study of line shapes, the average

$$g(\tau) = \langle \exp i[\eta(t) - \eta(t+\tau)]\rangle \tag{10.140}$$

is needed. It is apparent that this quantity is given by choosing $a_1 = 1$ and $a_2 = -1$ in the joint characteristic function. Special examples for various $b_\alpha(\xi)$ are considered in the next chapter. For a line shape for which the Gaussian approximation is valid, from Eq. (10.139),

$$g(\tau) = \langle \exp i[\eta(t) - \eta(t+\tau)]\rangle \approx \exp - [\psi(0) - \psi(\tau)] \quad (10.141)$$

where $\psi(\tau) = \langle \eta(t)\eta(t+\tau)\rangle$ is the correlation function for $\eta$. An alternate form is quite useful for a function $\eta$ for which

$$\eta(t+\tau) - \eta(t) = \int_t^{t+\tau} \dot{\eta}\, dt' \quad (10.142a)$$

This can be substituted into Eq. (10.139) for $\Phi_G(v)$ and an expansion of both sides of the equation in powers of $v$ yields

$$[\psi(0) - \psi(\tau)] = \tfrac{1}{2}\int_t^{t+\tau} dt' \int_t^{t+\tau} dt'' \langle \dot{\eta}(t')\dot{\eta}(t'')\rangle$$

$$= \int_0^{\tau} dt\, (\tau - t)\langle \dot{\eta}(0)\dot{\eta}(t)\rangle \quad (10.142b)$$

Since the process is assumed stationary, the time origin $t$ is arbitrary and is selected as $t = 0$. The double integral reduces to a single integral since $\langle \dot{\eta}(t')\dot{\eta}(t'')\rangle$ is a function of the time difference $t' - t''$ for a stationary random process. Figure 10.4 and Eq. (10.37) show a method of evaluation. If the process is of the form

$$\langle \dot{\eta}(t')\dot{\eta}(t'')\rangle = \alpha\, \delta(t' - t'') \quad (10.143a)$$

then the line shape follows from the Fourier transform of

$$g(\tau) = e^{-\alpha|\tau|} \quad (10.143b)$$

This method was used in the previous section for the oscillator output.

## 10.14 Random Processes

Consider the output of a recorder whose pen provides a tracing in time of the coordinate $x(t)$. Suppose some physical consideration limits the values of $x(t)$ to the range between the curves $\xi(t)$ and $\eta(t)$ shown in Fig. 10.10. Possible paths for $x(t)$ are shown in Figs. 10.10(a) and 10.10(b) and the character of these time-dependent processes are entirely different. The question immediately arises as to the type of probability distribution needed to distinguish between the different characters in the time evolution [5].

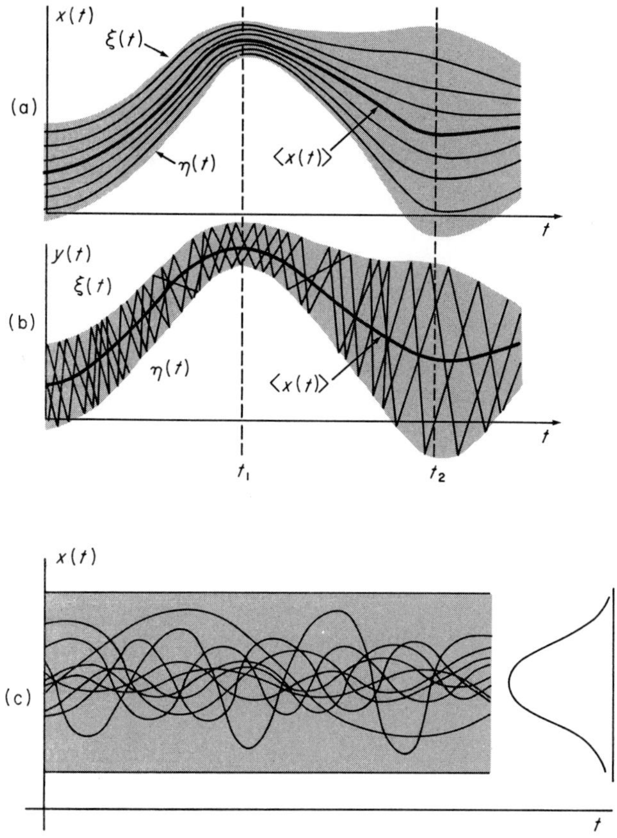


**Fig. 10.10** Possible trajectories $x(t)$ as a function of $t$ which lie between curves $\xi(t)$ and $\eta(t)$ with average value $\langle x(t)\rangle$. (b) Possible trajectories for $x(t)$ which lie between curves $\xi(t)$ and $\eta(t)$ and which have the same average value $\langle x(t)\rangle$ as the curves in (a). Note that more information than $\langle x(t)\rangle$ is needed to distinguish between the two sets of possible trajectories. (c) Possible trajectories for a stationary random variable $x(t)$. The profile $p(x; t)$ is shown at the right. $\langle x(t)\rangle$ implies a function which is the average of these trajectories. $\langle x(t)x(t + \tau)\rangle$ implies an average correlation for all of these trajectories. (Parts (a) and (b) from P. Beckman, "Elements of Applied Probability Theory," Fig. 4.1-3; Harcourt, New York, 1968.)

At a particular instant $t$, let $p(x; t)$ denote the probability density of the value of $x$ at time $t$. Then the distribution at time $t$ is given by the characteristic function

$$\varphi(a; t) = \langle e^{iax(t)}\rangle = \int_{-\infty}^{\infty} e^{iax}p(x; t)\, dx \tag{10.144}$$

The average value and dispersion are similar to Eqs. (10.129a) and (10.129b), but differ from these equations in that

$$\langle x(t)\rangle \qquad \text{and} \qquad \langle x^2(t) - \langle x(t)\rangle^2\rangle$$

are functions of time $t$. The two sets of curves in Fig. 10.10 can have the same moments and the same characteristic function. In order to distinguish between the two processes, it is necessary to examine the two-dimensional density

$$p(x_1, x_2; t_1, t_2)\, dx_1\, dx_2 \tag{10.145a}$$

or the probability that $x$ lies in the range $dx_1$ at time $t_1$ and $x$ lies in the range $dx_2$ at time $t_2$. The distribution is given by the two-dimensional characteristic function

$$\begin{aligned}\varphi(a_1, a_2; t_1, t_2) &= \langle \exp ia_1x_1 + ia_2x_2\rangle \\ &= \int_{-\infty}^{\infty} \exp(ia_1x_1 + ia_2x_2)\, p(x_1, x_2; t_1, t_2)\, dx_1\, dx_2 \end{aligned}\tag{10.145b}$$

and the moments of this function are discussed in Eqs. (10.132)–(10.135). The average value and dispersion are the same as those for the one-dimensional density $p(x; t)$, and the new quantity is the correlation $\langle x_1x_2\rangle$, which is given by Eq. (10.135), or

$$\langle x_1x_2\rangle = \langle x(t_1)x(t_2)\rangle = \int x_1x_2p(x_1, x_2; t_1, t_2)\, dx_1\, dx_2 \tag{10.145c}$$

and this function correlates the probable events at $t_1$ and $t_2$.

The $n$-dimensional distributions can be defined as the probable values of $x$ at times $t_1, \ldots, t_n$, or $p(x_1, x_2, \ldots, x_n;\ t_1, t_2, \ldots, t_n)$. This has a characteristic function

$$\varphi(a_1, \ldots, a_n) = \langle \exp i \textstyle\sum a_mx_m\rangle \tag{10.146a}$$

New cross-correlations will occur and describe the distinction between the two processes in greater detail. Finally, the time intervals can be selected infinitesimally close, and $a_1, \ldots, a_n$ becomes a continuous variable $a(t)$ over the interval $t_1$ to $t_n$ and the sum becomes an integral such that the difference between the two processes is completely described by

$$\varphi[a(t)] = \left\langle \exp i \int^t a(t')x(t')\, dt'\right\rangle \tag{10.146b}$$

Fortunately, most physical processes do not require this detailed description. The two-dimensional correlation function is adequate for many physical problems encountered. Even so, even this function is of greater generality than that used in the earlier part of this chapter.

Stationary random processes are defined as "strictly stationary" or stationary "in the narrow sense" if the probability density for the $n$-dimensional density is invariant to time translation $t_i \rightarrow t_i + \tau$. Thus the one-dimensional density does not depend on time,

$$p(x; t) = p(x; t + \tau) = p(x) \tag{10.147}$$

and the two-dimensional density depends only on the time difference $t_2 - t_1 = \tau$,

$$p(x_1, x_2; t_1, t_2) = p(x_1, x_2; t_1, t_1 + \tau) = p(x_1, x_2; \tau) \tag{10.148}$$

Hence the correlation function is dependent only on $\tau$, or

$$\langle x_1 x_2 \rangle = \langle x(t_1) x(t_2) \rangle = \langle x(t_1) x(t_1 + \tau) \rangle = \psi(\tau) \tag{10.149}$$

Usually, Eqs. (10.147) and (10.149) are adequate for physical processes which are stationary in the "wider sense." Most of the processes discussed earlier in this chapter are stationary in the "wider sense."

Average values over an ensemble of all possible realizations at a particular time $t$ have been used in the earlier discussion. The time average of $x(t)$,

$$\overline{x(t)} = \lim_{T \to \infty} \frac{1}{T} \int_0^T x(t)\, dt \tag{10.150}$$

is also of interest. A random process is denoted as ergodic if the probability that the ensemble average equals the time average is unity, that is,

$$P[\langle x(t) \rangle - \overline{x(t)}] = 1$$

A necessary and sufficient condition for a process to be ergodic is

$$0 = \lim_{T \to \infty} \frac{1}{T^2} \int_0^T dt_1 \int_0^T dt_2\, [\langle x(t_1) x(t_2) \rangle - \langle x(t_1) \rangle^2] \tag{10.151}$$

A sufficient condition for a stationary process to be ergodic is the existence of the integral

$$\int_0^\infty [\langle x(t) x(t + \tau) \rangle - \langle x \rangle^2]\, d\tau \tag{10.152}$$

A very lucid discussion of the aspects discussed in this section is given by Beckman [*5*].

Equation (10.146b) for the complete description of the process is of such generality as to raise questions regarding its use in physics. Feynman introduces expressions of this type in his discussion of path integrals in quantum mechanics [*17*] and gives examples of the evaluation of such integrals.

The space averages implied in Eq. (10.103a) are similar to the time averages discussed here. The ensemble average is a superposition of all possible arrangements for particles in volume $V$. The pair correlation function is independent of position in the volume in the same sense that the time correlation is independent of the time origin in interval $T$. Thus the space correlation is a function of $\mathbf{r} = \mathbf{r}' - \mathbf{r}''$, while the time correlation is a function of $\tau = t' - t''$ only.

### *10.14.1 Gaussian Random Variable x(t)*

If, as indicated earlier, the characteristic function of the random variables $x_m$ at times $t_m$ is written as

$$\begin{aligned}\Phi(a_1, \ldots, a_r) &= \left\langle \exp i \sum_{m=1}^{r} a_m x_m \right\rangle \\ &= \exp\left( -\tfrac{1}{2} \sum_{m}^{r} \sum_{n}^{r} a_m a_n \langle x_m x_n \rangle \right) \end{aligned} \tag{10.153}$$

and if the bilinear sum follows, the process is referred to as a Gaussian random process. The cumulant, or $\ln \Phi(a_1, \ldots, a_r)$, expansion terminates after the second-order term for a Gaussian random process. This implies that the moment terms are related by

$$\begin{aligned} \langle x_n \rangle &= 0 \\ \langle x_m x_n \rangle &\neq 0 \\ \langle x_m x_n x_p \rangle &= 0 \\ \langle x_m x_n x_p x_q \rangle &= 1 \cdot 3 \langle x_m x_n \rangle^2 \\ &\vdots \end{aligned} \tag{10.154}$$

for a Gaussian random process. Equation (10.153) can be generalized in the same fashion by transforming to a continuous index $t$ and the characteristic function of a Gaussian process is, in general,

$$\varphi[a(t)] = \left\langle \exp i \int_0^t a(t')x(t')\,dt' \right\rangle$$

$$= \exp\left[-\tfrac{1}{2}\int_0^t dt' \int_0^t dt''\, a(t')a(t'')\langle x(t')x(t'')\rangle\right] \quad (10.155)$$

For a stationary Gaussian process, this reduces to

$$\varphi[a(t)] = \exp\left[-\tfrac{1}{2}\int_0^t dt' \int_0^t dt''\, a(t')a(t'')\psi(t'-t'')\right] \quad (10.156)$$

where

$$\psi(t'-t'') = \langle x(t')x(t'')\rangle$$

is the usual correlation function. Kubo [*18*] uses this form in his study of line shapes. Changing $a(t)$ to $a(t) = 1$ yields a line shape correlation function $g(\tau)$ very similar to that given by Eq. (10.143b).

A method used by Rice [*4*] in his study of thermal noise and much earlier by Rayleigh and Gouy [*19*] in their discussion of blackbody radiation and white light is of considerable interest, and lends clarity to the various procedures used previously. Consider one of the many traces of $x(t)$ which are shown in Fig. 10.10. Since the interval of interest is of length $T$, assume that the phenomenon is periodic and expand it as a Fourier series expansion,

$$x(t) = \tfrac{1}{2}A_0 + \sum_{n=1}^{N} [A_n(\cos 2\pi\nu_n t) + B_n \sin 2\pi\nu_n t] \quad (10.157a)$$

where

$$\nu_n = \frac{n}{T} \quad (10.157b)$$

and

$$A_n = \frac{2}{T}\int_0^T dt\, x(t)\cos 2\pi\nu_n t \quad (10.157c)$$

The phenomenon is described by a characteristic set of frequencies or modes $\nu_n$. These modes are independent and there is no coupling between them. They form a new set of independent properties for the description. One can chose $N$ and $T$ arbitrarily large. Other traces can be expanded in terms of the same set of modes and it then makes sense to discuss the probability of the excitation of a particular mode or $A_n$. One can even consider the characteristic function of this particular mode, that is,

$$\varphi_n(a) = \langle e^{iaA_n} \rangle \tag{10.158}$$

Since the modes are independent, the characteristic function is merely the product of characteristic functions $\varphi_n$. The term $\langle A_n^2 \rangle$ is related to the correlation function by

$$\langle A_n^2 \rangle = \frac{4}{T^2} \int_0^T dt' \int_0^T dt'' \, \langle x(t')x(t'') \rangle \cos \omega_n t' \cos \omega_n t'' \tag{10.159}$$

A correlation function which describes a Gaussian random process has the properties

$$\begin{aligned} \langle x(t) \rangle &= 0 \\ \langle x(t')x(t'') \rangle &= 2D\, \delta(t' - t'') \\ \langle x(t_1)x(t_2) \cdots x(t_n) \rangle &= \sum_{\text{pair}} \langle x(t_1)x(t_i) \rangle \cdots \langle x(t_m)x(t_n) \rangle \\ &= 0 \qquad \text{for } n \text{ odd} \end{aligned} \tag{10.160}$$

It is just this property which cancels all higher terms in the cumulant expansion, or which in the moment expansion, permits all terms to be given as $\langle x(t')x(t'') \rangle$ to some power. For a particular mode, it then follows that

$$\langle \exp iaA_n \rangle = \exp[-\tfrac{1}{2}a^2 \langle A_n^2 \rangle]$$

where

$$\langle A_n^2 \rangle = \frac{4D}{T} \tag{10.161}$$

The probability that $A_n^2$ has a particular value of $A_n$ follows from the Fourier transform, and the form of the characteristic function indicates that the distribution is Gaussian.

A simple dynamic system with a single coordinate $q(t)$ responds to a single trace $x(t)$ shown in Fig. 10.10 if $x(t)$ denotes the driving force. Probability is involved since there is not sufficient information available for an external observer to decided ahead of time which trace will occur. If $x(t)$ is expanded in terms of modes with period $T$, these modes are independent in the Fourier sense and a single mode can be excited. Probability is involved since there is not sufficient information available to decide ahead of time which mode will be excited. A simple dynamic system can be designed to respond to a single mode and the average squared amplitude of the response will be proportional to the value given by Eq. (10.161) for a

Gaussian $x(t)$. There is an inclination to sum over the modes to find the response, and the answer is infinite. The same infinity would occur if one insisted that the simple system respond to all traces at the same time.

Exercise 10.15 (a) Show that $\langle A_n{}^2\rangle = \langle B_n{}^2\rangle$ for Eq. (10.160).

(b) Use the Fourier transform of the characteristic function to find the probability of the excitation of the amplitude $A_n$.

$$\text{Ans.}\quad P(A_n) = \frac{1}{(2\pi\langle A_n{}^2\rangle)^{1/2}} \exp\left[-\frac{A_n{}^2}{2\langle A_n{}^2\rangle}\right] \tag{10.162a}$$

(c) Find the joint probability of exciting amplitudes $A_1$, $A_5$, and $B_2$.

$$\text{Ans.}\quad P(A_1, A_5, B_2) = P(A_1)P(A_5)P(B_2)$$

(d) Show that

$$\langle A_n{}^2\rangle = \frac{2}{T}\,\psi(\nu_n) \tag{10.162b}$$

corresponds to the spectral density $2\psi(\nu)\,\Delta\nu$, where $\Delta\nu = 1/T$ and is in agreement with the earlier considerations as the sum is changed to an integral.

(e) Show that

$$\langle \exp iax(t)\rangle = \exp\left[-\tfrac{1}{2}a^2 \sum_n \langle A_n{}^2\rangle\right] = \exp[-\tfrac{1}{2}a^2\langle x(t^2)\rangle] \tag{10.162c}$$

Exercise 10.16 (a) Show that $D = \gamma/\beta$ or $R/\beta$ in the earlier considerations in this chapter.

(b) If the period of measurement is $T^*$, show that the average squared amplitude of a mode is

$$\langle A_n{}^2\rangle = \frac{4\gamma}{\beta T^*} \quad\text{or}\quad \frac{4R}{\beta T^*} \tag{10.163a}$$

Let $\Delta\nu = 1/T^*$ and show that the noise power is

$$\langle A_n{}^2\rangle = 4\gamma kT\,\Delta\nu \quad\text{or}\quad 4RkT\,\Delta\nu \tag{10.163b}$$

where temperature $T$ is introduced again. This is the noise power measured by a simple detector with a bandwidth $\Delta\nu$.

Exercise 10.17 (a) Discuss the temperature fluctuation of a small object which can come into thermal equilibrium with its surroundings.

Ans. Let $C$ be the heat capacity of a small thermal detector and $\gamma\, \Delta T$ the heat transfer to the surroundings. Let $F$ be the incident absorbed radiation in watts. Then

$$C\frac{d(\Delta T)}{dt} + \gamma\, \Delta T = F \tag{10.164}$$

and the response function is

$$G(\nu) = \frac{1}{i\omega C + \gamma}$$

The steady-state signal response is $\Delta T = F/\gamma$. Thermal noise also causes fluctuations in $\Delta T$ and by the development in Section 10.10, the fluctuation in energy is given by Eq. (10.86) and the fluctuation in temperature by Eq. (10.90b), or

$$\langle(\Delta T)^2\rangle = \frac{kT^2}{C}$$

and the spectral density of the noise source must be normalized to yield this fluctuation. Anticipating the answer,

$$\psi_n(\nu) = 2\gamma k T^2$$

and the spectral density $\theta(\nu)$ of the coordinate $\Delta T$ is given by

$$\theta(\nu) = \frac{2\gamma k T^2}{\gamma^2 + \omega^2 C^2}$$

The correlation function for the temperature fluctuation is

$$\langle \Delta T(t)\, \Delta T(t+\tau)\rangle = \frac{kT^2}{C} e^{-(\gamma/C)\tau} \tag{10.165}$$

and at $\tau = 0$ is in agreement with the fluctuation required by the canonical ensemble.

(b) A bolometer measures the incident radiation by measuring the change in resistance $R$ caused by the change in temperature $\Delta T$. Assume that this change in $\Delta R$ is measured by a narrowband detector with a bandwidth $\Delta\nu$ and then show that the minimum detectable signal increases as $(\Delta\nu)^{1/2}$ or as $(T^*)^{-1/2}$, where $T^*$ is the response time of the detector.

Ans. The minimum detectable signal must cause a temperature rise greater than the fluctuating noise, or

$$(\Delta T)_s^2 \geq \theta_n(\nu)\, \Delta\nu \simeq \frac{2kT^2}{\gamma} \Delta\nu$$

Since the incident flux and the temperature change are related by $\Delta T = F/\gamma$, the minimum detectable flux is given by

$$F_s = [(2kT^2\gamma)\,\Delta\nu]^{1/2} \tag{10.166}$$

By using detectors with longer measuring intervals, it is possible to increase the signal-to-noise ratio.

## References


1. E. Kappler, *Ann. Phys.* **31**, 377 (1938).
2. A. Einstein, *Ann. Phys.* **17**, 549 (1905); see also "Investigations on the Theory of Brownian Movement" (R. Furth, ed.) (translated by A. D. Cowper). Dutton, New York, 1926.
3. M. J. Perrin, "Brownian Movement and Molecular Reality." Taylor and Francis, London, 1910.
4. S. O. Rice, Mathematical analysis of random noise. *Bell Syst. Tech. J.* **23**, 282 (1944); **24**, 46 (1945).

4a. N. Wax, ed., "Selected Papers on Noise and Stochastic Processes." Dover, New York, 1954; see Beckman [5].

5. P. Beckman, "Elements of Applied Probability Theory." Harcourt, New York, 1968.
6. H. Nyquist, *Phys. Rev.* **32**, 110 (1928); J. Lawson and G. E. Uhlenbeck, "Noise Signals." McGraw-Hill, New York, 1950.
7. L. Landau and E. M. Lifshitz, "Statistical Physics." Addison-Wesley, Reading, Massachusetts, 1958.
8. H. B. Callen and R. F. Greene, *Phys. Rev.* **86**, 702 (1952); **88**, 1387 (1952).
9. S. R. de Groot and P. Mazur "Non-Equilibrium Thermodynamics," North Holland Publ., Amsterdam, 1962; R. Kubo, *J. Phys, Soc. Jap.* **12**, 570 (1957).
10. L. van Hove, *Phys. Rev.* **95**, 249 (1954).

10a. J. de Boer, *Rep. Progr. Phys.* **12**, 305 (1949).

10b. C. J. Pings, Structure of simple liquids by X-ray diffraction. *In* "Physics of Simple Liquids" (H. N. Temperley, ed.), pp. 389–445. Wiley, New York, 1968.

10c. D. McIntyre and J. V. Sengers, Study of fluids by light scattering. In "Physics of Simple Liquids" (H. N. Temperley, ed.), pp. 449–505. Wiley, New York, 1968.

10d. I. L. Fabelinskii, "Molecular Scattering of Light." Plenum, New York, 1968.

10e. H. C. van de Hulst, "Light Scattering by Small Particles." Wiley, New York, 1957.

10f. K. A. Stacey, "Light Scattering in Physical Chemistry." Academic Press, New York, 1956; M. Kerher, "The Scattering of Light." Academic Press, New York, 1969.

11. K. Shimoda, T. C. Wang, and C. H. Townes, *Phys. Rev.* **102**, 1308 (1956).
12. W. E. Lamb, Jr., *Phys. Rev. A* **134**, 1429 (1964).
13. M. O. Scully and W. E. Lamb, Jr., *Phys. Rev.* **159**, 208 (1967). (This reference includes references to related papers.)
14. C. V. Heer and R. D. Graft, *Phys. Rev. A* **140**, 1088 (1965).

14a. C. V. Heer, *Phys. Rev. Lett.* **17**, 774 (1966).

14b. M. Sargent, III, W. E. Lamb, Jr., and R. L. Fork, *Phys. Rev.* **164**, 436, 450 (1967).
14c. R. A. Settles and C. V. Heer, *Appl. Phys. Lett.* **12**, 350 (1968).
15. C. Hayashi, "Non-linear Oscillations in Physical Systems." McGraw-Hill, New York, 1964.
16. G. E. Uhlenbeck, An outline of statistical mechanics. *In* "Fundamental Problems in Statistical Mechanics" (E. G. D. Cohen, ed.), p. 1–29. North-Holland Publ., Amsterdam, 1968.
17. R. P. Feynman and A. R. Hibbs, "Quantum Mechanics and Path Integrals," p. 182. McGraw-Hill, New York, 1965.
18. R. Kubo, *In* "Fluctuation Relaxation and Resonance in Magnetic Systems " (D. ter Haar, ed.), pp. 23–68. Oliver & Boyd, Edinburgh, 1962.
19. Lord Rayleigh and G. Gouy, *Phil. Mag.* [5], **27**, 460 (1889).

# Stochastic Processes in Quantum Systems

# XI

## 11.1 Introduction

This chapter considers time-dependent phenomena in quantum mechanical systems. As an introduction, the elementary theory of the time-proportional transition rate from an initial state to a final state, which is caused by a time-dependent perturbation, is considered. This problem is then examined in terms of the two-state density matrix. Many of the concepts introduced in Chapter X for stationary random processes are used in this elementary treatment. The density matrix with phenomenological damping is used to discuss the response of an atomic or molecular system to a sinusoidal signal. In the unsaturated region, this yields the electric or magnetic susceptibility and if population inversion occurs, the electric polarization for simple laser phenomena. This approach includes some aspects of saturation by strong signals. In order to consider strong signals in greater detail, a unitary matrix is given for the evolution of the wavefunction of a two-state system which interacts with a saturating pulse. The stimulated electric polarization is used to discuss photon echoes and related phenomena.

Although each of the previous techniques can be used to solve specific problems, they do not incorporate damping in the proper manner and the phenomenological damping permits the system to relax to nonnatural states. The method of Wigner and Weisskopf is used to introduce damping in a simple manner and to discuss the natural width of an emission process. More exact and recent theories, which use the resolvent, are not introduced since these methods become quite cumbersome in most real problems. Spontaneous emission by an atom and recoilless emission, or the Mössbauer effect, are then discussed as special examples. As an illustrative exercise, the emission of radiation by a nucleus bound in a simple harmonic oscillator potential well is examined in considerable detail. Time-dependent annihilation–creation operators are introduced in this simple example and this permits a very satisfying approach for understanding these operators. Since the normal modes of a crystal are independent, this illustrative example permits the development of the Mössbauer effect for crystals and the scattering of x rays and neutrons by crystals in a single step.

Damping and the "golden rule" for transition rates are developed in greater detail and then are used for the study of line broadening and scattering. Line broadening by collisions in gases is examined by using the transition rate theory and the correlation function of the previous chapter. In the latter sections of the chapter, the inelastic scattering of x rays, neutrons, and optical radiation by the fluctuations in density, dielectric constant, and so on, which occur in gases, liquids, and solids, is developed. The development of the laser and the observation of the scattered radiation has made possible the observation of many internal properties of these macroscopic systems. With intense laser radiation, even stimulated scattering occurs.

Whenever possible, experimental examples are included with the development given in this chapter.

## 11.2 Elementary Theory of Transition Rates

One of the most fundamental problems in the quantum mechanics of real physical systems concerns the probability of a perturbation stimulating the system to change from an initial state to a final state. In its most direct form, it is assumed that the isolated system can be defined as having a Hamiltonian $H_0$ with a known set of energies $E_a$ and wavefunctions $| a)$. Since the wavefunctions and energies are known only approximately for atoms or molecules, the discussion is essentially limited to the one-particle aspect, where the particle is an atom or a molecule. An example of such a system is now considered.

Let the Hamiltonian of the one-particle system under observation be $H_0$ and the time-dependent perturbation be $U(t)$; then

$$H = H_0 + U(t) \tag{11.1a}$$

where

$$H_0 \mid a) = E_a \mid a) \tag{11.1b}$$

The evolution in time is given by the time-dependent Schrödinger equation

$$i\hbar \frac{\partial \psi}{\partial t} = H\psi \tag{11.2}$$

If the wave function is expanded as

$$\psi(t) = \sum \mid a)c_a \exp -iE_a t/\hbar \tag{11.3a}$$

direct substitution into Eq. (11.2) yields

$$i\hbar \dot{c}_a = \sum_{a'} (a \mid U \mid a')c_{a'} \exp(i\omega_{aa'}t) \tag{11.3b}$$

where $\hbar\omega_{aa'} = E_a - E_{a'}$. Further progress requires an experimental arrangement. Assume that the experimental arrangement prepares atoms in state $\mid a)$ and is able to measure the number in state $\mid b)$ at time $t$. With this assumption,

$$c_a = 1 \quad \text{and} \quad c_b = 0, \qquad t = 0 \quad \text{and} \quad b \neq a$$

If the time interval is short, Eq. (11.3b) can be integrated to yield

$$c_b \approx (i\hbar)^{-1} \int_0^t (b \mid U \mid a)e^{i\omega_{ba}t}\, dt \tag{11.4}$$

and the probability of finding the atom or particle in state $\mid b)$ is

$$P_{ab} = \mid c_b \mid^2 = \hbar^{-2} \int_0^t dt' \int_0^t dt'' \, (b \mid U(t') \mid a)(b \mid U(t'') \mid a)^* \times \exp[i\omega_{ba}(t' - t'')]$$

The mathematical techniques that were used to develop Eq. (10.37) and which are shown in Fig. 10.4 can be used to write this equation in terms of $t'$ and $\tau = t' - t''$. In any given experiment, it is not usually possible to make a meaningful measurement on a single particle. An average is taken over a large number of particles and the average measurement depends on the ensemble average of

$$\langle (b \mid U(t') \mid a)(b \mid U(t' - \tau) \mid a)^* \rangle$$

It is now assumed that this ensemble average is independent of the time origin, or is independent of $t'$, and is stationary. This assumption is necessary for the probability of the transition to be linear in $t$ and the transition rate to be independent of time. It implies that $U(t)$ cannot be defined and that this ensemble average is a stochastic function of the type discussed in Sections 10.3 and 10.14. Physical processes that cause $U(t)$ to be a stochastic function are spontaneous emission, the thermal influence on the interaction between molecules, collisions, and so on. Thus the average transition rate is given by

$$\begin{aligned} W_{ab} &= \lim_{t \to \infty} t^{-1} \langle P_{ab} \rangle \\ &= \hbar^{-2} \int_0^\infty d\tau \, \langle (b \mid U(t) \mid a)(b \mid U(t - \tau) \mid a)^* \rangle e^{i\omega_{ba}\tau} \\ &\quad + \text{complex conjugate} \end{aligned} \tag{11.5}$$

For the simple type of perturbation to be given in Exercise 11.1, the transition rate is proportional to the spectral density. Thus the response of the two-state atomic system to a set of random pulses with a spectral density $\psi(\nu)$ follows immediately from Eq. (11.6b) of Exercise 11.1. The interaction of an atom with an electric or magnetic field is of the form $-\mathbf{P} \cdot \mathbf{E}(t)$ or $-\mathbf{M} \cdot \mathbf{B}(t)$ and the matrix element is of the form of Eq. (11.6a). The transition rate is proportional to the spectral density of the electric or magnetic field experienced by the atom or molecule at frequency $\nu_{ab}$ and can be related to the intensity by Eq. (10.28).

Exercise 11.1 For many problems, the perturbation is of the form

$$(b \mid U(t) \mid a) \approx (b \mid U_0 \mid a) f(t) \tag{11.6a}$$

and can be separated from the matrix element. Show that the transition rate becomes

$$\begin{aligned} W_{ab} &= 2\hbar^{-2} \mid (b \mid U_0 \mid a) \mid^2 \int_0^\infty d\tau \, \psi(\tau) \cos 2\pi\nu_{ba}\tau \\ &= \hbar^{-2} \mid (b \mid U_0 \mid a) \mid^2 \psi(\nu_{ba}) \end{aligned} \tag{11.6b}$$

in terms of the correlation function and the spectral density. Here $\psi(\tau)$ is defined in terms of the perturbating field as $\langle f(t) f(t + \tau) \rangle = \psi(\tau)$. The definition of the $\delta$ function in terms of cosines is useful for this transformation:

$$4 \int_0^\infty d\tau \, (\cos 2\pi\nu\tau) \cos 2\pi\nu_{ab}\tau = \delta(\nu - \nu_{ab}) \tag{11.7}$$

## 11.3 Density Matrix and the Response of Atoms and Molecules to Perturbations

The stochastic arguments of the previous section are frequently discussed in terms of the density matrix. Again, the one-particle Hamiltonian and its solutions are assumed known. Using Eqs. (11.1a, b) and (11.3a, b), a bilinear operator $\sigma$ can be defined as

$$\sigma = \big|\, \psi(t))(\psi(t) \,\big| = \sum_{a,a'} |\, a)(a' \,|\, C_{a'}^* C_a \tag{11.8}$$

where $C_a = c_a \exp(-iE_a t/\hbar)$. Direct differentiation with respect to time yields the equation of motion of this operator as

$$i\hbar \frac{d\sigma}{dt} = H\sigma - \sigma H \tag{11.9a}$$

This equation has the simple solution for $U = 0$ of

$$\sigma(t) = e^{-iH_0 t/\hbar} \sigma(0) e^{iH_0 t/\hbar} = \sigma_{\mathrm{I}} \tag{11.9b}$$

It is convenient to use the interaction representation and define the $U = 0$ solution as $\sigma_{\mathrm{I}}$. Then, the equation of motion for $\sigma_{\mathrm{I}}$ is

$$i\hbar \frac{d\sigma_{\mathrm{I}}}{dt} = U_{\mathrm{I}}\sigma_{\mathrm{I}} - \sigma_{\mathrm{I}}U_{\mathrm{I}} \tag{11.10a}$$

where

$$U_{\mathrm{I}} = e^{iH_0 t/\hbar} U(t) e^{-iH_0 t/\hbar} \tag{11.10b}$$

Direct integration yields the integral equation

$$\sigma_{\mathrm{I}}(t) = \sigma_{\mathrm{I}}(0) + (i\hbar)^{-1} \int_0^t dt' \, (U_{\mathrm{I}}\sigma_{\mathrm{I}} - \sigma_{\mathrm{I}}U_{\mathrm{I}}) \tag{11.10c}$$

This expression can be iterated to yield another integral equation,

$$\begin{aligned} \sigma_{\mathrm{I}}(t) = \sigma_{\mathrm{I}}(0) &+ (i\hbar)^{-1} \int_0^t dt' \, [U_{\mathrm{I}}(t')\sigma_{\mathrm{I}}(0) - \sigma_{\mathrm{I}}(0)U_{\mathrm{I}}(t')] \\ &+ (i\hbar)^{-2} \int_0^t dt' \int_0^{t'} dt'' \, \{U_{\mathrm{I}}(t')[U_{\mathrm{I}}(t'')\sigma_{\mathrm{I}}(t'') - \sigma_{\mathrm{I}}(t'')U_{\mathrm{I}}(t'')] \\ &+ \text{hermitian conjugate}\} \end{aligned} \tag{11.11}$$

Further analysis requires a physical model. One model is to regard this particle as a member of an ensemble of particles and take a suitable ensemble average. This average will be denoted by changing $\sigma$ to $\varrho$ and this is the usual density matrix describing the process. An average over the single integral gives a zero contribution if the perturbation $U_{\mathrm{I}}(t')$ and $\sigma_{\mathrm{I}}(0)$ are statistically independent. The double integral can be transformed in the same manner as that used for Eq. (11.5) or Eq. (10.37). If the change in $\sigma_{\mathrm{I}}(t)$ is slow, then $\sigma_{\mathrm{I}}(0)$ can be used in the double integral and regarded as uncorrelated with the perturbation. This leaves the average of the term

$$\begin{aligned} \langle U_{\mathrm{I}}(t')U_{\mathrm{I}}(t'-\tau)\rangle &\neq 0, \qquad \tau < \tau_{\mathrm{c}} \\ &\approx 0, \qquad \tau > \tau_{\mathrm{c}} \end{aligned}$$

of primary interest and it is assumed that for times greater than a correlation time $\tau_{\mathrm{c}}$, this average is zero. It is also assumed that for $t' \ll \tau_{\mathrm{c}}$, the interaction is sufficiently weak that $\hbar^{-2}U^2\tau_{\mathrm{c}}^2 \ll 1$ and this contribution can be neglected for short times. It is further assumed that the average is independent of the choice of $t'$ for $t < \tau_{\mathrm{c}}$. With these assumptions, the integral can be taken over $t'$ and the double integral grows linearly in $t$. Since the average value is zero for $\tau > \tau_{\mathrm{c}}$, the integral over $\tau$ can be taken between zero and infinity. Then, the change in the density matrix which is stimulated by a stationary random perturbation is $t^{-1}[\varrho_{\mathrm{I}}(t) - \varrho_{\mathrm{I}}(0)]$, which is written as

$$\begin{aligned} \left(\frac{d\varrho_{\mathrm{I}}}{dt}\right)_{\mathrm{r}} \approx -\hbar^{-2} \int_0^\infty d\tau \, \langle U_{\mathrm{I}}(t)[U_{\mathrm{I}}(t-\tau)\varrho_{\mathrm{I}}(t) - \varrho_{\mathrm{I}}(t)U_{\mathrm{I}}(t-\tau)] \\ + \text{hermitian conjugate}\rangle \end{aligned} \tag{11.12}$$

This defines the average slope for the change in the density matrix. $\varrho_{\mathrm{I}}(t)$ rather than $\varrho_{\mathrm{I}}(0)$ is used on the right-hand side since it is assumed that $\varrho_{\mathrm{I}}$ does not change much during the interval $\tau_{\mathrm{c}}$.

Exercise 11.2 In order to compare this expression with the earlier development, let all particles start in state $|\,a)$ at $t = 0$, or

$$(a\,|\,\varrho_{\mathrm{I}}(0)\,|\,a) = 1 \qquad \text{and} \qquad (b\,|\,\varrho_{\mathrm{I}}(0)\,|\,b) = 0$$

and then show that

$$\frac{d(b\,|\,\varrho\,|\,b)}{dt} \approx W_{ab} \tag{11.13}$$

for the increase in the probability of being in state $|\,b)$ and is the same as the time-proportional transition rate which is given by Eq. (11.5).

### 11.3.1 Two-State Density Matrix as an Example

The two-level problem shown in Fig. 11.1 has a simple form and illustrates the salient features of the more general problem. Damping by random perturbations is considered first. Let $(a \mid \varrho \mid a)$ and $(b \mid \varrho \mid b)$ denote the probabilities of the particle being in states $\mid a)$ and $\mid b)$, respectively. If these are the only allowed states, then the sum of these two terms is unity. The change in these two states is given by Eq. (11.12), and

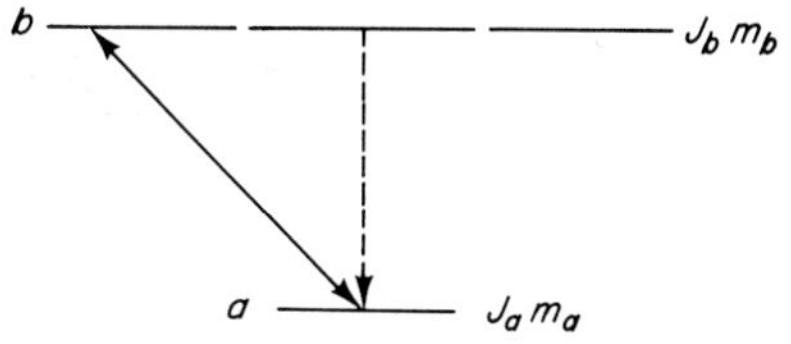


**Fig. 11.1** Typical designation of the angular momentum states of a nucleus or of an atom.

$$\frac{d(b \mid \varrho \mid b)}{dt} = -\frac{d(a \mid \varrho \mid a)}{dt} = \frac{(a \mid \varrho \mid a) - (b \mid \varrho \mid b)}{T_1} \tag{11.14a}$$

where

$$\frac{1}{T_1} = W_{ab} \tag{11.14b}$$

and $W_{ab}$ is given by Eqs. (11.5) and (11.13) and is the usual first-order transition rate.

The damping of the off-diagonal terms is divided into two parts. The nonadiabatic damping is due to stimulated transitions and is given by

$$\left[\frac{d(b \mid \varrho \mid a)}{dt}\right]_{\text{nonad}} \approx -(b \mid \varrho \mid a)\left[\frac{2}{\hbar^2}\int_0^\infty d\tau \, \langle (b \mid U(t) \mid a)(b \mid U(t-\tau) \mid a)^* \rangle e^{i\omega_{ba}\tau}\right] \tag{11.15a}$$

The contribution of the rapidly oscillating term at $2\omega_{ab}$ in Eq. (11.12) is omitted. The real part of the expression in the square brackets provides damping and comparison with Eq. (11.13) yields

$$\left[\frac{d(b \mid \varrho \mid a)}{dt}\right]_{\text{nonad}} = -\frac{(b \mid \varrho \mid a)}{T_1} \tag{11.15b}$$

The adiabatic term is due to perturbations which change the energy level spacings, or change $\omega_{ab}$. These perturbations do not cause transitions, but they do change the stochastic character of the off-diagonal elements. This is a new assumption, since in the initial problem, the spacing between energy levels $\omega_{ba}$ was assumed constant. The problem is now broadened to include the effects of interactions that cause $\omega_{ba}$ to become a function of time and the model is now phenomenological. Keeping only the diagonal elements in Eq. (11.12) yields

$$\left[\frac{d(b \mid \varrho \mid a)}{dt}\right]_{\mathrm{ad}} = -\frac{(b \mid \varrho \mid a)}{T_2'}$$

where

$$\frac{1}{T_2'} = \frac{1}{\hbar^2}\int_0^\infty d\tau \, \langle[(b \mid U(t) \mid b) - (a \mid U(t) \mid a)] \times [(b \mid U(t-\tau) \mid b) - (a \mid U(t-\tau) \mid a)]\rangle \tag{11.15c}$$

$T_2'$ is written in a form to illustrate that this term depends on the correlation of the frequency shift $\langle \delta\omega_{ba}(t) \, \delta\omega_{ba}(t+\tau)\rangle$. Thus for many problems, the damping of the off-diagonal elements is described by

$$\frac{d(b \mid \varrho \mid a)}{dt} = -\frac{(b \mid \varrho \mid a)}{T_2} \tag{11.16a}$$

where

$$\frac{1}{T_2} = \frac{1}{T_1} + \frac{1}{T_2'} \tag{11.16b}$$

A serious shortcoming of this semiclassical theory of relaxation is that the random perturbation drives the system to equal populations for both states. This corresponds to thermodynamic equilibrium at infinite temperature. In order to correct this defect, it is necessary to have $W_{ba} \neq W_{ab}$. Detailed balance at thermal equilibrium implies

$$W_{ba} = W_{ab} \exp\frac{-\hbar\omega_{ab}}{kT} \tag{11.17}$$

A simple stimulating perturbation induces as many up transitions as down transitions and spontaneous emission to the perturbing reservoir is necessary to obtain the given relationship. The quantized aspects are considered in a later section.

If a coherent signal $U_s$ is applied to the ensemble, it is usually assumed for many problems that the density matrix is of the form

$$\dot{\varrho} = (i\hbar)^{-1}[(H_0\varrho - \varrho H_0) + (U_s\varrho - \varrho U_s)] + \dot{\varrho}_r \tag{11.18a}$$

where

$$(b \mid \dot{\varrho}_{\mathrm{r}} \mid b) - (a \mid \dot{\varrho}_{\mathrm{r}} \mid a)$$

$$= - \frac{[(b \mid \varrho \mid b) - (a \mid \varrho \mid a)] - [(b \mid \varrho_0 \mid b)] - (a \mid \varrho_0 \mid a)]}{T_1} \qquad (11.18b)$$

and

$$(b \mid \dot{\varrho}_{\mathrm{r}} \mid a) = - \frac{(b \mid \varrho \mid a)}{T_2} \qquad (11.18c)$$

The previous development is altered to force the system to relax toward the distribution $\varrho_0$ and this is not inherent in the earlier equations.

**Exercise 11.3** From everyday experience, a gas at 300°K does not emit visible photons and the transition rate for excitation by collision $W_{ab} \approx 0$. Why does this imply that the rate of collisional deexcitation of a molecule $W_{ba} \approx 0$.

Ans. See Eq. (11.17)

### *11.3.2 Stimulated Absorption or Emission by a Sinusoidal Signal*

Assume that the coherent perturbation in Eq. (11.18a) is of the form

$$U_{\mathrm{s}}(t) = U_0 e^{-i\omega t} + U_0{}^{+} e^{+i\omega t} \qquad (11.19a)$$

Equation (11.18a) can be written as four equations, and with the more compact notation $\varrho_{ab} = (a \mid \varrho \mid b), \ldots,$ Eq. (11.18a) becomes

$$\dot{\varrho}_{ba} = -(i\omega_{ba} + T_2^{-1})\varrho_{ba} + (i\hbar)^{-1}(b \mid U_0 \mid a)e^{-i\omega t}(\varrho_{aa} - \varrho_{bb}) = \dot{\varrho}_{ab}^{*} \qquad (11.19b)$$

$$(\dot{\varrho}_{bb} - \dot{\varrho}_{aa}) = (i\hbar)^{-1}[(b \mid U_0 \mid a)e^{-i\omega t}\varrho_{ab} - \varrho_{ba}(a \mid U_0{}^{+} \mid b)e^{i\omega t}]$$
$$- T_1^{-1}[(\varrho_{bb} - \varrho_{aa}) - (\varrho_{bb}^0 - \varrho_{aa}^0)] \qquad (11.19c)$$

and

$$\varrho_{aa} + \varrho_{bb} = 1$$

The rotating wave approximation is used and all terms that would occur at frequencies of the order of $2\omega$ are omitted. Only the resonant-type terms are kept and $\omega_{ba} > 0$ is assumed. Expanding $\varrho_{ba}$ as

$$\varrho_{ba}(t) = \varrho_{ba}(\omega)e^{-i\omega t} + \text{complex conjugate} = \varrho_{ab}^{*}(t) \qquad (11.20a)$$

yields, upon substitution into Eqs. (11.19b, c),

$$\varrho_{ba}(\omega) = \frac{\hbar^{-1}(b \mid U_0 \mid a)(\varrho_{bb} - \varrho_{aa})}{(\omega - \omega_{ba}) + (i/T_2)} \tag{11.20b}$$

A steady-state solution implies $\dot{\varrho}_{aa} = 0$ and $\dot{\varrho}_{bb} = 0$ and then the shift induced by the signal in the diagonal element of the density matrix is

$$(\varrho_{bb} - \varrho_{aa}) = \frac{(\varrho_{bb}^0 - \varrho_{aa}^0)[1 + (\omega_{ba} - \omega)^2 {T_2}^2]}{1 + (\omega_{ba} - \omega)^2 {T_2}^2 + 2\hbar^{-2} \mid (b \mid U_0 \mid a) \mid^2 T_1 T_2} \tag{11.20c}$$

This steady-state solution has a long history and an equation of this type was introduced by Block for the analysis of nuclear magnetic resonance data [*1*]. It has found subsequent use in microwave resonance experiments and more recently in the analysis of lasers. This equation is unusual in that it introduces saturation effects. Even so, the restrictions used in deriving the relaxation times $T_1$ and $T_2$ must be given further consideration in many problems involving saturation.

An equation of this type, with perturbing fields at two frequencies and a nonlinear response of the medium, has been studied between the radio frequencies of nuclear resonance experiments and the optical frequencies which occur in optical laser experiments [*2*].

This two-state problem can immediately be applied to the perturbations of an atom or a molecule by an electric or magnetic field. The perturbation is then of the form

$$U = -\mathbf{P} \cdot \mathbf{E}(t) \quad \text{or} \quad -\mathbf{M} \cdot \mathbf{B}(t) \tag{11.21a}$$

In order to maintain the two-level character, it is almost necessary that the radiation be circularly polarized and directed along any nonzero, constant magnetic field. In zero magnetic field, linear polarization can be used. In Fig. 11.1, a transition stimulated by right circular polarization is shown. For magnetic interactions with the medium inside coils or waveguides, the magnetic problem is usually discussed in terms of linear polarization. The electric polarization induced by the incident radiation is given by

$$\mathbf{P}(t) = n \operatorname{tr} \mathbf{P}\varrho(t) = n[(a \mid \mathbf{P} \mid b)\varrho_{ba}(t) + \text{complex conjugate}] \tag{11.21b}$$

and for magnetic polarization by[1]

$$\mathbf{M}(t) = n \operatorname{tr} \mathbf{M}\varrho(t) = n[(a \mid \mathbf{M} \mid b)\varrho_{ba}(t) + \text{complex conjugate}] \tag{11.21c}$$

[1] The use of the density matrix for magnetic resonance phenomena is discussed by Abragam [*3a*] and by Slichter [*3b*].

where $n$ is the number of atoms per cubic meter and $\varrho_{ba}(t)$ is given by Eq. (11.20b); $b$ implies the state $j_b = 1$, $m_b = -1$ and $a$ implies $j_a = 0$, $m_a = 0$ in Fig. 11.1, but are more general and use all elements of the trace in the general case. Equation (11.21) assumes that the density matrix for an $n$-molecule system is $n$ times the density matrix for a one-molecule system, or $\varrho_n = n\varrho$. This need not be true since interactions between molecules become important and it may be necessary to use a many-body approach. Since dipole fields are long-range, it is necessary to use the local field in the simplest of problems. This local field was introduced in Section 7.10.

Absorption is usually discussed in the unsaturated region, and then $\varrho_{ba}(\omega)$ simplifies to

$$\varrho_{ba}(\omega) \approx \frac{\hbar^{-1}(b \mid U_0 \mid a)(\varrho_{bb}^0 - \varrho_{aa}^0)}{(\omega - \omega_{ba}) + (i/T_2)} \tag{11.22a}$$

If $\mathbf{E}(t)$ is expanded as

$$\mathbf{E}(t) = \hat{e}\_E(\omega)e^{-i\omega t} + \text{complex conjugate}$$

and $\mathbf{P}(t)$ is expanded in a similar manner, then

$$\begin{aligned} P(\omega) &\approx E(\omega)\frac{n\hbar^{-1} \mid (b \mid P_{-1} \mid a) \mid^2 (\varrho_{bb}^0 - \varrho_{aa}^0)}{(\omega - \omega_{ba}) + (i/T_2)} \\ &= \varepsilon_0\varkappa(\omega)E(\omega) \end{aligned} \tag{11.22b}$$

where $\varkappa$ is the electric susceptibility, defined by $D = \varepsilon E = \varepsilon_0(\varkappa + 1)E$. The electric susceptibility $\varkappa$ is complex,

$$\varkappa = \varkappa'(\omega) + i\varkappa''(\omega) \tag{11.23a}$$

and its components follow from Eq. (11.22b):

$$\varepsilon_0\varkappa'(\omega) = (n_b - n_a)\hbar^{-1} \left| (b \mid P_{-1} \mid a) \right|^2 \frac{\omega - \omega_{ba}}{(\omega_{ba} - \omega)^2 + (1/T_2)^2} \tag{11.23b}$$

$$\varepsilon_0\varkappa''(\omega) = (n_b - n_a)\hbar^{-1} \left| (b \mid P_{-1} \mid a) \right|^2 \frac{1/T_2}{(\omega_{ba} - \omega)^2 + (1/T_2)^2} \tag{11.23c}$$

A typical dispersion curve and an absorption or emission curve are shown in Fig. 11.2 and have Lorentz line shapes.

The same arguments apply to simple magnetic absorption, and with

$$M(\omega) = \mu_0^{-1}\chi(\omega)B(\omega)$$

and

$$\chi = \chi'(\omega) + i\chi''(\omega)$$

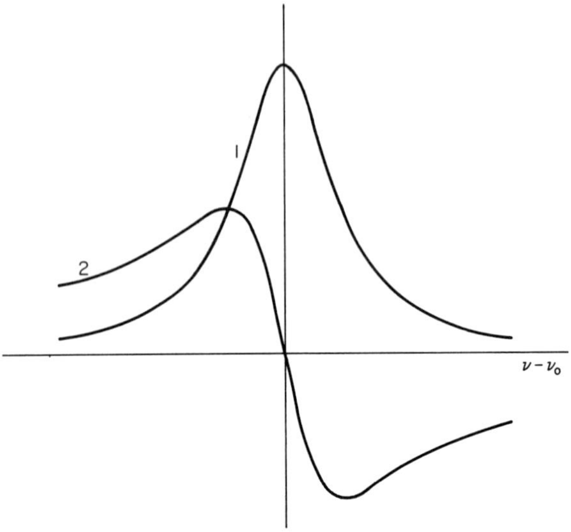


**Fig. 11.2** Typical (1) absorption and (2) dispersion curves for a Lorentzian line shape. $\varkappa''(\nu)$ is the absorption coefficient and $\varkappa'(\nu)$ the dispersion coefficient given by Eqs. (11.23c) and (11.23b), respectively.

Eqs. (11.23a–c) can be used by changing $\varkappa'$ to $\chi'$, $\varkappa''$ to $\chi''$, and $P_{-1}$ to $M_{-1}$.

If the energy levels are well separated and the interactions are weak, so that $\omega \gg T_2^{-1}$, then this can be regarded as a typical line shape. Other aspects can become important and some of these are discussed in subsequent sections.

The term $\varkappa'(\omega)$ can be related to the index of refraction in dilute systems, or in systems for which $\varkappa' \ll 1$, by

$$\frac{\varepsilon}{\varepsilon_0} = \varkappa' + 1 = \mu^2$$

Since

$$\frac{(\varepsilon/\varepsilon_0) - 1}{(\varepsilon/\varepsilon_0) + 2} = \frac{\bar{n}\alpha}{3}$$

for the Lorentz–Lorentz local field correction, for small $\varepsilon/\varepsilon_0$, $\varkappa' \approx \bar{n}\alpha$ and the above approximation can be used.

Note that the response in the unsaturated region has the same form as the response of a simple harmonic oscillator with damping. It differs in that the coefficient $\varkappa''$ is a function of $n_b - n_a$ and can give rise to stimulated emission or absorption. Only absorption occurs for a thermal distribution for which $n_b < n_a$.

In the unsaturated region, Eq. (11.22b) can be generalized to a susceptibility tensor. The component $P(\omega)$ in Eq. (11.22b) is along $\hat{e}_-$. For a more general $\mathbf{E}(t)$, the principle of superposition can be used and $\varrho_{ab}$ can be determined for each $\omega$. For a particular value of $\omega$, the vector field can be resolved along the $\hat{e}_\mathrm{M}$ and then $\varrho_{ba}$ can be determined for each value of $\hat{e}_\mathrm{M}$. Since only pairs of levels are connected by the perturbation, $\varrho(m_b m_a)$ is determined for each pair of levels connected by $\hat{e}_\mathrm{M}$ and then Eqs. (11.23a–c) are used for the polarization. This calculation yields

$$P_x(\omega) = \varepsilon_0 \varkappa_{xy}(\omega) E_y(\omega), \qquad \varkappa_{xy} = \hat{x} \cdot \left[\sum \varkappa_{MM} \hat{e}_M{}^* \hat{e}_M\right] \cdot \hat{y}$$

as one form of the susceptibility tensor $\varkappa_{xy}$. An electric or magnetic field can fix the axis of quantization and requires the tensor form. A magnetic field will split the energy levels shown in Fig. 11.1 and $\omega_{ab}$ will be different for right and left circular polarization. This frequency difference occurs in the equation for $\varkappa'$, and if a plane wave is traveling in the direction of the axis of quantization, the values of $\varkappa'$ are different for right circularly and left circularly polarized waves. This yields two indices of refraction $\mu_\mathrm{r}$ and $\mu_l$ and is a simple example of Faraday rotation.

In the unsaturated region, it is the line width

$$\Delta\nu_{1/2} = \frac{1}{\pi T_2}$$

which appears in the absorption and dispersion. In magnetic problems, the interactions between the magnetic dipoles produce a magnetic field at each atom which is the signal plus a fluctuating component due to the fields of the other dipoles. This is the random aspect which is present even in the absence of the signal and which gives rise to a line width. As interactions become stronger, this simple model must be improved to include all aspects of relaxation.

This model for two states has two relaxation times and is more nearly in accord with experimental absorption than a simple transition rate. Also, the macroscopic $\mathbf{P}(t)$ or $\mathbf{M}(t)$ can be used with Maxwell's equations and the properties of a coil or waveguide filled with the magnetic material or the properties of a laser cavity filled with lasing material have features that are related to the experimental systems.

For strong signals, saturation occurs in a manner which is also characteristic of many experimental systems. In Eq. (11.20c), the difference in the energy level occupation $\varrho_{bb} - \varrho_{aa}$ is driven toward zero as the strength of the signal increases. The off-diagonal elements of the density matrix $\varrho_{ba}$ tend toward zero and the oscillating component of the stimulated polarization tends toward zero. Saturation becomes important when the term

$$2\hbar^{-2} \left| (b \mid U_0 \mid a) \right|^2 T_1 T_2 \approx 1$$

in the denominator of Eq. (11.20c) becomes comparable to unity. It should be noted that the maximum effect of the stimulating signal is to equalize the population of the two states and the maximum stimulated absorption or emission per molecule is limited to

$$h\nu \frac{T_2}{2T_1}$$

Even so, very strong saturation is not included in the model. At sufficiently large fields, harmonics can be generated and it is necessary to keep the terms that change at $2\omega$, which were omitted in the rotating wave approximation.

The nature of the discussion often obscures the important features of the problem. In the macroscopic form of Maxwell's equations, the polarizations are introduced through

$$\mathbf{D} = \varepsilon_0 \mathbf{E} + \mathbf{P} \qquad \text{and} \qquad \mathbf{B} = \mu_0(\mathbf{H} + \mathbf{M})$$

For electric polarization, the electric field and the source $\mathbf{P}$ are related by

$$\text{curl curl } \mathbf{E} + \frac{1}{c^2} \frac{\partial^2 \mathbf{E}}{\partial t^2} = -\mu_0 \frac{\partial^2 \mathbf{P}}{\partial t^2} \tag{11.24a}$$

In the unsaturated region, a Fourier expansion can be used for both $\mathbf{E}$ and $\mathbf{P}$ and these two polar vectors are related by a susceptibility tensor for each $\omega$. For details concerning $\varkappa_{xy}(\omega)$, an equation of motion is needed for $\mathbf{P}$. In classical physics, the Lorentz Force Law

$$m\dot{\mathbf{v}} = e\mathbf{E} + e(\mathbf{v} \times \mathbf{B})$$

can be used to relate the motion of the electrons and the fields. In the Hamiltonian formalism, the equivalent of the Lorentz Force Law is the interaction energy

$$-\mathbf{P} \cdot \mathbf{E} = U(t)$$

and this is used to obtain the response of an atom or molecule to the field. Classical motion saturates as the relativistic correction for the electron mass becomes important. Saturation of **P** in the quantum mechanical problem occurs through the normalization, or $\varrho_{aa} + \varrho_{bb} = 1$.

In Eq. (11.24a), the field depends on the source, the response of the source depends on the stimulating field, and the problem is nonlinear. The opening sentence in Bloembergen's [*4*] discussion of nonlinear susceptibilities illustrates the historical aspects of the problem: "Nonlinear properties of Maxwell's constitutive relations, $D = \varepsilon(E)E$ and $B = \mu(H)H$, have been known from the beginning." The nonlinearity is a basic feature in the understanding of the structure of matter and in the usefulness of real physical systems. The response of **P** to a field **E** requires a knowledge of the physical system and only for problems that have the simplicity of almost independent atoms or molecules can a simple model be given. The two-level density matrix provides one simple model, and another simple model is given in the next section. Problems for which the independent-particle model are inadequate are beyond the scope of this text and detailed references are suggested.

There are two elementary ways to treat the equation relating the field and the polarization in the short-wavelength region. An oscillating dipole generates a field at a large distance $r$ from the source

$$\mathbf{E}_s = -\mathbf{P}\,\frac{\omega^2}{4\pi\varepsilon_0 c^2}\,\frac{e^{ikr}}{r} \tag{11.24b}$$

where the component of **P** transverse to $\hat{k}$ is used. This feature is used in scattering problems. A second procedure is to regard the radiation as along the $\hat{z}$ axis and $A$ as a slowly varying function of $z$ and $t$:

$$E(\mathbf{r}, t) \approx Ae^{i(kz-\omega t)} + \text{complex conjugate}$$

Retaining only large terms, Eq. (11.24a) is approximately

$$2ikc^2\,\frac{\partial A}{\partial z} + 2i\omega\,\frac{\partial A}{\partial t} \approx -\varepsilon_0^{-1}\omega^2 P \tag{11.24c}$$

where $P$ is the coefficient of $e^{-i\omega t}$ in $P(t)$ and is evaluated at the position of each molecule in a unit volume. For a cavity, the $\partial A/\partial z$ can be omitted and an equation similar to Eq. (10.110) is obtained. In an amplifier, $\partial A/\partial t$ can be omitted and the gain or loss per unit length and the index of refraction for a given frequency follow. Both time and space dependence can be important in some highly saturated systems and the analysis of the entire equation is of interest. This occurs in the problem of "self-induced transparency" as a pulse traverses a medium [5].

### *11.3.3 Response of an Atom to a Sequence of Saturating Pulses and Echoes*

Most of the previous discussion has depended on the perturbation being weak, and for a weak perturbation, the order in which a sequence of pulses occurs is not important. This information is lost. As the perturbation becomes stronger, the exact solution of Eq. (11.3b) is needed and a complete solution of the problem would follow from

$$i\hbar c_a(t) = \sum_{a'} \int_0^t (a \mid U \mid a')c_{a'}(t')[\exp i\omega_{aa'}t']\, dt'$$

This equation is nonlinear and a computer solution is needed for even a simple two-state problem. Some aspects of the nature of the problem can be seen by using a coherent perturbation of the type given by Eq. (11.19a). If terms that change at $2\omega$ are omitted, the remaining part of the problem has an exact solution. Limiting the discussion to two states $\mid a)$ and $\mid b)$, the evolution of the wave function in time is given by

$$\psi(\tau) = \mathscr{U}(t, t_0)\psi(t_0) \tag{11.25a}$$

The unitary matrix for this harmonic perturbation of duration $t - t_0$ is

$$\begin{aligned}\mathscr{U}(t, t_0) = {} & \mid b)(b \mid \left[\cos q\tau - \frac{i\Delta}{2q}\sin q\tau\right]e^{i\varphi_1} \\ & + \mid b)(a \mid \left[\frac{iu}{q}\sin q\tau\right]e^{i\varphi_2} \\ & + \mid a)(b \mid \left[\frac{iu^*}{q}\sin q\tau\right]e^{i\varphi_3} \\ & + \mid a)(a \mid \left[\cos q\tau + \frac{i\Delta}{2q}\sin q\tau\right]e^{i\varphi_4}\end{aligned} \tag{11.25b}$$

where

$$4q^2 = \Delta^2 + 4uu^*, \qquad \Delta = \omega_{ba} - \omega, \qquad t - t_0 = \tau$$

and the strength of the interaction is

$$(b \mid U \mid a) \approx -\hbar u e^{-i\omega t}$$

The other term in $(b \mid U \mid a)$ gives rise to a nonresonant term and is omitted. The phases are needed for some studies and are

$$\varphi_1 = -\left(\frac{E_b}{\hbar} - \frac{\Delta}{2}\right)\tau$$

$$\varphi_4 = -\left(\frac{E_a}{\hbar} + \frac{\Delta}{2}\right)\tau$$

$$\varphi_2 = -\left[\frac{E_b t}{\hbar} - \frac{E_a t_0}{\hbar} - \frac{\Delta(t + t_0)}{2}\right]$$

$$\varphi_3 = -\left[\frac{E_a t}{\hbar} - \frac{E_b t_0}{\hbar} + \frac{\Delta(t + t_0)}{2}\right]$$

This solution is quite different from the earlier solution and if the molecule is prepared in the initial state $\mid a)$ at time $t_0$ and an observation is made at time $t$, the probability that the molecule is in state $\mid b)$ is

$$P_{ab}(t) = \left|(b \mid \mathscr{U}(t, t_0) \mid a)\right|^2 = \frac{uu^*}{q^2} \sin^2 q(t - t_0) \qquad (11.26)$$

At resonance, the probability is equal to $\sin^2[(uu^*)^{1/2}(t - t_0)]$ and the probability depends on the strength $(uu^*)^{1/2}$ and the time interval $t - t_0$. For a $\pi$ pulse, the argument of the sine is given as

$$\pi = (uu^*)^{1/2}(t - t_0) = \left|\frac{U_0}{\hbar}\right|(t - t_0)$$

and the probability is zero. A $\pi/2$ pulse yields a probability of one if an observation is made at time $t$.

If no specific measurement is made, it is more convenient to discuss the problem in terms of the electric or magnetic polarization. This retains the wave features of the problem and many anomalous aspects that are characteristic of wave phenomena will be emphasized.

This polarization **P** or magnetization **M** of the molecule follows from

$$\mathbf{P}(t) = \operatorname{tr} \mathbf{P}\sigma(t) = (a \mid \mathbf{P} \mid b)(b \mid \sigma(t) \mid a) + \text{complex conjugate} \qquad (11.27a)$$

where

$$\sigma(t) = \mathscr{U}(t, t_0)\sigma(t_0)\mathscr{U}^+(t, t_0)$$

Direct evaluation for this problem for $\sigma(t_0)$ diagonal yields

$$\mathbf{P}(t) = (\sigma_a - \sigma_b)e^{-i\omega t}(a \mid \mathbf{P} \mid b)\left\{\frac{u}{4q^2}\,[\Delta(1 - \cos 2q\tau) + i2q \sin 2q\tau]\right\} + \text{complex conjugate} \tag{11.27b}$$

for the polarization. The polarization oscillates at a frequency $\omega$, but has a modulation in amplitude and phase proportional to $2q(t - t_0)$. Both amplitude and frequency modulation are introduced by saturating radiation. Saturation is a very natural phenomenon in quantum mechanical problems, since the absolute square of the wavefunction must be unity,

$$|\,\psi(t)\,|^2 = 1$$

This was apparent in the density matrix development by the condition $\varrho_{aa} + \varrho_{bb} = 1$. Equation (11.27b) is appropriate for a single, isolated molecule and an ensemble average is needed for the polarization of $N$ molecules. $\Delta$ may be different for each molecule in the ensemble and this is referred to as inhomogeneous broadening of the line. The Doppler shift causes this broadening in gases and inhomogeneities in crystals cause $\omega_{ba}$ to vary from site to site. If the interaction is weak and $\Delta$ is distributed over a broad range about $\omega_{ba}$, the quantity within the curly brackets in Eq. (11.27b) averages to zero unless $\Delta = 0$ as the ensemble average is taken. The information obtained is the same as that for the transition rate in Section 11.2.

#### 11.3.3.1 Echoes

A very dramatic experimental example of this approach is the experimental observation of "spin echoes" and "photon echoes." If a sequence of pulses is used as shown in Fig. 11.3, the wavefunction at some time later is

$$\psi(t) = \mathscr{U}_0(t, t_3)\,\mathscr{U}(t_3, t_2)\,\mathscr{U}_0(t_2, t_1)\,\mathscr{U}(t_1, 0)\psi(0) \tag{11.28a}$$

where the interaction is off between pulses and the wavefunction evolves as

$$\mathscr{U}_0(t, t_0) = \exp\left(-\frac{iH_0(t - t_0)}{\hbar}\right)$$

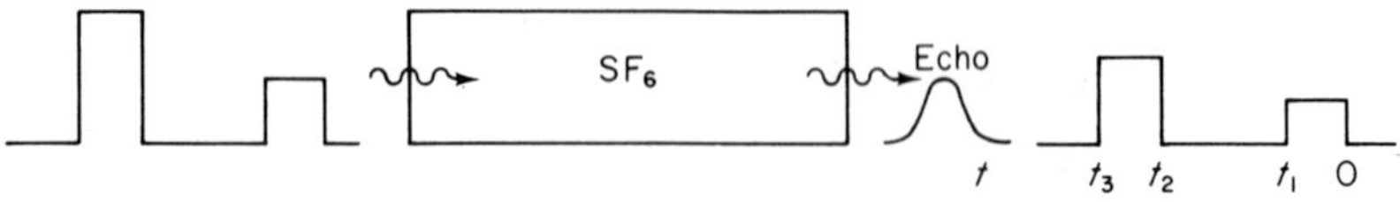


**Fig. 11.3** Photon echoes generated by $SF_6$. Two pulses of 10.6-$\mu$m radiation are incident on the absorbing $SF_6$. These two pulses stimulate the $SF_6$ molecules, which subsequently emit an echo pulse. Equation (11.28b) describes the stimulated polarization.

during this interval. The expression for the polarization becomes

$$\mathbf{P}(t) = (a \mid \mathbf{P} \mid b)(\sigma_a - \sigma_b)e^{-i\omega t}\left\{\frac{u^*}{4q^2}\,[\Delta(1 - \cos 2q\tau) - i2q \sin 2q\tau]\right\}$$
$$\times\left[\frac{1}{2}\left(\frac{u'}{q'}\right)^2(1 - \cos 2q'\tau')\right]e^{-i\Delta(t-t_3-t_2+t_1)}$$
$$+\text{complex conjugate} + \text{other terms} \tag{11.28b}$$

where the prime is for the second pulse. Again, in a collection of atoms, $\Delta = \omega_{ba} - \omega$ will be different for different atoms and an ensemble average over $\Delta$ is needed. An important term in this average is

$$\langle e^{-i\Delta(t-t_3-t_2+t_1)}\rangle$$

and several examples are discussed.

An early use of an equation of this type occurred in the discussion of atomic and molecular beams stimulated [*6*] by "separated oscillating fields." In the cesium beam atomic clock [*6*], a beam of cesium atoms effuses from an oven, is collimated by slits, and passes through a stimulating cavity which is oscillating at $\nu \approx 9192.68 \times 10^6 = \nu_{ba}$. This is the first pulse in Eq. (11.28a) and since the transition is between $(4, 0) \leftrightarrow (3, 0)$ states in the $(Fm)$ quantum numbers for the ground state of Cs, the transition is magnetic dipole and $P$ is replaced by $M$. The excitation interval $(t_1 - 0)$ is the time of passage through the cavity. During passage between cavities, there is no excitation and this is the interval $(t_2 - t_1)$. Passage through the second cavity occurs at $t_2$ and Eq. (11.28b) is appropriate for the magnetic polarization of the cesium atoms in the second cavity if the term $t - t_3$ is omitted and $\tau' = t - t_2$. The Doppler effect appears in the calculations as a difference in time intervals for each velocity group effusing from the oven and an average over the velocities yields a term in $M$ which gives maximum response in the second cavity when $\nu \approx \nu_{ba}$. The passage interval between the cavities favors $\Delta = 0$ in $\langle \exp i\Delta(t_2 - t_1)\rangle$ and yields a response line shape which is much narrower than the Doppler width.

A direct application of this equation occurs in experiments on "spin echoes" and the concept of "echoes" was introduced by Hahn [*7*] in 1950, or shortly after the introduction of nuclear and atomic resonance experiments. Again, magnetic transitions occur and $P$ is replaced by $M$. In these experiments, the sample of nuclear spins or atomic spins is subjected to a strong pulse of stimulating radiation with $\nu \approx \nu_{ba}$ during the interval $t_1 - 0 = \tau$. The strength and duration of the first pulse is selected so that

$$2q\tau = \tfrac{1}{2}\pi$$

A second pulse is applied at time $t_2$ with a duration $t_3 - t_2 = \tau'$ and a pulse strength

$$2q'\tau' = \pi$$

For an ensemble of atoms or molecules with different values of $\Delta$, the term $\langle \exp -i\Delta(t - t_3 - t_2 + t_1)\rangle$ is nonzero near $t - t_3 - t_2 + t_1 = 0$ and a pulse, or "echo," occurs. In nuclear resonance, the sample within the coils has a macroscopic magnetic polarization as long as this average is nonzero and this excites the coil so that energy is fed back into the circuit which previously had supplied the two pulses to the sample. These atoms or molecules emit coherently since they are excited by a magnetic field **B** which stimulates them in a coherent manner.

Photon echoes provide the example shown in Fig. 11.3 [*8*, *8a*]. The output of a laser is used to stimulate an absorbing material with two pulses. These are usually electric dipole transitions and the electric polarization expression is appropriate. Intense laser radiation is able to saturate the molecules in the absorbing material, or provide a $\pi/2$ first pulse ($2q\tau = \pi/2$) and a $\pi$ second pulse ($2q'\tau' = \pi$). An oscillating electric polarization $P$ at frequency $\nu$ is induced on each molecule and the scattered radiation at distance $r$ is proportional to

$$\mathbf{E}_s = -\mathbf{P}\frac{\omega^2}{4\pi\varepsilon_0 c^2}\frac{e^{ikr}}{r}$$

where the component of $P$ transverse to $\hat{k}$ is used. The molecules emit spontaneously as $P^2$, but this is not the interesting term. Since the molecules are excited by an almost plane incident wave, the scattering in the forward direction has this same amount of spatial coherence. If the average value of $\langle \exp -i\Delta(t - t_3 - t_2 + t_1)\rangle$ is nonzero and of the order of unity, the scattering in the forward direction has an amplitude comparable with the incidence pulses and an echo occurs when $t - t_3 - t_2 + t_1 = 0$. The formation of the ensemble average over $\Delta$ yields the line shape. In the initial experiments, very light pink ruby, or dilute $Cr^{3+}$ ions in $Al_2O_3$, was used for the absorbing material and ruby lasers were used for excitation. Other experiments have examined echoes [*8*, *8a*] from $SF_6$, and $CO_2$ lasers oscillating at 10.6 μm were used for excitation. Since for $J > 0$, the levels are degenerate and stimulation by elliptically polarized radiation can excite all states of $m_b$ and $m_a$, the analysis is considerably more complex. Anomalous polarization occurs in the echoes and original references are suggested.[2]

[2] Gordon *et al.* and Heer and Kohl [*8*] give theoretical discussions of anomalous polarization aspects. See also Abella *et al.* [*8a*].

It should be noted that no damping effects have been included in the discussion. An average over the inhomogeneous broadening implied by $\Delta$ yields a line shape, but this is not due to damping effects in the usual sense. One of the bad features of this simple and direct model is the absence of damping. A very naive model is often used and it is assumed that the oscillating component of the polarization is damped as $e^{-t/T_2}$. The effect of this decay will be quite pronounced and the echo will decrease in amplitude as the interval between pulses approaches $T_2$. This provides a technique for measuring the decay constant $T_2$.

The simple model discussed in this section has some aspects of the experiment in self-induced transparency. If $q\tau = \pi$ in Eq. (11.25b) or Eq. (11.27b), the molecule returns to its initial state and there is no appreciable probability of energy being stored by the molecule for subsequent emission. Since $\Delta$, and therefore $q$, is not the same for all molecules, $q\tau$ cannot equal $\pi$ for all molecules at the same time and more subtle aspects must be considered for this very interesting phenomenon [5].

## 11.4 Natural Line Width

If an atom is prepared in state $|\, b)$ and subsequently emits a photon and enters state $|\, a)$, the state has a finite lifetime and a distribution in energy or frequency for the photons emitted by an ensemble of atoms is expected. This problem can be examined in greater detail by quantizing both the atom and the radiation field, or by using a Hamiltonian

$$H = H_{\mathrm{a}} + H_{\mathrm{r}} + U \tag{11.29a}$$

where $H_{\mathrm{a}}$ is the Hamiltonian describing the atom, $H_{\mathrm{r}}$ the radiation field, and $U$ the interaction between the atom and the radiation field. $U$ is time-independent in this formulation. The wavefunction in the absence of the interaction $U$ is described by the product of the atomic wavefunction and the wavefunction for the radiation field, and is designated by

$$|\, \alpha) = |\, a) \,|\, r)$$

In the presence of the interaction, the change in the system is described by the Schrödinger equation, and replacing $a$ and $a'$ in Eq. (11.2a) by $\alpha$ and $\beta$ yields

$$i\hbar \dot{c}_\beta = \sum_\alpha (\beta \,|\, U \,|\, \alpha) c_\alpha \exp \frac{i(E_\beta - E_\alpha)t}{\hbar} \tag{11.29b}$$

Spontaneous emission is discussed with the following assumptions. At $t = 0$, the atom is excited to state $|\, b)$ and there are no photons in the radiation field and this state is denoted by $|\, \beta)$. Subsequently, there is a probability of the atom going to state $|\, a)$ and emitting a photon with momentum $\hbar\mathbf{k}$ and polarization $\hat{u}$, and this state is designated by $|\, \alpha)$. An infinite number of states $|\, \alpha)$ can occur and the modes in volume $L^3$ are discussed in Chapter II. The more general case considers more than one quantum in state $|\, \beta)$ and $|\, \alpha)$ and this feature is included in the following discussion. Thus with $|\, r)$ denoting the state of the radiation field,

$$\begin{aligned} |\, \beta) &= |\, b)\,|\, r), \qquad & E_\beta &= E_b + E_r \\ |\, \alpha) &= |\, a)\,|\, r'), \qquad & E_\alpha &= E_a + E_r + \hbar\omega_k \end{aligned} \tag{11.30a}$$

Direct substitution into Eq. (11.47) yields the infinite set of equations

$$\begin{aligned} i\hbar\dot{c}_\alpha &= (\alpha\,|\, U \,|\, \beta)c_\beta \exp i(\omega_k - \omega_{ba})t \\ i\hbar\dot{c}_\beta &= \sum_\alpha (\beta\,|\, U \,|\, \alpha)c_\alpha \exp -i(\omega_k - \omega_{ba})t \end{aligned} \tag{11.30b}$$

Following the procedure of Weisskopf and Wigner [9], the initial state is assumed to decay as

$$c_\beta(t) = e^{-\gamma t/2} \tag{11.31a}$$

Direct substitution into the equation for $c_\alpha$ and integration yield

$$c_\alpha = \frac{1}{i\hbar}(\alpha\,|\, U \,|\, \beta)\frac{\exp i[\omega_k - \omega_{ba} + (i\gamma/2)]t - 1}{i(\omega_k - \omega_{ba}) - (\gamma/2)} \tag{11.31b}$$

$\gamma$ can be found by substituting this quantity into Eq. (11.30b) or by observing that the probability of $|\, c_\alpha\,|^2$ for large $t$ is given by

$$|\, c_\alpha\,|^2 = \frac{1}{\hbar^2}\,|\,(\alpha\,|\, U \,|\, \beta)\,|^2 \frac{1}{(\omega_k - \omega_{ba})^2 + (\gamma/2)^2} \tag{11.32a}$$

For large $t$, the system is definitely in one of the $\alpha$ states and

$$\sum_\alpha |\, c_\alpha\,|^2 = 1 \tag{11.32b}$$

can be used to determine $\gamma$. Further details require the use of the more exact theory of resonance fluorescence [10].

In the Wigner–Weisskopf model, the probability of making a transition $\beta \to \alpha$ is

$$P_{\beta\alpha} = \frac{1}{\hbar^2}\left|\,(\alpha\,|\, U \,|\, \beta)\right|^2 \frac{1}{(\omega_k - \omega_{ba})^2 + (\gamma/2)^2} \tag{11.33}$$

Once the limiting process has been completed, an alternate form is useful. This Lorentz line shape corresponds to the integral

$$\frac{2}{\gamma}\int_0^\infty dt\, e^{-\gamma t/2}\cos(\omega_k - \omega_{ba})t$$

and the probability can be written with this substitution. One now chooses to write the term

$$e^{i\omega_{ba}t}(\alpha \mid U \mid \beta)(\beta \mid U \mid \alpha) = (\alpha \mid U(0) \mid \beta)(\beta \mid U(t) \mid \alpha)$$

where the partial interaction representation

$$U(t) = e^{iH_a t/\hbar}Ue^{-iH_a t/\hbar} \tag{11.34}$$

is used. Then the transition probability from state $\beta$ to $\alpha$ is given by the expression

$$\begin{aligned} P_{\beta\alpha} &= \gamma^{-1}\hbar^{-2}\int_0^\infty d\tau\,(\exp -\tfrac{1}{2}\gamma\tau - i\omega_k\tau)(\alpha \mid U(0) \mid \beta)(\beta \mid U(\tau) \mid \alpha) \\ &\qquad + \text{complex conjugate} \\ &= \gamma^{-1}W_{\beta\alpha} \end{aligned} \tag{11.35}$$

and is related to the transition rate by $\gamma$.

This theory is now applied to a specific problem of electric dipole radiation. The interaction has its usual form, with $U = -\mathbf{P}\cdot\mathbf{E}$, where the electric field is expanded as

$$\mathbf{E}(\mathbf{r}) = \sum_k i\left(\frac{\hbar\omega_k}{2\varepsilon_0 L^3}\right)^{1/2}[a_k\hat{u}_k(\exp i\mathbf{k}\cdot\mathbf{r}) - a_k^{+}\hat{u}_k^{*}(\exp -i\mathbf{k}\cdot\mathbf{r})] \tag{11.36}$$

Here $a_k$ and $a_k^+$ are the annihilation and creation operators for a photon in a mode with index $k$, or a photon in direction $\hat{k}$ with polarization $\hat{u}_k$ and energy $h\nu_k$. The matrix element is given by

$$\left|(\alpha \mid U \mid \beta)\right|^2 = \left|(j_a m_a \mid \mathbf{P}\cdot\hat{u}_k^{*} \mid j_b m_b)\right|^2 (n_k + 1_k)\frac{\hbar\omega_k}{2\varepsilon_0 L^3} \tag{11.37a}$$

for the atom in the upper state. $n_k$ is the number of photons already present in mode with index $k$, and $1_k$ is the spontaneous emission term and is omitted if the atom is in the lower state initially. This occurs since $a_k^+ \mid r)$ denotes a state with one more photon and $a_k \mid r)$ denotes a state with one less photon. Hence

$$\mid r') = (n_k + 1)^{-1/2}a_k^{+} \mid r)$$

is the state of the radiation field with one more photon of index $k$. The matrix element of the field operator $(r' \mid \mathbf{E} \mid r)$ can be formed in this manner.

For spontaneous emission, $\gamma$ is given by using only the spontaneous term in Eq. (11.37a) and substituting into Eq. (11.32b). This yields

$$\gamma = \sum_{k;m_a;m_b} \frac{1}{\hbar^2} \left| (j_a m_a \mid \mathbf{P} \cdot \hat{u}_k^* \mid j_b m_b) \right|^2 \frac{\gamma}{(\omega_k - \omega_{ba})^2 + (\gamma/2)^2} \frac{\hbar\omega_k}{2\varepsilon_0 L^3}$$

$$= \frac{16\pi^3}{(2J_b + 1)3h\lambda^3\varepsilon_0} \mid (J_a \|P\| J_b) \mid^2 \qquad (11.37\text{b})$$

which is the Einstein spontaneous emission coefficient and was given as Eq. (2.44). Again, the sum over $k$ is changed to an integral over $L^3\nu^2\, d\Omega d\nu/c^3$ and the sum over the two states of polarization. $\hat{u}_k$ is always transverse to $\hat{k}$. The spectral distribution of the spontaneously emitting atoms is given by using the continuous variable notation,

$$P(b \to a; \theta, \varphi, \nu, \hat{u})\, d\Omega\, d\nu = \left(\frac{3}{8\pi}\right)(j_a 1 m_a M \mid j_b m_b)^2 \mid \hat{e}_M^* \cdot \hat{u} \mid^2$$
$$\times [n(\theta, \varphi, \nu, p) + 1] g(\nu)\, d\Omega\, d\nu$$

where

$$g(\nu) = \frac{\gamma}{(\omega_k - \omega_{ba})^2 + (\gamma/2)^2} \qquad (11.38)$$

$P(b \to a; \theta, \varphi, \nu, \hat{u})$ is the probability of an atom emitting a photon into solid angle $d\Omega$ making direction $\hat{k}$, or $(\theta, \varphi)$, with the axis of quantization $\hat{z}$, into frequency range $d\nu$, and with polarization $\hat{u}$. A time-proportional transition rate similar to Eq. (2.50) can be derived by examining $d \mid c_\alpha \mid^2/dt$ and yields

$$W(b \to a; k) = \gamma P(b \to a; k) \qquad (11.39)$$

where $k$ implies a mode, or the continuous index $(\theta, \varphi, \nu, \hat{u})$. The sum over all $k$ must yield $d \mid c_\beta \mid^2/dt = -\gamma$ for the transition rate out of state $\beta$ and the above is in accord with this result.

### 11.4.1 *Modifications of the Natural Line Width by Motion*

The spectral distribution of an ensemble of spontaneously emitting atoms is given by $g(\nu)$ and this is again the Lorentz line shape. Even though an atom may emit a photon of distinct momentum $\hbar\mathbf{k}$, an experimental detector must examine the radiation from an ensemble of atoms to study the spectral distribution. Since this radiation has the spectral distribution given by $g(\nu)$ and since this spectral distribution is the same as that of a series of random pulses of the type given by Eq. (10.26a), the experimental detector cannot distinguish these photons from a series of random pulses of this type. It is for this reason that the "classical wave train" model is often a useful concept.

If the atoms are moving with velocities $\mathbf{v}$, then a Lorentz transformation is necessary to describe the observed frequency. The details of this transformation and the Doppler line shape for the emitted radiation are discussed in Chapter II and given in detail by Eqs. (2.54)–(2.56).

The frequency shift due to the atomic motion is

$$\frac{\nu - \nu_e}{\nu_e} \approx \frac{\mathbf{v} \cdot \hat{k}}{c} \tag{11.40}$$

and the emission of each atom is shifted by this amount. A sum over the atoms yields the Doppler line shape. If the emitting atom is confined to a region smaller than one wavelength, then the line can be made narrower than the Doppler width [*11*]. The hydrogen maser and the rubidium frequency standard use confinement within an enclosure to reduce the line width due to motion.

In any emission process, momentum is conserved, and

$$m\mathbf{v} = \hbar\mathbf{k} = \frac{h\nu}{c}\,\hat{k} \tag{11.41a}$$

is the recoil of an atom with mass $m$ when it emits a photon of energy $h\nu$ in direction $\hat{k}$. Recoil introduces a line shift of

$$\frac{\nu - \nu_e}{\nu_e} = -\frac{h\nu_e}{mc^2} \tag{11.41b}$$

This is only a frequency shift of one part in $10^{14}$ for the 1420-MHz line of the hydrogen maser, but for a 100-keV $\gamma$ ray which is used in Mössbauer experiments, the shift can be greater than the natural width of the line.

If the emitter undergoes a random motion during its lifetime, the line is broadened by the transverse Doppler effect and is proportional to the average square velocity,

$$\left| \frac{\nu - \nu_e}{\nu_e} \right| = \left\langle \frac{v^2}{c^2} \right\rangle \approx \frac{kT}{mc^2} \tag{11.42}$$

and for thermal motion is of the order of $kT/mc^2$. For a quantum oscillator in its ground state, $kT$ is replaced by $\frac{1}{2}h\nu_0$ and the line width is of the order of

$$\frac{h\nu_0}{2mc^2}$$

for an oscillator with frequency $\nu_0$.

## 11.5 Recoilless Emission and the Mössbauer Effect

Spontaneous emission from atoms or molecules or $\gamma$-ray emission by nuclei in solids can be treated with Eq. (11.35). The latter problem of $\gamma$-ray emission, or the Mössbauer effect, is selected for detailed examination. In the initial state $\beta$, the nucleus is excited, no photons are present in the radiation field, and the solid is in the state $b$. The final state has a photon in state $k$, the nucleus in a lower state, and the crystal in state $a$. The interaction of the nucleus with the radiation field is either a magnetic dipole or an electric quadrupole with a phase factor $\exp i\mathbf{k}\cdot\mathbf{r}$, where $\mathbf{r}$ is the distance of the nucleus from the observer. Thus the matrix element of interest is of the form

$$(\beta \mid U(\tau) \mid \alpha) \propto (b \mid \exp i\mathbf{k}\cdot\mathbf{r}_m(\tau) \mid a) \exp i\omega_e\tau$$

The polarization of the emitted photon and the change in nuclear states could be given in terms of a reduced matrix element and a Clebsch–Gordon coefficient similar to that used for atoms for magnetic dipole transitions, or $(F_b 2 m_b M \mid F_a m_a)$ for electric quadrupole transitions. If the initial states $m_b$ are equally probable, the emitted radiation is isotropic and unpolarized. Thus the probability of emitting a photon in direction $\hat{k}$ with energy $h\nu$ is

$$g(\hat{k}, \nu)\, d\Omega\, d\nu$$

where

$$g(\hat{k}, \nu) = \frac{1}{4\pi}\int_0^\infty d\tau \exp\left[-\frac{1}{2}\gamma\tau - i(\omega-\omega_e)\tau\right] \times\langle \exp -i\mathbf{k}\cdot\mathbf{r}(0) \exp i\mathbf{k}\cdot\mathbf{r}(\tau)\rangle + \text{complex conjugate} \tag{11.43}$$

For spontaneous emission by a nucleus of infinite mass, the term in angular brackets is unity. An integral over $d\Omega\, d\nu$ is also unity and this fixes the constant for emission by a single nucleus as $1/4\pi$. An integral over $d\nu$ yields $\delta(\tau)$. The quantity in angular brackets is a statistical average over the initial states $b$ of the crystal and a sum over the unobserved final states.

### 11.5.1 Illustrative Exercise

As an illustrative exercise, consider the radiation emitted by a nucleus of mass $m$ bound in a simple harmonic oscillator potential well, or the problem which is shown in Fig. 11.4. Then the canonical thermal average value of the quantity in angular brackets in Eq. (11.43) is

$$F(\tau) = \langle e^{-ikq(0)} e^{ikq(\tau)} \rangle = \frac{\operatorname{tr} e^{-\beta H} e^{-ikq(0)} e^{ikq(\tau)}}{\operatorname{tr} e^{-\beta H}} \tag{11.44}$$

for an ensemble of identical harmonic oscillators.

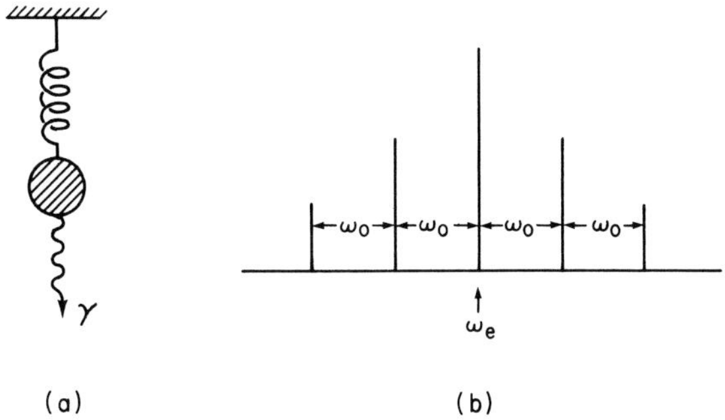


**Fig. 11.4** (a) An illustrative diagram of a $\gamma$ ray being emitted from a simple harmonic oscillator. (b) Central Mössbauer line $\omega_e$ and the displaced lines which correspond to the absorption or emission of one, two, or more vibrational quanta.

Assume that the Hamiltonian of the simple harmonic oscillator is

$$H_a = \frac{p^2}{2m} + \frac{1}{2} m\omega_0{}^2 q^2$$

with the commutation relation

$$qp - pq = i\hbar \tag{11.45}$$

Although any complete set of states can be used for determining the trace, it is convenient to choose the energy states. Then, with

$$a = \left(\frac{m\omega_0}{2\hbar}\right)^{1/2} q + \frac{i}{(2m\hbar\omega_0)^{1/2}} p \tag{11.46a}$$

the Hamiltonian becomes

$$H_a = \tfrac{1}{2}(aa^+ + a^+a)h\nu_0 \tag{11.46b}$$

and the commutation relation is

$$aa^+ - a^+a = 1 \tag{11.46c}$$

Not only is $e^{-\beta H}$ diagonal in this representation, but $q(\tau)$ can also be formulated in a simple manner. From Eq. (11.46a),

$$q(t) = \left(\frac{\hbar}{2m\omega_0}\right)^{1/2} [a(t) + a^+(t)] \tag{11.47}$$

where

$$a(t) = e^{iH_a t/\hbar} a e^{-iH_a t/\hbar} = ae^{-i\omega_0 t}$$

With

$$\varkappa^2 = k^2 \frac{\hbar}{2m\omega_0} = \left(\frac{\omega_e}{c}\right)^2 \frac{\hbar}{2m\omega_0} \tag{11.48}$$

the average value can be written as

$$\begin{aligned} F(\tau) &= \langle \exp -ikq(0) \exp ikq(\tau) \rangle \\ &= \langle \exp -i\varkappa(a + a^+) \exp i\varkappa[a(\tau) + a^+(\tau)] \rangle \end{aligned} \tag{11.49}$$

The identity

$$e^A e^B = e^{A+B} e^{[AB-BA]/2} \tag{11.50}$$

can be used to simplify the expression if $A$ and $B$ commute with $(AB - BA)$. This occurs for the $a$ and $a^+$ operators, and

$$F(\tau) = \langle \exp i[\alpha a + \alpha^* a^+] \rangle \exp(i\varkappa^2 \sin \omega_0 \tau) \tag{11.51}$$

where

$$\alpha = -\varkappa(1 - e^{-i\omega_0 \tau}) \tag{11.52}$$

It can be shown [*12*] for a thermal distribution that

$$\langle \exp i[\alpha a + \alpha^* a^+] \rangle = \exp[-\tfrac{1}{2}\alpha\alpha^*(2\bar{n} + 1)] \tag{11.53}$$

where $\bar{n} = 1/(e^{\beta h\nu} - 1)$ is the average number of quanta associated with the oscillator. The emission line shape follows from

$$\begin{aligned} g(\nu) = \int_0^\infty & d\tau \exp[-\tfrac{1}{2}\gamma\tau - i(\omega - \omega_e)\tau] \\ & \times \exp\{-\varkappa^2[(2\bar{n} + 1)(1 - \cos \omega_0 \tau) + i \sin \omega_0 \tau]\} \\ & + \text{complex conjugate} \end{aligned} \tag{11.54}$$

This is the exact solution for the emission line shape from a set of identical oscillators.

At zero temperature, or $\bar{n} = 0$, this expression for the line shape becomes

$$g(\nu) = \int_0^\infty d\tau \exp[-\tfrac{1}{2}\gamma\tau - i(\omega - \omega_e)\tau] \exp[-\varkappa^2(1 - \exp -i\omega_0\tau)] \quad (11.55a)$$

The time-dependent part of the exponential can be expanded as

$$\sum_u (\varkappa^2)^u \frac{e^{-i\omega_0 u\tau}}{u!} \quad (11.55b)$$

and yields a line shape

$$g(\nu) = (\exp -\varkappa^2) \sum_u \frac{\gamma}{(\omega - \omega_e + u\omega_0)^2 + (\gamma/2)^2} (\varkappa^2)^u \quad (11.55c)$$

The term on the right is the probability of observing a particular value of $\nu$. For $u = 0$, the line has its center at $\omega_e$ and has a width $\gamma$. This is referred to as recoilless emission and forms the Mössbauer line. For $u = 1$, a line occurs near $\omega = \omega_e - \omega_0$ and corresponds to the absorption of one quantum of vibrational energy by the oscillator in the emission process. In general, $u$ corresponds to the number of quanta absorbed by the oscillator and the displacement of the line to lower frequencies. Although the form of $g(\nu)$ is not simple, it is a normalized line shape and an integral over the frequency yields unity. The reduction in line intensity $\exp -\varkappa^2$ is compensated by an overall increase in line width, and the area under $g(\nu)$ remains unity.

If $\bar{n} \neq 0$ and the temperature is finite, then either absorption or emission of vibrational quanta can occur. The detailed form follows by including $\bar{n}$ in the discussion and repeating the discussion for zero temperature. Lines spaced $\omega_0$ apart will occur on each side of the infinite mass emission frequency $\omega_e$.

**Exercise 11.4** Use $Q = \text{tr} \exp(-\beta a^+ a h\nu_0)$ as the generating function for a simple harmonic oscillator to develop an expression for the average number of quanta $\bar{n}$ and $\langle n^2 \rangle$.

**Exercise 11.5** Estimate the strength of some lines for $kT = h\nu_0$ in Fig. 11.4(b).

**Exercise 11.6** Define the characteristic function as

$$\Phi(\xi) = \langle \exp[i\xi(\alpha a + \alpha^* a^+)] \rangle \quad (11.56a)$$

and show that the cumulant expansion, or the expansion of $\ln \Phi(\xi)$ about $\xi = 0$, yields

$$\ln \Phi(\xi) = -\tfrac{1}{2}\xi^2 \alpha\alpha^* \langle aa^+ + a^+a\rangle = -\tfrac{1}{2}\xi^2\alpha\alpha^*(2\bar{n} + 1) \quad (11.56\text{b})$$

for a thermal distribution. Note that all coefficients of odd powers of $\xi^m$ are zero and all even powers in the cumulant expansion cancel for $m > 2$. This is expected for a Gaussian random variable. For convenience in calculations with simple harmonic oscillator energy states with label $n$, we write

$$a^+ \mid n) = (n + 1)^{1/2} \mid n + 1); \qquad a \mid n) = n^{1/2} \mid n - 1); \qquad a^+a \mid n) = n \mid n) \quad (11.57)$$

It is also convenient to replace the exponential by $\cos + i \sin$, since the sine contains only odd powers and has an average value of zero. Note that $\langle n^2\rangle = 2\langle n\rangle^2 + \langle n\rangle$ for thermal radiation.

### *11.5.2 Mössbauer Effect in a Crystal Lattice*

The position of an emitting nucleus in a crystal lattice is given by

$$\mathbf{r}_n = \mathbf{R}_n + \mathbf{u}_n$$

where $\mathbf{R}_n$ is the fixed position of the unit cell and $\mathbf{u}_n$ is the displacement of the nucleus from this fixed position. The displacement $\mathbf{u}_n$ can be expressed as

$$\begin{aligned}\mathbf{u}_n(t) = \sum_q C_q\{a_q\hat{e}_q \exp i(\mathbf{q}\cdot\mathbf{R}_n - \omega_q t) \\ + a_q{}^+\hat{e}_q \exp -i(\mathbf{q}\cdot\mathbf{R}_n - \omega_q t)\}\end{aligned} \quad (11.58)$$

where

$$a_q a_{q'}^+ - a_{q'}^+ a_q = \delta_{q'q}$$

$q$ is an index describing the $3N$ independent modes of the crystal lattice in volume $V$, and $\hat{e}_q$ is the polarization vector. The annihilation and creation operators are treated in the same manner as that used for a single oscillator. Since the modes are *independent*, the previous development is quite easily generalized, and

$$\begin{aligned}&\langle[\exp -i\mathbf{k}\cdot\mathbf{r}_n(0)] \exp i\mathbf{k}\cdot\mathbf{r}_n(\tau)\rangle \\ &= \prod_q^N \langle \exp -i\varkappa_q[a_q + a_q{}^+] \exp\{i\varkappa_q[a_q(\tau) + a_q{}^+(\tau)]\}\rangle \\ &= \prod_q \exp\{-k^2C_q{}^2[(2\bar{n}_q + 1)(1 - \cos\omega_q\tau) + i \sin\omega_q\tau]\}\end{aligned} \quad (11.59)$$

Direct substitution into Eq. (11.43) yields the line shape $g(\nu)$ for the Mössbauer emission. In order to simplify the notation, the polarization vector $\hat{e}_q$ was chosen along $\hat{k}$ and the product with index $q$ is over the $N$ states with this polarization. If greater detail is needed, the specific polarization aspect $|\hat{k} \cdot \hat{e}_q|^2$ can be introduced for each of the three polarization states. Since the modes are independent, a detailed solution of the simple example provided a very rapid method for the general crystal. Here $\bar{n}_q$ is the average number of thermal quanta with energy $h\nu_q$.

The time-independent term in Eq. (11.59) is the Debye–Waller factor and can be approximated in terms of the Debye $\Theta$, which was introduced in Section 9.3. The coefficient $C_q$ can be found by noting that

$$N \sum m\omega_q^{\,2} C_q^{\,2} = \sum \tfrac{1}{2}h\omega_q$$

or

$$C_q = \left(\frac{\hbar}{2m\omega_q N}\right)^{1/2} \tag{11.60}$$

where $N$ is the number of unit cells per cubic meter. Many salient features are already apparent at $T = 0$ or $\bar{n}_q = 0$. Then,

$$\sum_q C_q^{\,2} = \frac{\hbar}{2mN} \sum_q \frac{1}{\omega_q} = \frac{(3/4m)\hbar^2}{k\Theta} \tag{11.61}$$

and the Debye–Waller factor $-(\omega_{\mathrm{e}}^{\,2}/c^2) \sum C_q^{\,2}$ becomes

$$2W = \frac{\hbar^2\omega_{\mathrm{e}}^{\,2}}{2mc^2} \frac{3}{2k\Theta} \tag{11.62a}$$

As indicated earlier, the peak line intensity is reduced by $e^{-2W}$ and this is a manifestation of recoil. The time-dependent term

$$\sum_q \frac{1}{\omega_q} e^{-i\omega_q\tau}$$

requires greater care, but the exponential can be expanded as a power series as was done in Eq. (11.55b), and retaining the first two terms yields

$$e^{-2W}\left[1 + \left(\frac{\omega_{\mathrm{e}}}{c}\right)^2 \sum_q C_q^{\,2} e^{-i\omega_q\tau} \cdots\right]$$

The leading term yields the same peak line shape as for a recoilless emission with a natural width $\gamma$. The second term gives the probability of the excitation of a lattice mode of frequency $\omega_q$ during the emission process. This gives rise to a broad and weak line shape with a peak near $\omega_e - \omega \approx k\Theta/\hbar$ below $\omega_e$ and none above.

At finite temperatures, $\bar{n}_q \neq 0$, and either absorption or emission of vibrational quanta can occur. As $\bar{n}_q$ increases, the broad, weak background above $\omega_e$ begins to increase in width. Any increase in width must be accompanied by a decrease in strength since the area under $g(\nu)$ remains unity. Thus the Debye–Waller factor for $T \ll \theta$ becomes

$$2W = k^2 \sum C_q^2(2\bar{n}_q + 1) \approx \frac{\hbar^2\omega_e^2}{2mc^2} \frac{3}{2k\Theta}\left[1 + \left(\frac{T}{\Theta}\right)^2\right] \qquad (11.62b)$$

If $k^2 = \omega_e^2/c^2$ is large and $T/\Theta$ is large compared to unity, then $\bar{n}_q \to kT/\hbar\omega_q$ and the expression for the line shape simplifies for small $\tau$. The large value of the coefficient can limit the region of interest to small $\tau$, and

$$(1 - \cos \omega_q\tau) \to \tfrac{1}{2}\omega_q^2\tau^2 \qquad \text{and} \qquad \sin \omega_q\tau \to \omega_q\tau$$

The sum over the coefficient of the $\tau^2$ term yields $\exp[-\frac{1}{2}\omega_e^2(kT/mc^2)\tau^2]$ and is the transverse Doppler broadening due to recoil and was discussed in Eq. (11.42). The $\tau^2$ dependence yields a Gaussian line shape. The sine term shifts the frequency and the center of the Gaussian is of the order of $\omega_e - \omega = k\Theta/\hbar$ below the Mössbauer line. The sharp Mössbauer line is always present and corresponds to the part of the line shape that occurs for large values of $\tau$.

The structure [*13*] of the $\gamma$-ray emission spectrum of emitters in two types of solids is shown in Fig. 11.5. The dashed line is for a recoil energy large compared to the Debye $\Theta$, with multiphonon processes dominant. The line shape is Gaussian. If the energy of the $\gamma$ ray is low and the temperature of the crystal is well below the Debye $\Theta$, the excitation of a few phonons occurs and the line shape has a structure which reflects the one-phonon, two-phonon, and so forth transitions. The sharp peak at $E_0 = h\nu_e$ is the zero-phonon, or Mössbauer, line. This line has the natural width and can be $10^{-10}$ of the background width. The contribution to the line shape above $E_0$ is due to the finite temperature of the lattice and indicates emission of thermal quanta.

The 8.4-keV $\gamma$ line of $^{169}$Tm and of many other emitters is so narrow that the hyperfine splitting is greater than the line width of the recoilless emission. Since the hyperfine splitting is a measure of the wavefunction of the s-state electrons at the nucleus and for nuclei with quadrupole moments, is a measure of the surrounding electric field gradients, the study of the splitting of the hyperfine states and the line shapes can be used for the study of local interactions in solids. The absorption spectrum of the 8.4-keV $\gamma$ radiation by thulium [*14*] in $TmCl_3 \cdot 6H_2O$ is shown in Fig. 11.6. The first excited state of $^{169}Tm$ has $I = \frac{3}{2}$ and a quadrupole moment. By the Wigner–Eckart Theorem, a $\frac{3}{2}$ state can be split by an electronic charge distribution with axial symmetry into $\pm\frac{3}{2}$ and $\pm\frac{1}{2}$ energy levels. The crystal symmetry of the crystal is so low that the crystal field splits the $^3H_6$ electronic

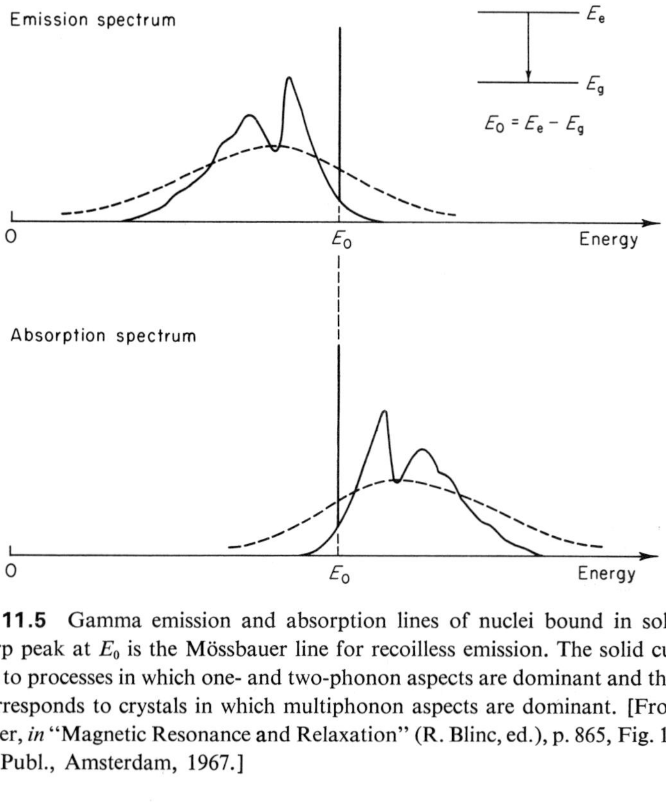


**Fig. 11.5** Gamma emission and absorption lines of nuclei bound in solids. The very sharp peak at $E_0$ is the Mössbauer line for recoilless emission. The solid curve corresponds to processes in which one- and two-phonon aspects are dominant and the dashed curve corresponds to crystals in which multiphonon aspects are dominant. [From R. L. Mössbauer, *in* "Magnetic Resonance and Relaxation" (R. Blinc, ed.), p. 865, Fig. 1; North-Holland Publ., Amsterdam, 1967.]

state, or $J = 6$ state, of $Tm^{3+}$ into 13 nondegenerate states. The separation of the two lowest electronic states is small and an analysis of the effect of crystal splitting must include the combined nuclear and electronic interactions with the crystal. The energy level structure is more complex and four rather than two twofold degenerate energy levels occur in the nuclear excited state. Four transitions are allowed and these appear as two lines on Fig. 11.6. The line intensity and position depend on the thermal distribution among the energy levels and this is apparent in the figure.

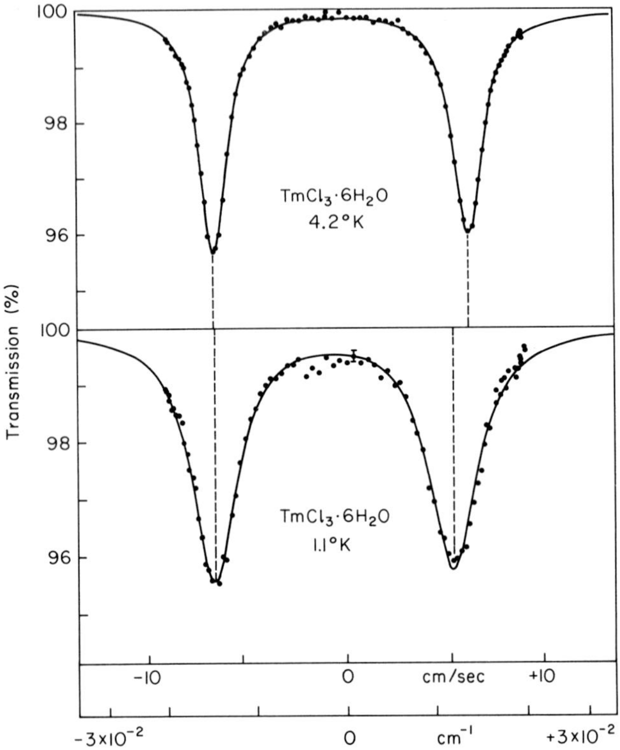


**Fig. 11.6** Absorption of the 8.4-keV $\gamma$ line by $^{169}$Tm in a crystal of $TmCl_3 \cdot 6H_2O$ at low temperatures. The figure shows a pseudoquadrupole shift which is due to the difference in population of the two participating electronic levels. [From M. J. Clauser, E. Kankeleit, and R. L. Mössbauer, *Phys. Rev. Lett.* **17**, 5 (1966), Fig. 2.]

## 11.6 Damping and the Golden Rule for Transition Rates

Equation (11.29b) is quite general and if a particular state $\beta$ is prepared at $t = 0$, or $c_\beta = 1$, the probability of a transition to state $\alpha$ is given by

$$P_{\beta\alpha} = \frac{1}{\hbar^2} | (\alpha | U | \beta) |^2 \left| \int_0^t dt\, e^{iyt} \right|^2$$

$$= \frac{1}{\hbar^2} | (\alpha | U | \beta) |^2 \left[ \frac{2(1 - \cos yt)}{y^2} \right] \tag{11.63}$$

where $y = (E_\alpha - E_\beta)/\hbar$. The Hamiltonian of the complete problem is independent of time and the interaction between the particle and the field

$U$ is independent of time and can be taken outside the integral. As noted earlier in this chapter, the term inside the square brackets grows as $t^2$ for small $yt$ and is a rapidly oscillating function as $t$ grows. This problem is different from the discussion of Eq. (11.4) in that the field has a large number of states with almost the same energy. The number of states in $L^3$ in a plane-wave expansion is $L^3k^2\,d\Omega\,dk/(2\pi)^3$ and for a known dependence $E(k)$, can be determined and is known as the density of states $\varrho(E)$,

$$N(E+dE)-N(E)=\varrho(E)\,dE \tag{11.64}$$

For photons, $E=h\nu$ and $k^2c^2=\omega^2$, and for free particles, $E=\hbar^2k^2/2M$. Exercise 11.7 discusses $N(E)$ for photons and particles. As $L^3\to\infty$, the density of states becomes large and it is no longer possible to distinguish experimentally between the states $\alpha$; that is, there is a limitation on the solid angle $d\Omega$ and energy $dE$ that can be measured by the physical size and response time of the detecting apparatus. Even if the system changes from a definite state $\beta$ to a definite state $\alpha$, this is not an experimentally possible measurement. Experimental measurements consider all possible states in the energy interval $dE$ and an average over these states is needed. Either the viewpoint that an ensemble of similar systems is examined or the viewpoint that the probability for all possible states of a single system in energy range $dE$ must be averaged can be used. In either case, the average

$$2\sum_{\Delta E_\alpha}\frac{1-\cos(E_\alpha-E_\beta)t/\hbar}{(E_\alpha-E_\beta)^2/\hbar^2}=\frac{1}{\hbar}\sum_{\Delta y}\frac{2(1-\cos yt)}{y^2} \tag{11.65a}$$

is of interest. If the argument of the cosine term is larger than 1 the cosine oscillates as the terms are summed and has an average value of zero. The sum is independent of time and grows as $1/y^2$ and gives rise to a negligible contribution. For small arguments of the cosine, Eq. (11.65a) has a peak of $t^2$ at $y=0$, or $E_\alpha=E_\beta$, and a width at half-maximum of $\Delta y=2\pi/t$. The approximate area under the curve is $\Delta y t^2$, or $2\pi t$, and this area grows linearly in time. Since the contribution outside the range $\Delta y=2\pi/t$ is negligible, the sum over $\Delta y$ is the same as integration between $\pm\infty$ and the result is in accord with the integral form given by Eq. (10.60). The transition probability becomes

$$[P_{\beta\alpha}]=\sum_{\Delta E_\alpha}P_{\beta\alpha}\approx\frac{1}{\hbar^2}\int dE_\alpha\,|(\alpha\,|\,U\,|\,\beta)|^2\,\varrho(E_\alpha)\left[\frac{2(1-\cos yt)}{y^2}\right] \tag{11.65b}$$

With the index $\alpha$ denoting all states with energy near $E_\alpha$, then

$$[P_{\beta\alpha}]=\left[\left(\frac{2\pi}{\hbar}\right)|(\alpha\,|\,U\,|\,\beta)|^2\,\varrho(E_\alpha)\right]t \tag{11.66a}$$

and the transition probability grows linearly in time. The important formula, called the "golden rule" in quantum mechanics, is the time-proportional transition rate $d[P_{\beta\alpha}]/dt$, or

$$[W_{\beta\alpha}] = \left(\frac{2\pi}{\hbar}\right) |(\alpha|U|\beta)|^2 \varrho(E_\alpha) \tag{11.67a}$$

The development here demonstrates the type of averaging which is necessary to obtain a transition probability proportional to time, or a constant transition rate. An alternate argument is that only aspects for large $t$ are measurable, and

$$\lim_{t\to\infty} \frac{2(1 - \cos yt)}{y^2} \to 2\pi t\, \delta(y)$$

so that

$$P_{\beta\alpha} \approx \left[\left(\frac{2\pi}{\hbar}\right) |(\alpha|U|\beta)|^2 \delta(E_\alpha - E_\beta)\right] t \tag{11.66b}$$

This yields the second form of the "golden rule":

$$W_{\beta\alpha} = \left(\frac{2\pi}{\hbar}\right) |(\alpha|U|\beta)|^2 \delta(E_\alpha - E_\beta) \tag{11.67b}$$

and the $\delta$ function implies a *sum over the final states.*

In forming the average, all states $\alpha$ within the range $2\pi/t = \Delta y = \Delta E_\alpha/\hbar$ contribute. This can be written as $\Delta E_\alpha t \approx h$, and, since the state $\beta$ was prepared at time $t = 0$ and the measurement made at time $t$, it is the energy statement of the Heisenberg uncertainty principle, $\Delta E\, \Delta t \approx h$. This implies that those transitions that conserve energy to within the range $\Delta E_\alpha = h/t$ contribute almost all of the transition rate. Here again, the linear increase with time of the probability gives rise to the constant transition rates which are so characteristic of the relaxation or damping observed in physical systems. Properly interpreted, averaging gives rise to this linear increase.

Once the limiting process that leads to Eq. (11.67b) is completed and a transition rate is defined, an alternate form is very useful. Let the states of $H_a$ be given by the labels $a$ and $b$ and the states of $H_r$ by $r$ and $r'$. Then, with

$$E_\beta = E_b + E_r, \qquad E_\alpha = E_a + E_{r'}$$

the $\delta$ function can be written as

$$\delta(E_\alpha - E_\beta) = \delta(h\nu_{r'r} - h\nu_{ba}) = h^{-1} \int_{-\infty}^{\infty} d\tau \exp[i(\omega_{ba} - \omega_{r'r})\tau] \tag{11.68a}$$

where

$$\delta(cx) = c^{-1}\,\delta(x) \tag{11.68b}$$

is used. It is convenient to introduce the partial interaction representation and write $U(t)$ in the manner used in Eq. (11.35). Then, the transition rate is given as

$$W_{\beta\alpha} = \hbar^{-2} \int_0^\infty d\tau\, \langle e^{-i\omega\tau}(\alpha \mid U(0) \mid \beta)(\beta \mid U(\tau) \mid \alpha)\rangle$$

$$+ \text{ complex conjugate} \tag{11.69}$$

where the angular brackets imply a sum over the final states $\alpha$ and a statistical average over the initial states $\beta$. When $H_a$ is used for the interaction representation for $U(t)$, $\omega = \omega_{r'r}$, and when $H_r$ is used for the interaction representation, $\omega = -\omega_{ba}$. In general, $H_a$ and $H_r$ can refer to any interacting systems with a time-independent interaction $U$. The time variable becomes apparent in the partial interaction representation. An ensemble average will often be used and the quantity within the brackets becomes a correlation function.

Exercise 11.7 Determine the density of states for radiation and for free particles in volume $L^3$.

$$\text{Ans.}\quad \varrho(E) = \frac{L^3 4\pi (h\nu)^2}{h^3 c^3} \tag{11.70a}$$

$$\varrho(E) = L^3 4\pi \frac{2^{1/2} m^{3/2} E^{1/2}}{h^3} \tag{11.70b}$$

Then show that Eq. (11.67a) yields the spontaneous transition rate given by Eq. (11.37b).

### *11.6.1 Damping and Statistical Mechanics*

In the earlier part of this chapter, a Hamiltonian describing a particle perturbed by a semiclassical interaction $U(t)$ was considered. It was further assumed that the matrix element $(b \mid U(t) \mid a) = (b \mid U \mid a)f(t)$ and then it was shown that the transition rate was proportional to the correlation function $\langle f(t)f(t+\tau)\rangle$. Such a system always relaxes to an equal population for all states and corresponds to equilibrium at infinite temperature. This fault was partially corrected in Eq. (11.18b) by forcing the system to relax toward the states described by the diagonal density matrix $\varrho_0$.

This criticism is corrected by considering a particle interacting with a quantized field. If the field and particle can be regarded as independent in the sense suggested by Eq. (11.29a) for the Hamiltonian, and if it makes sense to discuss the change in the atom from state $b$ to $a$ with the addition of a photon to the radiation field, then it immediately follows from Eq. (11.39) that

$$\frac{W(b \to a; k)}{W(a \to b; k)} = \frac{n_k + 1_k}{n_k} \tag{11.71a}$$

Here $n_k$ corresponds to the stimulated term and in the absence of spontaneous emission, would yield the same equal population as the semiclassical perturbation $f(t)$. Equation (11.39) is valid for an atom or molecule in any radiation field and is therefore valid in a thermal radiation field. Thermal equilibrium within an ensemble of similar particles requires the detailed balance argument that

$$N_b W(b \to a; k) = N_a W(a \to b; k)$$

and for thermal equilibrium,

$$e^{-\beta(E_b - E_a)} = \frac{N_b}{N_a} = \frac{W(a \to b; k)}{W(a \to b; k)} = \frac{n_k + 1}{n_k} \tag{11.71b}$$

This can be combined with Eq. (11.63) to yield

$$\bar{n}_k = \frac{1}{e^{\beta h\nu} - 1}$$

for the number of thermal quanta in a mode and is the usual Bose–Einstein formula for the number of quanta in the radiation field. This equation should be compared with Eq. (3.87b) for the ratio of the transition rates in a system with a single constraint.

Consider as an example the interaction of a particle with a field with an interaction of the form

$$U \to U_a{}^+ F + U_a F^+ \tag{11.72a}$$

with matrix elements

$$(\alpha \mid U \mid \beta) = (a \mid U_a \mid b)(r' \mid F^+ \mid r) \tag{11.72b}$$

$F$ is an annihilation operator and $F^+$ a creation operator and since $| r')$ is assumed to contain one more quantum than $| r)$, the matrix element $(r \mid F \mid r') = 0$. The $U_a$ is an operator describing the particle. A scalar interaction rather than a vector interaction like $-\mathbf{P}\cdot\mathbf{E}$ is used to simplify the discussion. It is further assumed that the initial state of the field is not precisely known and an average over the initial state $|\,r)$ is needed as well as a sum over the final state $|\,r')$. The transition rate with the sum and average is, from Eq. (11.69),

$$W_{ba} = 2\hbar^{-2}\,|\,(a\,|\,U_a\,|\,b)\,|^2 \int_0^\infty d\tau\; \psi_{ba}(\tau)\cos 2\pi\nu_{ba}\tau \qquad (11.73a)$$

where the correlation function is given by

$$\psi_{ba}(\tau) = \mathrm{tr}[\varrho_0 F(t)F^+(t-\tau)] + \text{complex conjugate} \qquad (11.73b)$$

The transition that corresponds to the absorption of one quantum is given by Eq. (11.73a) with $a$ and $b$ interchanged and with the correlation function

$$\psi_{ab}(\tau) = \mathrm{tr}[\varrho_0 F^+(t-\tau)F(t)] + \text{complex conjugate} \qquad (11.73c)$$

The reduction of Eq. (11.69) to the form just given will be discussed in Exercise 11.8. The correlation functions for emission and absorption of one quantum are no longer equal:

$$\psi_{ab}(\tau) \neq \psi_{ba}(\tau)$$

and the difference of these two correlation functions is expressed quite simply as

$$\psi_{ba}(\tau) - \psi_{ab}(\tau) = \mathrm{tr}\,\varrho_0[F(t)F^+(t-\tau) - F^+(t-\tau)F(t)] \neq 0 \qquad (11.74)$$

The difference $W_{ba} - W_{ab}$ is directly dependent on this difference. For operators $F$ and $F^+$ that do not commute, $FF^+ - F^+F \neq 0$, this expression is not equal to zero at $\tau = 0$. For large $\tau$, our earlier discussion in Chapters X and XI is applicable and the difference tends toward zero.

Since the correlation function $\psi_{ba}(\tau)$ and the spectral density $\psi_{ba}(\nu)$ are related by Eq. (10.33a), Eq. (11.73a) for the transition rate can be written in terms of the spectral density as

$$W_{ba} = \hbar^{-2}\,|\,(a\,|\,U_a\,|\,b)\,|^2\psi_{ba}(\nu_{ba}) \qquad (11.75)$$

The transition rate for absorption of one quantum has $a$ and $b$ interchanged and the spectral density for absorption $\psi_{ab}(\nu_{ba})$ follows from the transform of the correlation function, which is given by Eq. (11.73c). The spectral densities for absorption and emission are unequal,

$$\psi_{ba}(\nu_{ba}) \neq \psi_{ab}(\nu_{ba})$$

These expressions are now examined for a thermal field for which $\varrho_0$ is given by the canonical ensemble

$$\varrho_0 = e^{\beta(A-H_r)} \tag{11.76a}$$

$F$ is expanded in a set of modes $\eta_k$,

$$F = \sum_k a_k\eta_k \qquad \text{and} \qquad F^+ = \sum_k a_k^+\eta_k^* \tag{11.76b}$$

where typical modes are the plane-wave modes $L^{-3/2} \exp i\mathbf{k} \cdot \mathbf{r}$. The product $\eta_k\eta_k^*$ will occur throughout and it is replaced by $L^{-3}c_k$, where $c_k$ depends on the definition of $F$. Its value for the electric field operator is given by Eq. (11.36). The trace is reduced to a sum over the modes $k$ and the number of quanta $n_k$ in each mode. The Hamiltonian for the field is also expanded in modes, and

$$H_r = \sum_k H_k = \sum_k a_k^+a_kh\nu_k = \sum_k n_kh\nu_k \tag{11.76c}$$

The trace now reduces to

$$L^{-3}\sum_k c_k \sum_{n_k} (n_k \mid \varrho_0a_k(t)a_k^+(t-\tau) \mid n_k) \tag{11.76d}$$

and the evaluation of this expression will be discussed in Exercise 11.9. Direct substitution into Eq. (11.73c) yields a correlation function for emission of

$$\psi_{ba}(\tau) = L^{-3}\sum_k c_k(\bar{n}_k + 1) \cos 2\pi\nu_k\tau \tag{11.77a}$$

and this expression without the 1 is the correlation function for absorption $\psi_{ab}(\tau)$. The difference in correlation functions is

$$\psi_{ba}(\tau) - \psi_{ab}(\tau) = L^{-3}\sum_k c_k \cos 2\pi\nu_k\tau \tag{11.77b}$$

and is the expected result for spontaneous emission, since $aa^+ - a^+a = 1$ for each index $k$. Direct substitution into Eq. (11.73a) yields the spontaneous transition rate

$$\gamma = \langle[W_{ba}]\rangle - \langle[W_{ab}]\rangle$$

The spectral density for thermal electromagnetic radiation for each state of polarization is given by

$$\psi_{ab} = \left[\frac{1}{2}\left(\frac{h\nu}{2\varepsilon_0}\right)\bar{n}_\nu \frac{\nu^2}{c^3}\right] \tag{11.77c}$$

and can be compared with the energy density of thermal radiation of $\bar{n}h\nu^3/c^3$ given by Eq. (2.18). The appearance of $\varepsilon_0$ is necessary so that $\varepsilon_0 E^2$ has the dimensions of energy in the mks system of units.

It is often stated that the zero-point energy of $\frac{1}{2}h\nu$ per mode can be used to stimulate spontaneous emission. The zero-point energy can be used if it is used in the sense which is implied by Eq. (11.77c). The spectral density is greater for down transitions than for up transitions, but the density of states is the same for both up and down transitions.

If $c_k$ is appropriately chosen and $c$ is replaced by the velocity of sound, the discussion applies to phonons as quanta. The same arguments apply as long as the interaction is described by a quantized field. This is apparent in solid-state physics and is used to describe processes in which an excitation interacts with a field that has quanta such as plasmons or magnons.

Exercise 11.8 Show that Eq. (11.69) with an average over the initial state $|\, r)$ and a sum over the final state $|\, r')$ reduces for the $F$ dependence to

$$\sum_{rr'} P(r)(r \,|\, F \,|\, r')(r' \,|\, F^+ \,|\, r) \cos 2\pi\nu_{r'r}\tau$$
$$= \tfrac{1}{2}\sum_{rr'} P(r)(r \,|\, F(t) \,|\, r')(r' \,|\, F^+(t-\tau) \,|\, r) + \text{complex conjugate} \qquad (11.78\text{a})$$

where the interaction representation

$$F(t) = e^{iH_r t/\hbar} F e^{-iH_r t/\hbar} \qquad (11.78\text{b})$$

is used to replace the cosine term. $P(r)$ is the probability of the initial state $|\, r)$ for the field, and $P(r) = (r \,|\, \varrho_0 \,|\, r)$ in this diagonal representation. Then, find $\psi_{ba}(\tau)$.

Exercise 11.9 (a) Show that the matrix element in Eq. (11.76d) simplifies since the $a_k$ operators with different values of the index $k$ commute and the sum over $n_k$ for a particular value of $k$ yields

$$(\bar{n}_k + 1)e^{i\omega_k\tau} \qquad (11.79\text{a})$$

Then, show that the appropriate term for absorption has the 1 omitted.

(b) Show that the ratio or the spectral densities for emission and absorption of thermal quanta is given by

$$\frac{\psi_{ba}(\nu_{ba})}{\psi_{ab}(\nu_{ba})} = \exp\frac{h\nu_{bu}}{kT} = \frac{\bar{n}+1}{\bar{n}} \qquad (11.79\text{b})$$

## 11.7 Theory for Broadening of Spectral Lines in Gases

This section considers the broadening of spectral lines by random perturbations. These perturbing forces may be collisions in gases, the fluctuating field created by other magnetic dipoles upon a given dipole, and so on. Most of the development in this section is based on the joint characteristic function which was developed in Sections 10.13 and 10.14.

### *11.7.1 Elementary Theory*

A very elementary theory of impact broadening assumes that the transition rate for a stimulated effect follows from Eq. (11.4), and with a coherent perturbation of the form

$$U(t) = Ue^{-i\omega t} + \text{hermitian conjugate}.$$

the transition rate is

$$W_{ab} = \hbar^{-2} \,|\, (a \,|\, U \,|\, b) \,|^2 \lim_{t\to\infty} t^{-1} \left| \int_0^t dt' \exp i(\omega - \omega_{ab})t' \right|^2 \tag{11.80}$$

It is further assumed that the atoms or molecules undergo random collisions and each collision at time $t_m$ shifts the phase by $\varphi_m$. With this assumption, the integral from 0 to $t$ is

$$\begin{aligned}\int_0^t dt' \exp i(\omega - \omega_{ab})t' &\to \sum_m (\exp i\varphi_m) \int_{t_m}^{t_{m+1}} dt' \exp i(\omega - \omega_{ab})t' \\ &= \sum_m (\exp i\varphi_m) A_m\end{aligned}$$

In the Lorentz impact theory, $\varphi_m$ is considered a random variable which is large compared to 1 rad. The double sum which occurs in forming the absolute square reduces to a single sum:

$$\begin{aligned}\sum_{m,m'} [\exp i(\varphi_m - \varphi_{m'})] A_m A_{m'}^* &\approx \sum_m A_m A_m{}^* \\ &\approx \sum_m \frac{2[1 - \cos(\omega - \omega_{ab})\tau_m]}{(\omega - \omega_{ab})^2}\end{aligned}$$

since the angles $\varphi_m$ have random values. Here, $\tau_m = t_{m+1} - t_m$ is the time between collisions. The order in which the collisions occur is no longer important and the $\tau_m$ are random variables. It should also be noted that the transition rate is the sum of the rates for each interval. For random collisions with an average time between collisions of $T$, the probability that a collision occurs at $\tau = 0$ and the next collision occurs at $\tau$ is proportional to $T^{-1}e^{-\tau/T}$. The expected number of collisions in interval $t$ is $t/T$, and

$$\sum_m \to \frac{t}{T}\int_0^\infty d\tau \frac{1}{T} e^{-\tau/T} \frac{1-\cos \Delta\tau}{\Delta^2} \tag{11.81}$$

Direct integration yields the probability of stimulated emission or absorption and a transition rate of

$$W_{ab}(\nu) = \frac{1}{\hbar^2} |\,(a\,|\,U\,|\,b)\,|^2 \frac{2/T}{(\omega - \omega_{ab})^2 + (1/T)^2} \tag{11.82}$$

and has a Lorentzian line shape for the spectral density. This is a typical spectral distribution for an atom or molecule which undergoes a strong collision, or a collision which completely disrupts the phase between the coherent signal and the responding atom or molecule. Collisions which contribute to the viscosity of a gas are usually regarded as strong collisions. Equation (4.73) can be used to show that the average time between collisions is of the order of

$$\frac{1}{T} = n\langle V\sigma\rangle \tag{11.83a}$$

or $nV\sigma$ for hard spheres with a cross section $\sigma$. This time interval is pressure-sensitive and the line width between half-maxima,

$$(\Delta\nu)_{1/2} = \frac{1}{\pi T} \propto P \tag{11.83b}$$

is linear in pressure.

This simple theory can be used if the collisions are weak and random. The interval $T$ can be approximated by considering the interval during which the average square phase shifts by 1 rad, or the time interval in which the field and the atom become uncorrelated. If $\delta$ is the average random phase shift per collision and $\lambda$ collisions occur per second, then

$$1 \text{ rad} = \langle\varphi^2\rangle^{1/2} = \delta(\lambda T)^{1/2} \tag{11.84a}$$

and then the line width follows from

$$\frac{1}{T} = \lambda\delta^2 \tag{11.84b}$$

and gives the characteristic breadth of the spectral line.

### 11.7.2 Correlation Function and Line Shapes

In Section 11.4, the spontaneous emission of a photon by an atom or molecule was considered and resulted in Eq. (11.35) for the probability of the emission of a photon of frequency $\nu$. The general matrix element is given by Eq. (11.37a) and the probability that a molecule emits a photon in frequency interval $d\nu$, direction $d\Omega$, and polarization $\hat{u}$ is given by substituting directly into Eq. (11.35) and summing over interval $\Delta k$, or by multiplying by $L^3\nu^2\, d\Omega\, d\nu/c^3$. Since the initial quantum number $m_b$ is not known and the final-state quantum number $m_a$ is not measured, an average over $m_b$ and a sum over $m_a$ are needed. This is equivalent to ignoring the $m$ dependence and averaging over a random orientation of the axis of quantization. Only products like $P_xP_x$ will survive the average, and $P_xP_x = P_yP_y = P_zP_z$. With these assumptions, the emission is spherically symmetric and unpolarized. The probability of the molecule changing from state $b$ to $a$ or $j_b$ to $j_a$ and emitting a photon of frequency $\nu$ is

$$P_{ba}(\nu) = A \int_0^\infty d\tau\, e^{(-\gamma\tau - i\omega\tau)/2} (a \mid P_z(0) \mid b)(b \mid P_z(\tau) \mid a) + \text{complex conjugate} \tag{11.85}$$

Since the integral of $P_{ba}$ over $d\nu$ yields $\delta(\tau)$ and the integral must be unity, we have $A^{-1} = |\,(a \mid P_z \mid b)\,|^2$.

The $P_z(\tau)$ is in the interaction representation. If the atom undergoes collisions or perturbations for which the transition to other states can be ignored, the most that can happen during a collision is a shift in phase:

$$\mid b) \rightarrow e^{i\varphi} \mid b)$$

In order to examine the effect of a nonconservative interaction on the evolution in time of the states $\mid b)$ and $\mid a)$, Eqs. (11.1)–(11.3) are reconsidered. Let the interaction due to the collision be $U_c(t)$ with nonzero matrix elements $(a \mid U_c \mid a) \neq 0$, $(b \mid U_c \mid b) \neq 0$, and $(a \mid U_c \mid b) \neq 0$. Only two states are used and it is assumed that the molecule is in state $\mid b)$ initially. Let

$$c_b = e^{-i\varphi_b}$$

$c_a$ can be found by the integration of Eq. (11.3b). Direct substitution of $c_a$ into the differential equation for $c_b$ and a second integration yields

$$-i\varphi_b \approx (i\hbar)^{-1} \int_0^t dt' (b \mid U_c(t') \mid b) - \tfrac{1}{2}\Gamma t \tag{11.86}$$

Although the integral equation can be solved for the complex part of $\varphi_b$, it is easier to observe that the solution must be consistent with

$$-\frac{d\,|\,c_b\,|^2}{dt} = \frac{d\,|\,c_a\,|^2}{dt} = W_{ba} = \Gamma$$

where $W_{ba}$ is the transition induced by collisions and can be obtained from Eq. (11.5). Then the matrix element reduces to

$$(b\,|\,P_z(\tau)\,|\,a) = (b\,|\,P_z\,|\,a)(\exp i\omega_{ba}\tau)\exp -i(\varphi_b{}^* + \varphi_a)$$

Within the framework of the previous considerations, the probability of a system with one excited molecule emitting a photon of frequency $\nu$ is given by

$$P_{ba}(\nu) = \int_0^\infty d\tau\, e^{-[(\gamma/2)+\Gamma]\tau} e^{i(\omega_{ba}-\omega)\tau} \langle e^{i[\eta(t)-\eta(t-\tau)]}\rangle$$
$$+ \text{ complex conjugate} \qquad (11.87a)$$

The accumulated phase function $\eta(t)$ is defined as

$$\eta(t+\tau) - \eta(t) = (i\hbar)^{-1} \int_t^{t+\tau} dt'\, [(b\,|\,U_c(t')\,|\,b) - (a\,|\,U_c(t')\,|\,a)] \qquad (11.87b)$$

and $\Gamma$ is the nonradiative transition rate. In the ensemble average, $\eta(t)$ is independent of the time origin $t$. Equation (11.87a) will be used for the discussion of line shapes for molecules emitting from gases and for other line-shape effects for simple systems.

Anderson[3] introduced an equation of the form of Eq. (11.85) into the theory of pressure broadening. Kubo and Tomita [*15b*] introduced a similar expression into the study of magnetic resonance by using the magnetization $M(t)$ rather than the polarization $P(t)$. These equations have intuitive appeal since the emission of radiation appears to depend on the correlation function $\langle P(t)P(t+\tau)\rangle$. The system radiates until the correlation between the polarization at two different times is zero. The magnitude of the polarization is fairly well established by the matrix element, but the phase of the polarization is subject to changes due to collisions or other perturbations.

[3] See such references as Baranger [*15*], Anderson [*15a*], and Kubo and Tomita [*15b*].

### 11.7.3 Random Pulses and Line Shapes

Let the accumulated phase function be due to a series of random pulses,

$$\eta(t) = \sum_k \varphi(t - t_k)$$

The characteristic function of the joint distribution of $\eta(t)$ and $\eta(t + \tau)$ is given by Eq. (10.136) with $\varphi(\xi) = b(\xi)$ and a pulse rate of $\lambda$ per second. The joint distribution is valid for any choice of $a_1$ and $a_2$, and $a_1 = 1$ and $a_2 = -1$ are selected. The correlation function that describes this aspect of collisions or perturbations is

$$\begin{aligned} g(\tau) &= \langle \exp[i\eta(t) - i\eta(t - \tau)] \rangle \\ &= \exp\left( \lambda \int_{-\infty}^{\infty} d\xi \{\exp[i\varphi(\xi) - i\varphi(\xi - \tau)] - 1\} \right) \end{aligned} \tag{11.88a}$$

It should be noted that

$$| g(\tau) | \leq 1$$

and any divergences which may occur are introduced by the method of calculation.

In general, the pulses will not all have the same accumulated phase. Since the pulses are independent, the characteristic function is the product of the characteristic functions for the various pulses and the product of exponentials can be written as a sum. The line shape for random pulses with an accumulated phase function $\Delta_m$ and with an arrival rate $\lambda_m$ per second is given by the more general relationship

$$g(\tau) = \exp \sum_m \lambda_m \int_{-\infty}^{\infty} d\xi \{\exp[i\varphi_m(\xi) - i\varphi_m(\xi - \tau)] - 1\} \tag{11.88b}$$

The character of a given pulse is described by $d\varphi(t)/dt$ and the accumulated phase is

$$\varphi(\xi) = \int_{-\infty}^{\xi} dt \frac{d\varphi}{dt}$$

The difference of the accumulated phases at two different times is

$$\varphi(\xi) - \varphi(\xi - \tau) = \int_{\xi-\tau}^{\xi} dt \frac{d\varphi}{dt} \tag{11.89}$$

Some simple examples which occur in diverse fields are now considered.

Two interesting accumulated phase functions are shown in Fig. 11.7. The first, which is shown in Fig. 11.7(a), has a sudden jump of $\Delta$ at $\xi = 0$ in the accumulated phase and corresponds to the interaction described by $d\varphi/dt = \Delta\delta(t)$ or a $\delta$ function at $t = 0$. Its integral is the step function. The

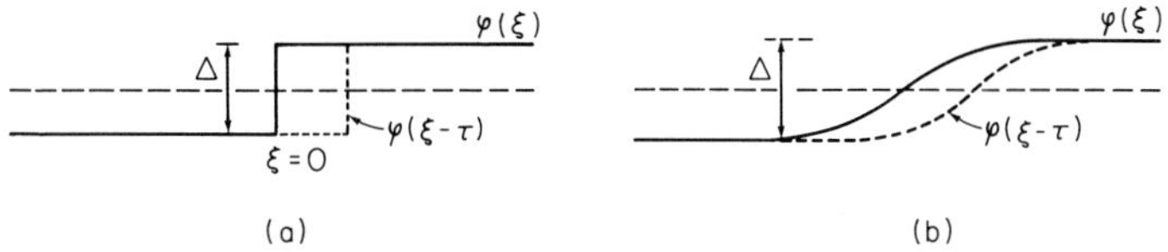


**Fig. 11.7** (a) Shift in phase of $\Delta$ during a sudden collision, or $\varphi = \Delta\delta(\xi)$. (b) Shift in phase of $\Delta$ during a collision described by $\varphi(\xi) = (\Delta/\pi)[(\tan^{-1}\alpha\xi) + (\pi/2)]$.

correlation function follows from Eq. (11.88a) by noting that the integrand is zero except for an interval of length $\tau$ and is

$$g(\tau) = \exp\{\lambda[(\exp i\Delta) - 1]\tau\} \tag{11.90a}$$

For random pulses of various accumulated phase shifts $\Delta_m$, the more general expression for the correlation function is

$$g(\tau) = \exp \sum_m \lambda_m[(\exp i\Delta_m) - 1]\tau \tag{11.90b}$$

Different approximations are now made. The *impact approximation* assumes that the phase shift $\Delta_m$ is large for almost all collisions, and then $\cos\Delta_m$ and $\sin\Delta_m$ become random variables between $\pm 1$. Their average values are zero and the correlation function reduces to

$$g(\tau) \to e^{-\lambda\tau} \tag{11.90c}$$

where $\lambda$ is the total number of pulses per second. This yields a Lorentzian line shape,

$$P_{ba}(\nu) = \frac{2A}{(\omega_{ba} - \omega + B)^2 + A^2} \tag{11.91}$$

where

$$A = (\tfrac{1}{2}\gamma + \Gamma) + \lambda \qquad \text{and} \qquad B = 0 \tag{11.90d}$$

This is in accord with the earlier development in which it was found that the width of the line increased linearly with the number of hard collisions per second. A hard collision is a collision for which the accumulated phase shift is greater than 1 rad. No line shift occurs in the impact approximation.

If all phase shifts $\Delta_m$ are small compared to 1 rad, then

$$\ln g(\tau) = \tau \sum \lambda_m (e^{i\Delta_m} - 1) \to \tau \sum \lambda_m (i\Delta_m - \tfrac{1}{2}\Delta_m{}^2 \cdots)$$
$$= \lambda(i\bar{\Delta} - \tfrac{1}{2}\langle\Delta^2\rangle)\tau \qquad (11.90e)$$

These weak interactions give rise to a Lorentzian line shape with both a line shift and line broadening,

$$A = (\tfrac{1}{2}\gamma + \Gamma) + \tfrac{1}{2}\lambda\langle\Delta^2\rangle \qquad \text{and} \qquad B = \lambda\langle\Delta\rangle$$

It should be noted that the number of weak collisions designated by $\lambda$ may far exceed the number of strong collisions denoted earlier by $\lambda$. The line broadening depends on the average square phase shift per collision and this can be a small quantity for very small phase shifts. The shift in the line depends on the average phase shift per collision and the shift can become appreciable.

Both hard and soft collisions occur and the line shift may be primarily due to soft collisions, while the line broadening is strongly influenced by the hard collisions. Thus, since the sum remains correct in the expansion, the collisions can be grouped as hard and soft, and then

$$\{\text{line shift}\} = \lambda_{\text{s}}\langle\Delta\rangle$$
$$\{\text{line-broadening increase}\} = \lambda_{\text{h}} + \tfrac{1}{2}\lambda_{\text{s}}\langle\Delta^2\rangle$$

This result is similar to Eq. (11.16b), for which the line width was given by $(1/T_2) = (1/T_1) + (1/T_2{}')$. Here, $T_1$ corresponds to the term $\frac{1}{2}\gamma + \Gamma$ and $T_2{}'$ to the increase in line broadening due to perturbations that perturb the radiating polarization.

The second function of interest is shown in Fig. 11.7(b):

$$\varphi(\xi) = \frac{\Delta}{\pi} \tan^{-1} \alpha\xi \qquad (11.92a)$$

The total accumulated phase shift as $\xi \to \infty$ is $\Delta$, but the phase shift can be either "fast" or "slow." The collision or perturbation is described by

$$\frac{d\varphi}{dt} = \frac{\alpha\Delta/\pi}{1 + \alpha^2 t^2}$$

Large $\alpha$ corresponds to a "fast" change near $t = 0$. The phase difference is

$$\varphi(\xi) - \varphi(\xi - \tau) = \frac{\Delta}{\pi} [\tan^{-1} \alpha\xi - \tan^{-1} \alpha(\xi - \tau)] \qquad (11.92b)$$

and is an even function about $\xi = \tau/2$. The two curves are separated by an interval of $\tau$. For large values of $\alpha\tau$, the phase difference is approximately $\pi$ for an interval for $\xi$ of length $\tau$ and the integral reduces to Eq. (11.90b) and the discussion given there is applicable.

For small $\alpha\tau$, the difference is approximately $\alpha\tau$ for an interval for $\xi$ of length $2/\alpha$ and the correlation function is

$$g(\tau) \approx \exp\left(\frac{2\lambda}{\alpha}\left\{\left[\exp\left(i\frac{\Delta\alpha}{\pi}\tau\right)\right] - 1\right\}\right) \tag{11.93a}$$

These two forms apply to different values of $\alpha$ for the same pulse shape. Since $\alpha$ and $\Delta$ may be different for each pulse, an average over these values may also be necessary.

If the collisions are sufficiently slow so that $2\lambda/\alpha$ becomes very large, then only small values of $\tau$ are of importance for the correlation function and the correlation function is given by the Gaussian

$$g(\tau) \approx \exp\left\{\frac{2\lambda}{\alpha}\left[i\frac{\alpha\Delta}{\pi}\tau - \frac{1}{2}\left(\frac{\alpha\Delta}{\pi}\right)^2\tau^2\right]\right\} \tag{11.93b}$$

Again, for $\lambda_m$ pulses with coefficients $\Delta_m$ and $\alpha_m$, the characteristic function is a product of characteristic functions and this product appears as a sum in the exponential and an average value can be used for the coefficients in Eq. (11.93b). The line shape is Gaussian:

$$g(\nu) = \left(\frac{\pi^3}{\lambda\alpha\Delta^2}\right)^{1/2} \exp\left\{-\frac{[\omega - \omega_{ba} - (2\lambda\Delta/\pi)]^2}{4\lambda\alpha\Delta^2/\pi^2}\right\} \tag{11.93c}$$

where the constants may be replaced by the average values $\langle\Delta\rangle$ and $\langle\alpha\Delta^2\rangle$. The line width is given by

$$2\pi\Delta\nu_{1/2} = \frac{4\Delta}{\pi}(\ln 2)^{1/2}(\lambda\alpha)^{1/2}$$

The line becomes Gaussian when the pulse rate $\lambda > 2\alpha$, which implies that line shapes become Gaussian when the important parts of the phase shifts of adjacent pulses overlap [*16*].

### *11.7.4 Spectral Line Broadening in Gases*

Physical interactions yielding the accumulated phase function $\varphi(\xi)$ have been avoided in the previous discussion. If it is assumed that in a gas the interaction energy causes the shift, then for a potential that varies as $r^{-n}$, the phase shift is

$$\hbar \frac{d\varphi}{dt} = \frac{C}{r^n} \tag{11.94a}$$

The phase is usually determined from a straight classical path, or

$$r^2 = b^2 + v^2 t^2 \tag{11.94b}$$

The impact parameter $b$ and the velocity $v$ become parameters in the theory. It is convenient to write the phase correlation as

$$\varphi(\xi) - \varphi(\xi - \tau) = \int_{\xi-\tau}^{\xi} dt \frac{d\varphi}{dt} = \Delta(\xi \tau b v) \tag{11.95}$$

and then replace the sum over $m$ in Eq. (11.88b) by the binary collision average which is given by Eq. (4.70) for the classical collision rate for binary encounters. Then, the correlation function for spectral line broadening in gases is given by

$$\ln g(\tau) = n\left(\frac{\mu}{2\pi kT}\right)^{3/2} \int_0^\infty 4\pi v^2\, dv \left[\exp\left(-\frac{\mu v^2}{2kT}\right)\right] \int_0^\infty v 2\pi b\, db \times \int_{-\infty}^{\infty} d\xi \{[\exp i\Delta(\xi b v \tau)] - 1\} \tag{11.96}$$

With the simple interaction energy given by Eq. (11.94a) and $n = 6$, the the phase shift for the van der Waals interaction is

$$\hbar\varphi(\xi) = C \int_{-\infty}^{\xi} \frac{dt}{(b^2 + v^2 t^2)^3} \approx \frac{3C}{8b^5 v} \left[\left(\tan^{-1} \frac{\xi v}{b}\right) + \frac{\pi}{2}\right] \tag{11.97a}$$

Only the arc tangent term changes sufficiently rapidly to be retained and the remaining terms in the expansion are omitted. The total phase shift during a collision as $r$ changes from $-\infty$ to $+\infty$ is

$$\varphi(+\infty) = \frac{\pi}{\hbar} \frac{3C}{8b^5 v} \tag{11.97b}$$

From Table 4.1, a typical value is $C = 10^{-77}$, and for thermal velocities, phase shifts of the order of $\varphi \approx \pi(10^4/b^5)$ radians occur, with $b$ in units of $10^{-10}$ m. These quite large and hard collisions cause line broadening which may be 10–100 times greater than expected from the viscosity cross section. Soft collisions occur for values of $b$ for which $\varphi < 1$. Since the phase shift falls off as $b^{-5}$, the cross section for soft collisions is not much larger than for hard collisions for the van der Waals type of interaction. The time duration of the collision is of the order of $b/v$, or of the order of $10^{-11}$–$10^{-12}$ sec for particles with thermal velocities. If the line width is much narrower, $\Delta\nu_{1/2} \ll 10^{11}$–$10^{12}$, then small values of $\tau \approx 10^{-11}$ are not of interest and $\varphi(\xi)$ can be replaced by $\varphi(+\infty)$, and then

$$\int_{-\infty}^{\infty} d\xi\ [e^{i\Delta(\xi b v \tau)} - 1] \rightarrow [e^{i\varphi(+\infty)} - 1]\tau$$

Direct substitution into Eq. (11.96) yields, for the integration over $b$,

$$\sigma(\nu) = 2\pi\left(\frac{3\pi C}{8\hbar v}\right)^{2/5} \int_0^{\infty} x\, dx \left[\left(\exp\frac{i}{x^5}\right) - 1\right] \qquad (11.98a)$$

With $y = x^{-5}$, the integral reduces to standard form, and with $\Gamma(n)$ as the gamma function, becomes

$$-\tfrac{1}{2}i^{-2/5}\Gamma(\tfrac{3}{5}) = -0.60 + i0.44 \qquad (11.98b)$$

The integral is readily estimated by observing that the sine and cosine are rapidly oscillating functions between 0 and 1 and the primary contribution is the integral of $-x\,dx$, and is approximately $-\frac{1}{2}$. The complex term $i \sin x^{-5}$ contributes between 1 and $\infty$ and its contribution is approximately $\frac{1}{3}i$. The first region corresponds to the region of hard collisions and provides the line broadening, and the second region corresponds to soft collisions and provides the line shift.

Without giving the details of the integration over the velocity, the correlation function can be written as

$$\ln g(\tau) = n\left[\bar{v}\left(\frac{3\pi C}{8\hbar\bar{v}}\right)^{2/5} 2^{2/5}\pi^{4/5}\Gamma\left(\frac{9}{5}\right)\Gamma\left(\frac{3}{5}\right)\left(-\cos\frac{\pi}{5} + i\sin\frac{\pi}{5}\right)\right]\tau$$

$$= -(n\bar{v}\sigma)\tau \qquad (11.99a)$$

$\bar{v} = (8kT/\pi\mu)^{1/2}$ is the average velocity for the colliding molecules. This equation is of the form $\lambda = n\bar{v}\sigma$, where $(3\pi C/8\hbar\bar{v})^{2/5}$ is the primary term determining an effective cross section. The cross section is complex and yields a line shift as well as line broadening. The line shape in the central part of the spectral line is given by

$$g(\nu) = \frac{2n\bar{v}\sigma_r}{(\omega - \omega_{ab} - n\bar{v}\sigma_i)^2 + (n\bar{v}\sigma_r)^2} \qquad (11.99b)$$

where $\sigma = \sigma_r + i\sigma_i$ follows from Eq. (11.99a). The line shift and line width are related for the van der Waals interaction by

$$\frac{\sigma_i}{\sigma_r} = -\tan\frac{\pi}{5}$$

and both are linear in the pressure. In the use of these equations $C$ is either positive or negative and its absolute value is used. The line shift has the sign of $C$ and is negative, or toward lower frequencies for attractive potentials and toward higher frequencies for repulsive potentials. The term $\frac{1}{2}\gamma + \Gamma$ is neglected here in the line shape.

Correlation times with $\tau < \Delta\nu_{1/2}^{-1}$ begin to examine the character of the collision itself. Thus the shape of the spectral line in the wings, or in the region in which $|\nu - \nu_{ba}| > \Delta\nu_{1/2}$, gives information regarding the character of the collision or the function $\varphi$. If only the wings of the line are considered, then only small values of $\tau$ are of interest, or values of $\tau$ for which

$$\tau < \Delta\nu_{1/2}^{-1}$$

In the region in which $\xi < b/v$, the difference in the arc tangents is approximately

$$\tan^{-1}\frac{\xi v}{b} - \tan^{-1}\frac{(\xi - \tau)v}{b} \approx \frac{v\tau}{b}$$

This difference tends to zero rapidly for $\xi > b/v$ and the integral can be taken between $\xi = \pm b/v$. With this approximation, the integral over $b$ and $\xi$ is

$$\begin{aligned} \sigma &= \int_0^\infty 2\pi b\, db \int_{-b/v}^{b/v} d\xi \left\{\exp\left[i\left(\frac{3C}{8\hbar b^6}\right)\tau\right] - 1\right\} \\ &= -\left(\frac{2\pi}{3}\right)\left(\frac{3C}{8\hbar}\right)\left[\frac{2\Gamma(\frac{1}{2})}{i^{1/2}}\left(\frac{\tau^{1/2}}{v}\right)\right] \end{aligned}$$

It is interesting to note that limiting the range on the $\xi$ introduces a $1/v$ dependence into the cross section and the form of the potential $1/r^n$ determines the exponent of $\tau$ as $\tau^{3/n}$. An average can now be taken over the velocity in Eq. (11.96) and yields $n$ times the above expression. This yields a line shape in the *wings of the line* of

$$g(\nu) = n\left(\frac{2\pi}{3}\right)\left(\frac{3}{8\hbar}\right)^{1/2} |\omega - \omega_{ab}|^{-3/2} \qquad \text{for} \quad |\omega - \omega_{ab}| \gg \Delta\omega_{1/2} \tag{11.100}$$

The falloff in the wings of the line as $|\omega - \omega_{ba}|^{-3/2}$ is one of the characteristics of the $r^{-6}$ dependence of the van der Waals interaction.

The cross section in Eq. (11.100) is proportional to $1/v$ and cancels the relative velocity in the binary collision average. For this reason, the line shape in the wings of the line is similar to that for atoms selected at random positions and is referred to as the *statistical* limit.

It is necessary to consider the difference between Eq. (11.100) and Eq. (11.93b). There is not a sufficiently large number of collisions that follow a $1/r^n$ law in a gas to warrant the type of expansion used for Eq. (11.93b). The range of $\tau$ in this example was limited by the requirement that the measurements be made in the wings of the line. A $\tau^2$ dependence is reached for a potential $1/r^{3/2}$, but the ensuing problem cannot be evaluated. Potentials of this type have a very long range, and the increase in the number of collisions with increasing $b$ is greater than the decrease in the phase function. The approximation method used for Eq. (11.93b) is needed for these long-range collisions.

Exercise 11.10 Show for values of $n > 2$ in Eq. (11.94a) that

$$\varphi(+\infty) = \frac{\pi}{\hbar}\left(\frac{C}{b^{n-1}v}\right)\Gamma(n-1)\frac{2^{2-n}}{[\Gamma(n/2)]^2} \qquad (11.101a)$$

for the accumulated phase shift. Show also that

$$\varphi(\xi) = \frac{1}{2}\varphi(\infty)\left[\left(\tan^{-1}\frac{\xi b}{v}\right) + \frac{\pi}{2}\right]$$

Exercise 11.11 Show that the falloff in the wings of the line is proportional to

$$\frac{1}{|\omega - \omega_{ab}|^{1+(3/n)}} \qquad (11.101b)$$

for a $1/r^n$ potential.

### 11.7.5 Some Experimental Results

Detailed measurements have been made [*17*] on the foreign-gas broadening and shift of the 0.7601-$\mu$m line of krypton. The experimental data are shown in Fig. 11.8 for various foreign gases. In accord with the earlier development in this chapter, the shift and broadening are linear in the number density or pressure. A linear dependence is expected in the central

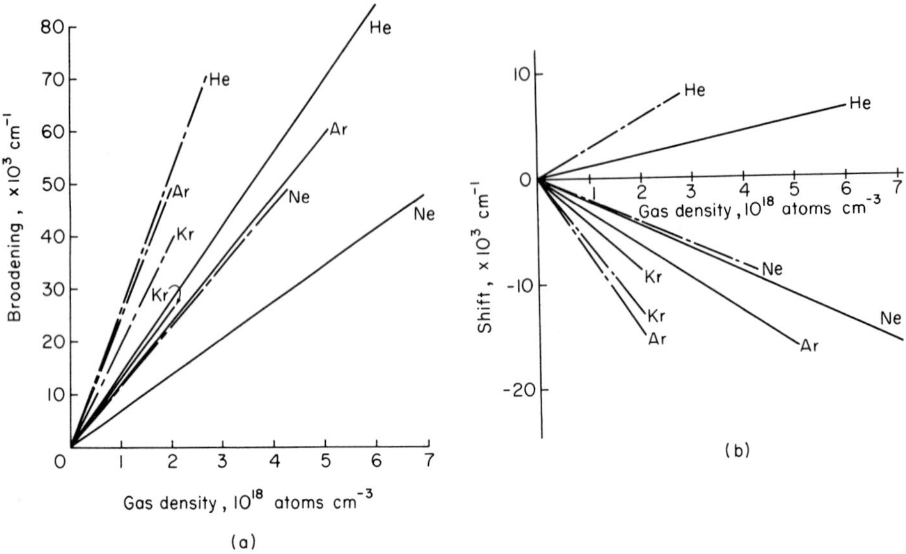


**Fig. 11.8** (a) Foreign-gas broadening of the krypton 0.7601-$\mu$m line at 80°K (solid curves) and 295°K (dashed curves) and various gas densities. Argon, neon, and helium are used as foreign gases. The broadening is given in units of $\text{cm}^{-1}$ and $10 \times 10^{-3}$ $\text{cm}^{-1}$ corresponds to a frequency width of $3 \times 10^8$ Hz. (b) Shift of the 0.7601-$\mu$m line of krypton by foreign gases at 80°K (solid curves) and 295°K (dashed curves). [From J. M. Vaughn and G. Smith, *Phys. Rev.* **166**, 17 (1969), Figs. 3 and 4.]

portion of the line for almost any model since it is a count of the number of binary collisions. In the same sense, the central part of the line has the Lorentzian line shape given by Eq. (11.91). The data can be interpreted in terms of the Lennard-Jones potential

$$\hbar \frac{d\varphi}{dt} = \frac{C_{12}}{r^{12}} - \frac{C_6}{r^6} \tag{11.102}$$

Then, from $\varphi(+\infty)$ for this potential, the equivalent of Eq. (11.94a) is developed for this accumulated phase function, and the ratio of the line shift to the line width is given in terms of

$$\left\{\frac{\text{line shift}}{\text{line width}}\right\} = \frac{B}{2A}$$

where $B$ and $A$ are the real and imaginary parts of the integral of the quantity $[\exp i(\alpha x^{-11} - x^{-5})] - 1$ in Eq. (11.96). In this model, the line shift can be either plus or minus and depends on the relative size of $\alpha$, which is proportional to $C_{12}/C_6^{11/5}$. Collisions of Kr with Ar and Ne have large van der Waals coefficients and the shift is negative, or toward the red. Krypton–helium collisions have a small van der Waals term, the hard-core repulsion is dominant, and the line shift is positive, or toward the violet. The quantitative interpretation of these data permits an evaluation of $C_6$ which is in good agreement with the data given in Table 4.1. Krypton–krypton collisions require some care since a resonance interaction can occur:

$$\mathrm{Kr} + \mathrm{Kr}^* \leftrightarrow \mathrm{Kr}^* + \mathrm{Kr}$$

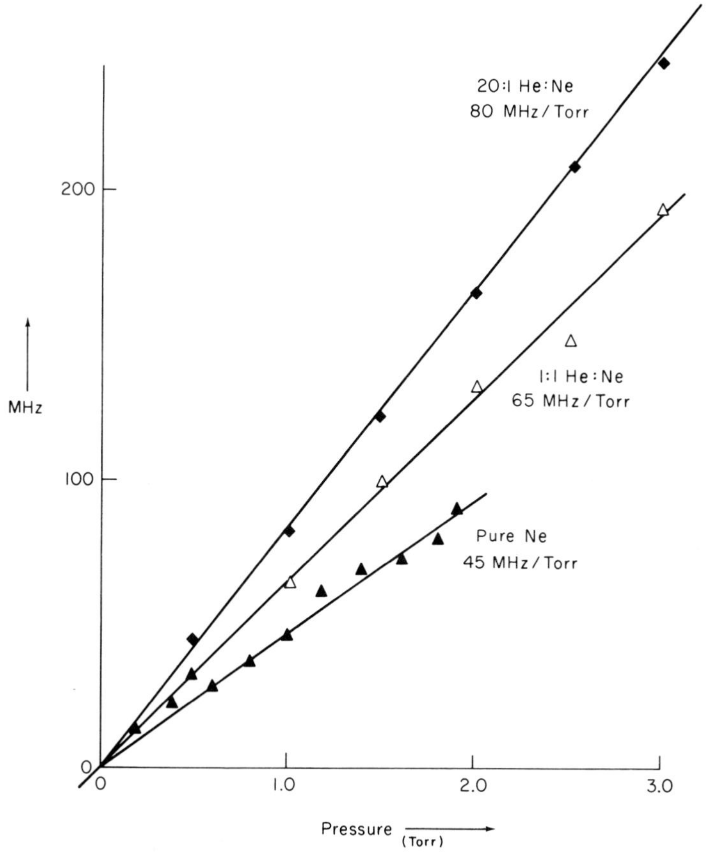


**Fig. 11.9** Pressure broadening increase in full Lorentz width of the 0.6328 μm red line in Neon, $2p^65s'[\frac{1}{2}]^0_1$ - - - $2p^53p'[1\frac{1}{2}]_2$. The zero pressure width is subtracted out and the curves give the pressure effect in pure Ne of 45 MHz/Torr, in a 1:1 mixture of He–Ne of 65 MHz/Torr, and in a 20:1 mixture of He–Ne of 80 MHz/Torr. (This figure was kindly furnished by J. W. Knutson Jr. and W. R. Bennett, Jr.; see also Knutson and Bennett [*18*].)

and the attractive potential is proportional to $1/r^3$. This is discussed in some detail by Vaughn and Smith [*17*].

Collision broadening has also been measured for the 0.6328-$\mu$m spectral line in the He–He laser [*18*]. Saturation of the atomic resonance and the occurrence of the Lamb [*19*] "dip" at the center of the Doppler line make this measurement possible. Experimental results are shown in Fig. 11.9.

The line shape for absorption or emission in gases and plasmas has a long history and it is not possible in a short discussion to give proper credit to the important contributions made by many workers. Review articles[4] summarize much of the prior experimental and theoretical research.

## 11.8 Scattering by Crystals, Liquids, and High-Pressure Gases

In Section 10.11, the scattering of electromagnetic waves by gases or liquids was treated by adding the spherically scattered waves at the detector. The signal was proportional to the square of the sum, and the single-center scattering and pair scattering were the quantities of interest. An approach which is very popular for the scattering of x-ray photons or neutrons by crystals is the use of the second form of the "golden rule," or in greater detail, Eq. (11.69). Inherent in the development of Eq. (11.69) was the assumption that the Hamiltonian described two systems that in some region of space interacted with one an other. It is now assumed that a particle is in volume $L^3$ and in almost plane-wave states with momentum $k$ and spin $s$. The scatterer is in state $b$. The transition rate to a state $k's'$ for the particle and a state $a$ for the crystal follows from Eq. (11.69). The states $\beta$ and $\alpha$ are summarized for photons and neutrons in Table 11.1.

**Table 11.1**

| | $\beta$ $(ksb)$ | $\alpha$ $(k's'a)$ |
|---|---|---|
| Photons | $E_\beta = E_r + E_b$ | $E_\alpha = E_{r'} + E_a$; $E_r - E_{r'} = \hbar\omega$ |
| Neutrons | $E_\beta = (\hbar^2 k^2/2M) + E_b$ | $E_\alpha = [\hbar^2 (k')^2/2M] + E_a$; $(\hbar^2/2M)[k^2 - (k')^2] = \hbar\omega$ |

[4] For review articles on line broadening, see Chen and Takeo and Breene [*20*]. Papers and reference to other literature are given by Smith *et al.* and Dillon *et al.* [*20a*].

A cross section can be related to the transition rate by

$$\sigma(ksb; k's'a) = \frac{W(ksb; k's'a)}{L^{-3}v} \tag{11.103}$$

where $L^{-3}v$ is the flux for one particle in $L^3$ in state $ks$. Scattering into an energy range $d\varepsilon$ and into solid angle $d\Omega$ is usually used and summing the transition rate over the final states $L^3(k')^2\, d\Omega\, dk'/(2\pi)^3$ permits a differential cross section $d^2\sigma/d\varepsilon\, d\Omega$ to be defined. For neutrons, the velocity $v = \hbar k/M$ and the energy $d\varepsilon$ follows from $\varepsilon = \hbar^2(k')^2/2M$. For photons, $v = c/\mu$, where $\mu$ is the index of refraction for frequency $\nu$ and $d\varepsilon$ follows from $\varepsilon = \hbar k'v$. Combining Eqs. (11.69) and (11.103) yields the expression for the differential cross section for scattering:

$$\frac{d^2\sigma}{d\varepsilon\, d\Omega} = B \int_0^\infty d\tau\, e^{-i\omega\tau} \left\langle \sum_a (ksb \mid U(0) \mid k's'a)(k's'a \mid U(\tau) \mid ksb) \right\rangle$$
$$+ \text{complex conjugate} \tag{11.104a}$$

where the brackets denote a statistical average over the initial states and a sum over the final states. The coefficient $B$ is given by

$$B = \begin{cases} \dfrac{4\pi^2 M^2}{h^5}\, \dfrac{k'}{k}, & \text{neutrons and electrons} \\[2ex] \dfrac{(k')^2}{h^3 v^2}, & \text{photons} \end{cases} \tag{11.104b}$$

Since each interaction term contains $L^{-3}$, the $L^6$ term in the numerator cancels, and $L = 1$ m is taken for convenience in the ensuing discussion.

### *11.8.1 Scattering Cross Section and Spectral Distribution*

The specific form of the interaction potential is required. The explicit form is not given here, but in the second quantization approach, it can be noted that the interaction potential must annihilate the particle in state $ks$ and create a particle in state $k's'$ and must contain terms like $a_{k'}^{\dagger}a_k$ in the potential energy of interaction $U$. A second approach is to regard the incoming wave as $\exp i\mathbf{k} \cdot \mathbf{r}$ and the outgoing wave as $\exp i\mathbf{k}' \cdot \mathbf{r}$ and therefore write $\mid k) = \exp i\mathbf{k} \cdot \mathbf{r}$. Both approaches are included by writing

$$(k'a \mid U(\tau) \mid kb) = \xi^3 U_0 \sum_n (a \mid \exp i\boldsymbol{\varkappa} \cdot \mathbf{r}_n(\tau) \mid b) \tag{11.105a}$$

where

$$\boldsymbol{\varkappa} = \mathbf{k} - \mathbf{k}' \tag{11.105b}$$

and $n$ is a sum over the scattering centers. The interaction occurs at the location of the scattering particle $\mathbf{r}_n(\tau)$ and the scattering substance is regarded as a sum of scattering centers

$$\xi^3 U_0 \, \delta(\mathbf{r} - \mathbf{r}_n)$$

where $\xi$ has the dimensions of length, so that the interaction has the dimension of energy. The spin index is omitted since its inclusion requires greater detail in the interaction potential than is given here. With this model, closure is possible over the final states with index $a$ for the scatterer, and

$$\frac{d^2\sigma}{d\varepsilon \, d\Omega} = (B\xi^6 U_0^{\,2}) S(\varkappa, \omega) \tag{11.106a}$$

where

$$S(\varkappa, \omega) = \int_0^\infty d\tau \, (\exp -i\omega\tau) \Big\langle \sum_{m,n} [\exp -i\boldsymbol{\varkappa} \cdot \mathbf{r}_m(0) \exp i\boldsymbol{\varkappa} \cdot \mathbf{r}_n(\tau)] \Big\rangle \tag{11.106b}$$

The coordinates $\mathbf{r}_m$ are operators and the angular brackets indicate a statistical average over the initial states [*21*].

If a grand canonical average is used and if $S(\boldsymbol{\varkappa}, \omega)$ is integrated over $\omega$, the resulting integral yields $\delta(\tau)$, or

$$S(\boldsymbol{\varkappa}) = \int \frac{d\omega}{2\pi} S(\varkappa, \omega) = \Big\langle \sum_{m,n} [\exp -i\boldsymbol{\varkappa} \cdot \mathbf{r}_m(0) \exp i\boldsymbol{\varkappa} \cdot \mathbf{r}_n(0)] \Big\rangle$$

$$= \{\text{see Eqs. (10.105b) and (10.105c)}\}$$

At the same instant of time, two coordinates commute and this expression reduces to Eqs. (10.105b), (10.105c), and (10.103b) for elastic scattering. The area under the scattering curve is the same as the strength of the elastic scattering for a given $\varkappa$. At $\varkappa = 0$, this integral is directly related to the compressibility and the pair correlation function $g(\mathbf{r})$.

The solution of the scalar wave equation in which the velocity of propagation is a function of position and time forms an illustrative and useful example of scattering. The scalar wave equation

$$\nabla^2\psi - \frac{1}{v^2} \frac{\partial^2\psi}{\partial t^2} = 0$$

can be rewritten as

$$\nabla^2\psi - \frac{1}{v_0^2}\frac{\partial^2\psi}{\partial t^2} = -\frac{\Delta\eta}{v_0^2}\frac{\partial^2\psi}{\partial t^2} \tag{11.107a}$$

where the spatial and time variation of the velocity is given by

$$\frac{\Delta\eta(\mathbf{r}, t)}{v_0^2} = \frac{1}{v_0^2} - \frac{1}{v^2}$$

It is now assumed that the incident wave and the solution of the homogeneous wave equation is

$$\psi_\mathrm{i} = \exp i(\mathbf{k}_a \cdot \mathbf{r} - \omega_a t)$$

Again, the first Born approximation is used and $\partial^2\psi/\partial t^2$ on the right-hand side of (11.107a) is replaced by $-\omega_a^2\psi_i$. With this approximation, the asymptotic solution at large $r$ for a particular Fourier component is

$$\psi_\mathrm{s}(\mathbf{r}, \omega) = -k_a^2\frac{\exp ik'r}{4\pi r}\int dt' \int d\mathbf{r}'\ \Delta\eta(\mathbf{r}', t')\exp i(\boldsymbol{\varkappa}\cdot\mathbf{r}' - \omega t') \tag{11.107b}$$

where

$$\boldsymbol{\varkappa} = \mathbf{k}_a - \mathbf{k}', \qquad \omega = \omega_a - \omega'$$

$\boldsymbol{\varkappa}$ is the change in the wave vector and $\omega$ is the change in frequency for this Fourier component. A detailed derivation is given in Exercise 11.12.

The spectral density of the scattered wave,

$$\sigma(\boldsymbol{\varkappa}, \omega) = \lim_{T\to\infty} T^{-1}\,|\,\psi_\mathrm{s}(r, \omega)\,|^2\, r^2$$

is dependent on the absolute square of the double integral. Only terms that *increase linearly in time and linearly in the volume of the sample are of interest.* This occurs if the ensemble average of the correlation function is a function only of the time difference $t' - t''$ and the coordinate difference $\mathbf{r} = \mathbf{r}' - \mathbf{r}''$,

$$\langle\Delta\eta(\mathbf{r}', t')\,\Delta\eta(\mathbf{r}'', t'')\rangle = \langle\Delta\eta(0, 0)\,\Delta\eta(\mathbf{r}, \tau)\rangle$$

Then, the cross section for scattering with a change in wave vector $\boldsymbol{\varkappa}$ and a change in frequency $\omega$ is given by

$$\sigma(\boldsymbol{\varkappa}, \omega) = \frac{\pi^2}{\lambda^4} V\int d\mathbf{r}\int_{-\infty}^{\infty} d\tau\,[\exp i(\boldsymbol{\varkappa}\cdot\mathbf{r} - \omega\tau)]\langle\Delta\eta(0, 0)\,\Delta\eta(\mathbf{r}, \tau)\rangle \tag{11.107c}$$

The integral of this cross section over $\omega$ yields $\delta(\tau)$ and reduces this expression for the cross section to the elastic cross section

$$\sigma(\boldsymbol{\varkappa}) = \int \sigma(\boldsymbol{\varkappa}, \omega) \frac{d\omega}{2\pi} = \{\text{Eq. (10.101a)}\}$$

The correlation function is a real function and a stationary random variable in space and time and possesses those attributes that were discussed in Sections 10.4 and 10.14,

$$\langle \Delta\eta(0, 0)\, \Delta\eta(\mathbf{r}, \tau)\rangle = \langle \Delta\eta(0, 0)\, \Delta\eta(-\mathbf{r}, -\tau)\rangle \xrightarrow[r>r_c]{} 0$$

This can be shown by using the equivalent expression

$$\langle \Delta\eta(\mathbf{r}', t')\, \Delta\eta(\mathbf{r}' + \mathbf{r}, t' + \tau)\rangle$$

and choosing new variables $\mathbf{r}' + \mathbf{r} = \mathbf{r}''$ and $t' + \tau = t''$. The expression is stationary with respect to the choice of $\mathbf{r}''$ and $t''$. By its definition, $\sigma(\boldsymbol{\varkappa}, \omega)$ is real, but it also possesses the properties of a spectral density that

$$\sigma(\boldsymbol{\varkappa}, \omega) = \sigma(-\boldsymbol{\varkappa}, -\omega) = \sigma^*(\boldsymbol{\varkappa}, \omega)$$

and only cosine transforms are needed. Although the sample volume is explicitly shown in the cross section, it is inherent in the use of the correlation function that the correlation function fall to zero for large distances and that $V$ be large compared to the correlation volume.

If the fluctuation in velocity of propagation is directly related to the fluctuation in number density

$$\Delta\eta = \frac{d\eta}{dn} \Delta n$$

then the phenomenon is completely described by the fluctuation in number density. Let the number density at time $t$ be defined by

$$n(\mathbf{r}, t) = \sum_m \delta[\mathbf{r} - \mathbf{r}_m(t)]$$

Then, the average number of particles per unit volume $\bar{n}$ follows from

$$\bar{N} = \bar{n}V = \int d\mathbf{r} \left\langle \sum_m \delta[r - r_m(t)] \right\rangle$$

and is a constant independent of time and position. The deviation from the average value is

$$\Delta n(\mathbf{r}, t) = n(\mathbf{r}, t) - \bar{n}$$

and its Fourier expansion is

$$\int d\mathbf{r}\, \Delta n(\mathbf{r}, t) \exp -i\boldsymbol{\varkappa} \cdot \mathbf{r} = \sum_{m} [\exp i\boldsymbol{\varkappa} \cdot \mathbf{r}_m(t)] - \bar{N}\, \delta(\boldsymbol{\varkappa})$$

Time is treated as a parameter in both the $\delta$ function and the Fourier transform. At each instant of time, it is possible to write an expression for $n(\mathbf{r}, t)$ even though there is no equation of motion which describes the evolution of $n$ from $t = 0$ to $t$. Again, it is argued that the correlation function is a stationary random variable in space and time, and then the correlation of the fluctuation is given by

$$V \int d\mathbf{r}\, (\exp i\boldsymbol{\varkappa} \cdot \mathbf{r}) < \Delta n(0, 0)\, \Delta n(\mathbf{r}, \tau)\rangle$$

$$= \left\langle \sum_{m,n} [\exp -i\boldsymbol{\varkappa} \cdot \mathbf{r}_m(0) \exp i\boldsymbol{\varkappa} \cdot \mathbf{r}_n(\tau)] \right\rangle - \bar{N}^2\, \delta(\boldsymbol{\varkappa}) \qquad (11.107\text{d})$$

The scattering cross section $\sigma(\boldsymbol{\varkappa}, \omega)$, the spectral distribution function $S(\boldsymbol{\varkappa}, \omega)$, and the scattering intensity ratio $I(\boldsymbol{\varkappa}, \omega)/I_0$ are related by

$$\sigma(\boldsymbol{\varkappa}, \omega) = \frac{\pi^2}{\lambda^4} \left(\frac{d\eta}{dn}\right)^2 S(\boldsymbol{\varkappa}, \omega) = \frac{I(\boldsymbol{\varkappa}, \omega)}{I_0} \qquad (11.107\text{e})$$

for $\varkappa \neq 0$ for this simple scalar wave equation with a velocity of propagation which fluctuates with the number density. The index of refraction of such an assembly is discussed in Section 11.8.

The term $\Delta\eta$ can be a function of other variables. According to the discussion in Section 10.10, the independently fluctuating thermodynamic variables at constant volume are

$$\Delta n \quad \text{and} \quad \Delta T \quad (\text{constant } V)$$

or

$$\Delta P \quad \text{and} \quad \Delta S \quad (\text{constant } V)$$

For a binary mixture, fluctuating variables are

$$\Delta n_{\text{A}},\ \Delta n_{\text{B}},\ \Delta T \quad \text{or} \quad \Delta P,\ \Delta S,\ \Delta c$$

where $c = n_{\text{A}}/(n_{\text{A}} + n_{\text{B}})$ for a solution. Some examples are considered in subsequent sections.

Van Hove [*21*] introduced and discussed the pair correlation function $G(\mathbf{r}, \tau)$ which is related to spectral distribution $S(\boldsymbol{\varkappa}, \omega)$ by

$$G(\mathbf{r}, \tau) = \frac{1}{N} \int [\exp +i(\boldsymbol{\varkappa} \cdot \mathbf{r} - \omega\tau)] S(\boldsymbol{\varkappa}, \omega) \frac{d\boldsymbol{\varkappa}\, d\omega}{(2\pi)^4}$$

$$= \frac{1}{N} \int (\exp +i\boldsymbol{\varkappa} \cdot \mathbf{r}) \left\langle \sum_{m,n}^{N} [\exp -i\boldsymbol{\varkappa} \cdot \mathbf{r}_m(0)] \exp i\boldsymbol{\varkappa} \cdot \mathbf{r}_n(\tau) \right\rangle \frac{d\boldsymbol{\varkappa}}{(2\pi)^3}$$

$$= \frac{1}{N} \left\langle \sum_{m,n}^{N} \int d\mathbf{r}'\, \delta(\mathbf{r} + \mathbf{r}_m(0) - \mathbf{r}')\, \delta(\mathbf{r}' - \mathbf{r}_n(\tau)) \right\rangle \tag{11.108}$$

In the interpretation of $S(\boldsymbol{\varkappa}, \omega)$, it should be noted that $\boldsymbol{\varkappa}$ gives the change in $\mathbf{k}$ and $\omega$ gives the change in energy between the incident and scattered particles. $G(\mathbf{r}, \tau)$ describes the correlation between the presence of a particle at coordinate $\mathbf{r}$ at time $\tau$ and the presence of a particle at $\mathbf{r} = 0$ at $\tau = 0$.

Exercise 11.12 Show Eq. (11.107b) is an asymptotic solution to the wave equation with a position- and time-dependent velocity of propagation.

*Ans.* Consider first the solution of the scalar wave equation

$$\nabla^2\psi - \frac{1}{c^2} \frac{\partial^2 \psi}{\partial t^2} = \varrho(\mathbf{r}, t) \tag{11.109a}$$

The Fourier integral time transform

$$\psi(\mathbf{r}, t) = \int \frac{d\omega}{2\pi} \psi(\mathbf{r}, \omega) e^{-i\omega t} \tag{11.109b}$$

is used to transform the time-dependent equation into the scalar Helmholtz equation,

$$\nabla^2\psi(\mathbf{r}, \omega) + \left(\frac{\omega}{c}\right)^2 \psi(\mathbf{r}, \omega) = \varrho(\mathbf{r}, \omega)$$

which has a solution

$$\psi(\mathbf{r}, \omega') = -\frac{1}{4\pi} \int d\mathbf{r}'\, \varrho(\mathbf{r}', \omega') \frac{\exp ik'R}{R}$$

$$\approx -\frac{1}{4\pi} \frac{\exp ik'r}{r} \int d\mathbf{r}'\, \varrho(\mathbf{r}', \omega') \exp -i\mathbf{k}' \cdot \mathbf{r}'$$

$$\approx -\frac{1}{4\pi} \frac{\exp ik'r}{r}$$

$$\times \int d\mathbf{r}' \int dt'\, \varrho(\mathbf{r}', t') \exp -i(\mathbf{k}' \cdot \mathbf{r}' - \omega' t') \tag{11.109c}$$

where $R = |\mathbf{r} - \mathbf{r}'|$ and $k^2c^2 = \omega^2$. It can be shown that $\psi(\mathbf{r}, \omega')$ is a solution by using $\nabla^2(1/R) = 4\pi\delta(\mathbf{R})$. The asymptotic solution for large $r$ follows from $|\mathbf{r} - \mathbf{r}'| = r - \hat{k} \cdot \mathbf{r}'$ in direction $\hat{k}$. With

$$\varrho(\mathbf{r}', t') = \frac{\Delta\eta}{c^2} \frac{\partial^2\psi}{\partial t^2} = -\frac{\omega_a^2}{c^2}(\Delta\eta)\psi_i$$

this solution yields Eq. (11.107b).

**Exercise 11.13** Show that Eq. (11.107c) can be used for the scattering of the de Broglie waves of electrons or neutrons with

$$k_a^2\, \Delta\eta(\mathbf{r}, t) \to \left(\frac{2M}{\hbar^2}\right) U(\mathbf{r}, t) \tag{11.109d}$$

Start with the time-dependent Schrödinger equation. Let

$$\hbar\omega = \frac{\hbar^2}{2M}[k_a^2 - (k')^2] \quad \text{and} \quad \boldsymbol{\varkappa} = \mathbf{k}_a - \mathbf{k}'$$

**Exercise 11.14** For nonmagnetic material, $\mu = \mu_0$, Maxwell's equation for the electric vector is

$$\frac{1}{\mu_0} \text{curl curl } \mathbf{E} + \frac{\partial^2 \mathbf{D}}{\partial t^2} = 0$$

Let div $\mathbf{E} \approx 0$ and then assume that

$$D_i = \varkappa\varepsilon_0 E_i + \varepsilon_0[\Delta\varepsilon_{ij}(\mathbf{r}, t)]E_j$$

Show that the correlation function in Eq. (11.107c) becomes

$$\langle \Delta\eta(0, 0)\, \Delta\eta(\mathbf{r}, \tau)\rangle \to \langle \Delta\varepsilon_{fg}(0, 0)\, \Delta\varepsilon_{ij}(\mathbf{r}, \tau)\rangle \tag{11.109e}$$

*Ans.* The form of equation is the same as Eq. (11.109c) with

$$\psi(\mathbf{r}, \omega') \to E_i(\mathbf{r}, \omega')$$

and

$$\varrho(\mathbf{r}', t') \to -k_a^2(\Delta\varepsilon_{ij})E_j$$

In order to ensure that div $\mathbf{E} = 0$, it is convenient to introduce the transverse polarization vectors $\hat{u}_k$ and $\hat{u}_{k'}$. Then the polarization is explicitly shown by writing

$$\sigma(\boldsymbol{\varkappa}, \omega, \hat{u}_k, \hat{u}_{k'}) = \frac{\pi^2}{\lambda^4} V \int d\mathbf{r} \int d\tau \, [\exp i(\boldsymbol{\varkappa} \cdot \mathbf{r} - \omega\tau)](\hat{u}_k{}^* \cdot \hat{x}_f)(\hat{u}_{k'} \cdot \hat{x}_g)$$
$$\times (\hat{u}_{k'}^* \cdot \hat{x}_i)(\hat{u}_k \cdot \hat{x}_j)\langle \Delta\varepsilon_{fg}(0, 0) \, \Delta\varepsilon_{ij}(\mathbf{r}, \tau)\rangle \qquad (11.109\text{f})$$

### 11.8.2 Classical Aspects

Choose $\mathbf{r}$ along $\hat{\varkappa}$ in Eq. (11.107b) and denote the displacement by $u$. Regard $u(0)$ and $u(\tau)$ as commuting variables. Then, the average

$$\left\langle \sum_{m,n} \exp i\varkappa[u_m(0) - u_n(\tau)] \right\rangle \qquad (11.110\text{a})$$

is the quantity of interest. Consider first the self-scattering term with $m = n$,

$$\langle \exp -i\varkappa[u(0) - u(\tau)]\rangle \qquad (11.110\text{b})$$

If the scattering centers are in a liquid, the number of interactions is large during the interaction interval and a Gaussian approximation would seem appropriate. Using Eq. (10.139), the quantity in (11.110b) is approximated by

$$\langle \exp -i\varkappa[u(0) - u(\tau)]\rangle = \exp\{-\tfrac{1}{2}\varkappa^2\langle[u(0) - u(\tau)]^2\rangle\} \qquad (11.110\text{c})$$

If diffusion is of primary importance, then from Eq. (10.59) for Brownian motion, the average value becomes

$$\exp -\varkappa^2 D\tau \qquad (11.110\text{d})$$

and yields an $S(\boldsymbol{\varkappa}, \omega)$ of

$$S(\boldsymbol{\varkappa}, \omega) = \frac{2D\varkappa^2}{\omega^2 + (D\varkappa^2)^2} \qquad (11.110\text{e})$$

with a line width of the order of $2\varkappa^2 D$. For neutrons at 300°K, a wave number $\varkappa = 10^{10}\ \text{m}^{-1}$ can occur at large-angle inelastic scattering. Similar values of $\varkappa$ can occur for x rays. The diffusion coefficient is of the order of $10^{-9}\ \text{m}^2/\text{sec}$ and $\varkappa^2 D = 10^{11}$. This suggests a correlation time of the order of $10^{-11}$ sec and a line broadening $\Delta\omega_{1/2} \approx 10^{11}$. Inelastic scattering with such small energy changes is difficult to observe with present experimental techniques. Larger changes in energy depend on the local correlated motions and are discussed in greater detail in the next section.

The velocity correlation function is not directly apparent in this development, but replacing $\eta(t)$ by $u(t)$ in Eqs. (10.141) and (10.142) yields the necessary relationship. The diffusion coefficient is often defined by the velocity correlation function as

$$D = \int_0^\infty \langle \dot{u}(0)\dot{u}(t)\rangle \, dt \tag{11.110f}$$

For a Gaussian random process for the velocities

$$\langle \dot{u}(t')\dot{u}(t'')\rangle = D\,\delta(t' - t'') \tag{11.110g}$$

and this aspect was discussed in relation to Eq. (10.160).

### *11.8.3 Crystal Scattering*

It was noted in the previous section that correlations at zero temperature or elastic and inelastic scattering yield information regarding the observable aspects of the scatterer. Some salient features of this problem can be examined by considering a crystal lattice with one scatterer per unit cell. The position of each cell is given by

$$\mathbf{r}_m = \mathbf{R}_m + \mathbf{u}_m$$

where $\mathbf{R}_m$ is the fixed position of the unit cell and $\mathbf{u}_m$ is the scatterer position as a function of time. The $\mathbf{R}_m$ do not depend on time and commute. Since relative coordinates are used, the $\mathbf{u}_m$ with different subscripts *need not commute.* A typical displacement for an atom in the crystal is given by Eq. (11.58). Let

$$\psi_{mn} = \mathbf{q} \cdot (\mathbf{R}_m - \mathbf{R}_n) - \omega_q \tau \tag{11.111}$$

be the relative phase between scattering centers. For convenience, choose $\varkappa$ along $\hat{e}_q$ and then, using the procedure that was used in developing Eq. (11.49) or Eq. (11.59) and the *independence of normal modes q*, the canonical average is

$$\begin{aligned}\Big\langle \sum_{m,n} [\exp - i\boldsymbol{\varkappa} \cdot \mathbf{r}_m(0) \exp i\boldsymbol{\varkappa} \cdot \mathbf{r}_n(\tau)] \Big\rangle &= \sum_{m,n} [\exp - i\boldsymbol{\varkappa} \cdot (\mathbf{R}_m - \mathbf{R}_n)] \\ &\quad \times \prod_q^N (\exp\{-\varkappa^2 C_q^2[(2\bar{n}_q + 1)(1 - \cos\psi_{mn}) - i \sin\psi_{mn}]\})\end{aligned} \tag{11.112a}$$

Again, the detailed analysis of a simple oscillator has provided a simple solution to the complex problem of a crystal. Scattering offers more information than Mössbauer emission since the relative particle positions as well as the single-particle positions are important. In general, $\varkappa_q = \varkappa(\hat{\varkappa} \cdot \hat{e}_q)$, and $C_q^2$ is replaced by $C_q^2 \mid \hat{\varkappa} \cdot \hat{e}_q \mid^2$ in the Eq. (11.112a) with $q$ summed over $3N$ modes.

The time-independent term has the same form as the Debye–Waller factor for recoilless emission, which was given by Eq. (11.62a). If only one-phonon processes are important, Eq. (11.112a) has the approximate expansion

$$(\exp -2W)[\exp -i\boldsymbol{\varkappa} \cdot (\mathbf{R}_m - \mathbf{R}_n)] \times \left\{1 + \sum_q \varkappa^2 C_q^{\,2}[(\bar{n}_q + 1) \exp i\psi_{mn} + \bar{n}_q \exp -i\psi_{mn}] + \cdots\right\} \tag{11.112b}$$

Terms like

$$\sum_{m,n} \exp[i(\boldsymbol{\varkappa} - \mathbf{q}) \cdot (\mathbf{R}_m - \mathbf{R}_n)] \to N\, \delta(\boldsymbol{\varkappa} - \mathbf{q} - \mathbf{s}) \tag{11.113}$$

occur, and when the sum over $m$ and $n$ is completed, it is small unless $\boldsymbol{\varkappa}$ or $\boldsymbol{\varkappa} \mp \mathbf{q}$ are reciprocal lattice vectors. This introduces a $\delta(\boldsymbol{\varkappa} \pm \mathbf{q} - \mathbf{s})$ term into the expansion and expresses the conservation of linear momentum when $\mathbf{s}$ is a reciprocal lattice vector. With these substitutions, $S(\boldsymbol{\varkappa}, \omega)$ is given by

$$S(\boldsymbol{\varkappa}, \omega) \approx Ne^{-2W}\Big\{\delta(\omega)\, \delta(\boldsymbol{\varkappa} - \mathbf{s}) + \sum_q \varkappa^2 C_q^{\,2}[(\bar{n}_q+1)\, \delta(\omega+\omega_q)\, \delta(\boldsymbol{\varkappa}-\mathbf{q}-\mathbf{s}) + \bar{n}_q\, \delta(\omega - \omega_q)\, \delta(\boldsymbol{\varkappa} + \mathbf{q} - \mathbf{s})] + \cdots\Big\} \tag{11.114}$$

The first term within the curly brackets corresponds to the elastic scattering of the incoming neutron or photon with a change in direction along a reciprocal lattice vector, or $\hat{\varkappa} = \hat{k} - \hat{k}' = \hat{s}$. The second set of terms corresponds to inelastic scattering and the excitation of the lattice by one quantum, $\omega = -\omega_q$, or the absorption of one quantum from the lattice, $\omega = \omega_q$. The term $\delta(\omega \pm \omega_q)$ expresses the conservation of energy. The product $\delta(\omega \pm \omega_q)\, \delta(\boldsymbol{\varkappa} \mp \mathbf{q} - \mathbf{s})$ expresses conservation of energy and of linear momentum for the combined system of scatterer and particle. For a given $\mathbf{k}$, the conservation of energy and linear momentum severely restrict the energies and directions of the scattered particle [*22, 22a*]. The $S(\boldsymbol{\varkappa}, \omega)$ can be interpreted as the probability that the momentum of the incoming particle changes momentum by $\hbar\boldsymbol{\varkappa}$ and energy by $\hbar\omega$. Multiple-phonon processes that contribute are included in Eq. (11.112a), but terms of order $\varkappa^4$, $\varkappa^6$, ... must be included in the series expansion for these processes.

Inelastic scattering of a very low-energy neutron is particularly interesting. Let $\mathbf{k} = 0$, and then only the absorption of phonons by the incident particle can occur for inelastic scattering. For one-phonon scattering, conservation of energy requires

$$\frac{(\hbar k')^2}{2M} = \frac{(\hbar \varkappa)^2}{2M} = \hbar\omega = \hbar\omega_q \tag{11.115}$$

and conservation of linear momentum requires $\mathbf{k}' = \boldsymbol{\varkappa}$. This equation relates the linear momentum of the neutron to its energy and at the same time the energy of the phonon to the wave vector $\boldsymbol{\varkappa}$ and indirectly $q$. Figure 11.10 shows the dependence of the $\omega_q$ on $\varkappa$ for the 100 and 110 directions in an aluminum crystal [*23*] at 300°C. Longitudinal and transverse modes are shown and $C_q^2$ must be modified to $C_q^2 \mid \hat{\varkappa} \cdot \hat{e}_q \mid^2$. The difference in the sound velocity of the longitudinal and transverse waves yields a curve for each polarization. Since $\boldsymbol{\varkappa}$ is along $\mathbf{q} - \mathbf{s}$, the transverse modes are apparent. This type of measurement gives a very direct means of determining the dispersion relationship $\omega(\mathbf{q})$ for phonons. If the unit cell contains more than one particle, then the dispersion relationship can be determined for the optical branch and the acoustic branch.

If in the neutron scattering experiment, the nucleus or the scatterer changes its state, the scattered wave is incoherent. The change in state of the individual scatterer destroys the phase and the ensuing expression is very similar to that for Mössbauer emission. Conservation of linear momentum is no longer important and the line shape depends on

$$[(\bar{n}_q + 1)\,\delta(\omega + \omega_q) + \bar{n}_q\,\delta(\omega - \omega_q)]$$

The scattering cross section for inelastic incoherent scattering provides a very direct means of measuring the density of states $g(\nu)$ for the phonon spectrum [*22, 22a, 24*] which was discussed in Section 9.3. The results of various experimental measurements of $g(\nu)$ for vanadium by means of incoherent neutron scattering are shown in Fig. 11.11.[5] Thermal neutrons were used for these experiments.

Other references [*24*] are suggested for scattering by binary solutions, crystals with defects, and so on.

[5] See Dolling and Woods [*22a*] for a discussion of the inelastic scattering of thermal neutrons and of Fig. 11.11.

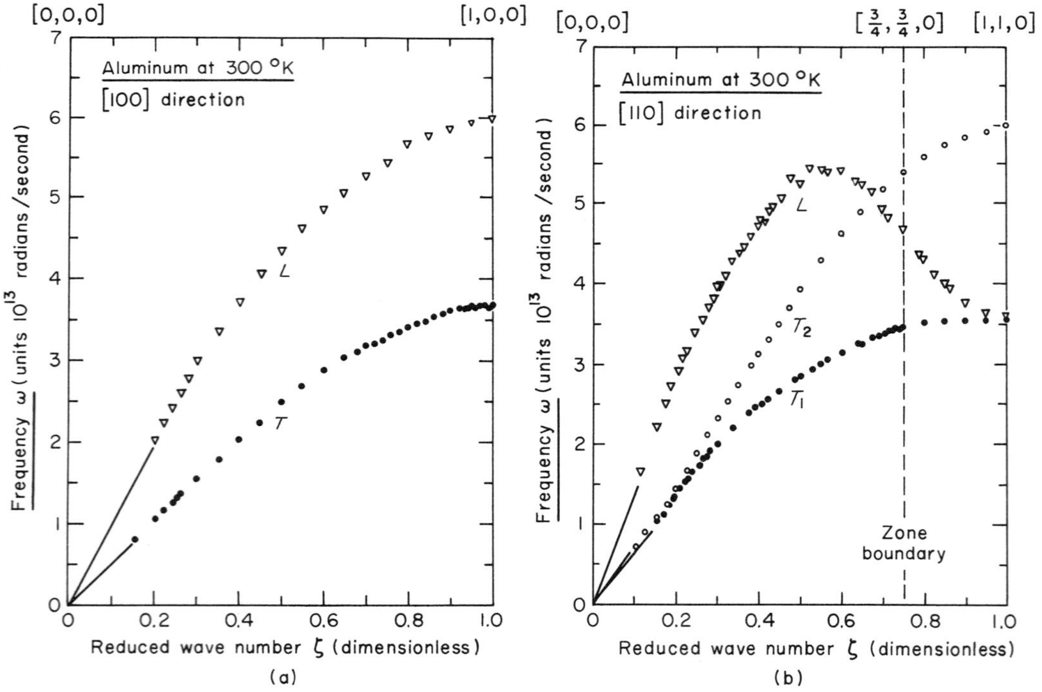


**Fig. 11.10** Measurement of the vibrational frequencies as a function of wave vector for the [100] and [110] directions in Al at 300°C by very low-energy neutrons. [J. L. Yarnell, J. L. Warren, and S. H. Koenig, Experimental dispersion curves for phonons in aluminum. In *Lattice Dyn., Proc. Int. Conf. Copenhagen, 1963* (R. F. Wallis, ed.), p. 60, Figs. 2 and 3. Pergamon, Oxford, 1963.]

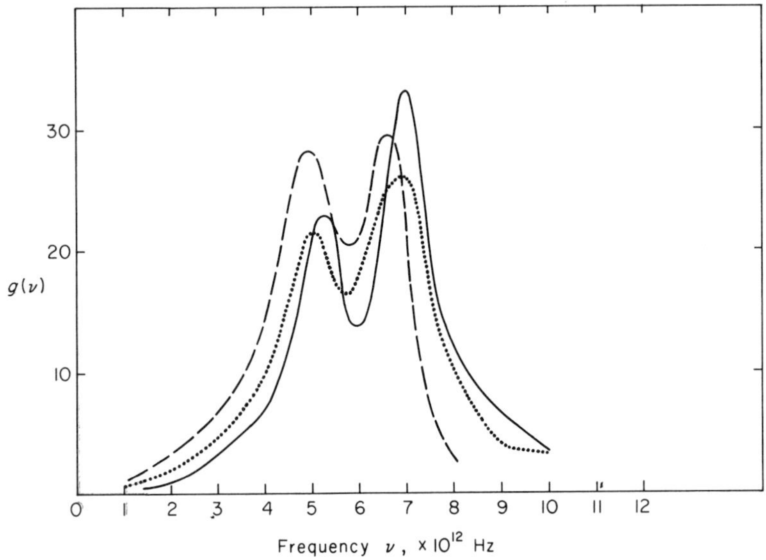


**Fig. 11.11** Density-of-state curves $g(\nu)$ as a function of $\nu$ for vanadium, measured by incoherent neutron scattering. (See Dolling and Woods [*22a*] for a discussion of these experimental curves.) [From G. Dolling and A. D. B. Woods, *in* "Thermal Neutron Scattering" (P. A. Egelstaff, ed.), p. 211, Fig. 5.8; Academic Press, New York, 1965.]

The intensity of the scattered radiation from a crystal lattice and its linear increase with $N$ require comment. As noted earlier, the integral of Eq. (11.112a) over $d\nu$ yields $\delta(\tau)$ and then the subsequent sum over $q$ cancels the Debye–Waller factor. The remaining double sum is the usual elastic scattering term for a crystal lattice:

$$\sum_{m,n} \exp i\boldsymbol{\varkappa} \cdot (\mathbf{R}_m - \mathbf{R}_n)$$

A crystal lattice has long-range order, and terms like

$$\sum_{m} \exp i\boldsymbol{\varkappa} \cdot (\mathbf{R}_m - \mathbf{R}_0) \approx N\,\delta(\boldsymbol{\varkappa} - \mathbf{s})$$

are linear in $N$ when $\boldsymbol{\varkappa}$ is a reciprocal lattice vector $\mathbf{s}$. This sum is a very sharp function and falls to zero for changes in $\boldsymbol{\varkappa} = \mathbf{s} + \Delta\boldsymbol{\varkappa}$ as small as $\Delta\varkappa \approx a/L$, where $a$ is the interatomic distance and $L$ is the crystal dimension. A highly parallel bundle of rays leaves the crystal when $\boldsymbol{\varkappa} = \mathbf{s}$ and the angular divergence of the beam $\delta\theta \approx 10^{-7}$–$10^{-8}$ rad for reasonably sized crystals. At $\boldsymbol{\varkappa} = \mathbf{s}$, the intensity is proportional to $N^2$, but in order to measure this $N^2$ intensity, the spread of the incident beam must be less than $\Delta k/k < 10^{-7}$ rad and the detector must be able to resolve $\Delta k'/k' < 10^{-7}$. An average over this spread in $k$ and $k'$ yields a double sum which increases linearly in $N$. An alternate viewpoint shows that the double sum has a maximum of $N^2$ and a width of $1/N$. The area under the curve is $N$ and is the measurable parameter. For these reasons, the averaging process is omitted and the linear increase given by Eq. (11.113) is used.

### *11.8.4 Liquid Scattering*

It was noted in Section 11.8.2 that scattering in liquids and dense gases would depend on the diffusion coefficient for correlation times longer than $10^{-11}$ sec. For shorter correlation times, local oscillatory motions can become important and the correlation in position given by Eq. (11.112a) should apply. The local order which exists in a liquid is assumed to be similar to that in a polycrystalline solid or in a powder sample and an average over all the orientations of individual regions is needed. For a given $\boldsymbol{\varkappa}$, an average of Eq. (11.112a) for single-phonon processes, over all crystal orientations, yields $S(\boldsymbol{\varkappa}, \omega)$ for a powder pattern. A granule of a powder has the same high degree of order as that of a crystal and the direction of the scattered radiation is given by $\boldsymbol{\varkappa} = \mathbf{s}$. For a given $\mathbf{s}$, this determines the angle $\theta$ between $\mathbf{k}$ and $\mathbf{k}'$. For inelastic scattering, the discussion following Eq. (11.114) can be used and conservation of energy and momentum requires $\delta(\omega \pm \omega_q)\,\delta(\boldsymbol{\varkappa} \mp \mathbf{q} - \mathbf{s})$. If the granule contains $N_\alpha$ unit cells and has a characteristic dimension $l = N_\alpha^{1/3}a$, then the angular width of the scattered beam is of the order of $a/l$, or $N_\alpha^{1/3}$, and is less than 1° for granules with $l \approx 100a$. Here, $l$ can be regarded as the maximum correlation distance, and different granules have arbitrary relative phases as well as a random spatial orientations of the reciprocal lattice vectors. The powder pattern is a ring with $\hat{k}$ as the symmetry axis and an experiment measures the magnitude $|\mathbf{s}|$ in elastic scattering. Inelastic scattering measures those values of $\omega_q$ for which $\varkappa$ lies in the range

$$s - q < \varkappa < s + q$$

for powder patterns.

Elastic scattering measurements give an average of a sequence of instantaneous configurations of the scatterer. For a liquid, the scattering sum is

$$\left\langle \sum_{m,n} \exp i\boldsymbol{\varkappa} \cdot (\mathbf{R}_m - \mathbf{R}_n) \right\rangle \to \left\langle \sum_{m,n} \frac{\sin \varkappa R_{mn}}{\varkappa R_{mn}} \right\rangle = \{\text{Eq. (10.104b)}\}$$

and only the magnitude of the distance between particles is important. In this average, $\boldsymbol{\varkappa}$ is held fixed and $\mathbf{R}_m - \mathbf{R}_n$ is regarded as having a random orientation in the sequence of configurations, or $\varkappa R_{mn} \cos \gamma$ is averaged over $\gamma$. If a particular particle is selected and then the probability of finding another particle at $R_{m0}$ is considered, it is apparent that for neighbors, next-nearest neighbors, and so on, the characteristic intermolecular distance is important. After a few average distances, the number of particles between $r$ and $r + dr$ is almost constant. The correlation function $g(r)$ in Eq. (10.104b) is the probability of finding a second particle at distance $r$. Since discreteness can be apparent for only short distances, the angular width is quite broad for the scattered pattern that has $\hat{k}$ as its symmetry axis.

Inelastic scattering implies that local order is maintained for a time interval longer than the period of vibration $\nu_q$, which is observed as the energy shift in the scattered beam. If the local structure has the same degree of order as a small granule, then the condition on the wave vector is the same as that for the granule. Shear modes as well as longitudinal modes are included in the development.

Inelastic scattering at a fixed angle of 60° between $\mathbf{k}$ and $\mathbf{k}'$ is shown in Fig. 11.12 for solid and liquid aluminum [*25*]. $S(\boldsymbol{\varkappa}, \omega)$ is quite similar

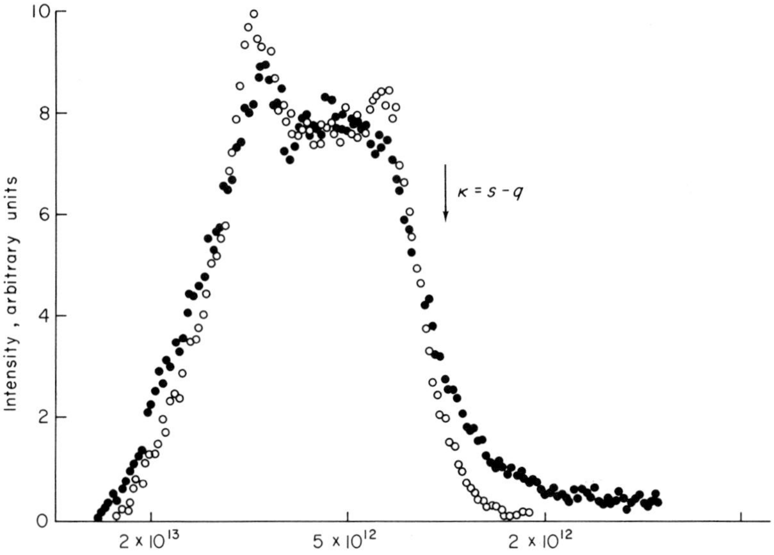


**Fig. 11.12** Scattering of $\lambda = 4 \times 10^{-10}$ m neutrons by solid and liquid aluminum at 630°C and 677°C at a fixed angle of 60° between $\hat{k}$ and $\hat{k}'$. Open circles are for solid Al and dots are for liquid Al. [From P. Schofield, *in* "Physics of Simple Liquids" (H. N. Temperley, ed.), p. 599, Fig. 4; Wiley, New York, 1968.]

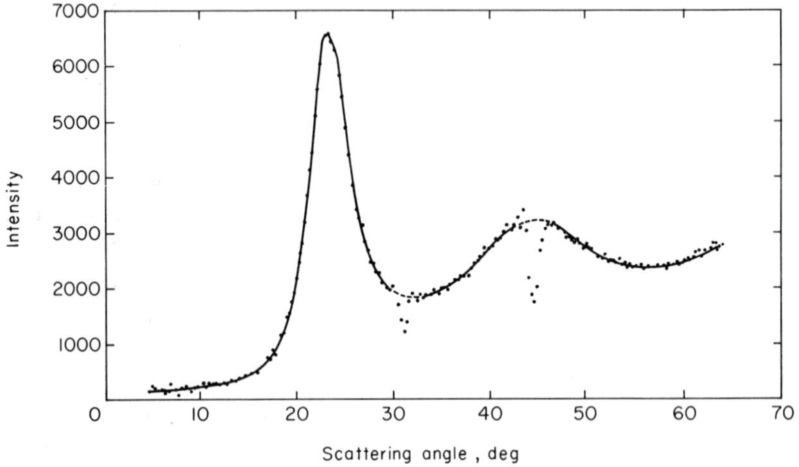


(a)

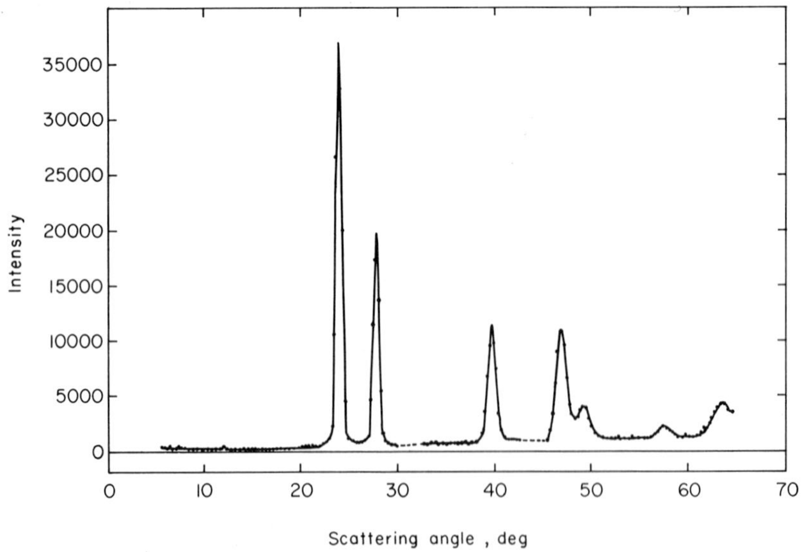


(b)

**Fig. 11.13** Diffraction patterns of $1.06_4$-Å neutrons for (a) liquid (26°K) and (b) solid (4.2°K) neon. [From D. G. Henshaw, *Phys. Rev.* **111**, 1470 (1968), Figs. 3 and 6.]

in both the liquid and solid states. A sharp decrease in intensity is predicted by the model at $\varkappa = s - q$ and this is apparent in the line shape. The range in frequencies is from $3\times 10^{12}$ to $2\times 10^{13}$ Hz, and the peak in the frequency spectrum occurs near the value $\nu = k\Theta/h$. The angular dependence of the scattering for solid and liquid neon [*26*] is shown in Fig. 11.13.

The general correlation function given by Eq. (11.106b) is applicable in general. Both classical diffusion and polycrystalline scattering are special examples. For very short correlation times, the polycrystalline aspect of the liquid occurs, and for long correlation times, or times greater than $10^{-11}$ sec, the diffusion aspect is expected [*27*].

### *11.8.5 Scattering of Optical Photons*

Equations (11.106b) and (11.107c) are for scalar point scatters and the correlation between scattering centers. The spin of the neutron and the intrinsic angular momentum or polarization of the photons have been ignored. Equation (11.104a) can be used for the scattering of polarized photons. If $\hat{u}_k$ is the polarization state of the incident photon [see Eqs. (2.51) and (2.52) for linear and circular polarization] and $\hat{u}_{k'}$ is the polarization of the scattered photon, then the perturbation $U$ must annihilate a photon with index $k$ and create a photon with index $k'$, where $k$ is a four-index symbol implying $\hat{k}$, $h\nu$, and $\hat{u}_k$ or $s$. The perturbation must contain field terms like

$$a_{k'}^{\dagger}\hat{u}_{k'}^{*}(\exp -i\mathbf{k}\cdot\mathbf{r})a_k\hat{u}_k(\exp i\mathbf{k}\cdot\mathbf{r})$$

where $\hat{u}_{k'}$ and $\hat{u}_k$ are transverse to $\hat{k}'$ and $\hat{k}$, respectively. This requires an electronic configuration for the atom, molecule, or unit cell in a crystal that is described by a polarizability tensor $\alpha_{xy}$. This is the only type of term that couples the "in" and "out" electromagnetic waves when the wavelength is large compared to the highly correlated distribution of electrons about point $\mathbf{r}$. Thus the matrix element of interest has the form

$$(k'a \mid U(\tau) \mid kb) = \sum_m (a \mid (\hat{u}_{k'}^{*}\cdot\hat{x}_i)\alpha_{ij}(\tau)(\hat{x}_j\cdot\hat{u}_k)\exp i\boldsymbol{\varkappa}\cdot\mathbf{r}_m(\tau) \mid b)$$

where the notation implies a summation over the repeated index for the polarizability tensor and a sum over $m$ for the atoms, molecules, or unit cells in a crystal; $b$ and $a$ are quantum numbers that describe the scattering system; and $\alpha_{ij}$ is a tensor operator of the type $T_m^{(2)}$, $T_m^{(1)}$, $T_0^{(0)}$ and the Wigner–Eckart theorem can be used to determine its matrix elements. Closure can be used over the final states $| a)$ and with brackets implying a thermal average over the initial states, the scattering is described by

$$S(\boldsymbol{\varkappa}, \omega, \hat{u}_k, \hat{u}_{k'}) = \int_0^\infty d\tau \exp -i\omega\tau \Big\langle \sum_{m,n} [(\hat{u}_k{}^* \cdot \hat{x}_f)\alpha_{fg}^*(0)(\hat{x}_g \cdot \hat{u}_{k'}) \exp i\boldsymbol{\varkappa} \cdot \mathbf{r}_m(0)] \times [(\hat{u}_{k'}^* \cdot \hat{x}_i)\alpha_{ij}(\tau)(\hat{x}_j \cdot \hat{u}_k) \exp i\boldsymbol{\varkappa} \cdot \mathbf{r}_n(\tau)] \Big\rangle \tag{11.116}$$

The quantity $\langle \alpha_{fg}\alpha_{ij} \rangle$ indicates that the scattering is dependent on a fourth-order correlation tensor. The scalar product, which shows the explicit form of the vector products, is used since $\hat{u}_k$ is transverse to $\hat{k}$ and $\hat{u}_{k'}$ to $\hat{k}'$. If $\hat{k} = \hat{X}_3$ defines a space-fixed axis, then $\hat{u}_k$ is described by $\hat{X}_1$ and $\hat{X}_2$ and $\hat{u}_{k'}$ by $\hat{\theta}$ and $\hat{\varphi}$ for $\hat{k}' = \hat{r}$. Here the $\hat{x}_i$ denote unit vectors in any appropriate reference frame and may be body-fixed for molecules. For molecules and atoms, $\alpha_{ij}$ may commute with $\mathbf{r}_m$ when $\mathbf{r}_m$ is chosen as the center of mass. The form of $S$ is similar to Eq. (11.107b) or (11.109f) with $\Delta\eta$ replaced by the tensor form $\Delta\varepsilon_{ij}$. This is also the expected form for the scattering and follows from the elementary concept that the stimulated polarization of the molecule is given by $P_i = \alpha_{ij}E_j$ and the radiated energy is proportional to $P^2$. Usually, the electric dipole radiation is the dominant interaction and $\alpha_{ij}$ is the usual polarizability tensor. In the most general case, $\alpha_{ij}$ connects axial as well as polar vectors and $P_i = \alpha_{ij}E_j + \beta_{ij}H_j$ and $M_i = \chi_{ij}H_j + \gamma_{ij}E_j$ must be considered as possible sources for the scattered radiation.

### 11.8.5.1 Brillouin and Rayleigh Scattering

Assume that the polarizability tensor is diagonal, $\alpha_{ij} = \alpha\delta_{ij}$, and therefore is invariant to rotation of the coordinate system. The average value of

$$\Big\langle \sum_{m,n} [\exp -i\boldsymbol{\varkappa} \cdot \mathbf{r}_m(0) \exp i\boldsymbol{\varkappa} \cdot \mathbf{r}_n(\tau)] \Big\rangle$$

determines the scattering. The wave number $k = 2\pi\mu/\lambda$ is of the order of $2\pi \times 10^6\ \mathrm{m}^{-1}$ and is $10^{-4}$ of the x-ray and thermal-neutron wave numbers. Equation (11.114) can be used for the one-phonon approximation. Since reciprocal lattice vectors are of the order of $10^{10}\ \mathrm{m}^{-1}$, $\mathbf{s} = 0$, and scattering requires

$$\mathbf{k} - \mathbf{k}' = \boldsymbol{\varkappa} = \mathbf{q} \quad \text{or} \quad \begin{cases} \boldsymbol{\varkappa} = \pm\mathbf{q}, & \omega = \mp\omega_q \\ \varkappa = 0, & \omega = 0 \end{cases} \tag{11.117a}$$

This is again an expression for the conservation of linear momentum and energy. The line at lower frequency corresponds to the excitation of a lattice vibration and a decrease in frequency for the scattered photon. This term has $(\bar{n}_q + 1)$ as its coefficient and is referred to as the Stokes line. The term with $\bar{n}_q$ as its coefficient is referred to as the anti-Stokes line and corresponds to the absorption of one quantum from the lattice and an increase in the frequency of the emitted photon. At sufficiently low temperatures, only the Stokes line occurs. The unshifted term corresponds to a term similar to that which occurred in Mössbauer emission and represents scattering in the forward direction.

This expression for the conservation of linear momentum in scattering limits the sum over the index $q$. The maximum value of $q$ is of the order of $2k$ for scattering at 180° and

$$\frac{q}{k} \approx \frac{v_q}{c/\mu} \approx 10^{-5}$$

The energy shift upon scattering is given by $\omega = \omega_q$ and the acoustic frequency is of the order of $10^9$–$10^{10}$ Hz. Thus $k \approx k'$ for the optical wave or

$$\varkappa = \frac{4\pi\mu}{\lambda} \sin\frac{\theta}{2} = q$$

and the frequency shift upon scattering with $\hbar q v_q = \hbar\omega_q$ is given by

$$\nu = 2\nu_0 \frac{\mu v_q}{c} \sin\frac{\theta}{2} \tag{11.117b}$$

$\nu_0$ is the frequency of the optical radiation, $v_q$ the sound velocity in direction $\hat{q}$ or $\hat{\varkappa}$, $c$ the velocity of light, and $\mu$ the index of refraction of the medium.

Brillouin light scattering has been used to determine the hypersonic sound velocities $v_q$ for $\nu_q \approx 10^{10}$ Hz in ammonium chloride, $NH_4Cl$. The longitudinal sound velocity [*28*] along the (110) direction is shown in Fig. 11.14. The hypersonic velocity corresponds to $\nu_q = 1.8 \times 10^{10}$ Hz and the ultrasonic velocity to $\nu_q = 10^7$ Hz. At $T_\lambda = -30.55$°C, the single crystal of $NH_4Cl$ undergoes a second-order lambda transition.

Scattering by gases forms a particularly interesting study of the use of the correlation function. For time intervals sufficiently short, the motion of the particle is given by the classical trajectory

$$\mathbf{r}(\tau) - \mathbf{r}(0) = \mathbf{v}\tau$$

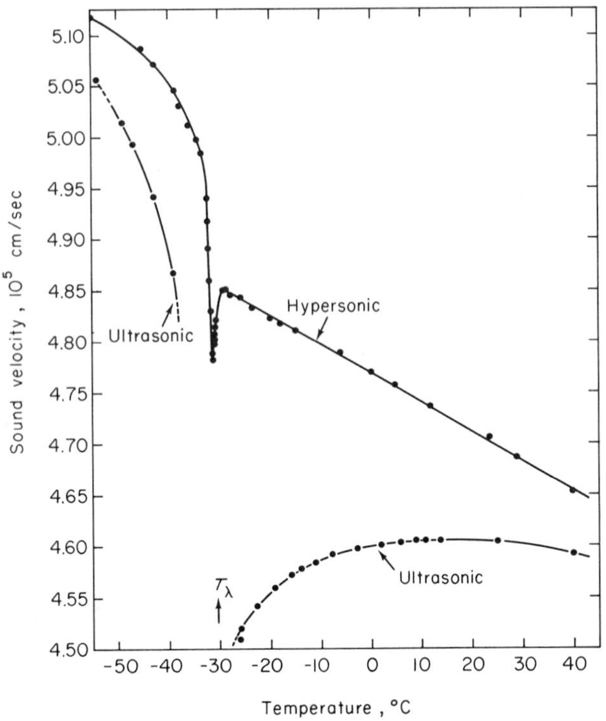


**Fig. 11.14** Temperature dependence of the longitudinal sound velocity in the [110] direction for a single crystal of $NH_4Cl$. The lambda transition occurs at $-30.55$°C. Hypersonic data are obtained from the Brillouin laser light scattering. [From P. D. Lazay, J. H. Lunacek, N. A. Clark, and G. B. Benedek, *in* "Scattering Spectra of Solids" (G. B. Wright, ed.), p. 595, Fig. 2; Springer-Verlag, Berlin and New York, 1969.]

between collisions. If $\lambda$ is the wavelength of the incident radiation and $l$ is the average path of a gas particle, then Doppler broadening of the spectral line is the dominant feature for $l > \lambda/2\pi$ and $\tau < l/v$. The motions of different particles are not correlated and the Boltzmann average over velocities yields the line shape

$$\left\langle \sum_{m,n} [\exp -i\boldsymbol{\varkappa} \cdot \mathbf{r}_m(0) \exp i\boldsymbol{\varkappa} \cdot \mathbf{r}_n(\tau)] \right\rangle \approx \bar{N}\langle \exp i\boldsymbol{\varkappa} \cdot \mathbf{v}\tau \rangle$$

$$= \bar{N} \exp -\left[ (\varkappa\tau)^2 \frac{kT}{2m} \right] \tag{11.118a}$$

or a spectral density of

$$S(\boldsymbol{\varkappa}, \omega) = \bar{N}\left(\frac{2\pi m}{\varkappa^2 kT}\right)^{1/2} \exp\left(-\frac{\omega^2}{2\varkappa^2 kT/m}\right) \tag{11.118b}$$

This expression can be compared to Eq. (2.54) for the Doppler broadening for spontaneous emission with $\omega \to (\omega - \omega_0)$ and $\varkappa = 2\pi/\lambda$. The line width for scattering depends on $\varkappa = (4\pi/\lambda)\sin(\theta/2)$ and is greatest for the backward, or $\theta = 180°$, scattering and negligible in the forward direction. At 60°, the line shape is the same as that for spontaneous emission. For $\varkappa = 0$, the special techniques used for elastic scattering in Section 10.11.1 are needed.

As the gas density increases or the average path becomes less than one wavelength, $l < \lambda/2\pi$, the particle changes direction during the scattering event and the average or diffusive motion is needed. One approximation is to assume that the particle motions are uncorrelated. Following the discussion in Section 11.8.2 for diffusive motion, we have

$$\left\langle \sum_{m,n} [\exp -i\boldsymbol{\varkappa}\cdot\mathbf{r}_m(0) \exp i\boldsymbol{\varkappa}\cdot\mathbf{r}_n(\tau)] \right\rangle \approx \bar{N}\exp(-\varkappa^2 D\tau) \qquad (11.119a)$$

and the spectral density is

$$S(\boldsymbol{\varkappa}, \omega) = \frac{2D\varkappa^2}{\omega^2 + (D\varkappa^2)^2}, \qquad l < \frac{\lambda}{2\pi} \qquad (11.119b)$$

The line width changes from the Doppler width to the narrower width [*11*]

$$\Delta\omega_{1/2} = 2\varkappa^2 D \qquad (11.119c)$$

For gases, the diffusion coefficient is given by $D \approx \frac{1}{2}l\bar{v} \approx \bar{v}/2n\sigma$ and the decrease in line width is inversely proportional to pressure. Although diffusion is the measurement of the position of a given particle, it implies the existence of other particles and these other particles must be included.

The spectral distribution $\sigma(\boldsymbol{\varkappa}, \omega)$ for inelastic scattering by a gas is shown in Fig. 11.15 for deuterium gas at various pressures [*29*]. At low pressures, the line has the Doppler shape. With increasing pressure, the central Rayleigh line and the two Brillouin lines appear. These lines become narrower with increasing pressure. Again, the Brillouin lines are a manifestation of the long-range correlation implied by the existence of acoustic waves.

A phenomenological method is used to treat these correlations in gases and liquids. For electromagnetic radiation, the scattering is due to the fluctuation of the dielectric constant $\varepsilon$ and in Eq. (11.107c), $\Delta\eta = \Delta\varepsilon$. In the earlier discussion of crystals, a canonical ensemble was used for the average. A grand canonical ensemble is used when the volume is constant and the number density varies; then, $\Delta n$ and $\Delta T$ can be used as the independent fluctuating variables. For this discussion, it is assumed that only density fluctuations are important and

$$\Delta\varepsilon \approx \left(\frac{\partial\varepsilon}{\partial n}\right)_{\mathrm{T}} \Delta n = \Delta\eta \tag{11.120a}$$

The term $(\partial\varepsilon/\partial n)_N \, \Delta T$ is regarded as small and is omitted. Thus

$$\Delta n = \Delta\left(\frac{N}{V}\right) = -n\,\Delta V$$

Acoustic waves are one manifestation of correlation in gases and liquids, and occur as changes in pressure at constant entropy. Pressure and entropy are independent fluctuating variables, and

$$\Delta\varepsilon \approx -n\frac{\partial\varepsilon}{\partial n}\left[\left(\frac{\partial V}{\partial P}\right)_S \Delta P + \left(\frac{\partial V}{\partial S}\right)_{\mathrm{P}} \Delta S\right] \tag{11.120b}$$

It is now assumed that the fluctuations occurring in gases and in liquids have these same thermodynamic coefficients, and that

$$\langle\Delta\varepsilon(0,0)\,\Delta\varepsilon(\mathbf{r},\tau)\rangle = A\langle\Delta P(0,0)\,\Delta P(\mathbf{r},\tau)\rangle + B\langle\Delta S(0,0)\,\Delta S(\mathbf{r},\tau)\rangle \tag{11.120c}$$

Direct substitution of this correlation function into Eq. (11.107c) yields $\sigma(\boldsymbol{\varkappa}, \omega)$. Here, $A$ and $B$ are the coefficients of $(\Delta P)^2$ and $(\Delta S)^2$ in $(\Delta\varepsilon)^2$. In accord with the discussion in Section 10.10,

$$\langle\Delta S\,\Delta P\rangle = 0$$

and changes in pressure and entropy are treated as independent.

The correlation functions for pressure and for entropy are obtained from the phenomenological equations of motion. Fluctuations in pressure follow from the wave equation for acoustic waves, which was developed in Section 5.8,

$$\nabla^2(\Delta P) - \frac{1}{v^2}\frac{\partial^2(\Delta P)}{\partial t^2} = 0 \tag{11.120d}$$

or

$$\Delta P(\mathbf{r}, t) = \sum_q A_q \exp i(\mathbf{q}\cdot\mathbf{r} - \omega_q t + \varphi_q)$$

where

$$q^2 v^2 = \omega_q^{\,2}$$

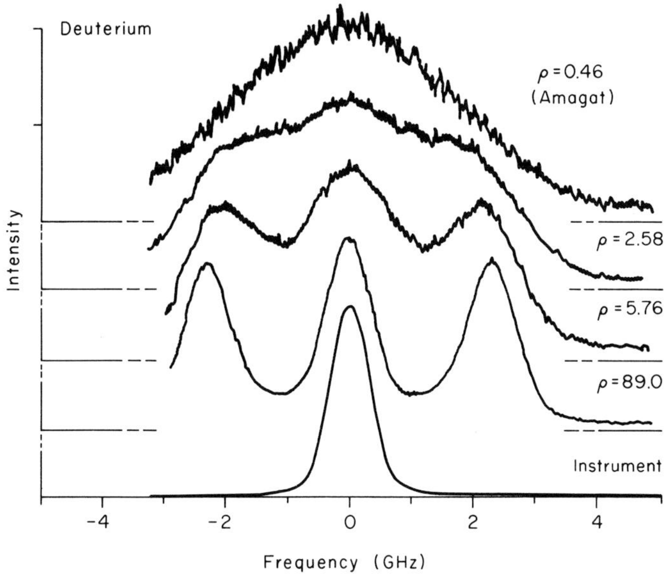


**Fig. 11.15** Rayleigh–Brillouin spectrum of deuterium [*29*] for scattering at 90° for 20 mW of 0.6328-$\mu$m laser radiation. With increasing density, the line shape changes from the Doppler line shape to the Stokes and anti-Stokes components of the Brillouin line and the central Rayleigh line. [The density in amagats is $\varrho = (M/V_{\mathrm{NTP}})(V_{\mathrm{NTP}}/V)$ and is a measure of the volume per particle.] [From E. H. Hara and A. D. May, *Int. Conf. Quantum Electron., 6th, 1970,* p. 195.]

For an expansion for volume $L^3$, $q_x = 2\pi n_x/L$, and so on. The sum includes values of $q$ up to $q = 2\pi/\lambda_a \approx 1/l$, or acoustic wavelengths greater than the average path between collisions, $\lambda_a > 2\pi l$. The $\Delta P(\mathbf{r}, t)$ is a stationary random variable if $\varphi_q$ is a random variable and the correlation function for acoustic oscillations is of the form

$$\langle \Delta P(0,0)\, \Delta P(\mathbf{r}, \tau)\rangle = \sum_q \langle A_q{}^2\rangle \exp i(\mathbf{q}\cdot\mathbf{r} - \omega_q\tau) + \text{complex conjugate} \tag{11.120e}$$

Direct substitution into Eq. (11.107c) yields

$$\delta(\boldsymbol{\varkappa} \pm \mathbf{q})\, \delta(\omega \pm \omega_q)$$

and gives the same frequency shift and angular dependence as Eq. (11.117b) for the Brillouin lines. All directions of $q$ are equally probable and with $\boldsymbol{\varkappa} - \mathbf{q}$, the wave vector $\boldsymbol{\varkappa}$ is also isotropic. This does not imply that $\mathbf{k}' = \mathbf{k} + \boldsymbol{\varkappa}$ is isotropic. The Brillouin line for a gas does not appear until the largest permitted value of $q$ reaches $\varkappa$ and the threshold is approximately $(4\pi/\lambda)\sin(\theta/2) < 1/l$ and is inversely proportional to the pressure. The width of the Brillouin line depends on the damping of the acoustic wave. Linearization of the Navier–Stokes equation (5.62) yields a damping term

$$\frac{\eta}{nmv^2}\,\frac{\partial[\nabla^2(\Delta P)]}{\partial t} \tag{11.121a}$$

which should be included with the acoustic wave equation. Equation (11.120e) becomes a solution when multiplied by the damping term

$$\exp\left[-\frac{q^2\eta}{2nm}\,|\,\tau\,|\right] \tag{11.121b}$$

and yields $\sigma_{\mathrm{B}}(\boldsymbol{\varkappa}, \omega)$ for the Brillouin lines with a Lorentzian line shape having a line width about $\omega = \pm\omega_q$ of

$$\Delta\omega_{1/2} = \frac{\varkappa^2\eta}{nm} \tag{11.121c}$$

Since $\eta \approx \frac{1}{2}nm\bar{v}l$ and $l \approx 1/n\sigma$, the Brillouin line width is inversely proportional to pressure.

The central Rayleigh line, which is due to fluctuations in entropy, can be determined from the equation for the change in energy at constant pressure via heat flow:

$$C_{\mathrm{P}}\frac{\partial T}{\partial t} = K\nabla^2 T \tag{11.122a}$$

with $T\,dS = C_{\mathrm{P}}\,dT$, the equation for entropy change is approximately

$$\frac{\partial(\Delta S)}{\partial t} \approx \frac{K}{C_{\mathrm{P}}}\nabla^2(\Delta S) \tag{11.122b}$$

or

$$\Delta S \approx \sum_q B_q \exp\left[-q^2\left(\frac{K}{C_{\mathrm{P}}}\right)t\right]\exp i(\mathbf{q}\cdot\mathbf{r} + \varphi_q) + \text{complex conjugate} \tag{11.122c}$$

The correlation function is approximately

$$\langle\Delta S(0,0)\,\Delta S(\mathbf{r},\tau)\rangle \approx \sum_q \langle B_q^{\,2}\rangle \exp\left[-q^2\left(\frac{K}{C_{\mathrm{P}}}\right)|\,\tau\,|\right]\exp i\mathbf{q}\cdot\mathbf{r} + \text{complex conjugate} \tag{11.122d}$$

and for $\sigma_R(\boldsymbol{\varkappa}, \omega)$, this correlation function yields a Lorentzian line shape with $\boldsymbol{\varkappa}$ along $\mathbf{q}$ and isotropic. The line width about $\omega = 0$ is

$$\Delta\omega_{1/2} = 2\varkappa^2 \frac{K}{C_P} \tag{11.122e}$$

$Km/\eta C_V \approx 2.5$ for many gases and the line width for the central Rayleigh line is comparable to that for the Brillouin wings.

An integral of the scattering cross section $\sigma(\boldsymbol{\varkappa}, \omega)$ over $dv$ yields $\delta(\tau)$. The ensuing integral with $\boldsymbol{\varkappa} = \mathbf{q}$ reduces to an expression for $\langle(\Delta P)^2\rangle$ and $\langle(\Delta S)^2\rangle$ and the area under the Rayleigh line as compared to the area under the two Brillouin lines is given by the Landau–Placzek relation

$$\frac{\sigma_R(\varkappa)}{\sigma_B(\varkappa)} = \frac{C_P - C_V}{C_V} \tag{11.123}$$

This can be shown by evaluating in detail the ratio $B\langle(\Delta S)^2\rangle/A\langle(\Delta P)^2\rangle$ and will be discussed in Exercise 11.18. It always seems somewhat remarkable that scattering cross sections are related to thermodynamic quantities like heat capacity, isothermal compressibility, and so on.

It should be noted that this phenomenological theory for inelastic scattering by gases and liquids is incomplete. It does not include the short time intervals that give rise to Doppler broadening. As noted earlier, this is the region in which the optical wavelength $\lambda$ is less than the average path $l$. For Brillouin scattering, the optical and acoustic wavelengths are equal and this becomes the same as the criteria for the use of the equations of gas motion, which were developed in Chapter V. An equivalent statement is that the thermodynamic variables lose their validity when the variation in one average path is large. It has been suggested [*30*] that the transition from the kinetic to the hydrodynamic regime occurs for a given wavelength when

$$\lambda > 4\pi l \tag{11.124}$$

and this corresponds to limiting the sum in Eq. (11.120e) to values of $q < l$.

Lasers have made scattering experiments with optical photons readily accessible. The results for the spectrum of light scattered at 90° by water [*31*] are shown in Fig. 11.16. The central or unshifted line is the Rayleigh scattering of the entropy fluctuations and of any residual small particles in the water. The two shifted components represent the Brillouin scattering, which is due to the pressure fluctuations, or to the one-phonon term in Eq. (11.114) for the longitudinal acoustic waves. The width of the lines is the instrumental width and the resolution is not adequate to show this.

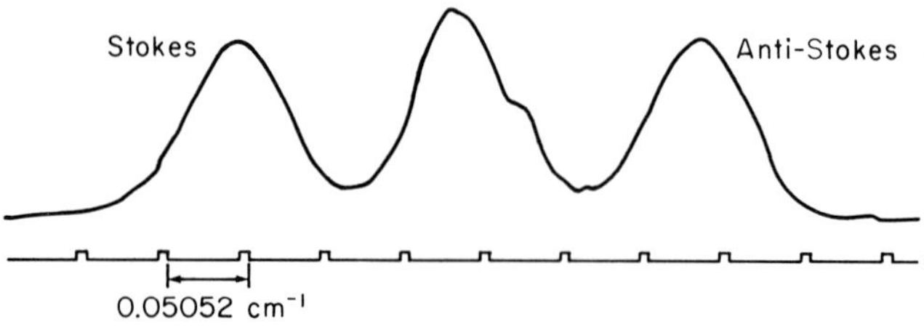


**Fig. 11.16** Brillouin and Rayleigh scattering by water. The unshifted central peak at the laser frequency of 0.6328 $\mu$m is due largely to Rayleigh scattering by residual particles. [From G. Benedek, J. B. Lastovka, K. Fritsch, and T. Greytak, *J. Opt. Soc. Amer.* **54**, 1284 (1964), Fig. 1.]

**Exercise 11.15** The frequency shift in water for scattering at 90° for the red 0.6328-$\mu$m line of the He–Ne laser is $4.33 \times 10^9$ Hz. Compare this result for the hypersonic velocity of sound with the ultrasonic sound velocity of 1491 m/sec.

**Exercise 11.16** Show for $\hat{k} = \hat{X}_3$ and $\hat{k}' = \hat{X}_2$ that $\hat{X}_1$ is the only possible polarization of the observed scattered radiation and is in accord with statement that radiation scattered at 90° by isotropic scatterers is linearly polarized.

**Exercise 11.17** Show that the probability of polarization $\hat{u}_{k'}$ from an assembly of isotropic scatterers

$$\Delta\varepsilon_{ij} = \delta_{ij}\,\Delta\varepsilon$$

follows from

$$(\hat{u}_k^* \cdot \hat{X}_f)(\hat{X}_f \cdot \hat{u}_{k'})(\hat{u}_{k'}^* \cdot \hat{X}_i)(\hat{X}_i \cdot \hat{u}_k)$$

where $\hat{u}_k$ is the polarization of the incident radiation along $\hat{k}$ and $\hat{u}_{k'}$ is the measured polarization for scattered radiation along $\hat{k}'$.

Show that for $\hat{k} = \hat{X}_3$ and $\hat{k}'$ along $\hat{r}$, the polarization of the incident radiation can only be a linear combination of $\hat{X}_1$ and $\hat{X}_2$ and the polarization of the scattered radiation can only be a linear combination of $\hat{\theta}$ and $\hat{\varphi}$. (See Section 2.8 for linear and circular polarization vectors.) Find the ratio $\sigma_\theta/\sigma_\varphi$ for $\hat{u}_k = \hat{X}_1$.

$$\text{Ans.} \quad \sigma_\theta/\sigma_\varphi = (\hat{X}_1 \cdot \hat{\theta})^2/(\hat{X}_1 \cdot \hat{\varphi})^2 = (\cos^2\theta\cos^2\varphi)/\sin^2\varphi$$

Exercise 11.18 Show that Eq. (11.123) follows from $B\langle(\Delta S)^2\rangle/A\langle(\Delta P)^2\rangle$. Use Section 10.10 to show that $\langle(\Delta S)^2\rangle = kC_P$ and $\langle(\Delta P)^2\rangle = -kT(\partial P/\partial V)_S$. The expression $T\,dS = C_V\,dT + T(\partial P/\partial T)_S\,dP$ with $dT$ written in terms of $S$ and $P$ can be used to find

$$\left(\frac{\partial V}{\partial S}\right)_P = \frac{C_P - C_V}{C_P(\partial P/\partial T)_V}$$

$$\frac{(\partial V/\partial P)_S}{(\partial V/\partial S)_P} = -\frac{C_V}{T(\partial P/\partial T)_V}$$

#### 11.8.5.2 Molecular Raman Scattering

The transition from the general state $b$ to the general state $a$ for the scattering system includes the rotational and vibrational states of the molecules, and the polarizability tensor $\alpha_{ij}$ describes the scattering in general terms. For a free molecule, $\alpha_{ij}$ commutes with the center-of-mass coordinate and this simple system is described in terms of the free-particle translational quantum numbers and the rotational and vibrational quantum numbers. During molecular collisions, this simple separation is not possible, but the discussion of Section 11.7 can be used for collisional line broadening. For a diatomic molecule, the quantum numbers describing the rotational and vibrational states are $vJM$ and Raman scattering can occur between states

$$vJM \rightarrow v'J'M'$$

In the interaction representation, the time dependence becomes apparent through

$$(v'J'M' \mid \alpha_{ij}(\tau) \mid vJM) = (v'J'M' \mid \alpha_{ij}(0) \mid vJM) \times \exp \frac{i[E(v'J'M') - E(vJM)]\tau}{\hbar} \tag{11.125a}$$

and Eq. (11.116) will have scattering peaks at

$$\omega = \frac{E(v'J'M') - E(vJM)}{\hbar} \tag{11.125b}$$

Correlation effects between molecules have been neglected in this approximation and the scattering is just $N$ times the scattering by a single molecule. Linear molecules have more vibrational states and symmetric rotators require three quantum numbers to describe their rotational states.

In the frequency domain, the shift in frequency of the scattered light for Raman scattering measures the energy spectrum of the rotational and vibrational states of the molecule. In the absence of external electric or magnetic fields, the energy, and therefore the frequency shift, is independent of the index $M$. Polarization of the scattered radiation does depend on the change in $J$ and on the change in $M$. Since $\alpha_{XY}$ is a tensor operator of the type $T_m^{(2)}$ or $T_0^{(0)}$, the matrix elements depend on the Clebsch–Gordon coefficient $(J2Mm \mid J'M')$ or $(J0M0 \mid JM')$ and the general selection rule on $J$ is $\Delta J = 0, \pm 1, \pm 2$ and on $M$ it is $M + m = M'$. The axis of quantization is in the space-fixed frame and convenient choices are along the electric vector for linearly polarized radiation and along the direction $\hat{k}$ of the incident wave for unpolarized radiation. If the molecules are not oriented and the orientation is not measured after scattering, an average over the initial values of $M$ and a sum over the final values $M'$ are needed.

An average over the equally probable initial states $M$ is seldom used and the average is performed by observing that this is equivalent to a random choice for the orientation of the space-fixed axes. Random orientation is equivalent to

$$\langle \alpha_{XX}^2 \rangle = \langle \alpha_{YY}^2 \rangle = \langle \alpha_{ZZ}^2 \rangle = \alpha^2 + \frac{4}{45}\gamma^2 \tag{11.126a}$$

and

$$\langle \alpha_{XY}^2 \rangle = \langle \alpha_{YZ}^2 \rangle = \langle \alpha_{ZX}^2 \rangle = \frac{1}{15}\gamma^2 \tag{11.126b}$$

Two functions which are invariant to the choice of axes are the trace of the second-order tensor,

$$\sum_i \alpha_{ii} = 3\alpha \tag{11.127a}$$

and, for the fourth-order tensor,

$$\tfrac{1}{2}\sum_{i,j} (3\alpha_{ij}\alpha_{ij} - \alpha_{ii}\alpha_{jj}) = \gamma^2 \tag{11.127b}$$

Taking the average values of $(3\alpha)^2$ and $\gamma^2$ yields their relationship with $\langle \alpha_{XX}^2 \rangle$ and $\langle \alpha_{XY}^2 \rangle$ given in Eqs. (11.126a) and (11.126b).

If the incident radiation is unpolarized, then

$$\hat{u}_k = \hat{X}\cos\delta + \hat{Y}\sin\delta$$

where $\delta$ is a random phase angle and can be used to describe the unpolarized radiation. This limits the tensor products of interest to $\alpha_{Xi}\alpha_{Xj}$ and $\alpha_{Yi}\alpha_{Yj}$

after the average over $\delta$. This is apparent in Eq. (11.116) by observing that only the terms $\cos^2 \delta$ and $\sin^2 \delta$ survive the average. If the observation of the scattered radiation is along $\hat{k}' = \hat{Y}$, then $i$ and $j$ are limited to $X$ or $Z$. For experimental data, the scattering ratio

$$\frac{S(\varkappa, \omega, u, \hat{Z})}{S(\varkappa, \omega, u, \hat{X})} = \frac{2\langle \alpha_{XZ}^2 \rangle}{\langle \alpha_{XX}^2 \rangle + \langle \alpha_{XZ}^2 \rangle} = \frac{6\gamma^2}{45\alpha^2 + 7\gamma^2} \tag{11.128}$$

is defined as the degree of depolarization of the scattered radiation for unpolarized incident radiation. $S(\varkappa, \omega, u, \hat{X})$ is the intensity scattered at $90°$, or $\hat{k} = \hat{Z}$, $\hat{k}' = \hat{Y}$, for unpolarized incident radiation and polarization $\hat{X}$ and frequency shift $\omega$ for the observed radiation.

For linear molecules, the symmetry axis is in the body-fixed reference frame and the two constants needed to describe the polarization of the molecule are $\alpha_\perp = \alpha_{xx} = \alpha_{yy}$ and $\alpha_\parallel = \alpha_{zz}$. Space-fixed and body-fixed coordinate frames are related by the rotation matrix

$$\hat{X} = \sum R_{Xx}(\varphi\theta)\hat{x}$$

where the sum is over $x, y, z$. The polarizability tensor transforms as

$$\alpha_{XY} = \sum R_{Xx} R_{Yy} \alpha_{xy}$$

Since the trace of the tensor $\alpha_{ij}$ is invariant under rotation, the quantity $2\alpha_\perp + \alpha_\parallel$ is an invariant and the polarization of a linear molecule reduces to

$$\alpha_{XY} = \alpha_\perp \delta_{XY} + (\alpha_\parallel - \alpha_\perp) R_{Xz} R_{Yz} \tag{11.129a}$$

All necessary terms follow from $R_{Xz} = \sin\theta\cos\varphi$, $R_{Yz} = \sin\theta\sin\varphi$, and $R_{Zz} = \cos\theta$ and these angles should not be confused with the orientation of $\hat{k}'$. The term $\delta_{XY}$ has no angular dependence and can be treated as a tensor operator $T_0^{(0)}$ which has selection rules $\Delta J = 0$ and $\Delta M = 0$. This term gives rise to the undisplaced Rayleigh line and has the polarization aspects discussed in the previous section for an isotropic polarizability. Pure rotational Raman scattering requires a nonzero value of $(\alpha_\parallel - \alpha_\perp)$ and its angular dependence depends on the product of two rotation operators $R_{Xz} R_{Yz}$ which is a spherical harmonic $Y_m^{(2)}$ or a tensor operator $T_m^{(2)}$. For integral $J$ values, the matrix element is zero unless $\Delta J = 0, \pm 2$. In terms of the previous notation,

$$\alpha = 2\alpha_\perp + \alpha_\parallel \quad \text{and} \quad \gamma^2 = (\alpha_\perp - \alpha_\parallel)^2 \tag{11.129b}$$

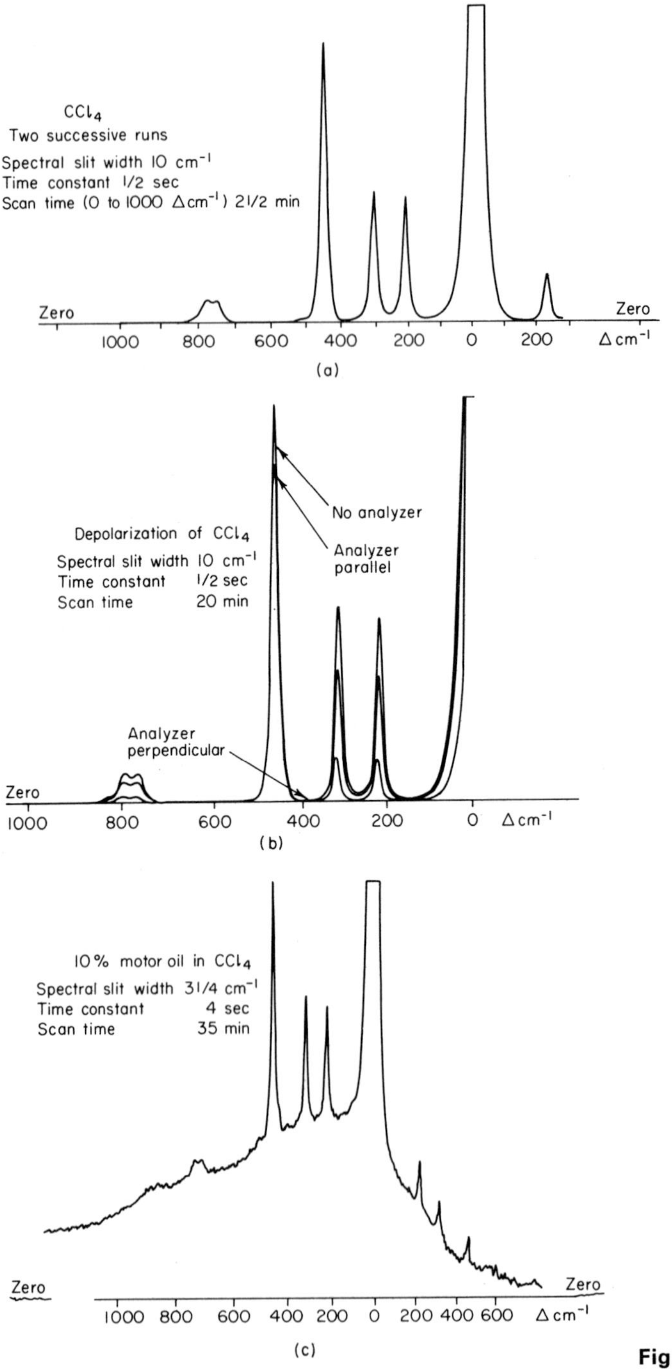

CCl4
Two successive runs
Spectral slit width 10 cm⁻¹
Time constant 1/2 sec
Scan time (0 to 1000 Δcm⁻¹) 21/2 min
Zero
Zero
1000 800 600 400 200 0 200 Δcm⁻¹
(a)
Depolarization of CCl4
Spectral slit width 10 cm⁻¹
Time constant 1/2 sec
Scan time 20 min
No analyzer
Analyzer parallel
Analyzer perpendicular
Zero
1000 800 600 400 200 0 Δcm⁻¹
(b)
10% motor oil in CCl4
Spectral slit width 31/4 cm⁻¹
Time constant 4 sec
Scan time 35 min
Zero
Zero
1000 800 600 400 200 0 200 400 600 Δcm⁻¹
(c)


**Fig. 11.17**

The highly directional and very narrow bandwidth of optical lasers has provided an excellent source for Raman spectroscopy. The vibrational Raman spectrum [*32*] of $CCl_4$ is shown in Fig. 11.17. A 4-mW 0.6328-$\mu$m (red) laser is used as a source. Scan times of the order of minutes were used. In the molecular frame of reference, the polarizability tensor is a function of the vibrational coordinates $Q_i$, that is, $\alpha_{xy}(Q_1, \ldots, Q_n)$, and under the perturbation of an electric field, terms linear in $Q_i$ occur. These lead to vibrational Raman lines with $\Delta v = \pm 1$. Polarization aspects of the Rayleigh and Raman lines are shown in Fig. 11.17(b). In Fig. 11.17(c), motor oil is added to the $CCl_4$ and the characteristic vibrational Raman lines are quite evident above the background. General surveys of developments in Raman spectroscopy are available in the literature [*32*].[6]

### 11.8.5.3 Plasmon Scattering

In an electron gas or an electron gas imbedded in a fixed array of positive charges, the scattering of a photon is proportional to the term

$$\left\langle \sum_{m,n} [\exp -i\boldsymbol{\varkappa} \cdot \mathbf{r}_m(0) \exp i\boldsymbol{\varkappa} \cdot \mathbf{r}_n(\tau)] \right\rangle$$

where the electron coordinates are given by $\mathbf{r}_m$. As noted earlier, this is proportional to the fluctuation in the number density of electrons and to the fluctuation in the dielectric constant. Long-range Coulomb forces give rise to a correlation between the particles, and quantized collective modes are needed for a description.

The order of magnitude of the frequency of oscillation of these collective modes follows from the following classical argument. Let the electron gas be displaced by $x$ so that the polarization $P = Nex$. If the plasma is contained between two slabs, the polarization gives rise to a depolarization

[6] Porto [*33*] presents a very readable review of Raman scattering in solids. Pershan and Lacina [33*a*] and Wolff [*33b*] provide excellent reviews of these scattering topics.

---

**Fig. 11.17** (a) Vibrational Raman lines [*32*] of $CCl_4$. (b) Depolarization of vibrational lines of $CCl_4$, where "analyzer perpendicular" implies the measurement of the $\hat{Z}$ component of polarization of the radiation scattered at 90°, or $\hat{k}' = \hat{Y}$. The incident radiation is along $\hat{k} = \hat{Z}$. (c) Vibrational Raman lines of $CCl_4$ with motor oil added to the $CCl_4$. [Parts (a)-(c) from J. R. Ferraro, Advances in Raman instrumentation and sampling techniques, in "Raman Spectroscopy" (H. A. Szymanski, ed.), Chapter 2, pp. 44–81, Figs. 13, 20, and 18, respectively; Plenum, New York, 1967.]

field $E_d = -P/\varepsilon_0$. The equation of motion of an electron in the plasma is given by

$$M\ddot{x} = eE_d = -\frac{Ne^2x}{\varepsilon_0}$$

so that the resonant frequency is

$$\omega_p^{\ 2} = \frac{Ne^2}{M\varepsilon_0}$$

One can further show that the dielectric constant is given by

$$\frac{\varepsilon}{\varepsilon_0} = 1 - \frac{\omega_p^{\ 2}}{\omega^2}$$

for frequency $\omega$.

The Coulomb term, which is due to the fluctuation in local number or charge density, can be combined with the elastic properties of the medium to yield a Hamiltonian for the electron motion. This can be quantized to yield a dispersion relation of the form $\omega_q^{\ 2} = \omega_p^{\ 2} + bq^2$. This quantized excitation is referred to as a "plasmon" and the energy of the plasmon is $\hbar\omega_q$.

Again the discussion leading to Eq. (11.117a) can be repeated. Now, the lowest-energy plasmon has energy $\hbar\omega_p$ and momentum $\hbar q$. Only $\mathbf{s} = 0$ is of interest and the frequency of the scattered wave is given by Eq. (11.117b) with a suitably chosen plasmon velocity $v_q$. Two lines occur with a frequency displacement $\omega_q$ for the central line, where

$$\varkappa = \frac{4\pi}{\lambda_0} \sin\frac{\theta}{2} = q = \pm\frac{\omega_q}{v_q}$$

Since the optical wavelength in the medium and $v_q$ for the plasmons may not be easily determinable quantities, $(4\pi v_q/\lambda_0)\sin(\theta/2)$ is the experimental parameter in the theory that is related to $\omega_q$. Scattering of ruby laser light by a gaseous hydrogen plasma [*34*] is shown in Fig. 11.18. The central line is the Rayleigh scattering by the hydrogen and the displaced lines are the scattering by the collective excitations of the plasma. Collective motions are apparent for the small-angle scattering at $\theta = 13.5°$.

### 11.8.5.4 Stimulated Scattering

In principle, there is stimulated scattering associated with each spontaneous process. The previous discussion has assumed that the scattering event annihilated one photon with index $k$ and created a photon of index $k'$, and in annihilation–creation operator notation was proportional to the

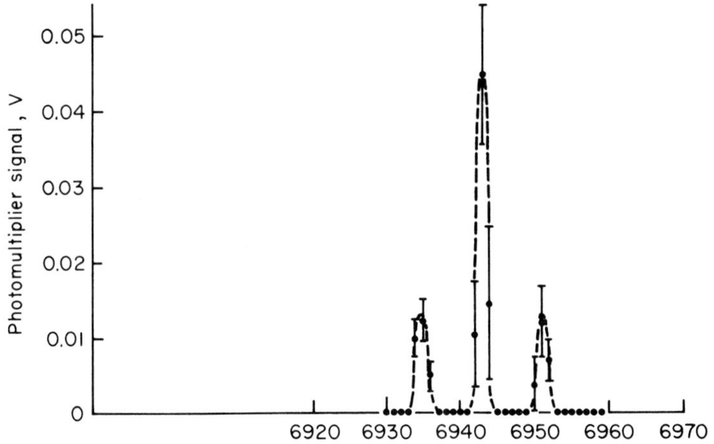


**Fig. 11.18** Rayleigh and Brillouin light scattering from a gaseous hydrogen plasma at the ruby laser frequency. The scattering was observed at the small angle of 13.5°. [From S. Ramsden and W. Davies, *Phys. Rev. Lett.* **16**, 303 (1966), Fig. 2A.]

matrix element of

$$a_{k'}^{\dagger} \hat{u}_k^* \exp -i\mathbf{k}' \cdot \mathbf{r}\; a_k \hat{u}_k \exp i\mathbf{k} \cdot \mathbf{r}$$

In our earlier discussion, it was assumed that $a_{k'}^{\dagger}$ operated on the vacuum state and created photons of index $k'$ which were characteristic of spontaneous emission. For intense radiation fields, Eq. (11.116) must be multiplied by

$$(n_{k'} + 1_{k'})$$

and the probability of a photon in state $k'$ is enhanced by the number already present in state $k'$, that is, $n_{k'}$, and must give rise to stimulated scattering. Fluctuations in the background medium that cause a variation in the dielectric constant $\Delta\varepsilon_{ij}$ have been discussed in the previous sections.

If

$$\Delta\varepsilon_{ij} = \delta_{ij}\, \Delta\varepsilon$$

then the primary scattering effects are due to the fluctuations of $\Delta\varepsilon$ that are due to pressure and entropy in a single-component system. In a binary system, fluctuations in concentration can occur. The stimulated scattering effects occur whenever spontaneous effects occur and the following fluctuations cause the following effects:

$\Delta P$ Stimulated Brillouin scattering

$\Delta S$ Stimulated Rayleigh scattering

In binary systems, the stimulated Rayleigh scattering is related to the diffusive phenomenological equations for temperature and concentration:

$$\frac{\partial(\Delta T)}{\partial t} = K \nabla^2(\Delta T) + \text{coupling terms} \qquad (11.130a)$$

and

$$\frac{\partial(\Delta c)}{\partial t} = D \nabla^2(\Delta c) + \text{coupling terms} \qquad (11.130b)$$

$\Delta S$ will contain terms which depend on $\Delta T$ and $\Delta c$, and thermal as well as concentration-stimulated Rayleigh scattering can occur. Equations (11.120e) and (11.121b) can be used for $\Delta P$, while Eq. (11.122c) is the appropriate equation for $\Delta T$ and $\Delta c$. Coupling terms must be included in the detailed analysis [*35*].

For molecules, rotational and vibrational transitions can occur and stimulated Raman scattering is observed. Each process which gives rise to a nonzero value of the correlation function $\langle \Delta\varepsilon(0, 0)_{fg}\, \Delta\varepsilon(\mathbf{r}, \tau)_{ij} \rangle$ and stimulates the scattering of light can also cause stimulated emission. Table 11.2

**Table 11.2**

Scattering and Stimulated Scattering Phenomena

| Type of scattering | Research application |
|---|---|
| Rayleigh | Phase transitions; atmospheric propagation; critical opalescence |
| Brillouin | Phase transitions; velocity of sound and damping at hypersonic frequencies |
| Rotational Raman | Molecular structure of gases |
| Vibrational Raman | Molecular structure and molecular force constants; chemical identification |
| Phonon | Phase transitions and lattice properties in solids |
| Multiple-phonon | Anharmonic aspects of crystals and critical points in the Brillouin zone |
| Polariton | Coupling of phonons and light in the infrared |
| Magnon | Magnon dispersion relations; magnon–magnon interaction; spin waves |
| Plasmon | Properties of gaseous, liquid, and solid plasmons; coupling of the plasma with lattice modes |
| Electronic | Electronic energy levels of ions in solids |
| Landau-level | Cyclotron resonance and damping of nearly free electrons in solids; spin-flip processes |
| Field-induced Raman | Lattice deformation induced by an electric field; "forbidden" Raman effects |

lists some of the many stimulated scattering phenomena which can be observed with high-power laser excitation.

The stimulated Raman scattering from the spin-flip [*36*] process for the conduction electrons in InSb is selected as an example. In a strong magnetic field, the energy of the carriers is quantized into Landau levels with separation $g\mu_B B$, where $B$ is the magnetic field, $\mu_B$ the Bohr magneton, and $g$ the effective gyromagnetic ratio of the electrons in InSb. The $g$ value is quite large and varies from 48 at $B = 0$ to 35 at $B = 10$ W/m². Figure 11.19 shows the scattering of the 10.6-μm radiation of a $CO_2$ laser. Since the initial and final states of the scatterer have opposite spin quantum numbers, the Raman process is referred to as "spin-flip." The magnetic field is 5.2 W/m² and the scattered photon has a wavelength of 11.89 μm. In this experiment, the magnetic field $B$ is along $\hat{z}$, the 10.6-μm incident photon is

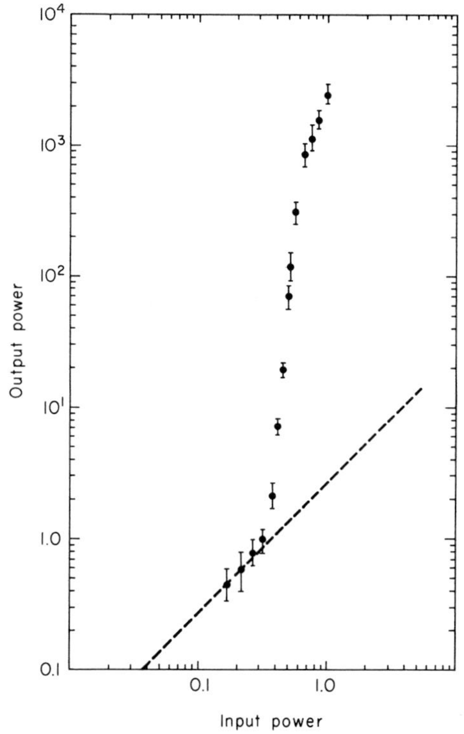


**Fig. 11.19** Variation of the spin-flip, Raman-scattered radiation at 11.89 μm as a function of the input power of the incident 10.59-μm radiation [*36*]. The magnetic field is 5.2 W/m². The exponential increase in scattered intensity corresponds to the onset of stimulated scattering. The dashed line has a slope of 1. [From C. K. N. Patel and E. D. Shaw, *Phys. Rev. Lett.* **24**, 451 (1970), Fig. 1.]

along the $\hat{y}$ direction, and the scattered photon at 11.89 $\mu$m is along $\hat{x}$. At low input power, the scattered radiation is linearly proportional to the input power. An exponential increase in the stimulated radiation occurs at threshold.

#### 11.8.5.5 Index of Refraction and Multiple Scattering

Most of the previous discussion has been concerned with the single scattering event by an inhomogeneous aspect of the medium. Consider again the scattering by an array of scatterers at position $\mathbf{r}_n$ with scattering amplitude $f$. For a single scattering center, this is the solution to the scalar Helmholtz equation

$$\nabla^2\psi + k_0^2\psi = -f\psi\, \delta(\mathbf{r} - \mathbf{r}_m) \tag{11.131}$$

which is given in Exercise 11.19 as

$$\psi(\mathbf{r}) = \psi_{\mathrm{i}}(\mathbf{r}) + \frac{1}{4\pi} f\psi(\mathbf{r}_m) \frac{e^{ik_0R_m}}{R_m} \tag{11.132}$$

where $\psi_{\mathrm{i}}$ is a solution to the homogeneous equation and $R_m = |\,\mathbf{r} - \mathbf{r}_m\,|$. Since this equation is linear, the solution for a group of scatterers is given by the superposition of the solutions, and

$$\psi(\mathbf{r}) = \psi_{\mathrm{i}}(\mathbf{r}) + \frac{1}{4\pi} f \sum_{1}^{N} \psi_m(\mathbf{r}_m) \frac{e^{ik_0R_m}}{R_m} \tag{11.133a}$$

$\psi_m(\mathbf{r}_m)$ is the field at the $m$th particle which is due to the incident field and all the contributions from all the other scattering centers, that is,

$$\psi_m(\mathbf{r}_m) = \psi_{\mathrm{i}}(\mathbf{r}_m) + \left(\frac{1}{4\pi}\right) f \sum_{n}{}' \psi_n(r_n) \frac{e^{ik_0R_{mn}}}{R_{mn}} \tag{11.133b}$$

These equations are quite general, but they miss an essential point. If the sample is homogeneous, only the overall shape causes scattering. Shape effects can be eliminated by considering an infinite medium or a thin sheet of infinite area, but the homogeneous aspect remains. In a Fourier expansion of

$$\sum_m \delta(\mathbf{r} - \mathbf{r}_m)$$

the homogeneous aspect is the small $\varkappa$ or dc term.

Let $\bar{n}$ be the average density of particles:

$$\bar{n}V = \int d\mathbf{r} \sum_m \delta(\mathbf{r} - \mathbf{r}_m)$$

and then write the wave equation *in the medium* as

$$\nabla^2\psi + k^2\psi = -f\left[\sum_m \delta(\mathbf{r} - \mathbf{r}_m) - \bar{n}\right]\psi = -f\,\Delta n(\mathbf{r})\psi \qquad (11.134)$$

where

$$k^2 = k_0^2 + \bar{n}f$$

Propagation in the medium is described by a new wave vector **k** and scattering is described by the deviation $\Delta n(\mathbf{r})$ from the homogeneous background. If the scattering due to $f\,\Delta n(\mathbf{r})$ is small, then this is a good approximation to the new wave vector and can be regarded as the index of refraction of the medium, that is,

$$k^2 = \mu^2 k_0^2 = k_0^2\left(1 + \frac{\bar{n}f}{k_0^2}\right) \qquad (11.135)$$

Equation (11.134) remains valid for an ensemble average, and $\langle\ \rangle$ can be placed about each term. If $\langle \Delta n(\mathbf{r})\rangle$ is a stationary random variable, then $\langle \Delta n(\mathbf{r})\rangle = 0$, and the scalar Helmholtz equation becomes

$$\nabla^2\langle\psi\rangle + k^2\langle\psi\rangle = 0$$

and makes the homogeneous aspect apparent. The cross section for elastic scattering is given by Eq. (10.101a),

$$\sigma(\boldsymbol{\varkappa}) = \left(\frac{1}{4\pi}\right)^2 f^2 V \int d\mathbf{r}\,(\exp -i\boldsymbol{\varkappa}\cdot\mathbf{r})\langle \Delta n(0)\,\Delta n(\mathbf{r})\rangle \qquad (11.136)$$

where $\boldsymbol{\varkappa} = \mathbf{k} - \mathbf{k}'$ *implies wave vectors* **k** *and* **k**$'$ *in the medium.* Scattering depends on the correlation of fluctuations in number density $\langle \Delta n(0)\,\Delta n(\mathbf{r})\rangle$.

For a periodic array such as that which occurs in crystals, the ensemble average of $\langle \Delta n(\mathbf{r})\rangle \neq 0$. At the Bragg angles, the lattice periodicity becomes quite apparent. Even so, $k$ for the medium should be used and $f\psi_m(\mathbf{r}_m)$ is replaced by $f\,\Delta n(\mathbf{r}_m)\psi_m(\mathbf{r}_m)$ in Eq. (11.133a). For single scattering in the sample, this is the same as the earlier discussion in this section. For multiple scattering at the Bragg angles, the sample size is important. The amplitude at the Bragg angles grows exponentially with passage through the sample, but whenever the exponential region is reached, multiple scattering effects must be included and the intensity is some fraction of the incident intensity.

**Exercise 11.19** The arguments used in deriving Snell's Law apply at the boundary of two media described by Eq. (11.135) and $\mu_1 \sin\theta_1$

$= \mu_2 \sin \theta_2$. Total reflection occurs for $\theta = \pi/2$ or $\sin \theta_1 = \mu_2/\mu_1$. Assume that the scattering length for neutrons by Ni is $f = 1.03 \times 10^{-14}$ m. Find the angle for total reflection.

Exercise 11.20 Water is filled with bubbles of radius $r_0$ and it is assumed that the scattering cross section is $\sigma = 4\pi r_0^2 = 4\pi f^2$. Find the index of refraction for water with $n$ bubbles per cubic meter. Discuss conditions under which such an effect would occur and be observable.

Exercise 11.21 Under what conditions will a fog have an index of refraction for sound waves?

## References


1. F. Block, *Phys. Rev.* **70**, 460 (1946).
2. N. Bloembergen and Y. R. Shen, *Phys. Rev. A* **133**, 37 (1964).

3a. A. Abragam, "The Principles of Nuclear Magnetism." Oxford Univ. Press, London and New York, 1961.

3b. C. P. Slichter, "Principles of Magnetic Resonance." Harper, New York, 1962.

4. N. Bloembergen, "Nonlinear Optics." Benjamin, New York, 1965.
5. S. L. McCall and E. L. Hahn, *Phys. Rev.* **183**, 457 (1969); P. K. Cheo and C. H. Wang, *Phys. Rev. A* **1**, 225 (1970).
6. N. F. Ramsey, "Molecular Beams." Oxford Univ. Press (Clarendon), London and New York, 1956; H. Lyons, *Ann. N. Y. Acad. Sci.* **55**, 811 (1952).
7. E. L. Hahn, *Phys. Rev.* **80**, 580 (1950).
8. J. P. Gordon, C. H. Wang, C. K. N. Patel, R. E. Slusher, and W. J. Tomlinson, *Phys. Rev.* **179**, 294 (1969); C. V. Heer and R. H. Kohl, *Ibid.* **1**, 693 (1970).

8a. I. D. Abella, N. A. Kurnit, and S. R. Hartman, *Phys. Rev.* **141**, 391 (1966).

9. V. Weisskopf and E. Wigner, *Z. Phys.* **63**, 54 (1930); **65**, 580 (1930).
10. W. Heitler, "Quantum Theory of Radiation." Oxford Univ. Press (Clarendon), London and New York, 1954; A. Messiah, "Quantum Mechanics," Vol. II. North-Holland Publ., Amsterdam, 1962; M. L. Goldberger and K. M. Watson, "Collision Theory." Wiley, New York, 1964.
11. R. H. Dicke, *Phys. Rev.* **89**, 472 (1953); see also **93**, 99 (1954).
12. A. Messiah, "Quantum Mechanics," Vol. I. North-Holland Publ., Amsterdam, 1962.
13. R. L. Mössbauer, Recoilless absorption of gamma rays and studies of nuclear hyperfine interactions in solids. *In* "Magnetic Resonance and Relaxation" (R. Blinc, ed.), pp. 864–880. North-Holland Publ., Amsterdam, 1967.
14. M. J. Clauser, E. Kankeleit, and R. L. Mossbauer, *Phys. Rev. Lett.* **17**, 5 (1966).
15. M. Baranger, "Atomic and Molecular Processes," Chapter 13. Academic Press, New York, 1962.

15a. P. W. Anderson, *Phys. Rev.* **76**, 647 (1949).

15b. R. Kubo and K. Tomita, *J. Phys. Soc. Jap.* **9**, 888 (1953).
16. R. Kubo, "Fluctuation Relaxation and Resonance in Magnetic Systems," pp. 23–68. Oliver & Boyd, Edinburgh, 1962.
17. J. M. Vaughn and G. Smith, *Phys. Rev.* **166**, 17 (1969).
18. J. W. Knutson, Jr. and W. R. Bennet, Jr., *Bull. Am. Phys. Soc.* **16**, 594 (1971); R. H. Cordover and P. A. Bonczyk, *Phys. Rev.* **188**, 696 (1969).
19. W. E. Lamb, Jr., *Phys. Rev. A* **134**, 1429 (1964).
20. S. Chen and M. Takeo, *Rev. Mod. Phys.* **29**, 20 (1957); R. G. Breene, *Ibid.* **29**, 94 (1957).
20a. E. W. Smith, J. Cooper, and C. R. Vidal, *Phys. Rev.* **185**, 140 (1969); T. A. Dillon, E. W. Smith, and J. Cooper, *Phys. Rev. A* **2**, 1839 (1970).
21. L. van Hove, *Phys. Rev.* **95**, 249 (1954).
22. G. Placzek and L. van Hove, *Phys. Rev.* **93**, 1207 (1954); see C. Kittel, "Quantum Theory of Solids," p. 377. Wiley, New York, 1963.
22a. G. Dolling and A. D. B. Woods, *in* "Thermal Neutron Scattering" (P. A. Egelstaff, ed.), p. 193. Academic Press, New York, 1965.
23. J. L. Yarnell, J. L. Warren, and S. H. Koenig, *Lattice Dyn., Proc. Int. Conf. Copenhagen, 1963*, p. 57 (R. F. Wallis, ed.). Pergamon, Oxford, 1963.
24. M. A. Krivoglaz, "Theory of X-Ray and Thermal-Neutron Scattering by Real Crystals." Plenum, New York (1969); H. N. Temperley, J. S. Rawlinson, and G. S. Rushbrooke, eds., "Physics of Simple Liquids." Wiley, New York, 1968.
25. K. Skold and K. E. Larson, The neutron scattering technique and relaxation phenomena in liquids. *In* "Magnetic Resonance and Relaxation" (R. Blinc, ed.), p. 78. North-Holland Publ., Amsterdam, 1967.
26. D. G. Henshaw, *Phys. Rev.* **111**, 1470 (1958).
27. P. Schofield, Experimental knowledge of correlation functions in simple liquids. *In* "Physics of Simple Liquids," (H. N. V. Temperley, J. S. Rowlinson, and G. S. Rushbrooke, eds.), p. 563. Wiley, New York, 1968.
28. P. D. Lazay, J. H. Lunacek, N. A. Clark, and G. B. Benedek, *in* "Light Scattering Spectra of Solids" (G. B. Wright, ed.), pp. 593–602. Springer-Verlag, Berlin and New York, 1969.
29. E. H. Hara and A. D. May, *Int. Conf. Quantum Electron. 6th, 1970*, pp. 194–195; see also E. H. Hara, A. D. May, and H. F. P. Knaap, *Can. J. Phys.* **49**, 420 (1971).
30. S. Yip and M. Nelkin, *Phys. Rev. A* **135**, 1241 (1964); J. P. Boon and P. Deguent. *Phys. Rev. A* **2**, 2542 (1970).
31. G. Benedek, J. B. Lastovka, K. Fritsch, and T. Greytak, *J. Opt. Soc. Amer.* **54**, 1284 (1964).
32. H. A. Szymanski, ed., "Raman Spectroscopy." Plenum, New York, 1967; see J. R. Ferraro, Chapter 2 for Fig. 11.17.
33. S. P. S. Porto, Light scattering with laser sources. *In* "Light Scattering Spectra of Solids" (G. B. Wright, ed.), pp. 1–24. Springer-Verlag, Berlin and New York, 1969.
33a. P. S. Pershan and W. B. Lacina, Raman scattering in mixed crystals. *In* "Light Scattering Spectra of Solids" (G. B. Wright, ed.), pp. 439–454. Springer-Verlag, Berlin and New York, 1969.
33b. P. A. Wolff, Light scattering from solid state plasmas. *In* "Light Scattering Spectra of Solids" (G. B. Wright, ed.), pp. 273–284. Springer-Verlag, Berlin and New York, 1969.

34. S. Ramsden and W. Davies, *Phys. Rev. Lett.* **16**, 303 (1966); see also D. E. Evans and P. G. Carolan, *Phys. Rev. Lett.* **25**, 1605 (1970).
35. N. Bloembergen, W. H. Lowdermilk, M. Matsuoka, and C. S. Wang, *Phys. Rev. A* **3**, 404 (1971); *Phys. Rev. Lett.* **25**, 1476 (1970); J. Boon and P. Deguent, *Phys. Rev. A* **2**, 2542 (1970).
36. C. K. N. Patel and E. D. Shaw, *Phys. Rev. Lett.* **24**, 451 (1970).

# APPENDIX A: Fundamental Constants

| Quantity | Symbol | Value[a] | |
|---|---|---|---|
| Standard gravitational acceleration | $g$ | 9.8062 | m $sec^{-2}$ |
| Standard atmosphere (760 Torr) | | 101,323 | N $m^{-2}$ |
| Gravitational constant | $G$ | $6.6732 \times 10^{-11}$ | N $m^2$ $kg^{-2}$ |
| Standard volume of ideal gas | $V_0$ | 22.4136 | $m^3$ $kmole^{-1}$ |
| Ice point | | 273.15 | °K |
| Avogadro's number | $\mathcal{N}$ | $6.022169 \times 10^{26}$ | $kmole^{-1}$ |
| Gas Constant ($R_0/N$) | $R_0$ | 8314.34 | J $kmole^{-1}$ $°K^{-1}$ |
| Boltzmann Constant | $k$ | $1.380622 \times 10^{-23}$ | J $°K^{-1}$ |
| Atomic mass unit | amu | $1.66053 \times 10^{-27}$ | kg |
| Electron rest mass | $m_e$ | $9.109558 \times 10^{-31}$ | kg |
| Velocity of light | $c$ | $2.997925 \times 10^8$ | m $sec^{-1}$ |
| Electron charge | e | $1.602191 \times 10^{-19}$ | C |
| Planck's constant | $h$ | $6.626196 \times 10^{-34}$ | J sec |
| Stefan–Boltzmann constant ($2\pi^5 k^4/15h^3c^2$) | $\sigma$ | $5.66961 \times 10^{-8}$ | W $m^{-2}$ $°K^4$ |
| First radiation constant ($8\pi hc$) | $c_1$ | $4.992579 \times 10^{-24}$ | J m |
| Second radiation constant ($hc/k$) | $c_2$ | $1.438833 \times 10^{-2}$ | m °K |
| Bohr magneton ($e\hbar/2m_e c$) | $\mu_B$ | $9.27096 \times 10^{-24}$ | J $T^{-1}$ |
| Fine structure constant ($\mu_0 c^2/4\pi$)($e^2/\hbar c$) | $\alpha^{-1}$ | 137.03602 | |
| Bohr radius | $a_0$ | $5.2917715 \times 10^{-11}$ | m |
| Rydberg constant | $R$ | $1.09737312 \times 10^7$ | $m^{-1}$ |
| Faraday constant ($Ne$) | $F$ | $9.648670 \times 10^7$ | C $kmole^{-1}$ |

[a] Values are taken from B. N. Taylor, W. H. Parker, and D. N. Langenberg, *Rev. Mod. Phys.* **41**, 375 (1969).

# APPENDIX B: Gaussian Integrals

| $n$ | $\int_0^\infty x^n \exp -ax^2\, dx$ | |
|---|---|---|
| 0 | $\frac{1}{2}\left(\frac{\pi}{a}\right)^{1/2}$ | |
| 1 | | $\frac{1}{2a}$ |
| 2 | $\frac{1}{4}\left(\frac{\pi}{a^3}\right)^{1/2}$ | |
| 3 | | $\frac{1}{2a^2}$ |
| 4 | $\frac{3}{8}\left(\frac{\pi}{a^5}\right)^{1/2}$ | |
| 5 | | $\frac{1}{a^3}$ |
| 6 | $\frac{15}{16}\left(\frac{\pi}{a^7}\right)^{1/2}$ | |
| 7 | | $\frac{3}{a^4}$ |
| $n$ | $I_{n+2} = -\frac{dI_n}{da}$ | |

# Index

## A



## B

## C

**D**

## H



## I



## J



## K



## L

## T

## U



## V